# Lecture Notes in Physics

Editorial Board

R. Beig, Wien, Austria
W. Beiglböck, Heidelberg, Germany
W. Domcke, Garching, Germany
B.-G. Englert, Singapore
U. Frisch, Nice, France
P. Hänggi, Augsburg, Germany
G. Hasinger, Garching, Germany
K. Hepp, Zürich, Switzerland
W. Hillebrandt, Garching, Germany
D. Imboden, Zürich, Switzerland
R. L. Jaffe, Cambridge, MA, USA
R. Lipowsky, Potsdam, Germany
H. v. Löhneysen, Karlsruhe, Germany
I. Ojima, Kyoto, Japan
D. Sornette, Nice, France, and Zürich, Switzerland
S. Theisen, Potsdam, Germany
W. Weise, Garching, Germany
J. Wess, München, Germany
J. Zittartz, Köln, Germany

## The Lecture Notes in Physics

The series Lecture Notes in Physics (LNP), founded in 1969, reports new developments in physics research and teaching – quickly and informally, but with a high quality and the explicit aim to summarize and communicate current knowledge in an accessible way. Books published in this series are conceived as bridging material between advanced graduate textbooks and the forefront of research and to serve three purposes:

- to be a compact and modern up-to-date source of reference on a well-defined topic

- to serve as an accessible introduction to the field to postgraduate students and nonspecialist researchers from related areas

- to be a source of advanced teaching material for specialized seminars, courses and schools

Both monographs and multi-author volumes will be considered for publication. Edited volumes should, however, consist of a very limited number of contributions only. Proceedings will not be considered for LNP.

Volumes published in LNP are disseminated both in print and in electronic formats, the electronic archive being available at springerlink.com. The series content is indexed, abstracted and referenced by many abstracting and information services, bibliographic networks, subscription agencies, library networks, and consortia.

Proposals should be sent to a member of the Editorial Board, or directly to the managing editor at Springer:

Christian Caron
Springer Heidelberg
Physics Editorial Department I
Tiergartenstrasse 17
69121 Heidelberg / Germany
christian.caron@springer.com

D. Benest
C. Froeschle
E. Lega (Eds.)

# Topics in Gravitational Dynamics

Solar, Extra-Solar and Galactic Systems

Editors

Daniel Benest
Claude Froeschle
Elena Lega
OCA - Nice
Observatoire de la Cote d'Azur
B.P. 4229
06304
Nice CX 4, France
benest@obs-nice.fr
claude@obs-nice.fr
elena@obs-nice.fr

D. Benest, C. Froeschle and E. Lega (Eds.), *Topics in Gravitational Dynamics: Solar, Extra-Solar and Galactic Systems*, Lect. Notes Phys. 729 (Springer, Berlin Heidelberg 2007), DOI 10.1007/ 978-3-540-72984-6

Library of Congress Control Number: 2007941801

ISSN 0075-8450
ISBN 978-3-540-72983-9 Springer Berlin Heidelberg New York

This work is subject to copyright. All rights are reserved, whether the whole or part of the material is concerned, specifically the rights of translation, reprinting, reuse of illustrations, recitation, broadcasting, reproduction on microfilm or in any other way, and storage in data banks. Duplication of this publication or parts thereof is permitted only under the provisions of the German Copyright Law of September 9, 1965, in its current version, and permission for use must always be obtained from Springer. Violations are liable for prosecution under the German Copyright Law.

Springer is a part of Springer Science+Business Media
springer.com
© Springer-Verlag Berlin Heidelberg 2007

The use of general descriptive names, registered names, trademarks, etc. in this publication does not imply, even in the absence of a specific statement, that such names are exempt from the relevant protective laws and regulations and therefore free for general use.

Typesetting: by the authors and Integra using a Springer LATEX macro package
Cover design: eStudio Calamar S.L., F. Steinen-Broo, Pau/Girona, Spain

Printed on acid-free paper      SPIN: 12072378       5 4 3 2 1 0

# Preface

This book is the outcome of the 2006 Aussois Winter School: "Open problems in Celestial Mechanics". In this school, we have reviewed the state of the art in the theory and applications of dynamical systems with particular interest in planetary and galactic dynamics. Recent developments have been achieved in two directions: (1) the application of Nekhoroshev's theorem to the motion of planets and asteroids and (2) the dynamical origin of the diversity of the extra-solar planetary systems.

The aims of the 2006 Aussois Winter School and consequently those of this book are:

- To provide a review of the dynamics of quasi-integrable Hamiltonian systems from the viewpoint of Nekhoroshev's theorem and its implications for slow diffusion processes.
- To introduce the dynamics of weakly dissipative systems.
- To review the dynamics of small bodies in the solar system.
- To discuss the dynamical origin and the diversity of the extra-solar planetary systems.
- To highlight special problems in galactic dynamics.

## Acknowledgments

The editors thank with great pleasure the sponsorship of the O.C.A. (Observatoire de la Côte d'Azur) and of the Département Cassiopée, together with the CNRS (the french "Centre National de la Recherche Scientifique"), through its Continuing Education Department ("Formation Permanente") and its staff, with a special mention to Mrs Jocelyne Gosselin.

U.N.S.A./C.N.R.S./O.C.A. Observatoire de Nice, February 2007

*Daniel Benest*
*Claude Froeschlé*
*Elena Lega*

# Bibliography

1. Benest, D. and Froeschlé, C. (eds), 1990. *Les méthodes modernes de la Mécanique Céleste [Modern Methods of Celestial Mechanics]*, Editions Frontieres (C36).
2. Benest, D. and Froeschlé, C. (eds), 1992. *Interrelations Between Physics and Dynamics for Minor Bodies in the Solar System*, Editions Frontieres (C49).
3. Benest, D. and Froeschlé, C. (eds), 1994. *An Introduction to Methods of Complex Analysis and Geometry for Classical Mechanics and Non-Linear Waves*, Editions Frontieres.
4. Benest, D. and Froeschlé, C. (eds), 1995. *Chaos and Diffusion in Hamiltonian Systems*, Editions Frontieres.
5. Benest, D. and Froeschlé, C. (eds): *Impacts on Earth*, Lect. Notes in Phys., **505**. Springer, Heidelberg (1998).
6. Benest, D. and Froeschlé, C. (eds), 1998. *Analysis and Modelling of Discrete Dynamical Systems – with Applications to Dynamical Astronomy*, Advances in Discrete Mathematics and Applications, **1**, Gordon and Breach.
7. Benest, D. and Froeschlé, C., 1999, At the frontiers of chaotic dynamics of gravitational systems. Special Issue of *Celest. Mech. Dyn. Astron.* **72**(1–2).
8. Benest, D. and Froeschlé, C. (eds): *Singularities in Gravitational Systems – Applications to Chaotic Transport in the Solar System*, Lect. Notes Phys., **590**. Springer, Heidelberg (2002).
9. Benest, D. and Froeschlé, C., and Lega, E. (eds), 2005. *Hamiltonian Systems and Fourier Analysis – New prospects for Gravitational Dynamics*, Advances in Astronomy and Astrophysics, **9**, Cambridge Scientific Publishers.

# Contents

**1 An Overview on the Nekhoroshev Theorem**
*Massimiliano Guzzo* .................................................. 1
1.1 Introduction ...................................................... 1
1.2 Classical Integrable Systems ...................................... 3
1.3 Quasi–Integrable Systems and Hamiltonian Perturbation Theory .. 7
1.4 Inside the Nekhoroshev Theorem ................................. 13
1.5 Exponential Stability in Degenerate Systems ..................... 19
References ............................................................ 26

**2 Diffusion in Hamiltonian Quasi-Integrable Systems**
*Elena Lega, Claude Froeschlé and Massimiliano Guzzo* ................ 29
2.1 Introduction ..................................................... 29
2.2 Arnold's Diffusion ............................................... 31
2.3 A Generalization of Arnold's Diffusion ........................... 36
2.4 The Fast Lyapunov Indicator ..................................... 38
2.5 The Model Problem .............................................. 40
2.6 The FLI for Detecting the Geometry of Resonances .............. 41
2.7 Local and Global Arnold's Diffusion in the Convex Case ......... 43
2.8 Diffusion in Non-Convex Systems ................................ 51
2.9 Conclusion ...................................................... 62
References ............................................................ 63

**3 Weakly Dissipative Systems in Celestial Mechanics**
*Alessandra Celletti* .................................................. 67
3.1 Introduction ..................................................... 67
3.2 The Dissipative Standard Map .................................... 68
3.3 Techniques for the Numerical Investigation of the Dynamics ...... 70
3.4 Adiabatic Invariants of the Pendulum ............................ 76
3.5 A Paradigm from Celestial Mechanics: The Spin-Orbit Model .... 77
3.6 The Restricted, Planar, Circular, Three-Body Problem ........... 83
3.7 Periodic Orbits for Non-Autonomous and Autonomous Systems ... 84
References ............................................................ 88

## 4 Connectance and Stability of Dynamical Systems
*Marina Cosentino, Dimitri Laveder, Elena Lega and Claude Froeschlé* .. 91
4.1 Introduction ................................................. 91
4.2 Connectance in Ecology: A Historical Perspective............... 93
4.3 Connectance in Economics...................................... 94
4.4 Linear Stability and Connectance: Basic Definitions............ 96
4.5 The Diagonal Dominance Criterion and its Applicability ........ 97
4.6 The Stability Recovery Phenomenon: Some Simple Criteria ......100
4.7 The Influence of the Matrix Size .............................106
4.8 Conclusions..................................................107
References .......................................................108

## 5 Chaotic Diffusion of Asteroids
*Kleomenis Tsiganis* .............................................111
5.1 Introduction .................................................111
5.2 Slow Dispersion of the Trojan Swarms .........................115
5.3 Resonant Populations of Unstable Asteroids ...................117
5.4 Chaotic Chronology of Asteroid Families ......................123
5.5 Analytical Models of Chaotic Diffusion in MMRs ...............127
5.6 Chaotic Transport in the Main Belt: A Numerical Survey........138
5.7 Conclusions—Discussion .......................................145
References .......................................................148

## 6 An Overview of the Rotations of Planets in the Solar System
*Jean Souchay*....................................................151
6.1 Introduction .................................................151
6.2 Theories of the Motion of Rotation of a Celestial Body .......153
6.3 Rotation of the Earth ........................................168
6.4 The Rotation of the Planets ..................................186
6.5 Conclusion ...................................................203
References .......................................................204

## 7 On the Stability of Extra-Solar Planetary Systems
*Elke Pilat-Lohinger and Rudolf Dvorak* ..........................209
7.1 Introduction .................................................209
7.2 Multi-planet Systems .........................................211
7.3 Binary Systems ...............................................214
7.4 Terrestrial Planets in Extra-Solar Planetary Systems..........222
7.5 Theoretical Considerations ...................................224
7.6 Gl 777 A .....................................................225
7.7 The **G4** Terrestrial Planet in HD108874.....................225
7.8 Concluding Remarks for Possible Terrestrial Planets ..........227
References .......................................................230

## 8 Origin Theories for the Eccentricities of Extrasolar Planets
*Fathi Namouni* .................................................. 233
8.1 Introduction .................................................. 233
8.2 Eccentricity Observations ..................................... 235
8.3 Eccentricity Origin Theories .................................. 236
8.4 Jet-Induced Excitation ........................................ 240
8.5 Concluding Remarks ........................................... 253
References ........................................................ 254

## 9 Methods for the Study of the Dynamics of the Oort Cloud Comets I: Modelling the Stellar Perturbations
*Marc Fouchard, Christiane Froeschlé, Hans Rickman and Giovanni B. Valsecchi* ........................................................ 257
9.1 Introduction .................................................. 257
9.2 Modelling the Stellar Pertubations ............................ 258
9.3 Stellar Encounters ............................................ 260
9.4 The Dynamical Models ......................................... 261
9.5 Results ....................................................... 265
9.6 Conclusions ................................................... 271
References ........................................................ 271

## 10 Methods for the Study of the Dynamics of the Oort Cloud Comets II: Modelling the Galactic Tide
*Marc Fouchard, Christiane Froeschlé, Sławomir Breiter, Roman Ratajczak, Giovanni B. Valsecchi and Hans Rickman* .......... 273
10.1 Introduction ................................................. 273
10.2 Models of Galactic Tide Effects on Cometary Orbits ........... 274
10.3 Comparisons Between the Different Models .................... 292
10.4 Hybrid Integrators ........................................... 293
10.5 Conclusion ................................................... 294
References ........................................................ 295

## 11 Special Features of Galactic Dynamics
*Christos Efthymiopoulos, Nikos Voglis[†] and C. Kalapotharakos* ......... 297
11.1 Introduction ................................................. 297
11.2 Basic Notions ................................................ 299
11.3 The Statistical Mechanical Approach: Violent Relaxation ..... 322
11.4 The Orbital Approach: Global Dynamics and Self-Consistent Models of Galaxies ........................................... 347
11.5 The $N$-Body Approach ........................................ 364
References ........................................................ 380

**12 On the Difficulty to Foresee Solar Cycles:
A Non-Deterministic Approach**
*Jean-Pierre Rozelot* ................................................. 391
12.1 Introduction ................................................... 391
12.2 Solar Cycle Activity ........................................... 393
12.3 Tools .......................................................... 398
12.4 Some Results .................................................. 400
12.5 Portrait of the Solar Activity Cycle in the Phases Space ......... 402
12.6 Conclusion .................................................... 404
References ......................................................... 405

**Index** ............................................................ 409

# 1

# An Overview on the Nekhoroshev Theorem

Massimiliano Guzzo

Dipartimento di Matematica Pura ed Applicata, Università degli Studi di Padova, via Trieste 63, 35121 Padova, Italy
guzzo@math.unipd.it

**Abstract.** The Nekhoroshev theorem has provided in the last decades an important framework to study the long-term stability of quasi-integrable dynamical systems. While KAM theorem predicts the stability of invariant tori for non–resonant motions, the Nekhoroshev theorem provides exponential stability estimates for resonant motions. The first part of this chapter reviews the mechanisms which are at the basis of the proof of Nekhoroshev theorem.

The application of Nekhoroshev theorem to systems of interest for Celestial Mechanics encounters difficulties due to the presence of the so–called proper degeneracy, or super–integrability, of the Kepler problem. For this reason, for concrete systems such as planetary systems or asteroids of the main belt, the application of the Nekhoroshev theorem requires modifications of the theory. At variance with the non-degenerate case, these systems can have chaotic diffusion in relatively short times. The second part of this chapter reviews the problems related to degenerate systems, and describes the mechanism of production of chaotic motions in short times.

## 1.1 Introduction

Classical Mechanics emphasizes the importance of a special class of mechanical systems which are called integrable systems. Integrability means essentially that all solutions are analytically determined. In the modern Hamiltonian picture integrable systems are characterized by a number of independent first integrals in involution with respect to the Poisson brackets equal to the number of the degrees of freedom and with compact connected level sets, and the Liouville–Arnol'd theorem guarantees that the equations of motion can be solved by quadratures; the phase–space is fibrated by invariant tori; the Hamiltonian flow is a linear drift on an invariant torus; it is possible to introduce symplectic action–angle conjugate variables $(I, \varphi) = (I_1, \ldots, I_n, \varphi_1, \ldots, \varphi_n) \in \mathbb{R}^n \times \mathbb{T}^n$ such that the Hamilton function depends only on the actions $I$. The properties of the system which are relevant for the long–term stability are determined mainly by the actions. The Kepler problem and the Euler–Poinsot rigid body are integrable, so that the Liouville–Arnol'd theorem provides a modern geometric interpretation to the classical solutions of these important problems. However, integrable systems are a crude approximation of real problems: to describe the motion of a planet or of an asteroid around a star we cannot neglect the influence of the other planets; the motion of an artificial satellite around a planet cannot be modeled only by the Kepler problem; to describe the rotation of celestial bodies we cannot treat them as isolated rotating bodies and so on. However, these problems can be considered and studied as *perturbations* of integrable systems. As a matter of fact, many problems of real interest for different fields of physics, such as celestial mechanics, dynamical astronomy, statistical physics, plasma physics and particle accelerators, can be studied by regarding them as perturbations of suitable integrable systems, and are called quasi–integrable systems. To be definite, we write the Hamiltonian of a perturbation of an integrable system in the following form:

$$H_\varepsilon(I,\varphi) = h(I) + \varepsilon f(I,\varphi) , \qquad (1.1)$$

where $(I,\varphi) \in V \times \mathbb{T}^n$, $V \subseteq \mathbb{R}^n$ is open and bounded; $f$ is analytic; $\varepsilon$ is a small parameter. In general, the Hamilton equations of (1.1) have solutions which can be explicitly written only when $\varepsilon = 0$, namely in the integrable case: the actions $I$ stay constant while the angles $\varphi$ circulate at uniform speed $\omega(I) = \partial h(I)$. Instead, when $\varepsilon \neq 0$, not only one is not able in general to provide explicit solutions, but also understanding the stability and diffusion properties of the actions becomes a non–trivial problem. Typically, even small perturbations of integrable systems can produce local oscillations as well as a a slow drift of the actions and, after certain time, these drifts can cumulate in such a way to drive the system into very different physical state. This phenomenon is called diffusion of the actions, and is usually very difficult to study.

For small perturbations the system (1.1) falls within the range of celebrated perturbation theories which provide fundamental answers to the problem of the long–term stability, such as the KAM [3, 33, 42] and Nekhoroshev theorems [44].

The KAM theorem (which applies to (1.1) provided that $h$ satisfies a 'non–degeneracy' condition) states that the solutions, for small $|\varepsilon|$, remain quasi–periodic for a set $S_\varepsilon \subseteq V \times \mathbb{T}^n$ of initial data with large relative volume: Vol $S_\varepsilon \geq O(1 - \sqrt{|\varepsilon|})$. In particular, the actions have at most $\sqrt{\varepsilon}$-small bounded quasi–periodic variations. Therefore, any initial condition in this set can be considered to produce a perpetually stable motion.

However, even though the stability set $S_\varepsilon$ has large relative measure, its complement can be open, dense, and if the number of degrees of freedom is larger than two also connected; thus, the possibility for solutions to wander in the complement of $S_\varepsilon$, such as the so called Arnol'd diffusion, can prevent any result of stability over an infinite time.

As a consequence, if one wants to study the evolution of the actions with initial data in an open set, possibly the whole action–domain $V$, one has to abandon the usual concept of stability (which refers to infinite times), but one can try to produce asymptotic estimates, i.e. bounds:

$$|I(t) - I(0)| \leq r(\varepsilon)$$

which are valid for times $|t| \leq T(\varepsilon)$ such that $r \to 0$ and $T \to \infty$ as fast as possible in the limit $\varepsilon \to 0$. In the case $h$ satisfies a technical property called steepness, the Nekhoroshev theorem [44, 45] provides the celebrated exponential estimate:

$$r(\varepsilon) = r_0 \varepsilon^a, \quad T(\varepsilon) = t_0 \exp(\varepsilon_0/\varepsilon)^b$$

where $r_0$, $t_0$, $a$, $b$ and $\varepsilon_0$ are suitable positive constants independent of $\varepsilon$ and of $I(0)$. In addition to these important asymptotic estimates, the geometric construction which is the core of the proof of Nekhoroshev theorem provides a global picture which is useful to understand the dynamics of the system, relating the geometric properties of the integrable approximation $h(I)$ to the possible way the diffusion of the actions can take place in the phase space. For this reason, understanding the concepts of the proof of the theorem is a key for understanding the possible type of dynamics which can be found in a quasi–integrable system, including diffusive motions.

The chapter is organized as follows: Section 1.2 is devoted to integrability and basic examples and definitions. Section 1.3 is devoted to a review of important results of Hamiltonian perturbation theory. The fourth section is devoted to a detailed description of the geometric argument which is at the basis of the proof of the Nekhoroshev theorem. The fifth section is devoted to degenerate systems.

## 1.2 Classical Integrable Systems

An integrable system is characterized by a set of independent first integrals, in involution with respect to the Poisson brackets, at least equal to the number of the degrees of freedom. The Liouville–Arnol'd theorem [2] completely describes the features of the motions of such systems.

**Liouville–Arnol'd Theorem.** *Let $M$ be a symplectic manifold with two–form $\omega$ and dimension $2n$. Let the $n$ functions $F_j : M \longrightarrow \mathbb{R}$, $j = 1, \ldots, n$ be pairwise in involution with respect the Poisson brackets, i.e.*

$$\{F_i, F_j\} = 0, \quad \forall i, j = 1, \ldots, n, \tag{1.2}$$

*and be such that the level sets*

$$\mathcal{N}_z = \{x \in M : F_j(x) = F_j(z), \forall j = 1, \ldots, n\}, \tag{1.3}$$

are compact and connected for any $z \in M$. Assume there exists $z_* \in M$ such that $\mathrm{d}F_1, \ldots, \mathrm{d}F_n$ are linearly independent on $\mathcal{N}_{z_*}$. Then, there exists a neighbourhood $U$ of $z_*$ such that

(i) $\mathcal{N}_z$ is diffeomorphic to the $n$–dimensional torus $\mathbb{T}^n$ for any $z \subset U$;
(ii) denoting $\mathcal{V} = \cup_{z \in U} \mathcal{N}_z$, there exists a diffeomorphism

$$a : \mathcal{V} \to V \times \mathbb{T}^n$$
$$z \longmapsto a(z) = (I(z), \varphi(z)) \tag{1.4}$$

with $V \subseteq \mathbb{R}^n$ open, such that

$$\omega_{|\mathcal{V}} = \mathrm{d}I \wedge \mathrm{d}\varphi , \tag{1.5}$$

and the $n$ functions $\tilde{F}_j = F_j \circ a^{-1}$, $j = 1, \ldots, n$, depend only on $I_1, \ldots, I_n$.

Thus, if an Hamiltonian system with Hamilton function $H$ has first integrals $F_1, \ldots, F_n$ satisfying the hypotheses of the Liouville–Arnol'd theorem (usually one chooses $H = F_1$), then there exist action–angle coordinates $(I, \varphi)$ such that $H \circ a^{-1}(I, \varphi) = h(I)$, for some $h$, and therefore the Hamilton equations take the form

$$\frac{\mathrm{d}I}{\mathrm{d}t} = 0$$
$$\frac{\mathrm{d}\varphi}{\mathrm{d}t} = \omega(I) , \quad \omega = \frac{\partial h}{\partial I} , \tag{1.6}$$

with trivial solutions $I(t) = I(0)$ and $\varphi(t) = \omega(I(0))t + \varphi(0)$. Any motion of an integrable system is thus a linear drift on the torus $I = I(0)$, at constant velocity $\omega(I(0))$.

The Liouville–Arnol'd theorem applies in particular to the Kepler system and to the Euler–Poinsot rigid body:

**The Kepler system.** The system is defined by $M = \{(p,q) \in \mathbb{R}^6$ such that $q \neq 0\}$ and

$$H = \sum_{j=1}^{3} \frac{p_j^2}{2} - \frac{1}{r} , \quad r^2 = \sum_{j=1}^{3} q_j^2 . \tag{1.7}$$

The energy $F_1 = H$, the modulus $F_2 = \|q \times p\|$ ($\|\ \|$ denotes the euclidean norm) and any of the spatial components of the angular momentum, such as $F_3 = (q \times p)_3$, satisfy the hypotheses of the Liouville–Arnol'd theorem in the subset $M_0$ of $M$ given by

$$M_0 = \{(p,q) \in M : H < 0, \ 0 < \|q \times p\| \neq |(q \times p)_3|\} \setminus \{(p,q) :$$
$$\|p\| = \|q\|^{-\frac{1}{2}} \text{ and } p \cdot q \neq 0\} , \tag{1.8}$$

consisting of all Keplerian elliptic motions, with the exception of the circular ones and those lying on the reference plane $q_3 = 0$. A set of possible action–angle variables are the well known Delaunay elements $(L, G, \Theta, l, g, \theta)$ defined by $L = \sqrt{-1/2H}$, $G = F_2$, $\Theta = F_3$, $l$ is the mean anomaly while $g$ is the argument of the pericentre and $\theta$ is the longitude of the ascending node of the Keplerian ellipse that the point is describing. With such action–angle variables the Hamiltonian is $h(L) = -1/2L^2$, whose solutions are $l(t) = l(0) + (1/L(0)^3)t$, while $L, G, \Theta, g$ and $\theta$ stay constant. Therefore, all the solutions are periodic because of the existence of other two first integrals independent of $F_1, F_2$ and $F_3$, which are the two angles $g, \theta$. In the generic central motion problems all solutions are quasi-periodic with two frequencies because of the existence of four first integrals (e.g. the energy and the three components of the angular momentum): the fourth integral forces the motion to take place in invariant tori of dimension two, instead of dimension three. In the Kepler problem a further first integral, related to the Runge–Lenz vector, reduces the number of frequencies to one.

**The Euler–Poinsot rigid body.** The Euler–Poinsot rigid body is a rigid body with a fixed point, free of external torque. The configuration space of the system is SO(3), and so the Hamiltonian phase–space is the cotangent bundle $T^*SO(3)$ with its natural symplectic two–form. It is well known that $T^*SO(3)$ is trivial and can be identified with $SO(3) \times \mathbb{R}^3$ in such a way that if $x \in T^*SO(3)$, then $(\mathcal{R}(x), M(x)) \in SO(3) \times \mathbb{R}^3$ represents the matrix configuration $\mathcal{R}$ and the angular momentum $M$ in a body principal reference frame; correspondingly, the Hamilton function becomes

$$H(x) = \frac{1}{2} \sum_{i=1}^{3} \frac{M_i(x)^2}{a_i} , \qquad (1.9)$$

where $a_1, a_2, a_3$ are the three principal moments of inertia. We will refer for simplicity to the symmetric case $a_1 = a_2 \neq a_3$, for which the angular momentum vector in space and its projection $L$ over the symmetry axis of the body are integrals of motion. Thus, denoting by $G$ the euclidean norm of $M$ and by $J$ the projection of $M$ over any fixed direction in space (say the one determined by a unit vector $E$, so that $J = M \cdot \mathcal{R}E$) one has a set of three integrals of motion $G, L, J$ which satisfy the hypotheses of the Liouville–Arnol'd theorem in the set

$$\Sigma(E) = \{x \in T^*SO(3) \text{ such that } G(x) > 0 , \ |L(x)| < G(x)$$
$$\text{and} |J(x)| < G(x)\}. \qquad (1.10)$$

The Andoyer–Deprit variables $(G, L, J, g, l, j)$ are action–angle variables: the actions $G, L, J$ are defined as above, while the definition of the angles $g, l, j$ depends on the choice of a fixed reference frame determined by the orthonormal basis $e_x, e_y, e_z = E$ and a principal basis $e_1, e_2, e_3$. Precisely, $j$ is the angle among the vectors $e_x$ and $e_z \times M$; $g$ is the angle among the vectors $e_z \times M$ and $M \times e_3$; $l$ is the angle among $M \times e_3$ and $e_1$ ([1, 19]; see also [34]).

The Hamiltonian, which turns out to depend only on the actions $G, L$, takes the form

$$h(G,L) = \frac{G^2}{2u_1} + \eta \frac{L^2}{2a_3} , \quad \eta = \frac{a_1 - a_3}{a_1} . \tag{1.11}$$

The solutions of the Hamilton equations are

$$g(t) = g(0) + \omega_1 t , \quad \omega_1 = \frac{G(0)}{a_1}$$

$$l(t) = l(0) + \omega_2 t , \quad \omega_2 = \eta \frac{L(0)}{a_3} , \tag{1.12}$$

and describe classical Euler–Poinsot precession: the motion is the composition of a uniform rotation about its symmetry axis with frequency $\omega_2$ and a uniform precession of this axis about the fixed direction of the angular momentum in space with frequency $\omega_1$. As in the Kepler case, the motions are quasi–periodic with a number of frequencies which is strictly smaller than the degrees of freedom. Again, this is a consequence of the fact that there exists another integral of motion independent of $G, L, J$.

As we have seen, the Kepler and the Euler–Poinsot problems are special in the class of integrable systems, because have a number of independent first integrals larger than the number of the degrees of freedom. Such systems will be called degenerate or super–integrable, and the fibration in invariant $\mathbb{T}^n$ tori introduced by the Liouville–Arnol'd theorem does not reflect the geometry of motions. The following version of the Liouville–Arnol'd theorem [46] better describes the integrability properties of degenerate systems:

**Liouville–Arnol'd Theorem** (degenerate case). *Let $M$ be a symplectic manifold of dimension $2d$ with symplectic form $\omega$. Assume there exist $2d - n$ real functions $F_1, \ldots, F_{2d-n}$ defined on $M$, with $n \leq d$, such that their differentials are everywhere linearly independent and the first $n$ of them are in involution with all others:*

$$\{F_i, F_j\} = 0 , \quad i = 1, \ldots, n , \quad j = 1, \ldots, 2d - n . \tag{1.13}$$

*Assume also that the level sets*

$$\mathcal{N}_z = \{x \in M : F_j(x) = F_j(z), \forall j = 1, \ldots, 2d - n\} \tag{1.14}$$

*are compact and connected for any $z \in M$. Then, for any $z \in M$*
*(i) $\mathcal{N}_z$ is diffeomorphic to the $n$–dimensional torus $\mathbb{T}^n$;*
*(ii) there exist a neighbourhood $U$ of $z$ and a diffeomorphism*

$$\begin{aligned} a : \mathcal{V} &\longrightarrow V \times \mathbb{T}^n \\ z &\longmapsto ((I, p, q), \varphi) , \end{aligned} \tag{1.15}$$

*where $\mathcal{V} = \cup_{z \in U} \mathcal{N}_z$ and $V \subseteq \mathbb{R}^{2d-n}$ is an open set, such that*

$$\omega_{|\mathcal{V}} = dI \wedge d\varphi + dp \wedge dq . \tag{1.16}$$

Moreover, for any function $H$ in involution with all $F_1, \ldots, F_{2d-n}$ there exists a function $h(I)$ such that $H \circ a^{-1}((I, p, q), \varphi) = h(I)$ depends only on the actions $I$.

We will refer to the $(I, \varphi, p, q)$ as to the generalized action–angle coordinates, to the $(I, \varphi)$ as to the non-degenerate variables and to the $(p, q)$ as to the degenerate ones.

## 1.3 Quasi–Integrable Systems and Hamiltonian Perturbation Theory

A quasi–integrable Hamiltonian system is a perturbation of an integrable Hamiltonian $h$ of the form

$$H_\varepsilon(I, \varphi) = h(I) + \varepsilon f(I, \varphi) , \tag{1.17}$$

where the $(I, \varphi)$ are action–angle coordinates for $h$ defined in some domain $V \times \mathbb{T}^n$, with $V \subseteq \mathbb{R}^n$, and $\varepsilon$ is a small parameter. In general, for any small but non-vanishing $\varepsilon$, the term $\varepsilon f$ breaks the symmetry properties of the system, and one no longer expects for the existence of a set of first integrals of $H_\varepsilon$ satisfying the hypotheses of the Liouville–Arnol'd theorem. In such a situation, it is not obvious at all whether, and in which sense, the behaviour of the perturbed system remains similar to the integrable one. The study of such behaviour had been indicated by Poincaré as *le problème général de la dynamique* [57].

One of the aims of the Hamiltonian perturbation theory is to look for the longest possible time–scale on which the actions stay almost constant. The most relevant results in such a perspective are the KAM and Nekhoroshev theorems.

The KAM theorem [3, 33, 42] proves the integrability of the system in a closed subset of $V \times \mathbb{T}^n$ of large measure. A possible formulation of the theorem is the following:

**KAM Theorem.** *Let the Hamiltonian (1.17) be analytic in a complex neighbourhood of $V \times \mathbb{T}^n$, and let $h$ satisfy the non–degeneracy condition $\det \partial^2 h(I) \neq 0$ at all $I \in V$. Then, there exist positive constants $\varepsilon_0, a_1, a_2$ and $a_3$, independent of $\varepsilon$, such that for any $\varepsilon < \varepsilon_0$ one can find:*

*(a) a near to identity smooth canonical transformation*

$$\begin{aligned} \mathcal{C}_\varepsilon : V' \times \mathbb{T}^n &\longrightarrow V \times \mathbb{T}^n \\ (I', \varphi') &\longmapsto (I, \varphi) \end{aligned} \tag{1.18}$$

*with $V' \subseteq \mathbb{R}^n$ open;*
*(b) a set $V_\varepsilon \subseteq V \cap V'$;*
*(c) a smooth function $h'_\varepsilon(I')$ defined on $V' \times \mathbb{T}^n$;*

*satisfying*

(i) $\mathrm{Vol}(V \backslash V_\varepsilon) \leq a_1 \sqrt{\varepsilon}$ ;

(ii) $\|I - I'\| \leq a_2 \sqrt{\varepsilon}$, $\|\varphi - \varphi'\| \leq a_3 \sqrt{\varepsilon}$;

(iii) $H_\varepsilon \circ C_\varepsilon =_{V_\varepsilon} h'_\varepsilon$, where $=_{V_\varepsilon}$ means equality of the functions at the two members, and of their derivatives, whenever $I' \in V_\varepsilon$.

As a consequence of the above theorem, it is defined a set $S_\varepsilon = C_\varepsilon(V_\varepsilon \times \mathbb{T}^n) \subseteq V \times \mathbb{T}^n$ of large relative Lebesgue measure made of invariant tori. In particular, any motion $(I(t), \varphi(t))$ with $(I(0), \varphi(0)) \in S_\varepsilon$ satisfies

$$\|I(t) - I(0)\| \leq 2a_2 \sqrt{\varepsilon} \text{ for any } t \in \mathbb{R} . \tag{1.19}$$

The set made of invariant tori is constructed as the complement of a neighbourhood of all resonances. Precisely, for any integer vector $k \in \mathbb{Z}^n$ one excludes a neighbourhood of the resonant manifold:

$$R_k = \{I \in V : \text{ such that } k \cdot \omega(I) = 0\}$$

according to a Diophantine law:

$$|k \cdot \omega(I)| \geq \frac{\gamma_0 \sqrt{\varepsilon}}{|k|^\tau} \tag{1.20}$$

where $\gamma_0 > 0$ is a suitable positive constant and $\tau > n - 1$.

For concrete systems, it is very difficult to know if a given initial condition is on a KAM torus or not. In fact, the purely analytic estimates of constants $\varepsilon_0, a_1, a_2, a_3$, as well as the Diophantine constant $\gamma_0$, can be very pessimistic. The improvement of these estimates, especially in connection with systems of real interest, is not a trivial problem, and has been afforded in the last decades by refining the proof of the theorem and by the use of algebraic manipulators (see [15, 16, 17, 39]).

Different ways of understanding if a given initial condition is on a KAM torus are based on numerical integrations. In particular, the frequency analysis of a solution directly checks if a numerically computed solution is quasi–periodic and which are the frequencies [35, 36].

In this chapter we are more concerned with motions in the set which is the complement of the set $S_\varepsilon$ made of KAM tori, which we call Arnol'd web, in view of relations (1.20). The theorem does not exclude the possibility for the solutions with initial conditions in the complement of $S_\varepsilon$ to wander in it already with speed of order $\varepsilon$, as it happens in the well known example [44]:

$$H = \frac{I_1^2}{2} - \frac{I_2^2}{2} - \varepsilon \sin(\varphi_1 + \varphi_2) , \tag{1.21}$$

which has some special solutions with the actions moving at a speed of order $\varepsilon$:

$$I_1(t) = \varepsilon t\,, \qquad I_2(t) = \varepsilon t$$
$$\varphi_1(t) = \frac{1}{2}\varepsilon t^2\,, \quad \varphi_2(t) = -\tfrac{1}{2}\varepsilon t^2\,. \tag{1.22}$$

The solutions of system (1.21) can be easily written because the Hamiltonian depends on only one harmonic. However, for future convenience, we analyse the motions of the actions in term of line of fast drift. Precisely, when $\varepsilon = 0$ any motion of the Hamiltonian $h = \frac{I_1^2}{2} - \frac{I_2^2}{2}$ is quasi–periodic with frequencies $(I_1, -I_2)$. When $\varepsilon \neq 0$ the motion of the actions is restricted to a line (called line of fast drift) parallel to the vector $(1,1)$ and passing through the initial value $(I_1(0), I_2(0))$. Moreover, energy conservation adds the constraint:

$$|\Delta h| = |h(I(t)) - h(I(0))| \leq 2\varepsilon\,,$$

so that all motions outside the asymptotes $I_1 = \pm I_2$ are bounded, while the motions on the asymptote can drift with speed of order $\varepsilon$.

Instead, the Hamiltonian:

$$H = \frac{I_1^2}{2} + \frac{I_2^2}{2} - \varepsilon \sin(\varphi_1 + \varphi_2)\,, \tag{1.23}$$

which is very similar to (1.21), has all the motions which are bounded, because the level sets of $h$ are circles.

For generic perturbations, i.e. with more than one independent harmonics, and/or for a higher number of degrees of freedom, analysing the stability properties of the actions in the complement set of $S_\varepsilon$ is not trivial as in the case of these two special examples. However, the Nekhoroshev theorem provides geometric conditions which are sufficient to prevent the possibility of these drifts of the actions in the phase space (up to exponentially long times). The more general geometric condition in Nekhoroshev theorem is the so called steepness. In concrete applications, it is not always easy to test if a given function satisfies the steepness condition because the condition is written in an implicit form. However, some explicit conditions which are sufficient for steepness are the following ones [44, 45]:

- *convexity:*
$$h'' u \cdot u = 0 \quad \Rightarrow \quad u = 0$$
$h = \frac{I_1^2}{2} - \frac{I_2^2}{2}$ is not convex, but $h = \frac{I_1^2}{2} + \frac{I_2^2}{2}$ is convex.
- *quasi-convexity:*
$$(h' \cdot u = 0\,, \ h'' u \cdot u = 0) \Rightarrow u = 0\,,$$

so that convexity is required only in the plane orthogonal to the frequency vector, which contains the spaces of fast drift at exact resonances. The typical example of function which is is not convex, but is quasi–convex is $h = \frac{I_1^2}{2} + \frac{I_2^2}{2} + I_3$.

- *three–jet non–degeneracy:*

$$\left(h' \cdot u = 0 \,,\, h''u \cdot u = 0 \,,\, \sum_{i,j,k} \frac{\partial^3 h}{\partial I_i \partial I_j \partial I_k} u_i u_j u_k = 0\right) \Rightarrow u = 0\,.$$

The cubic function

$$h = \frac{I_1^2}{2} + \frac{I_2^2}{2} + \frac{I_3^3}{6}$$

is not convex, not quasi–convex but satisfies the three–jet condition.

More recently, weaker general conditions have been considered, such as the so called rational convexity or rational steepness [27, 54] which is extensively described in Chap. 2, while the exponential stability around elliptic equilibrium points has been proved under the so called directional quasi–convexity ([7], see also [20, 23])

For all the integrable systems in these classes it holds the following theorem:

**Nekhoroshev Theorem.** *Let us consider Hamiltonian (1.17) and suppose that it is analytic in a complex neighbourhood of $V \times \mathbb{T}^n$.*

*If $h$ is steep, there exist positive constants $\varepsilon_0, a, b, t_0, r$ such that for any $\varepsilon < \varepsilon_0$, and for any motion $(I(t), \varphi(t))$ with $(I(0), \varphi(0)) \in V \times \mathbb{T}^n$ it is*

$$|I(t) - I(0)| \leq r\varepsilon^a \qquad (1.24)$$

*for any time $t$ satisfying:*

$$|t| \leq t_0 \exp\left(\frac{\varepsilon_0}{\varepsilon}\right)^b . \qquad (1.25)$$

The value of the constants $a, b$ depends on the steepness properties of $h$. The best estimates for the generic steep case are given in [53]. Instead, the Nekhoroshev stability for perturbations of convex function has been studied in [10, 38, 56]. The convex case corresponds, among the steep functions, to the case characterized by the best stability estimates, such as $a = b = 1/(2n)$. Moreover, dealing with convexity instead of steepness simplifies many technical calculations, so that in the literature the Nekhoroshev theorem is usually illustrated in connection with convexity. However, while convexity is a strong hypothesis on the Hamiltonian $h$, the steepness condition is, in a given sense, generic for analytic functions [45, 55]. As a consequence, the steep case seems to be important in view of the study of complicated realistic physical problems, which can likely be expected to satisfy the generic steepness condition rather than the stronger convexity assumption. As a matter of fact, steepness played an important role in two applications of Nekhoroshev theorem to Celestial Mechanics, namely the stability of the asteroids in the Main Belt [29, 41] and the stability of L4–L5 Lagrangian equilibrium points [7].

For convex systems, Guzzo et al. [56] obtained the following estimates:

**Nekhoroshev Theorem** (convex case). *Let the Hamiltonian (1.17) be analytic in a complex neighbourhood of $V \times \mathbb{T}^n$ and let $h$ be $m$-convex, i.e*

$$\partial^2 h(I) u \cdot u \geq m u \cdot u \tag{1.26}$$

*for any $u \in \mathbb{R}^n$, at any $I \in V$. There exist $\varepsilon_0, a_0, t_0, \varepsilon_* > 0$ such that if $\varepsilon < \varepsilon_0$, then any motion $(I(t), \varphi(t))$ satisfies*

$$\|I(t) - I(0)\| \leq a_0 \varepsilon^{\frac{1}{2n}} \tag{1.27}$$

*for any time $t$ such that*

$$|t| \leq t_0 \exp\left(\frac{\varepsilon_*}{\varepsilon}\right)^{\frac{1}{2n}}. \tag{1.28}$$

*If the complex neighbourhood of $V \times \mathbb{T}^n$ where (1.17) is analytic has the form $V_{\rho_I} \times \mathbb{T}^n_{\rho_\varphi}$, where*

$$\begin{aligned} V_{\rho_I} &= \{I \in \mathbb{C}^n : \text{there exists } I_0 \in V \text{ with } \|I - I_0\| \leq \rho_I\} \\ \mathbb{T}^n_{\rho_\varphi} &= \{\varphi \in (\mathbb{C}/2\pi\mathbb{Z})^n : |\Im \varphi_i| \leq \rho_\varphi \text{ for any } i = 1, \ldots, n\} \end{aligned}, \tag{1.29}$$

*then possible values for $\varepsilon_0, t_0, a_0$ and $\varepsilon_*$ are*

$$\varepsilon_0 = \frac{m \rho_I^2}{2^{10} \left(11 \frac{M}{m}\right)^{2n}}, \quad a_0 = \rho_I \left(11 \frac{M}{m}\right)^{-1}$$

$$t_0 = \left(11 \frac{M}{m}\right)^2 \frac{\rho_\varphi}{\Omega}, \quad \varepsilon_* = \left(\frac{\rho_\varphi}{6}\right)^{2n} \varepsilon_0, \tag{1.30}$$

*where $M$ is such that $\|\partial^2 h(I) u\| \leq M \|u\|$ for any $u \in \mathbb{R}^n$ and any $I \in V_{\rho_I}$, and $\Omega = \sup_{I \in V_{\rho_I}} \|\partial h(I)\|$.*

As already remarked, the main features of Nekhoroshev theorem are the exponential dependence (1.28) of the stability time on $\varepsilon$ and its validity for all initial data in an open set. A direct consequence is that, when Nekhoroshev theorem applies, large displacements of the actions are possible only after times much longer than (1.28). For real systems, this seems very effective: at variance with the KAM theorem, one does not need to know if the initial condition is resonant or not, and if $\varepsilon$ is suitably small the exponentially long time (1.28) can be so long to be considered practically infinite (for example when it exceeds the lifetime of our solar system). In the practice, things are more complicate. As in the case of KAM theorem the analytic estimates of the constants $\varepsilon_0, \varepsilon_*$ is very inefficient, so that an effective estimates of the stability time require a lot of analytic work, including algebraic manipulators (see for example [25, 26]). An alternative approach is based on numerical integrations and on the Lyapunov exponents theory, and is explained in detail in Chap. 2.

A difficulty which arises when attempting to apply Nekhoroshev theorem to real problems is related to the degeneracy of the Kepler and Euler–Poinsot systems. Nekhoroshev [44] proved a version of his theorem which applies to perturbations of degenerate systems, i.e. to Hamiltonians of the form

$$H_\varepsilon(I,\varphi,p,q) = h(I) + \varepsilon f(I,\varphi,p,q) \ , \qquad (1.31)$$

with $(I,\varphi) \in U \times \mathbb{T}^n$ and $(p,q) \in \mathcal{B}$ where $U \subseteq \mathbb{R}^n$ and $\mathcal{B} \subseteq \mathbb{R}^{2(d-n)}$. Assuming that $h$ is convex with respect to the actions $I$ and that $f$ is analytic in a complex neighbourhood of $U \times \mathbb{T}^n \times \mathcal{B}$, one gets the stability bound (1.27) on the motion of the actions $I$ for times $|t| \leq \min\{T_N, T_e\}$, where $T_N$ is an exponentially long time and $T_e$ is the escape time of $(p,q)$ from their domain $\mathcal{B}$. We note that no estimate on the motion of the degenerate variables was provided, and so $T_e$ can be in principle much smaller than $T_N$, possibly of order $\varepsilon^{-1}$.

One of the most interesting degenerate systems is represented by the planets of our solar system. The integrable approximation is obtained by neglecting the interactions among the planets, so that it is essentially the sum of decoupled Kepler problems, for which we can introduce the Delaunay action–angle variables:

$$\begin{aligned} L_j &= \mu_j \sqrt{k(m_0 + m_j)a_j}, & l_j &= M_j \\ G_j &= L_j \sqrt{1 - e_j^2}, & g_j &= \omega_j \\ H_j &= G_j \cos i_j, & h_j &= \Omega_j, \end{aligned} \qquad (1.32)$$

where $a_j$ denotes the semi–major axis of the $j$th planet, $e_j$ denotes its eccentricity, $i_j$ its inclination, $M_j$ its mean anomaly, $\omega_j$ the argument of the pericenter, $\Omega_j$ the longitude of the ascending node and $m_j$ its mass ($m_0$ denotes the mass of the Sun). With these variables, the complete Hamiltonian of the system has the quasi–integrable form:

$$H_\varepsilon = \sum_{j=1}^{N} h(L_j) + \varepsilon f(L, G, H, l, g, h) \ , \qquad (1.33)$$

where $h(L_j)$ is the Hamiltonian of the Keplerian part related to the $j$th planet. This problem is degenerate because $H_0 = \sum_{j=1}^{N} h(L_j)$ depends only on the actions $L_1, \ldots, L_N$, so that a direct application of Nekhoroshev theorem allows one to prove that, if $\varepsilon$ is suitably small, the action variables $L_j$ (i.e. the semi–major axes of the planets) stay almost constant up to exponentially long times, provided in that time the perturbation remains bounded (i.e. provided the system remains far from close approaches; sufficient conditions are provided in [44, 52]). However, as it is usual for non–degenerate cases, more detailed information on the dynamics can be obtained by the analysis of the normal form obtained after the average of the non–resonant combinations of the mean anomalies, which is also called secular problem. In fact, secular planetary theories describe the motion of the planets from Mercury to Neptune by means of equations obtained by averaging the mean anomalies [12, 35]. Therefore, the averaged semi–major axis of the planets are constant, and the secular system has an elliptic equilibrium point corresponding to a motion in which all the orbits are circular and lie on the same plane. The linearized system around the equilibrium point has quasi–periodic solutions, which are known as the Laplace–Lagrange solutions of the Solar System. Numerical investigations reveal that that this linear approximation is suitable to describe the motion of the planets up to times of order $10^7$yr: on the one hand, secular resonances are responsible of the chaotic motions of the inner planets [35, 37], while the chaotic motion of the outer planets [58] has been explained by means of three–planets mean motion resonances [31, 32, 43].

## 1.4 Inside the Nekhoroshev Theorem

In this section, we review the basics facts of the proof of Nekhoroshev theorem because they provide detailed informations on the motions, besides the celebrated asymptotic exponential estimate. Indeed, the geometric construction introduced in the course of the proof is rich, and can be used to describe in great detail the local behaviour of the actions, which turns out to be rather different from point to point in the action space. In order to bring this structure into light, we sketch it in the simplest steep case, i.e. the case of convex function $h(I)$.

The starting point is that, by means of a near to identity canonical transformation, one can remove from the quasi–integrable Hamiltonian (1.17) an harmonic term $\varepsilon f_k(I)e^{ik\cdot\varphi}$ only outside a small neighbourhood of the *resonant surface* defined by $k\cdot\omega(I)=0$, with $\omega=\partial h$. But, since $h$ is convex, the set of resonant surfaces related to all integer vectors $k\in\mathbf{Z}^n\backslash 0$ is dense in the action space $\mathbf{R}^n$. Therefore, in any given open set it is not possible to remove from the perturbation an infinite number of harmonics. This fact, which prevents in general the integrability of the system, was pointed out by Poincaré [57]. Nekhoroshev observed that, in spite of this essential obstruction to the integrability of the system, the long-term behaviour of the solutions can be studied removing only a finite number of harmonics $\varepsilon f_k(I)e^{ik\cdot\varphi}$, precisely those with order $|k|=\sum_{i=1}^n |k_i|$ up to a given threshold $K$. In fact, because of the analyticity of $f$, the harmonics with $|k|\geq K$ can be estimated to be smaller than $\exp(-K\rho_\varphi)$ ($\rho_\varphi$ is a positive parameter such that the Hamiltonian is analytic for $|\mathrm{Im}\varphi_i|\leq\rho_\varphi$, see (1.29)). As a consequence, if one chooses $K$ suitably large (for example $K\sim\varepsilon^{-\frac{1}{2n}}$), these terms turn out to be very small and they can determine large deviations of the actions only after long times of order $\exp(K\rho_\varphi)$. We focus therefore our attention on the harmonics with order $|k|\leq K$. To perform an average of these low order harmonics, we cover the action space with the following sets (see Fig. 1.1):

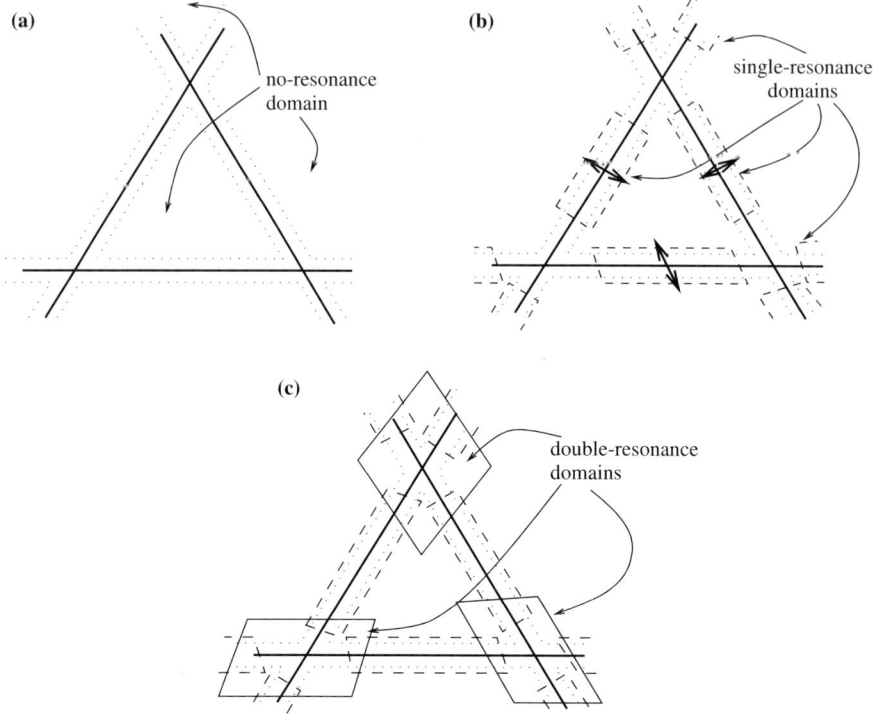

**Fig. 1.1.** Sketch of the geometric construction of Nekhoroshev theorem in the case $n = 3$. The pictures represent two–dimensional sections of the three–dimensional frequency space. The *solid lines* are the resonant manifolds; *dotted lines* delimit the non–resonant domain; *dashed boxes* delimit the single-resonance domains; *continuous boxes* delimit the double-resonance domains

- a *no–resonance domain*, which is the set of points which are far enough from all the resonant surfaces up to order $K$;
- the *single–resonance domains*, which are characterized by the presence of only one resonant surface of order smaller than $K$;
- the *double-resonance domains*, which are centred around the crossings of two single–resonance surfaces;

and so on up to the $(n-1)$–dimensional resonant domains. Any of these domains is characterized by being far from all resonant manifolds, except a finite number. For any set of independent integer vectors $k$ with $|k| \leq K$ generating a lattice $\Lambda$, we denote by $D_\Lambda$ the resonant domain related to $\Lambda$. The lattice $\Lambda$ can have dimension at most $n-1$ for the $(n-1)$–dimensional resonant domains. Therefore, in $D_\Lambda$, one can average the non–resonant low order harmonics, thus obtaining the normal form Hamiltonian

$$h(I) + \varepsilon g_\Lambda(I,\varphi) + \varepsilon \exp\left(-K\frac{\rho_\varphi}{2}\right) r(I,\varphi) \tag{1.34}$$

where the Fourier expansion of $g_\Lambda$ contains only harmonics related to integer vectors of $\Lambda$. The normal forms (1.34) can then be used to provide the confinement of the actions for exponentially long times. More precisely

- in the no–resonance domain (the one with $\Lambda = 0$), one can eliminate all the harmonic terms of the perturbation $\varepsilon f$ of order smaller than $K$ so that the behaviour of the system is essentially governed by an effective normal form Hamiltonian of type

$$h(I) + \varepsilon g_0(I) + \varepsilon \exp\left(-K\frac{\rho_\varphi}{2}\right) r_0(I,\varphi) \tag{1.35}$$

which, apart from an exponentially small remainder, looks like an integrable one. Therefore, in the no–resonance domain, the actions experience essentially only a slow diffusion forced by $\exp\left(-K\frac{\rho_\varphi}{2}\right) r_0$;

- in any single–resonance domain one can eliminate all the harmonics $\varepsilon f_k(I)e^{ik\cdot\varphi}$ of order smaller than $K$ except those with $k$ parallel to a resonant vector $\nu$, so that the normal form Hamiltonian is of type

$$h(I) + \varepsilon g_\nu(I, \nu \cdot \varphi) + \varepsilon \exp\left(-K\frac{\rho_\varphi}{2}\right) r_\nu(I,\varphi) \tag{1.36}$$

Apart from the exponentially small remainder, the above Hamiltonian is still integrable because it depends only on one angle, but in its motions the actions are no longer fixed. However, because $\dot{I} \sim \nu$, they can move only along the direction $\nu$, which is usually called the *fast drift direction* (indicated by arrows in Fig. 1.1b). The crucial point of the construction is that the convexity hypothesis on $h$ guarantees that the fast drift direction is transversal to the resonant surface. This fact produces the confinement: indeed, following indefinitely the fast drift direction, the motion would enter the no–resonance domain. But this is impossible, since in the no–resonance domain the actions are essentially fixed, as explained above. On the other hand, transversal motion with respect to the fast drift direction can be forced only by the remainder, so that it is exponentially slow;

- in the double-resonance domains the normal form Hamiltonian has two independent resonant terms $k_1$, $k_2$ of order smaller than $K$. Then, the motion of the actions can take place in any direction of the plane spanned by $k_1$ and $k_2$. However, due to the convexity of $h$, these motions are bounded: again, if the actions moved far enough from the double resonance, they would enter by construction either the no–resonance domain or one of the single resonance domains. But this is impossible, since in the no–resonance domain the actions are essentially fixed, and in the single-resonance domains they can change only along the fast drift direction.

The same reasoning applies to resonant domains of higher order, up to $n-1$. In conclusion, neglecting the exponentially small remainder, for each initial condition, the motion is confined within one of the resonance domains. As a consequence, the actions can change at most by a quantity equal to the radius of the resonant domains in the fast drift direction, whose largest value, corresponding to the $(n-1)$–resonant domains, is proportional to $1/K$. So, by increasing $K$ such a radius decreases, but at the same time the number of resonances to be taken into account increases as well, so that one must stop when $K$ is so large that all resonances with their resonant domains do not leave place for the no–resonance domain. The largest value one can choose turns out to be $K = (\varepsilon_*/\varepsilon)^{1/2n}$, with suitable $\varepsilon_*$.

Finally, the exponentially small remainder can force diffusion in the action space, but only with exponentially small speed. Then the result concerning bounded motions will be true only up to exponentially long times. We underline again that a fundamental role in the above construction is played by convexity. Indeed, if $h$ is not convex, the fast drift linear space can be parallel to the resonant surface, and so the actions could move indefinitely on it without leaving the resonant domain (as it happens for Hamiltonian (1.21)).

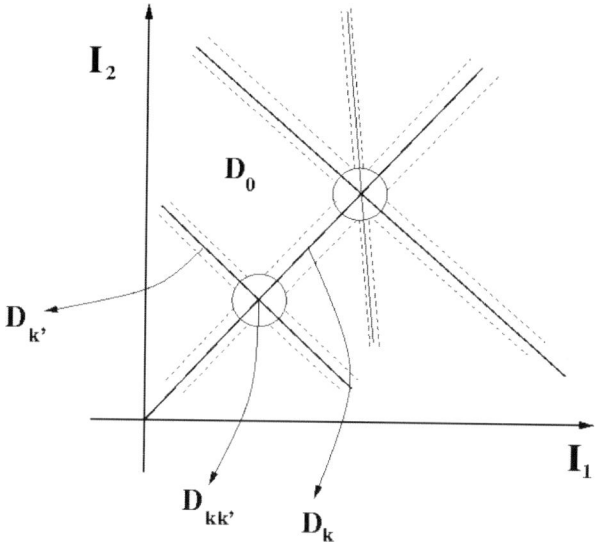

**Fig. 1.2.** Sketch of the geometry of resonances for Hamiltonian (1.37): the *lines* represent the single resonances; the crossing of two independent resonances $k_1 I_1 + k_2 I_2 + k_3 = 0$ $k'_1 I_1 + k'_2 I_2 + k'_3 = 0$ represent the double resonances. The domain $D_0$ is the non–resonant domain, while $D_k, D_{k'}$ are the single resonant domains related to the integer vectors $k$ and $k'$ respectively; $D_{k,k'}$ is the resonant domain related to the two dimensional lattice generated by $k$ and $k'$

The argument providing the confinement of the actions described above was introduced by Nekhoroshev in [44] for steep functions, and it is sometimes called the 'trap argument'. The detailed proof in the convex case, for which there exists also an alternative approach based on energy conservation [11, 24, 38, 56] can be found in [10].

As an illustration of this construction, we describe the set of resonances, resonant domains and the dynamics in the case of Hamiltonian:

$$H = \frac{I_1^2}{2} + \frac{I_2^2}{2} + I_3 + \frac{\varepsilon}{\cos(\varphi_1) + \cos(\varphi_2) + \cos(\varphi_3) + 4} \quad . \tag{1.37}$$

The integrable approximation:

$$h = \frac{I_1^2}{2} + \frac{I_2^2}{2} + I_3$$

is quasi–convex, and the perturbation is analytic, therefore the Nekhoroshev theorem applies to $H$. The frequency vector related to $h$ is

$$\omega = (I_1, I_2, 1) \, ,$$

and therefore for any $k \in \mathbb{Z}^3$ its resonance is the line in the $(I_1, I_2)$ plane defined by

$$k_1 I_1 + k_2 I_2 + k_3 = 0 \, .$$

The double resonances are the points at the crossings of at least two of these lines. The situation is schematically represented in Fig. 1.2.

The details of the dynamics of any initial condition $(I_0, \varphi_0)$ up to times which are small compared to the exponentially long time (1.28) depend on which resonant domain contains the actions $I_0$. Precisely, in the non–resonant domain the motion is conjugated to a system in which the actions are constant (apart from eventual exponentially slow drifts); in the single-resonant domain the motion is conjugated to a system in which the actions are confined to the line of fast drift orthogonal to the resonant line (apart from eventual exponentially slow drifts; see Fig. 1.3); in the double resonant domains the motion is confined near the center of the resonance (see Fig. 1.3).

Instead, on times which are of order or longer than the exponentially long time (1.28) the actions can drift along the resonant lines, visiting different resonant domains (see Fig. 1.4).

Numerical computations of the geometry of resonances and of the diffusion properties of quasi–integrable systems very similar to this one is described in details in Chap. 2.

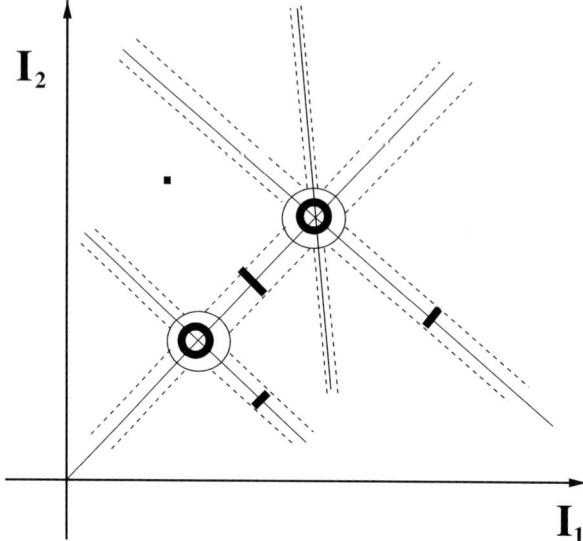

**Fig. 1.3.** Schematic representation of the dynamics of Hamiltonian (1.37) up to times which are small compared to the exponentially long time (1.28). In the non-resonant domain, the motion is conjugated to a system in which the actions are constant (the *bold points*); in the single-resonant domain the motion is conjugated to a system in which the actions are confined to the line of fast drift orthogonal to the resonant line (the *bold segments*); in the double-resonant domains the motion is confined near the centre of the resonance (the *bold circles*)

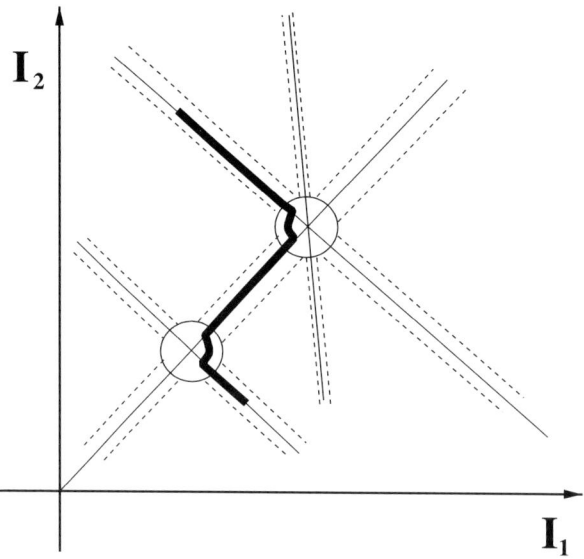

**Fig. 1.4.** Schematic representation of the dynamics of Hamiltonian (1.37) up to times which are longer than the exponentially long time (1.28): the actions can drift along the resonant lines, visiting different resonant domains

## 1.5 Exponential Stability in Degenerate Systems

A degenerate integrable system has a number of independent first integrals which is strictly larger than the number $d$ of the degrees of freedom. The additional first integrals force the motion to take place always in invariant manifolds of dimension strictly lower than $d$, which turn out to be lower dimensional tori: in correspondence, some frequencies are null and the number of the actions appearing in $h$ is strictly lower than $d$. For a detailed description of the geometry of these systems, which are called also 'non–commutatively integrable', see [18, 40, 46]; for geometric considerations about perturbation theory for degenerate systems see [21, 22].

The degenerate version of the Liouville–Arnol'd theorem recalled in Sect. 1.2 provides 'generalized action–angle coordinates' $(I, \varphi, p, q)$ with $(I, \varphi) \in U \times \mathbb{T}^n$, $U \subseteq \mathbb{R}^n$, and $(p, q) \in \mathcal{B} \subseteq \mathbb{R}^{2(d-n)}$, such that the two form is $dI \wedge d\varphi + dp \wedge dq$ and the Hamiltonian depends only on the $n$ actions $I$: $H(I, \varphi, p, q) = h(I)$. A perturbation of a degenerate system has therefore the form

$$H_\varepsilon(I, \varphi, p, q) = h(I) + \varepsilon f(I, \varphi, p, q) \,. \tag{1.38}$$

As it was observed by Nekhoroshev [44], if $h$ is steep, then Nekhoroshev theorem applies to (1.38), providing the usual bound for the motion of the actions $I$, but only for a time which is the smaller between an exponentially long time and the escape time of $(p, q)$ from the domain $\mathcal{B}$. An escape is possible because the motion of the so called degenerate variables $(p, q)$ is bounded in principle, in lack of other informations, only by the a priori estimate $(\dot{p}, \dot{q}) = O(\varepsilon)$; thus, the coordinates $(p, q)$ can experience variations of order 1 already on time scales of order $\varepsilon^{-1}$. Therefore, to provide accurate descriptions of the motions of the degenerate variables, one has to look at the structure of the perturbation with respect to these variables. To do this, we recover the definition of resonances given in Sect. 1.4 for the fast angles $\varphi$: we still define the resonant domains in the space of the actions $I$ with respect to the frequency vector $\omega = \frac{\partial h}{\partial I}$. Therefore, in the non–resonant domain the normal form is

$$h(I) + \varepsilon g_0(I, p, q) + \varepsilon \exp\left(-K\frac{\rho_\varphi}{2}\right) r(I, \varphi, p, q) \,. \tag{1.39}$$

If we neglect the exponentially small remainder the normal form

$$h(I) + \varepsilon g_0(I, p, q)$$

does not depend on the angles $\varphi$, so that the actions $I$ are constants of motion and can be regarded as parameters. Therefore, the degenerate variables $(p, q)$ satisfy the Hamilton equations of the reduced Hamiltonian: $\varepsilon g_0(I(0), p, q)$, which defines the secular problem. In many physically interesting situations, the secular problem is integrable or quasi–integrable. In such a case it is possible to choose the $(p, q) \in V \times \mathbb{T}^m$ (with $V \subseteq \mathbb{R}^m$ open) to be action–angle coordinates and to individuate a second small parameter $\eta$ such that the secular Hamiltonian takes the quasi–integrable form

$$g_0 = K_0(I,p) + \eta K_1(I,p,q) \ .$$

In the integrable case it is obviously $\eta = 0$. The stability of the degenerate variables is therefore reduced to the study of the stability of the quasi–integrable system $g_0$.

Instead, in the resonant domain related to an integer lattice $\Lambda$, the normal form Hamiltonian is

$$h(I) + \varepsilon g_\Lambda(I,\varphi,p,q) + \varepsilon \exp\left(-K\frac{\rho_\varphi}{2}\right) r(I,\varphi,p,q) \ . \quad (1.40)$$

where the Fourier expansion of $g_\Lambda$ with respect to the angles $\varphi$ contains only harmonics related to integer vectors in $\Lambda$. In this case, even neglecting the exponentially small remainder, the actions $I$ are not constants of motion, so that we cannot reduce the dynamics to the study of the secular Hamiltonian. This is the case of planetary systems in mean motion resonances and of asteroids in mean motion resonances with the planets.

It turns out that with suitable hypotheses (which are the hypotheses used by Arnol'd in [4] to study non–resonant motions of degenerate systems with KAM theory) a Nekhoroshev stability result for the degenerate and non–degenerate variables can be proved also for these resonant cases, except for a set of resonances of logarithmically small order. Such a fact has been developed for the first time in [29] for the asteroid belt problem, then in [30] in a more general situation and is reviewed below.

Instead, inside the resonances of logarithmically small order the degenerate variables can indeed experience chaotic motions in short time.

An argument which explains the production of chaotic motions in very low order resonances for systems with one degenerate and one non–degenerate degree of freedom was given by Neishtadt in [50] and used to explain chaos in the 3–1 mean motion resonance of the restricted three body problem [47, 51, 60]. This argument is reviewed below.

Some specific degenerate systems satisfy conservation laws which allow one to prove stability of the degenerate variables. For example, Nekhoroshev studied the planetary problem proving an exponential stability result for the semi–major axis, eccentricities and inclinations of the planets ([44], see also [52]) with a combined use of the Nekhoroshev theorem and the so–called Laplace argument, valid only for very small eccentricities. In the perturbed Euler–Poinsot rigid body the degenerate variables determine the motion in space of the angular momentum vector, which, in the resonant case can perform large chaotic motions in small time [5, 6, 8]. However, for gyroscopic motions, this is prevented by the analytic properties of the perturbation [9]. Other examples can be found in [7, 13, 14].

### 1.5.1 Nekhoroshev Stability Outside Very Small Order Resonances

The results described in this section were proved in [30], and apply to many physically interesting situations, such as the asteroid belt problem [29, 41].

We consider degenerate systems with Hamiltonian in the form (1.38). The main hypothesis concerns the average of $H$ with respect to the angles $\varphi$:

$$\mathcal{K}(I,p,q) = h(I) + \frac{\varepsilon}{(2\pi)^n} \int_0^{2\pi} \cdots \int_0^{2\pi} f(I,\varphi,p,q) \mathrm{d}\varphi_1 \ldots \mathrm{d}\varphi_n, \qquad (1.41)$$

which, only in this section, we will call the secular Hamiltonian. Following [1], we suppose that the secular Hamiltonian is integrable or quasi–integrable, so that it is possible to choose the $(p,q) \in V \times \mathbb{T}^m$ (with $V \subseteq \mathbb{R}^m$ open) to be action–angle coordinates and to individuate a second small parameter $\eta$ such that

$$\mathcal{K} = h(I) + \varepsilon(K_0(I,p) + \eta K_1(I,p,q)) \qquad (1.42)$$

with $K_0$ and $K_1$ analytic. In the integrable case it is obviously $\eta = 0$.

A degenerate system which has an integrable secular Hamiltonian is the spatial restricted three–body problem with small values of the mass-ratio; the Hamiltonian of the system describing the motion of an asteroid in the main belt perturbed by the planets is quasi–integrable (see [59] and also [29, 41]). In the former case $\eta = 0$, in the latter $\eta$ is of the order of the eccentricity and inclination of the orbits of the perturbing planets. Such situations are typical of Celestial Mechanics problems, where one often deals with different small parameters.

To apply a Nekhoroshev-like result we also assume that $h(I)$ is a steep function of the $I$, while $h(I) + \varepsilon K_0(I,p)$ is steep functions of all the actions $I, p$. For simplicity, in the following we restrict to the convex case, but the result suitable extends also to the steep case (which is necessary for the asteroid belt case).

As anticipated, in [30] it is proved the following theorem.

**Theorem 1.1.** *Let $H$ be as in (1.38) with secular Hamiltonian (1.41) be as in (1.42) analytic in a complex neighbourhood of $(I, p, \varphi, q) \in U \times V \times \mathbb{T}^n \times \mathbb{T}^m$, where $U \subseteq \mathbb{R}^n$, $V \subseteq \mathbb{R}^m$ are open sets. Let $h(I)$ be a convex function of the $I$ and $h(I) + \varepsilon K_0(I,p)$ be a convex functions of all the actions $I, p$. There exist positive constants $\varepsilon_0, a, b, c, d$ such that if $\varepsilon \leq \varepsilon_0$, then any motion of the system with initial condition satisfying*

$$\left| \nu \cdot \frac{\partial h}{\partial I}(I(0)) \right| > \varepsilon^a \quad \text{for any} \quad \nu \in \mathbb{Z}^n \setminus 0 \quad \text{with} \quad |\nu| = \sum |\nu_i| \leq b \ln \frac{1}{\varepsilon}$$

*has the actions $I(t), p(t)$ which remain close to their initial value for any time $t$ such that*

$$|t| \leq \exp \frac{d}{\max\{\varepsilon, \eta\}^c}.$$

For possible values of the constants $\varepsilon_0, a, b, c, d$ we refer to the paper [30]. The degenerate actions are therefore proved to be almost constant for exponentially long times if the initial action $I(0)$ is taken outside a small neighbourhood of the resonant manifolds of order logarithmically small. On the contrary, the result does not apply near the very low order ($|\nu| \sim 1$) resonant manifolds.

For asteroids in the belt case the result can be stated as follows [19, 20].

**Theorem 1.2.** *Let $n_j, e_j, i_j$, $j = 1, \ldots, 4$, denote the frequencies of the mean motion, the eccentricity and the inclination of the outer planets, from Jupiter to Neptune. Let $\varepsilon$ denotes the mass of Jupiter. There exist positive constants $c_1, c_2, \varepsilon_0, e_0, i_0$ such that if $\varepsilon \leq \varepsilon_0$, $e_j \leq e_0$ and $i_j \leq i_0$ for any $j \leq 4$, then for any motion $L(t), G(t), \Theta(t), g(t), l(t), \theta(t)$ with $L(0)$ satisfying*

$$\left|\nu_0 \frac{1}{L(0)^3} + \sum_{k=1}^{4} \nu_k n_k\right| \geq \varepsilon^{c_1} \tag{1.43}$$

*for any $(\nu_0, \nu_1, \ldots, \nu_4) \in \mathbb{Z}^5$ with $\sum_{k=0}^{4} |\nu_k| \leq c_2 \ln(1/\varepsilon)$, the semi–major axis, the eccentricity and the inclination of the asteroid are exponentially stable (the stability time grows exponentially as a suitable inverse power of $\varepsilon, e_1, \ldots, e_4, i_1, \ldots, i_4$).*

Therefore, outside some mean motion resonances of logarithmically small order (among which there are those associated with the Kirkwood gaps, and possibly many others, see [48]), if the masses, the eccentricities and the inclinations of the planets are suitably small, then the semi–major axis, the eccentricity and the inclination of the asteroid are exponentially stable. However, as explained in detail in [41], one does not expect that the thresholds $\varepsilon_0, e_0, i_0$ are uniform in the phase–space. Therefore, in principle, we do not know whether outside the low order mean motion resonances the entire phase space is exponentially stable or not, and consequently the search for exponentially stable objects has to be done with reference to specific regions of the phase–space (see [28, 49]).

### 1.5.2 Chaos and Order in the Very Low Order Resonances

When the reduction of point (1.1) cannot be done, the degenerate variables can experience chaotic motions in times which grow as an inverse power of $\varepsilon$. An argument which explains the production of chaotic motions in very low order resonances for systems with one degenerate and one non–degenerate degree of freedom was given by Neishtadt in [50] and used to explain chaos in the 3–1 mean motion resonance of the restricted three body problem [47, 51, 60]. To illustrate this argument, following [50], we consider a model Hamiltonian of the form:

$$H = \frac{I^2}{2} - \varepsilon[a(p,q)\cos\varphi + b(p,q)] \tag{1.44}$$

with $I, p \in \mathbb{R}$, $\varphi, q \in \mathbb{S}^1$.

For times $t \ll \frac{1}{\varepsilon}$, the degenerate variables $p, q$ are almost constant, so that it holds the adiabatic invariant approximation:

$$h = \frac{I^2}{2} - \varepsilon\, a(p,q) \cos\varphi\,, \quad p, q = \text{const}\,. \tag{1.45}$$

In this approximation, the action–angle variables $(I, \varphi)$ have equations of motion of a pendulum-like Hamiltonian, where the amplitude of the pendulum is the constant $a(p,q)$. Formally, outside the separatrices of the pendulum, i.e. outside the curves defined by

$$\varepsilon a(p,q) = \frac{I^2}{2} - \varepsilon\, a(p,q) \cos\varphi\,, \tag{1.46}$$

one can define an adiabatic invariant $\mathcal{I}$ of the system, whose first approximation is given by the action variable of the pendulum, i.e.

$$\mathcal{I} = \frac{1}{2\pi} \int_{h=\text{const}} I \, d\varphi\,. \tag{1.47}$$

More precisely, there exists a canonical change of variables:

$$(I, \varphi, p, q) \longmapsto (\mathcal{I}, \phi, p', q')$$

where the $(p', q')$ are near to the degenerate variables $(p, q)$, conjugating the Hamiltonian (1.44) to a function of the form:

$$\mathcal{H}(\mathcal{I}, \phi, p', q') = \hat{h}(\mathcal{I}, p', q') + \ldots \tag{1.48}$$

where the dots denote terms of higher order with respect to $\varepsilon$ (the order depends on the adiabatic approximation). The change of variable is of course defined only outside a suitable neighborhood of the separatrices (1.46).

We now describe the motions with fixed energy $\varepsilon H_0$ and fixed initial angle $\varphi = \varphi_0$ (of course, everything is valid by changing the values of $H_0, \varphi_0$). For any initial condition $p_0, q_0$ of the degenerate variables $p, q$, two situations are possible:

1. The motion with initial conditions $\varphi_0, p_0, q_0$ and $I_0 = \pm\sqrt{\varepsilon}\sqrt{H_0 + a(p_0, q_0)}$ is such that it never hits the separatrix (1.46). Therefore, the adiabatic approximation can be used to describe the motion and the function $\mathcal{I}$ is an approximate integral of motion. But, because the function $\mathcal{I}$ depends on the energy of the pendulum $\frac{I^2}{2} - \varepsilon a(p,q) \cos\varphi$ and on the variables $(p, q)$, the equation $\mathcal{I} = \text{const}$ defines a curve $\gamma(\mathcal{I})$ in the $(p, q)$ plane, and the motion of the degenerate variables is confined near this curve. We can now control if the hypothesis that the motion never hits the separatrix (1.46) is satisfied: the equation of the separatrix (1.46) defines the curve in the plane $(p, q)$, that we denote by $\Gamma$:

$$a(p, q) = H_0 - b(p, q)\,. \tag{1.49}$$

The motion never hits the separatrix (1.46) if the curve $\gamma(\mathcal{I})$ does not intersect the curve $\Gamma$.

Therefore, in the plane $p, q$ all motions with initial conditions on a curve $\gamma(\mathcal{I})$ which does not intersect the curve $\Gamma$ remain confined near the curve $\gamma(\mathcal{I})$, and there is not diffusion of the degenerate variables in the $(p, q)$ plane.

2. The second possibility is that a solution with initial conditions $\varphi_0, p_0, q_0$ and $I_0 = \pm\sqrt{\varepsilon}\sqrt{H_0 + a(p_0, q_0)}$ hits the separatrix (1.46). This can happen only if the initial conditions belong to a curve $\gamma(\mathcal{I})$ which intersects the curve $\Gamma$. In this case, the motion of the $p_t, q_t$ are confined near the curve $\gamma(\mathcal{I})$ only till it intersects the curve $\Gamma$: near the intersection the adiabatic approximation is not valid, so that the value of the adiabatic invariant can change a little. When the motion leaves the curve $\Gamma$ it is on a curve $\gamma(\mathcal{I}_1)$, with $\mathcal{I}_1$ which can be slightly different from $\mathcal{I}$. The cumulations of many of these changes in the value of the adiabatic invariant, due to many 'encounters' of the motions with the curve $\Gamma$, can produce chaotic diffusion of the degenerate variables $(p, q)$. It is evident that the region of the chaotic motions in the plane of the degenerate variables is the region defined by all $\gamma(\mathcal{I})$ which intersect $\Gamma$ in some points.

As an illustration of this mechanism of production of chaotic motions we discuss the case

$$H = \frac{I^2}{2} - \varepsilon[a(p, q)\cos\varphi + b(p, q)]$$

with

$$a(p, q) = \frac{p^2}{2} - \cos q + 2 \quad , \quad b(p, q) = p^2 + \cos q \, .$$

The one-dimensional system with Hamiltonian:

$$h = \frac{I^2}{2} - \varepsilon a \cos\varphi$$

has the hyperbolic equilibrium point $(I, \varphi) = (0, \pi)$ and separatrices characterized by $\frac{I^2}{2} - \varepsilon a \cos\varphi = \varepsilon a$. We now fix a value of energy $\varepsilon H_0$ and the initial angle $\varphi_0$. In the plane $(p, q)$ the separatrix has equation

$$H_0 = a(p, q) - b(p, q) \, ,$$

which in the present case is

$$H_0 = -\frac{p^2}{2} - 2\cos q + 2 \, .$$

All motions whose projection on the plane $p, q$ encounters this curve can be chaotic, the other are regular. The situation is represented in Fig. 1.5: on the top it is represented the dynamics in the $(p, q)$ plane. The bold curve represents the curve $\Gamma$. On the bottom we represent the level curves $\gamma(\mathcal{I})$ at

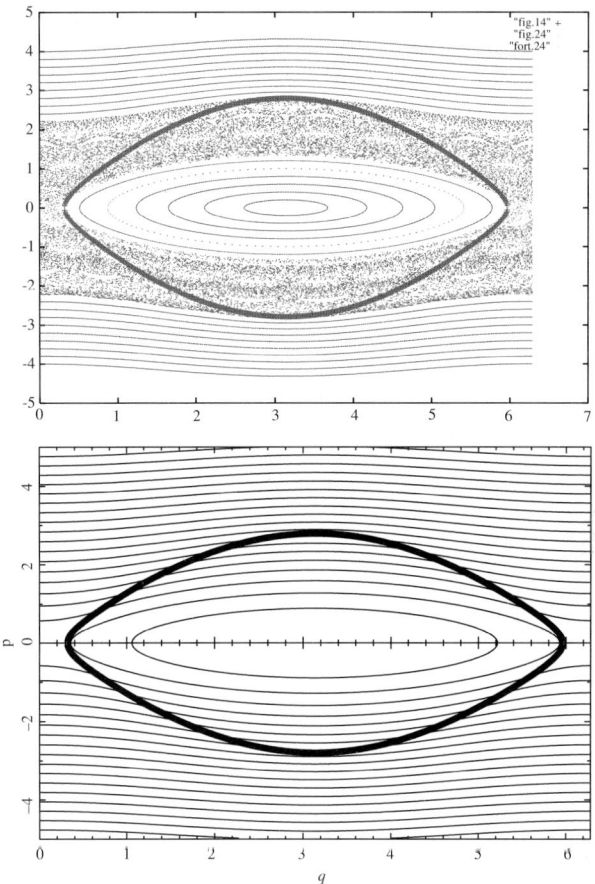

**Fig. 1.5.** On the top: representation of the projection of the dynamics on the $(p,q)$ plane with the condition $|\varphi| < 0.05$. The phase portrait corresponds to $H_0 = 0.1$. The *bold curve* represents the separatrix curve $\Gamma$. On the bottom we represent the level curves $\gamma(\mathcal{I})$ at different values of $\mathcal{I}$. The dynamics of the degenerate variables is regular in the region of the $(p,q)$ plane where the curves $\gamma(\mathcal{I})$ do not intersect $\Gamma$. Instead, where the curves $\gamma(\mathcal{I})$ intersect $\Gamma$ the motions of the degenerate variables is chaotic

different values of $\mathcal{I}$, as well as the curve $\Gamma$. It is evident that the dynamics of the degenerate variables is regular in the region of the $(p,q)$ plane where the curves $\gamma(\mathcal{I})$ do not intersect $\Gamma$. Instead, where the curves $\gamma(\mathcal{I})$ intersect $\Gamma$ the motions of the degenerate variables is chaotic.

# References

1. Andoyer, H., 1923 *Cours de Mécanique Céleste.* Gauthier–Villars, Paris.
2. Arnol'd, V. I., 1976, *Méthodes Mathématiques de la Méchanique Classique.* MIR, Moscow.
3. Arnol'd, V. I., 1963 Proof of a theorem by A. N. Kolmogorov on the invariance of quasi-periodic motions under small perturbations of the Hamiltonian. *Russ. Math. Surv.* **18**(9).
4. Arnol'd, V. I., 1963, Small denominators and problems of stability of motion in classical and celestial mechanics. *Russ. Math. Surveys* **18**, 85–191.
5. Benettin, G. and Fassò, F., 1996, Fast rotations of the rigid body: A study by Hamiltonian perturbation theory. Part I. *Nonlinearity* **9**, 137–186.
6. Benettin, G., Fassò, F., and Guzzo, M., 1997, Fast rotations of the rigid body: A study by Hamiltonian perturbation theory. Part II: Gyroscopic rotations. *Nonlinearity* **10**, 1695–1717.
7. Benettin, G., Fassò, F., and Guzzo, M., 1998, Nekhoroshev-stability of L4 and L5 in the spatial restricted three–body problem. *Regular Chaotic Dyn.*, **3**(3).
8. Benettin, G., Cherubini, A. M., and Fassò F., 2002, Regular and chaotic motions of the fast rotating rigid body: A numerical study. *Discrete Cont. Dyn. Syst., Series B* **2**, 521–540.
9. Benettin, G., Fassò, F., and Guzzo, M., 2005, Long–term stability of proper rotations and local chaotic motions in the perturbed Euler rigid body. *Regular Chaotic Mech.*, **11**, 267–284.
10. Benettin, G., Galgani, L., and Giorgilli A., 1985, A proof of Nekhoroshev's theorem for the stability times in nearly integrable Hamiltonian systems. *Cel. Mech.* **37**, 1.
11. Benettin, G., and Gallavotti, G., 1986, Stability of motions near resonances in quasi–integrable Hamiltonian systems. *J. Stat. Phys.* **44**, 293–338.
12. Bretagnon, P., 1974, Termes à longues périodes dans le système solaire. *Astron. Astrophys.* **30**, 141–154.
13. Biasco, L., Chierchia, L., 2005, Exponential stability for the resonant d'Alembert model of celestial mechanics. *Discrete Contin. Dyn. Syst.* **12**(4), 569–594.
14. Biasco, L., Chierchia, L., and Treschev, D., 2006, Stability of nearly integrable, degenerate Hamiltonian systems with two degrees of freedom. *J. Nonlinear Sci.* **16**(1), 79–107.
15. Celletti A., and Chierchia L., 1995, A constructive theory of Lagrangian tori and computer-assisted applications. Dynamics Reported, New Series, Springer Verlag 4.
16. Celletti A., and Chierchia L., 1997, On the stability of realistic three-body problems. *Comm. Math. Phys.* **186**(2), 413–449.
17. Celletti, A., Giorgilli, A., and Locatelli U., 2000, Improved estimates on the existence of invariant tori for Hamiltonian systems. *Nonlinearity* **13**(2), 397–412.
18. Dazord, P., and Delzant, T., 1987, Le problème général des variables actions–angles. *J. Diff. Geom.* **26**, 223–251.
19. Deprit, A., 1967, Free rotations of a rigid body studied in phase plane. *Am. J. Phys.* **55**, 424–428.
20. Dullin, H. and Fassó F., 2004, An algorithm for detecting Directional Quasi-Convexity. *BIT—Numer. Math.* **44**, 571–584.

21. Fassò, F., 1995, Hamiltonian perturbation theory on a manifold. *Cel. Mech. Dyn. Astron.* **62**, 43.
22. Fassò, F., 2005, Superintegrable Hamiltonian systems: Geometry and perturbations. *Acta Appl. Math.* **87**, 93–121.
23. Fassò, F., Guzzo, M., and Benettin, G., 1998, Nekhoroshev-Stability of Elliptic Equilibria in Hamiltonian Systems, *Commun. in Math. Phys.* **197**, 347–360.
24. Gallavotti, G., 1986, Quasi–integrable Mechanical systems. In Osterwalder, K. and Stora, R. (eds). *Critical phenomena, Random systems, Gauge theories*, North Olland, Amsterdam.
25. Giorgilli, A., Delshams, A., Fontich, E., Galgani, L., and Simó, C., 1989, Effective stability for a Hamiltonian system near an elliptic equilibrium point, with an application to the restricted three body problem. *J. Diff. Eq.* **77**, 167–198.
26. Giorgilli, A. and Skokos, C., 1997, On the stability of the Trojan asteroids. *Astron. Astrophys.* **17**, 254–261.
27. Guzzo, M., Lega, E., and Froeschlé, C., 2006, Diffusion and stability in perturbed non-convex integrable systems. *Nonlinearity* **19**, 1049–1067.
28. Guzzo, M., Knezevic, Z., and Milani, A., 2002, Probing the Nekhoroshev Stability of Asteroids. *Celest. Mech. Dyn. Astroc.* **83**(1–4).
29. Guzzo, M. and Morbidelli, A., 1997, Construction of a Nekhoroshev like result for the asteroid belt dynamical system. *Cel. Mech. Dyn. Astron.* **66**, 255–292.
30. Guzzo, M., 1999, Nekhoroshev stability of quasi–integrable degenerate Hamiltonian systems. *Regular Chaotic Dyn.* **4**(2).
31. Guzzo, M., 2005, The web of three–planets resonances in the outer Solar System. *Icarus* **174**(1), 273–284.
32. Guzzo, M. 2006, The web of three-planet resonances in the outer solar system II: A source of orbital instability for Uranus and Neptune. *Icarus* **181**, 475–485.
33. Kolmogorov A. N., 1954, On the conservation of conditionally periodic motions under small perturbation of the hamiltonian. *Dokl. Akad. Nauk. SSSR* **98**, 524.
34. Kozlov, V. V., 1974, Geometry of "action–angle" variables in Euler–Poinsot problem. *Vestn. Moskov. Univ., Math. Mekh.* **29**, 74–79.
35. Laskar, J., 1990, The chaotic motion of the solar system. A numerical estimate of the size of the chaotic zones. *Icarus* **88**, 266–291.
36. Laskar, J., Froeschlé, C., and Celletti, A., 1992, The measure of chaos by the numerical analysis of the fundamental frequencies. Application to the standard mapping. *Physica D* **56**, 253.
37. Laskar, J., 1996, Large scale chaos and marginal stability in the Solar System. *Celest. Mech. Dyn. Astron.* **64**(1/2), 115–162.
38. Lochak, P., 1992, Canonical perturbation theory via simultaneous approximation. *Russ. Math. Surv.* **47**, 57–133.
39. Locatelli, U. and Giorgilli, A., 2000, Invariant tori in the secular motions of the three–body planetary system. *Cel. Mech. Dyn. Astron.* **78**(1–4), 47–74.
40. Mischenko, A. S. and Fomenko, A. T., 1978, Generalized Liouville method of integration of Hamiltonian systems. *Funct. Anal. Appl.* **12**, 113–121.
41. Morbidelli, A. and Guzzo, M., 1997, The Nekhoroshev theorem and the asteroid belt dynamical system. *Cel. Mech. Dyn. Astron.* **65**, 107–136.
42. Moser, J., 1958, On invariant curves of area-preserving maps of an annulus. *Comm. Pure Appl. Math.* **11**, 81–114.
43. Murray, N. and Holman, M., 1999, The origin of chaos in the Outer Solar System. *Science* **283**, 1877–1881.

44. Nekhoroshev, N. N., 1977, Exponential estimates of the stability time of near-integrable hamiltonian systems. *Russ. Math. Surveys* **32**, 1–65.
45. Nekhoroshev, N. N., 1979, Exponential estimates of the stability time of near-integrable Hamiltonian systems, 2. *Trudy Sem. Petrovs.* **5**, 5–50.
46. Nekhoroshev, N. N., 1972, Action–angle variables and their generalizations. *Trudy Moskow Mat. Obsc.* **26**, 181–198 (Russian). English transl. in *Moskow Math. Soc.* **26**, 180–198.
47. Neishtadt, A. I., 1987, Jumps of the adiabatic invariant on crossing the separatrix and the origin of the 3:1 Kirkwood gap. *Dokl. Akad. Nauk SSSR* **295**, 47–50 (Russian). English transl. in *Sov. Phys. Dokl.* **32**, 571–573.
48. Nesvorny, D. and Morbidelli, A., 1998, Three–body mean motion resonances and the chaotic structure of the asteroid belt. *Astron. J.* **116**, 3029–3037.
49. Nesvorny, D. and Morbidelli, A., 1998, An analytic model of three–body mean motion resonances. *Celest. Mech. Dyn. Astron.* **71**, 243–271.
50. Neishtadt, A. I., 1987, On the change in the adiabatic invariant on crossing a separatrix in systems with two degrees of freedom. *Prikl. Matem. Mekhan* **51**(5) 750–757. *PMM USSR* **51**(5), 586–592.
51. Neishtadt, A. I., and Sidorenko, V. V., 2004, Wisdom system: Dynamics in the adiabatic approximation. *Cel. Mech. Dyn. Astron.* **90**(3–4), 307–330.
52. Niederman, L., 1996, Stability over exponentially long times in the planetary problem. *Nonlinearity* **9**(6), 1703–1751.
53. Niederman, L., 2004, Exponential stability for small perturbations of steep integrable Hamiltonian systems. *Ergodic Theory Dyn. Syst.* **24**(2), 593–608.
54. Niederman, L. 2007, Prevalence of exponential stability among nearly integrable Hamiltonian systems. *Ergodic Theory Dyn. Syst* **27**, 905–928.
55. Niederman, L., 2006, Hamiltonian stability and subanalytic geometry. *Ann. Inst. Fourier (Grenoble)* **56**(3), 795–813.
56. Poschel, J., 1993, Nekhoroshev estimates for quasi–convex hamiltonian systems. *Math. Z.* **213**, 187.
57. Poincaré, H. 1982, *Les méthodes nouvelles de la mécanique celeste*. Gauthier-Villars, Paris.
58. Sussman, G. J. and Wisdom, J., 1992, Chaotic evolution of the Solar System. *Science* **257**, 56–62.
59. Williams, J. G., 1969, Secular perturbations in the solar system. Ph.D dissertation, University of California, Los Angeles.
60. Wisdom, J., 1983, Chaotic behavior and the origin of the 3/1 Kirkwood gap. *Icarus* **56**, 51–74.

# 2

# Diffusion in Hamiltonian Quasi-Integrable Systems

Elena Lega[1], Claude Froeschlé[1] and Massimiliano Guzzo[2]

[1] Observatoire de la Côte d'Azur, B.P.4229, 06304 Nice cedex 4
   elena@obs-nice.fr
[2] Dipartimento di Matematica Pura ed Applicata
   guzzo@math.unipd.it

**Abstract.** The characterization of diffusion of orbits in Hamiltonian quasi-integrable systems is a relevant topic in dynamics. For quasi-integrable Hamiltonian systems a possible model for global diffusion, valid for perturbation larger than a critical value, was given by Chirikov; while for smaller perturbation the Nekhoroshev theorem leave the possibility of exponentially slow diffusion along a peculiar the Arnold's web. We have studied this problem using a numerical approach. The aim of this chapter is to give the state of the art concerning the detection of slow Arnold's diffusion in quasi-integrable Hamiltonian systems.

## 2.1 Introduction

The characterization of mechanisms for diffusion of orbits in quasi–integrable Hamiltonian systems and symplectic maps is a relevant topic for many fields of physics, such as celestial mechanics, dynamical astronomy, statistical physics, plasma physics and particle accelerators.

In 1979 Chirikov [5] described a possible model for global drift valid when the perturbation is greater than some critical value. Chirikov's model has so far been successfully used to describe diffusion in systems from different fields of physics (see, e.g [33]). One of the reasons of the broad detection of the Chirikov's diffusion is that its typical times fall within the simulation abilities of modern computers as far back as the seventies. For smaller perturbations the systems fall within the range of celebrated perturbation theories such as KAM [2, 24, 32] and Nekhoroshev theorems [35], which leave the possibility for a drift only on a subset of the possible dynamical states with peculiar topology, the so–called Arnold web, and force diffusion times to be at least exponentially long with an inverse power of the norm of the perturbation. The theoretical possibility of drift in slightly perturbed systems has been first shown in 1964 by Arnold [3] for a specific system, and is commonly called Arnold diffusion.

Being interested to applications to specific systems, and in particular to systems of interest for physics, we have used a numerical approach which, avoiding theoretical difficulties, measures directly the quantitative features of eventual long term diffusion. In 2003 we numerically detected a very slow local diffusion confined to the Arnold web [28] in a model perturbed system satisfying both the KAM and Nekhoroshev hypothesis. In that work we have numerically measured a diffusion coefficient showing that it decreases faster than a power low of the perturbing parameter and in agreement with the Nekhoroshev theorem.

The Arnold's mechanism is related to a specific "ad hoc" model and the generalization of the mechanism to generic quasi-integrable systems is still an open problem. One of the main difficulties is the so-called "large gap" problem (LGP hereafter), which is related to low order resonance crossings. A new approach to solve the LGP has been recently introduced in [7] to prove the existence of diffusion in the so-called a priori unstable systems. Instead, the LGP has not been yet solved for generic quasi-integrable systems.

Therefore, we found it interesting to explore the possibility for orbits of a generic system to globally diffuse, i.e. to explore macroscopic regions of the space, in a model problem in which the "gap problem" is present. In 2005 we have provided numerical evidence both on quasi–integrable Hamiltonian systems and symplectic maps of a relevant phenomenon of global diffusion of orbits occurring on the Arnold web [17]. More precisely, we have shown that a set of well chosen initial conditions practically explores the whole web and, in the case of maps, the process behaves as a global diffusion.

These results concern systems satisfying the hypothesis of Nekhoroshev theorem (for mapping see [21, 25, 26]) However, most interesting systems (for example, the system describing the motion of an asteroid in the Main Belt of our solar system, see [19, 31]) do not satisfy the hypotheses of the Nekhoroshev theorem, in its standard formulation, because they are represented by Hamiltonian which do not satisfy a suitable geometric condition called "steepness" (see Chap. 1 for the definition). For example, this happens when the Hamiltonian is properly degenerate, i.e. it does not depend on some action variables. For many of these systems the degeneracy can be, in some sense, removed by perturbation techniques adapted to the system ([1, 4, 19, 20], see Chap. 1). However there are non-steep functions even among the non–degenerate functions $h$. We find the quadratic non–convex functions among the simplest non–degenerate functions which are not steep in some points. Our work [18] was dedicated to the numerical investigation of the real possibility of diffusion of the actions in times much smaller than the exponential Nekhoroshev estimate for these quasi–integrable systems. We have found that, except for very special non–convex functions, for which the effect of non-convexity concerns low order resonances, the diffusion coefficient decreases faster than a power law (and possibly exponentially) of the perturbation's norm. According to the theory, we have found that the diffusion coefficient as a function of the perturbation's norm decreases slower than in the convex case.

Our aim here is to give in a self-consistent paper the state of the art concerning the detection of slow Arnold-like diffusion in quasi-integrable Hamiltonian models. In Sect. 2.2 we recall Arnold's model and try to numerically detect diffusion on it. In Sect. 2.3 we give a more general definition of Arnold's diffusion. A brief recall of the numerical tool used to discriminate the dynamics of the orbits is provided in Sect. 2.4. In Sect. 2.5 we introduce our model problem. We discuss results about local and global Arnold's diffusion on a suited Hamiltonian model in Sect. 2.6. In Sect. 2.7 we show the diffusion phenomenon in a more general case in which the convexity hypothesis is released. Conclusion is provided in Sect. 2.8.

## 2.2 Arnold's Diffusion

In a fundamental paper Arnold [3] proposed a mechanism for extremely slow diffusion for a specific Hamiltonian model. He considered the Hamiltonian:

$$H(I_1, I_2, \phi_1, \phi_2, t) = \frac{1}{2}I_1^2 + \varepsilon(\cos\phi_1 - 1) + \frac{1}{2}I_2^2 + \mu(\cos\phi_1 - 1)(\sin\phi_2 + \cos t) \tag{2.1}$$

where $I_1, I_2 \in \mathbb{R}$ and $\phi_1\, \phi_2 \in \mathbb{T}$ and $\varepsilon$ and $\mu$ are small parameters satisfying the relation $0 < \mu << \varepsilon < 1$. For $\mu = 0$ the Hamiltonian is the sum of a pendulum plus a rotator. Thus $\dot{I}_2 = 0$, $\dot{\phi}_2 = I_2 = \omega = const.$ and the change of $I_1$ and $\phi_1$ is described by the dynamics of the pendulum with hyperbolic equilibrium point $I_1 = \phi_1 = 0$. The $(I_1, I_2, \phi_1)$ phase space is foliated in pendula as illustrated in Fig. 2.1. Then, in the complete five-dimensional phase space the manifold $I_1 = \phi_1 = 0$, $I_2 = const.$ is a two-dimensional hyperbolic torus with

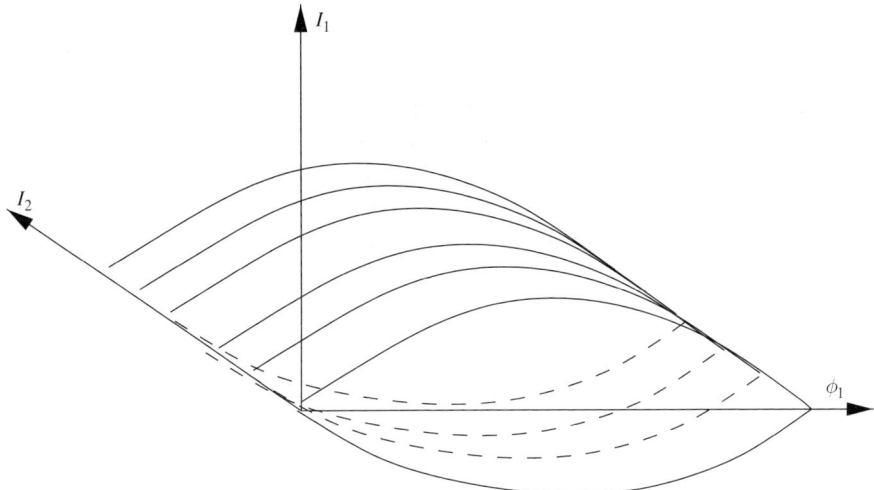

**Fig. 2.1.** The phase space of the unperturbed Arnold's model is foliated in pendula

a one-dimensional stable and unstable manifolds, which is called, for figurative reasons a *whiskered torus*. Changing the value of $I_2$ one has a one-dimensional family of two-dimensional whiskered tori. The departing whisker of any torus forms one manifold together with its arriving whisker.

Let us now consider the case $\mu \neq 0$. All whiskered tori persist under the $\mu$-perturbation ($\dot{I}_1 = \dot{\phi}_1 = 0$ if $I_1 = \psi_1 = 0\ \forall I_2, \phi_2, t$). Moreover, the $\mu$-perturbation causes the stable and unstable manifold of each whiskered torus to intersect creating homoclinic lobes which are not confined in a plane $I_2 = const.$ since the perturbation is $\phi_2$ dependent. In his paper Arnold proved that, for sufficiently small $\mu$ the homoclinic lobes of neighboring whiskered tori intersect allowing diffusion along the resonant line $I_1 = 0$. The time scale of this diffusion turns out to be exponentially small with $\varepsilon$ (more precisely scaling as $\exp(-1/\sqrt(\varepsilon)))$.

An essential ingredient for the Arnold's mechanism is that the $\mu$- perturbation preserves the whiskered tori. As soon as the perturbation is not "ad hoc", i.e. when a more general perturbing function is considered, then the tori for which the frequency $I_2 = \omega$ suitably near to rational are destroyed. Therefore the family of whiskered tori is interrupted by gaps whose size can be of order $O(\mu^{1/2})$. The gaps may be larger than the size of the intersections of the stable and unstable manifolds and this interrupts the heteroclinic chain driving diffusion. A lot of literature exists around the so called *large gap problem* (LGP). An extensive bibliography is available in [7] together with the introduction of a geometric mechanism overcoming the LGP in a so called a priori unstable system with $\varepsilon$ of order 1 and a generic $\mu$ perturbation.

Being interested in applications to specific systems, we have used a numerical approach which, avoiding theoretical difficulties, measures directly the quantitative features of eventual long-term diffusion.

### 2.2.1 About the Numerical Detection of Diffusion

Since the work of Arnold's a great debate about the detection of Arnold's diffusion can be found in the literature. The first numerical studies of chaotic motion in systems with at least three degrees of freedom where performed by Froeschlé and collaborators. In the early seventies Froeschlé [9, 10, 11] studied numerically the number of isolating integrals in systems with three degrees of freedom with a technique involving the projections of thin slices of the four-dimensional surface of section onto two-dimensional planes and calculation of the maximal Lyapunov exponent. The Froeschlé four dimensional map as well as a related mapping known as the "Billiard problem" have then been widely used for numerical studies of the diffusion process ([30, 36] Chap. 6, [29, 37]). With a similar mapping model [6] measures an exponential dependence of the diffusion coefficient on the perturbing parameter describing even the resonance overlap region. Also [27] reports the numerical detection of an orbit of the 4dimensional Froeschlé standard map leaving the crossing between the 1/6 and 1/5 resonance and spending a quite long time along the 1/5 resonance.

Also [8] shows exponentially slow diffusion of orbits of the four-dimensional standard map going into a large chaotic domain. More recently diffusion has been detected by [14] in a system with marginal overlapping of resonances with a geometrical resemblance to Arnold's diffusion.

Since a lot of literature concerns the four-dimensional Froeschlé's mapping (or related mappings) we remark that this system is not quasi-integrable in the sense stated by Arnold. We recall that the original definition of Arnold diffusion is, in the words of Arnold: "to give an example of a system of 5-dimensional phase space which satisfies KAM theorem for quasi-integrable systems but is nonstable" [3]. Let us recall the equations of the Froeschlé four-dimensional map [9, 10]:

$$M = \begin{cases} y_{n+1} = y_n + x_n & (\mathrm{mod}\, 2\pi) \\ x_{n+1} = x_n + k_1 \sin(y_{n+1}) + \varepsilon \sin(y_{n+1} + t_{n+1}) & (\mathrm{mod}\, 2\pi) \\ t_{n+1} = z_n + t_n & (\mathrm{mod}\, 2\pi) \\ z_{n+1} = z_n + k_2 \sin(t_{n+1}) + \varepsilon \sin(y_{n+1} + t_{n+1}) & (\mathrm{mod}\, 2\pi) \end{cases} \quad (2.2)$$

This system is not quasi-integrable unless $k_1 = k_2 = \varepsilon$. It is interesting to observe the difference on the geometry of resonances when considering the quasi-integrable case $k_1 = k_2 = \varepsilon = 0.01$, an intermediate case with $k_1 = k_2 = 0.2$, $\varepsilon = 0.01$ and $k_1 = k_2 = 0.8$, $\varepsilon = 0.01$ as in [37]. Figure 2.2 shows, for a set of $500 \times 500$ initial conditions regularly spaced in $x$, $z$, their corresponding dynamical behavior coded with gray-scale. The integration time is $n=10,000$ for the top panel and $n = 1000$ for the middle and bottom one. The initial angles are set to $\pi$. The numerical tool used is explained in Sect. 2.4. Purple color stands for the presence of KAM tori, while the darker color violet to black correspond to regular resonant motion and the yellow color reveal the presence of a chaotic zone. It appears clearly that the large region of invariant KAM tori of the top panel, is still present in the middle panel while has completely disappeared in the bottom panel being replaced by regular resonant motion and chaotic regions. Following the Arnold's definition the diffusion will be driven by Arnold's mechanism in the case of Fig. 2.2 top. For Fig. 2.2 middle it is interesting to observe that we can have at least an Arnold's like diffusion, i.e. a diffusion supported by a web of resonances merged into a large volume of invariant tori, although the system is not quasi-integrable. We remark that in the case of Fig. 2.2 bottom the diffusion will be driven by Chirikov overlapping of resonances. Some of the papers about the diffusion on a four dimensional map (or related maps) correspond to this last case: the diffusion detected is certainly not Arnold's diffusion.

### 2.2.2 Detection of Diffusion Using Arnold's Model

For our numerical investigations we start first trying to detect a diffusion along the resonant line $I_1 = 0$ of the Arnold's model. Figure 2.3 shows, for a set of $400 \times 400$ initial conditions regularly spaced in $I_1$, $I_2$, their corresponding

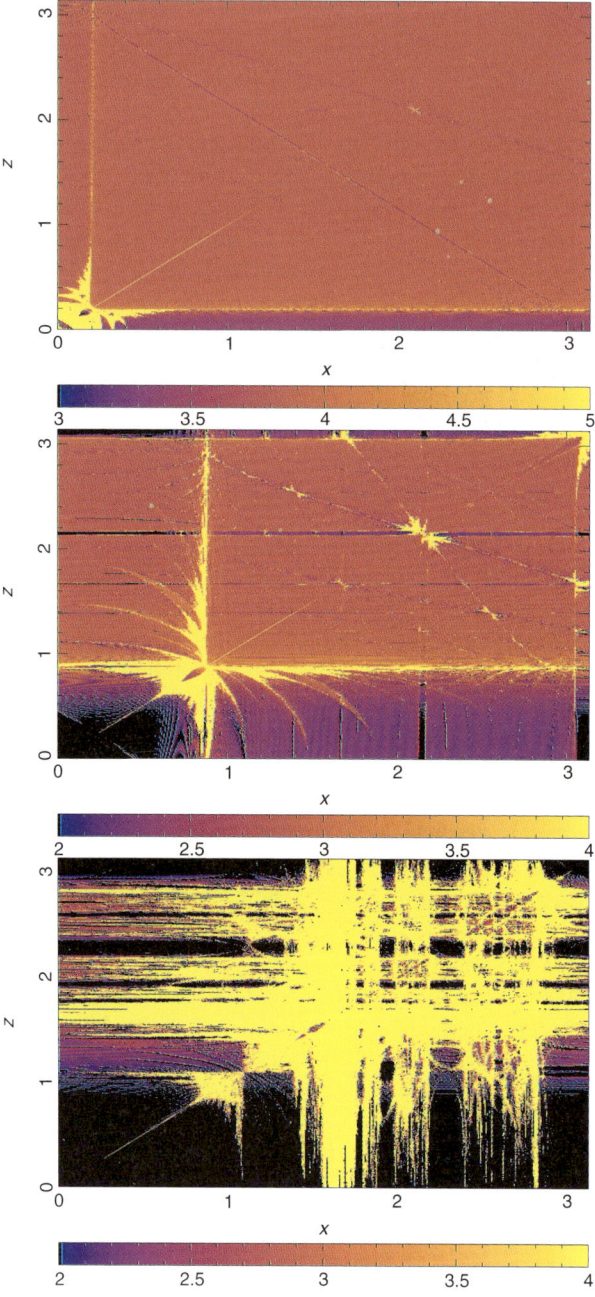

**Fig. 2.2.** The Arnold's web detected using the method of the Fast Lyapunov Indicator (see Sect. 2.4) for the four-dimensional Froeschlé map (2.2) with $\varepsilon = 0.01$ **top panel**: $k_1 = k_2 = 0.01$, **middle panel**: $k_1 = k_2 = 0.2$, **bottom panel**: $k_1 = k_2 = 0.8$. Purple color corresponds to the presence of KAM tori, while darker color correspond to regular resonant motion and yellow color reveal the presence of a chaotic zone

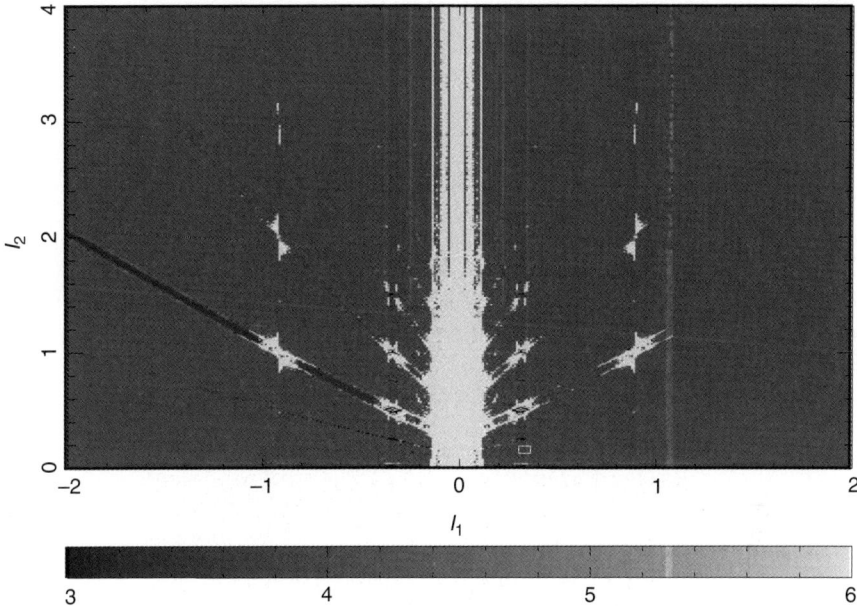

**Fig. 2.3.** The Arnold's web detected using the method of the Fast Lyapunov Indicator (see Sect. 2.4) for the Arnold's model of (2.1) with $\varepsilon = 0.1$ and $\mu = 0.01$. The large *gray region* in the picture stands for the presence of KAM tori, while *darker color* correspond to regular resonant motion and *white color* reveal the presence of a chaotic zone

dynamical behavior coded with gray-scale, for the Arnold's model with $\varepsilon = 0.1$ and $\mu = 0.01$. The numerical tool used is explained in Sect. 2.4. The initial angles are set to zero. The integration time is $t = 5000$.

The resonant line $I_1 = 0$ appears clearly as well as some low order resonances. We recall that the unperturbed Arnold's web of Hamiltonian (2.1) is $k_1 I_1 + k_2 I_2 + k_3 = 0$ with $(k_1, k_2, k_3) \in \mathbb{Z}$.

We have tried to measure a diffusion coefficient as if the phenomenon was Brownian like. This is in fact a very crude assumption, and moreover we are submitted to computational limitations: we can not take a very large number of initial conditions since we need long integration times. Although these difficulties, we say that we observe diffusion if the average evolution of the mean squared distance from the initial conditions grows linearly with time. More precisely, denoting with $I_2^{(j)}(0)$, $j = 1, \ldots, N$ the initial conditions of a set of $N$ orbits and with $I_2^{(j)}(t)$ the corresponding values at time $t$ we computed the quantity:

$$S(t) = \frac{1}{N} \sum_{j=1}^{N} [I_2^{(j)}(t) - I_2^{(j)}(0)]^2 \tag{2.3}$$

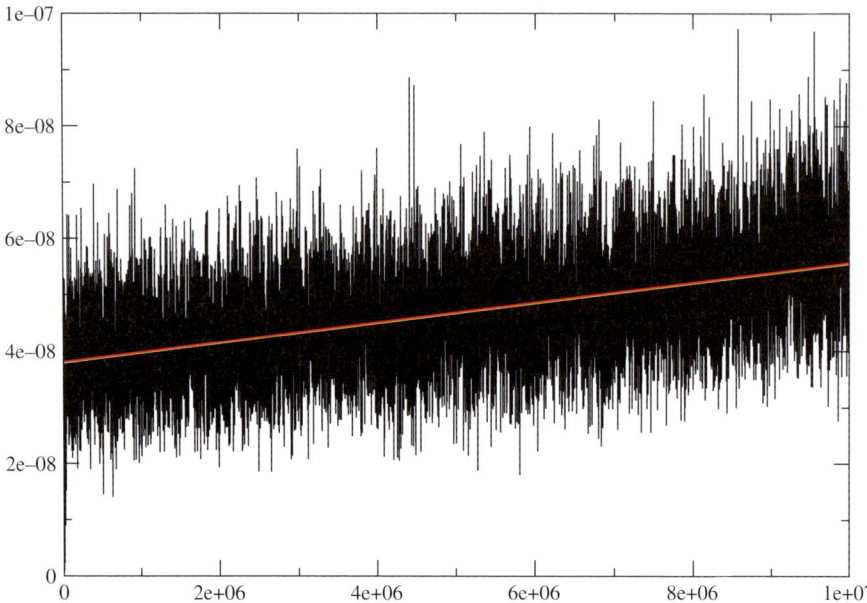

**Fig. 2.4.** Evolution with time of the quantity $S(t)$ for a set of 100 initial conditions taken along the resonant line $I_1 = 0$. The corresponding diffusion coefficient is $D = 10^{-15}$

In Fig. 2.4 it appears clearly that, on average, $S(t)$ grows linearly with time for $\varepsilon = 0.1$, $\mu = 0.01$ and $N = 100$. The initial actions are $I_1^{(j)} = 0.011$, $3 < I_2^{(j)}(0) < 3.01$, $j = 1, \ldots 100$, the initial angles have been set to zero. On a total time of $10^7$ we observed a slow diffusion characterized by a diffusion coefficient $D$, the slope of $S(t)$, $D = 1.5 \times 10^{-15}$.

When decreasing the value of $\varepsilon$ the CPU time needed to observe diffusion with this model becomes rapidly larger than a reasonable time for numerical experiences. Therefore it was not possible to quantify the relation between the diffusion coefficient and the perturbing parameter. Let us remark that without this relation it is not possible to say that we detected Arnold diffusion, we can only say that we observed a very slow diffusion along the resonant line $I_1 = 0$ decreasing as a power law of the perturbing parameter. For $\varepsilon$ lower than 0.1 we could not observe diffusion up to a time $t = 10^7$.

## 2.3 A Generalization of Arnold's Diffusion

It is interesting to extend the phenomenon to more generic models, not only from the mathematical point of view but even from the physical interesting point of the possibility of a slow diffusion process in quasi-integrable generic Hamiltonian systems.

However, the numerical examples of detection reported at the beginning of the previous section show that there are different points of view about Arnold's diffusion in the literature. It is therefore important to define clearly what we consider a generalization of the Arnold's diffusion. We generalize Arnold's diffusion to an exponentially slow diffusion process occurring on a system which satisfies both the KAM and Nekhoroshev theorem as we did in [17] (see Chap. 1 for the Nekhoroshev theorem). It will be useful to list the definitions used through the chapter.

(i) *We strictly refer to quasi–integrable systems*, i.e. to Hamiltonian systems with Hamilton functions of the form

$$H(I,\phi) = h(I) + \varepsilon f(I,\phi) , \tag{2.4}$$

or to symplectic maps:

$$I = I' + \frac{\partial S}{\partial \phi}(I',\phi)$$
$$\phi' = \phi + \frac{\partial S}{\partial I}(I',\phi) \tag{2.5}$$

where

$$S(I',\phi) = h(I') + \varepsilon f(I',\phi) \tag{2.6}$$

with action–angle variables $(I,\phi)$ defined on the open bounded domain $B \times \mathbb{T}^n \subseteq \mathbb{R}^n \times \mathbb{T}^n$.

(ii) *The functions $h, f$ are such that the Hamiltonian system (2.4) and the map (2.5) satisfy the hypotheses of both KAM and Nekhoroshev theorems (for the KAM and Nekhoroshev theorems for quasi–integrable maps see [21, 25, 26]) for suitably small $\varepsilon$.* It is sufficient that $h$ and $f$ are analytic and $h$ satisfies a suitable geometric condition, such as quasi–convexity or steepness.

(iii) *We consider values of the perturbing parameter $\varepsilon$ so small that both KAM and Nekhoroshev theorems apply.* This implies that the phase space is almost filled with a set of invariant tori $\mathcal{K}$. Any motion with initial condition on $\mathcal{K}$ is perpetually stable, so that instability can occur only on the complementary set of $\mathcal{K}$, called Arnold web. Moreover, the Nekhoroshev theorem implies that any eventual instability of the actions occurs only on very long times, which increase exponentially with a positive power of $1/\varepsilon$.

(iv) We will investigate the possibility that the actions $I$ explore macroscopic regions of a given action domain $B$. By Nekhoroshev theorem this is non-trivial and can require very long times as soon as $\varepsilon$ is small (so that condition (iii) is satisfied and the diameter of the set $B$ is bigger than the local fast oscillations of the actions which are allowed by the theorem). To fix ideas, *we say that a motion $(I(t),\phi(t))$ is "unstable" if there exists a time $t$ such that*

$$\|I(t) - I(0)\| \geq \frac{\text{diam } B}{2} \tag{2.7}$$

It is evident that for small $\varepsilon$ the existence of unstable motions is not a trivial fact as soon as diam $B > \sqrt{\varepsilon}$.

(v) We say that the $N$ motions $(I^{(j)}(t), \phi^{(j)}(t))$, $j = 1, \ldots N$, *diffuse in the action space* if the quantity $S(t)$ defined in (2.3), i.e. the average evolution of the squared distance of the actions from their initial value grows linearly with time. In other words there exists a constant $D > 0$ such that

$$S(t) \sim D\, t \tag{2.8}$$

for all $t$.

(vi) *Any diffusion of motions for a system satisfying conditions 1, 2 and 3 will be called Arnold diffusion.* For instance, the model studied in [37] is not in the form (2.6) and therefore the slow diffusion detected is not Arnold diffusion in the sense stated above.

## 2.4 The Fast Lyapunov Indicator

The different diffusion mechanisms are strictly related to the geometry of resonances: the Chirikov's one is characterized by resonance overlapping, while in the case of Arnold's diffusion resonances are arranged as a regular web (the so–called Arnold web) and the phase space is filled by a large number of invariant tori. A precise numerical detection of the Arnold web is possible with the Fast Lyapunov Indicator whose definition is related to the Lyapunov exponent theory.

When computing the Lyapunov Characteristics Indicator (LCI) the attention is focused on the length of time necessary to get a reliable value of their limit, while very little importance has been given to the first part of the computation. Actually, this part was considered as a kind of transitory regime depending, among other factors, on the choice of an initial vector of the tangent manifold.

In 1997 Froeschlé et al. [13] have remarked that the intermediate value of the largest LCI (which was called fast Lyapunov Indicator: FLI), taken at equal times for chaotic, even weakly chaotic, and ordered motion, allows to distinguish between them. It turns out that the FLI allows also to distinguish among ordered motions of different origins, like resonant and non resonant motion [16]. This is not possible with the LCI, which tends to zero when $t$ goes to infinity in both cases.

Given a map $M$ from $\mathbb{R}^n$ to $\mathbb{R}^n$, an initial condition $\boldsymbol{x}(0) \in \mathbb{R}^n$, and an initial vector $\boldsymbol{v}(0) \in \mathbb{R}^n$ of norm one, let us define the FLI function $\text{FLI}(\boldsymbol{x}(0), \boldsymbol{v}(0), T)$, $T$ belonging to $\mathbb{R}^+$, as

$$\text{FLI}(\boldsymbol{x}(0), \boldsymbol{v}(0), T) = \sup_{0 < t \leq T} \log \|\boldsymbol{v}(t)\|, \tag{2.9}$$

where $\boldsymbol{v}(t)$ is given by the system

$$\begin{cases} \boldsymbol{x}(t+1) = M(\boldsymbol{x}(t)) \\ \boldsymbol{v}(t+1) = \frac{\partial M}{\partial \boldsymbol{x}}(\boldsymbol{x}(t))\,\boldsymbol{v}(t) . \end{cases} \quad (2.10)$$

The same definition holds for a continuous flow defined by a set of differential equations:

$$\tfrac{d}{dt}\boldsymbol{X} = \boldsymbol{F}(\boldsymbol{X}) \ , \quad \boldsymbol{X} = (x_1, x_2 \ldots x_n) \quad (2.11)$$

where $F$ from $\mathbb{R}^n$ to $\mathbb{R}^n$ is a suitable regular function. The evolution $\boldsymbol{v}(t)$ of any vector $\boldsymbol{v}(0) \in \mathbb{R}^n$ can be obtained by integrating the linear variational equations:

$$\frac{d\boldsymbol{v}}{dt} = \left(\frac{\partial \boldsymbol{F}}{\partial \boldsymbol{X}}\right)\boldsymbol{v} \ . \quad (2.12)$$

In the specific case of interest here, i.e. the study of quasi-integrable Hamiltonian systems of the form (2.4) for any initial condition $(I(0), \varphi(0))$ and any initial tangent vector $(v_I(0), v_\varphi(0))$, the FLI at time $t$ is

$$\log \|(v_I(t), v_\varphi(t))\| \ . \quad (2.13)$$

In order to kill non significative fluctuations of (2.13), in formula (2.9) we have considered the supremum of the logarithm of the norm of the tangent vector. A running average could also have been used. Actually, as far as the mathematical development is concerned, we drop these averaging procedures, which however are useful in numerical computations. For $\varepsilon = 0$ it is evidently

$$v_I^0(t) = v_I(0) \ , \quad v_\phi^0(t) = v_\phi(0) + \frac{\partial^2 h}{\partial^2 I}(I(0))v_I(0)t \ .$$

If $\varepsilon$ is small we can estimate the evolution of $\|\boldsymbol{v}\|$ with hamiltonian perturbation theory. Following [16], if the initial condition is on a KAM torus then the norm $\|v^\varepsilon(t)\|$ satisfies

$$\|v^\varepsilon(t)\| = \left\|\frac{\partial^2 h}{\partial^2 I}(I(0))v_I(0)\right\| t + O(\varepsilon^\alpha t) + O(1) \ , \quad (2.14)$$

with some $\alpha > 0$. As a consequence, the FLI takes approximately the value of the unperturbed case on all KAM tori. Instead if the initial condition is on a regular resonant motion then it is [16]

$$\|v^\varepsilon(t)\| = \|C_\Lambda \Pi_{\Lambda^{\mathrm{ort}}} v_I(0)\| t + O(\varepsilon^\beta t) + t O(\rho^2) + O(\sqrt{\varepsilon}t) + O\left(\frac{1}{\sqrt{\varepsilon}}\right) \quad (2.15)$$

with some $\beta > 0$, $\Lambda^{\mathrm{ort}}$ being the linear space orthogonal to an integer lattice $\Lambda$ (the integer lattice $\Lambda \subseteq \mathbb{Z}^n$ defines the resonance, for details see [16] and Chap. 1, Sect. 1.4) and $C_\Lambda$ is a linear operator depending on the resonant lattice $\Lambda$ and on the initial action $I(0)$.

It is important to remark that the FLI on regular resonant motions is different at order O(1) from the unperturbed case on regular resonant motions. In fact the linear operator $C_\Lambda \Pi_{\Lambda^{ort}}$ is different from the Hessian matrix of $h$ at order O(1), i.e. $C_\Lambda \Pi_{\Lambda^{ort}}$ does not approach $\frac{\partial^2 h}{\partial I^2}$ as $\varepsilon$ approaches to zero. In this way we detect the presence of the resonances because the value of the FLI is different from the uniform value assumed on the KAM tori. Finally, for initial conditions on chaotic resonant motions the FLI is higher (since the tangent vectors growth exponentially with time) than the value characterizing KAM tori. As a consequence the resonance structure of the phase space can be detected computing the FLI with the same $v(0)$ and the same time interval $t$ on a grid of regularly spaced initial conditions.

## 2.5 The Model Problem

In the following of the paper we consider either the quasi-integrable model Hamiltonian studied in [12, 16] or the map studied in [12, 17] and in [18].

The hamiltonian is

$$H = \frac{I_1^2}{2} + \frac{I_2^2}{2} + I_3 + \varepsilon\, f \quad,\quad f = \frac{1}{\cos(\phi_1) + \cos(\phi_2) + \cos(\phi_3) + 3 + c}\,, \quad (2.16)$$

where $I_1, I_2, I_3 \in \mathbb{R}$ and $\phi_1, \phi_2, \phi_3 \in \mathbb{T}$ are canonically conjugate, $\varepsilon$ is a small parameter and $c > 0$.

The quasi–integrable symplectic map is

$$\begin{aligned}\phi_1' &= \phi_1 + I_1 \quad,\quad \phi_2' = \phi_2 - \alpha I_2 \\ I_1' &= I_1 + \varepsilon\frac{\partial f}{\partial \phi_1}(\phi_1',\phi_2') \quad,\quad I_2' = I_2 + \varepsilon\frac{\partial f}{\partial \phi_2}(\phi_1',\phi_2')\end{aligned} \qquad (2.17)$$

where $f = 1/(\cos(\phi_1)+\cos(\phi_2)+2+c)$, with $c>0$. the parameter $\alpha$ determines if the integrable approximation is convex ($\alpha < 0$) or not ($\alpha > 0$). In the convex case, at small $\varepsilon$, the KAM and Nekhoroshev theorems [21, 25, 26] apply to this kind of maps. The crucial parameters to set in the models are $\varepsilon$ and the value of constant $c$ appearing in the denominator of the perturbation. In fact, any Fourier harmonic $\varepsilon f_k = \varepsilon \int f(\phi) \exp(-i \sum_i k_i \phi) d\phi$ of the perturbing function $f$ is proportional to $\varepsilon$ and decreases (asymptotically) exponentially with $c \sum_i |k_i|$. The values of the two parameters must be balanced so that $\varepsilon$ is smaller than the critical value for Chirikov diffusion (determined with numerical methods, see [12, 16]) and the value of $c$ is not too large so that harmonics with large order $\sum_i |k_i|$ produce measurable effects on the finite time scale of our numerical computations. We found suitable value for our experiments $c = 1$ for the Hamiltonian and $c = 2$ for the map. We choose such a perturbing functions because the Fourier spectrum of the perturbation contains all harmonics at order $\varepsilon$. Any Hamiltonian or map satisfying this requirement would be equivalent while a simple trigonometric potential could not be sufficient to detect numerically the diffusion.

## 2.6 The FLI for Detecting the Geometry of Resonances

The geometry of resonances of (2.16) and (2.17) can be conveniently represented in the two–dimensional plane $I_1, I_2$. Indeed, each point on this plane individuates univocally the frequency of an unperturbed torus, and the resonances are all the straight lines $k_1 I_1 + k_2 I_2 + k_3 = 0$ for (2.16), $k_1 I_1 - \alpha k_2 I_2 + 2\pi k_3 = 0$ for (2.17), with $(k_1, k_2, k_3) \in \mathbb{Z}^3 \backslash 0$.

Of course, the set of all resonances is dense on the plane. However, one can expect that resonant orbits surround each resonance line up to a distance which decreases at most as $\sqrt{\varepsilon}/|k|$. Figure 2.5 shows the FLI chart of (2.16) for $t = 4000$ for a grid of $500 \times 500$ initial conditions regularly spaced on the action plane for $\varepsilon = 0.003$ (the other initial conditions are $\phi_1 = 0$, $\phi_2 = 0$, $\phi_3 = 0$, $I_3 = 1$, the initial tangent vector is $(v_{\phi_1}, v_{\phi_2}, v_{\phi_3}) = (0.5(\sqrt{5}-1), 1, 1)$ and $(v_{I_1}, v_{I_2}, v_{I_3}) = (1, 1, 1)$).

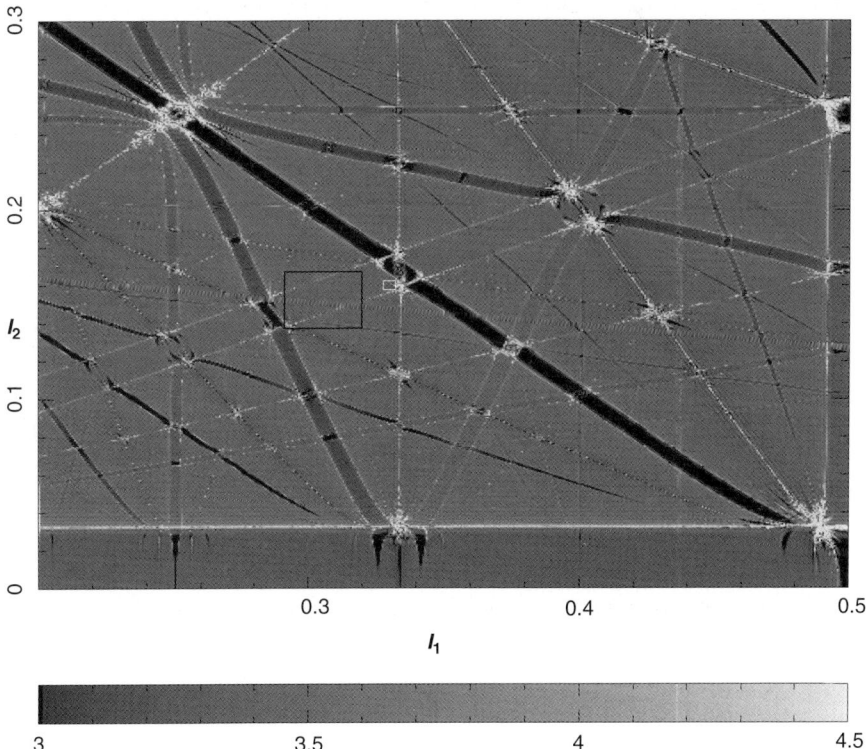

**Fig. 2.5.** FLI values computed at $t = 4000$ on a grid of $500 \times 500$ initial conditions, for Hamiltonian (2.16), regularly spaced on the action axes $I_1$ and $I_2$ for $\varepsilon = 0.003$. The other initial conditions are $I_3 = 1$, $\phi_1 = \phi_2 = \phi_3 = 0$, $(v_{\phi_1}, v_{\phi_2}, v_{\phi_3}) = (0.5(\sqrt{5}-1), 1, 1)$ and $(v_{I_1}, v_{I_2}, v_{I_3}) = (1, 1, 1)$. The gray scale range from black (FLI$\leq 3$) to white (FLI$\geq 4.5$). After [28]

The FLI is reported with a gray scale: the dark lines correspond to regular resonant motions while the white lines correspond both to chaotic resonant motions or to regular orbits very close to a separatrix. The orbits having a FLI value of about $\log(t)$ constitute the background of KAM tori. In the following analysis we focus the attention on a neighborhood of the resonance $I_1 = 2I_2$ (box in Fig. 2.5).

Figure 2.6 shows the Arnold web for the map (2.17) with $\varepsilon = 0.1$. The FLI has been computed on a set of $500 \times 500$ initial conditions regularly spaced in the intervals: $0 < I_1 < \pi$ and $0 < \alpha I_2 < \pi$ for the convex case $\alpha = -1$ (Fig. 2.6, left) and for the non convex one: $\alpha = 4$ (Fig. 2.6, right). The initial angles are chosen equal to zero and the integration time is $t = 1000$ iterations.

The white lines correspond to the chaotic part of the resonances while the gray background corresponds to the set of invariant tori characterized by a FLI value of about $\log t = 3$. Colors going from gray to black stay for regular resonant motions.

Let us remark that in the case of Fig. 2.5 and of Fig. 2.5, left, the phase space is filled with KAM tori while in Fig. 2.6, right the fraction of invariant tori is drastically reduced although resonance crossing do not overlap. These FLI charts provide the important information that for the considered values of $\varepsilon$ the three systems are in the Nekhoroshev regime although in the non convex case the system is close to the transition to the Chirikov one. The transition

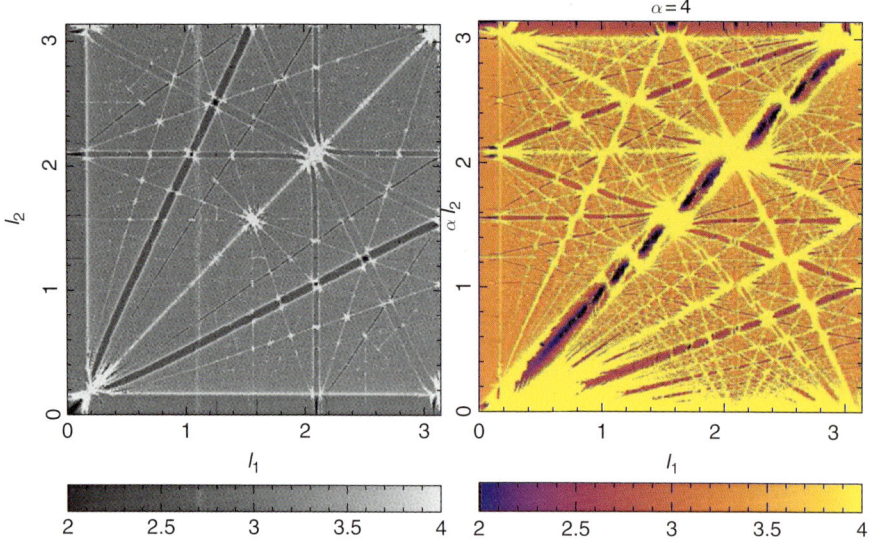

**Fig. 2.6.** Detection of the Arnold web for the convex case: $\alpha = -1$ (**left**) and for the non convex case $\alpha = 4$ (**right**). The perturbation parameter is $\varepsilon = 0.1$ and the integration time is $t = 1000$ iterations. The FLI values close to $\log t = 3$ correspond to invariant tori, higher values show the presence of chaotic orbits and lower values correspond to the regular part of resonances

between the two regime was studied in detail using the FLI in [12] and results were refined in [16] using a sophisticated method of Fourier representation of the Nekhoroshev theorem introduced in [15].

## 2.7 Local and Global Arnold's Diffusion in the Convex Case

### 2.7.1 Convexity and Diffusion

Before discussing diffusion on this kind of models it is instructive to illustrate the relation between the hypothesis of convexity (condition (ii) in Sect. 2.3 and Chap. 1) and diffusion. At this purpose let us start from a two degree of freedom quadratic convex Hamiltonian:

$$H = \frac{I_1^2}{2} + \frac{I_2^2}{2} + \varepsilon f(\varphi_1, \varphi_2) .$$

Such a Hamiltonian is iso-energetically non-degenerate and KAM theorem prevents the diffusion of the actions. Analyzing more closely the dynamics, the resonant normal form for a generic resonance: $k_1 \dot\varphi_1 + k_2 \dot\varphi_2$ is defined near the line of the action plane:

$$k_1 I_1 + k_2 I_2 = 0 , \qquad (2.18)$$

and has the form

$$\tilde{H} = H_0 + \varepsilon \exp - \left(\frac{\varepsilon_0}{\varepsilon}\right)^b r(I, \varphi)$$

with

$$H_0 = \frac{I_1^2}{2} + \frac{I_2^2}{2} + \varepsilon u(I, k_1 \varphi_1 + k_2 \varphi_2) .$$

The dynamics of $H_0$ can move the actions only on the line parallel to the vector: $(k_1, k_2)$, usually called line of fast drift (see Fig. 2.7, top), which is perpendicular to the resonant line (2.18). Therefore, there cannot be a diffusion along the resonance. Figure 2.7, bottom shows the FLI chart for a two degree of freedom quadratic convex hamiltonian system with perturbing function as in (2.16). We have integrated a set of 20 orbits with initial conditions on the line $I_1 = I_2$ with $I_1(0) = I_2(0)$ taken in a small neighborhood of 0.2. The points of the orbits intersecting the FLI section, i.e. having $|\phi_1| + |\phi_2| \leq 0.05$, have being plotted in black on the picture. It appears clearly that the only possible motion occurs perpendicularly to the resonant line.

When the number of degrees of freedom becomes greater than two then only an exponentially small remainder can force an exponentially slow diffusion along the resonant line.

This is the mechanism underlying the exponential stability predicted by the Nekhoroshev theorem.

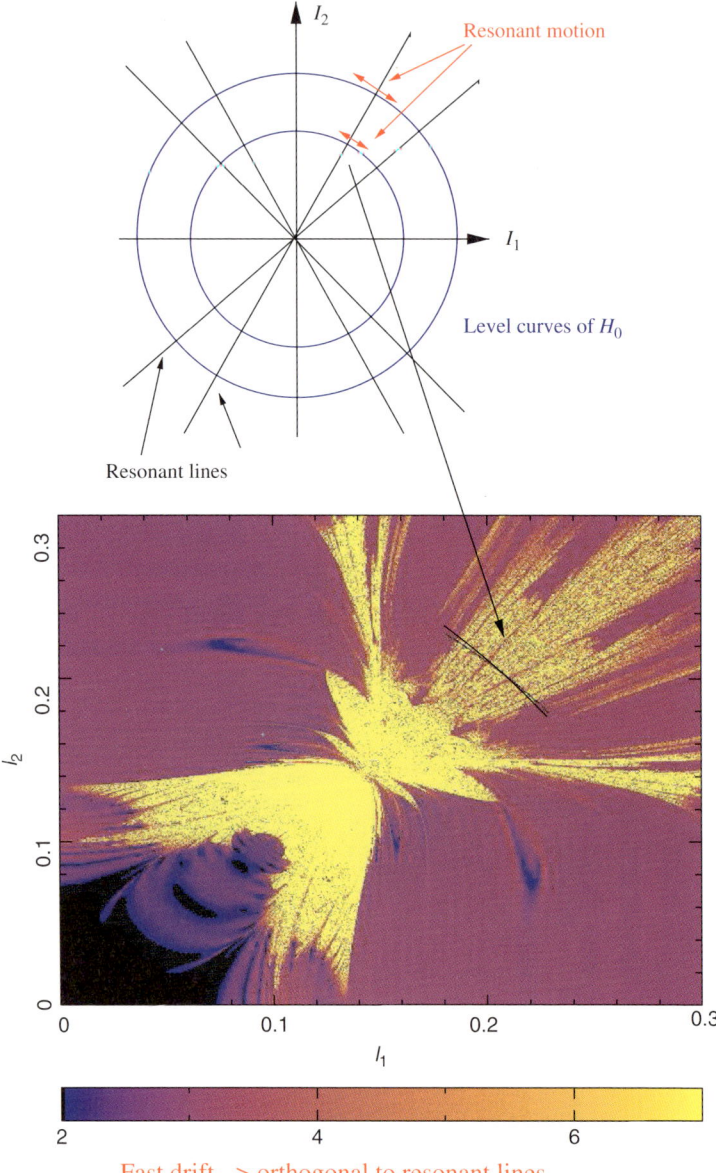

**Fig. 2.7.** (**top**) sketch of the fast drift direction with respect to resonant lines in a two degrees of freedom quadratic convex hamiltonian system. (**bottom**) FLI chart for a two degree of freedom quadratic convex hamiltonian system with perturbing function as in (2.16). The black points are the intersection of 20 orbits with the FLI section, more precisely with the section $\sigma = |\phi_1| + |\phi_2| \leq 0.05$. The motion on the fast drift direction appears clearly

## 2.7.2 Local Arnold's Diffusion

The aim of the following numerical experience is to measure this kind of exponentially slow diffusion as we did in [28]. Fig. 2.8 shows enlargements of the FLI chart of the action space around $I_1 = 0.3$, $I_2 = 0.15$ for different values of $\varepsilon$. Figure 2.8 (top) correspond to the box drawn in Fig. 2.5 for $\varepsilon = 0.003$. In these pictures the region between the two yellow lines is the resonance associated to $I_1 - 2I_2 = 0$, and the two yellow lines correspond to its hyperbolic border where diffusion is confined. These charts provide us the possibility of choosing initial conditions in this hyperbolic border and following their evolution by considering only the points of the orbits which intersect the section of the phase $S = \{(I_1, I_2) \in \mathbb{R}^2, I_3 \in R, \phi_i = 0, i = 1, 2, 3\}$. Of course this can be done numerically by selecting the points satisfying the conditions $\sigma = |\phi_1| + |\phi_2| \leq 0.05$, $\phi_3 = 0$ (reducing the tolerance 0.05 reduces only the number of points on the section, but does not change their diffusion properties). Let us remark that in such a way we minimize all projection effects and fast quasi-periodic movements. What remains is a very slow drift along the border of the resonance.

We have taken a set of 100 initial conditions in the small interval $0.303 \leq I_1 \leq 0.304$, $0.143 \leq I_2 \leq 0.144$ corresponding to orbits of the FLI charts having FLI values larger than $1.2 \log(t)$, i.e. corresponding to chaotic orbits in the border of the resonance. Such initial conditions are chosen far from the more stable crossing with other resonances. We have plotted on the FLI charts all the points in the double section described above. Let us remark that such points will appear on both side of the resonance (in fact the two yellow lines are connected by an hyperbolic region in the six-dimensional phase space).

Figure 2.8(top, left) shows the successive intersections with $\sigma \leq 0.05$, $\phi_3 = 0$, up to a time $t = 10^7$, while Fig. 2.8(top, right) extends to $t = 10^8$. Diffusion along the resonant line appears clearly, with speed of the fastest orbits of about $10^{-11}$, even for this value of the perturbing parameter, which is one order of magnitude lower than the threshold for transition to Chirikov regime. Figure 2.9 shows the phenomenon for one orbit of the previous sample. The values of $I_1(t) - 2I_2(t)$ are on average constant and close to zero, ensuring that there is no diffusion transversal to the resonant line (Fig. 2.9(left)). The variation of $2I_1(t) + I_2(t)$ (Fig. 2.9(right)), shows a very slow diffusion along the resonance. We have decreased $\varepsilon$ and observed diffusion along the resonance with smaller and smaller speed, up to $\varepsilon = 0.001$.

Increasing $\varepsilon$ to 0.007 (Fig. 2.8(middle, left) for $t = 10^6$, Fig. 2.8(middle, right) for $t = 2.4\,10^7$) the situation does not change, except for the speed of the fastest orbits, which is about $10^{-9}$. Instead, for $\varepsilon = 0.02$ (Fig. 2.8(bottom, left) for $t = 1.6\,10^4$, Fig. 2.8(bottom, right) for $t = 5\,10^5$) there is still diffusion along the resonance, but the higher order resonances intersecting the main one become evident and consequently the region of diffusion extends a little also on the direction transversal to the resonance. This happens because we are approaching the transition value to Chirikov regime. The speed of the drift

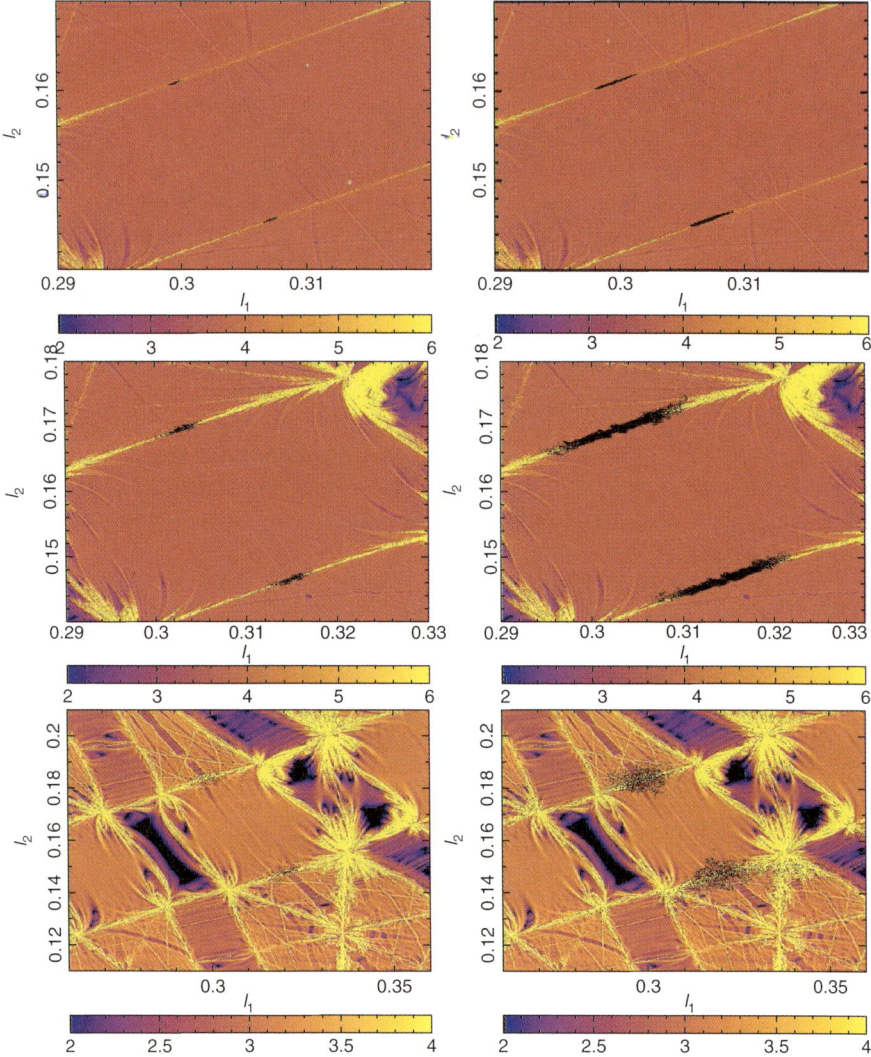

**Fig. 2.8.** Diffusion along the resonant line $I_1 = 2I_2$ for $\varepsilon = 0.003$ (**top**), $\varepsilon = 0.007$ (**middle**), $\varepsilon = 0.02$ (**bottom**) of a set of 100 initial conditions taken in the hyperbolic border of the resonance in the interval $0.303 < I_1 < 0.304$ and $0.143 < I_2 < 0.144$. The *black points* are the intersections of the orbits on the double section $|\phi_1| + |\phi_2| \leq 0.05$, $\phi_3 = 0$. The integration times are respectively $t = 10^7$ (*top, left*), $t = 10^8$ (**top, right**), $10^6$ (**middle, left**), $2.4\,10^7$ (**middle, right**), $1.6\,10^4$ (**bottom, left**), $5\,10^5$ (**bottom, right**). The color scale range from black to yellow. After [28]

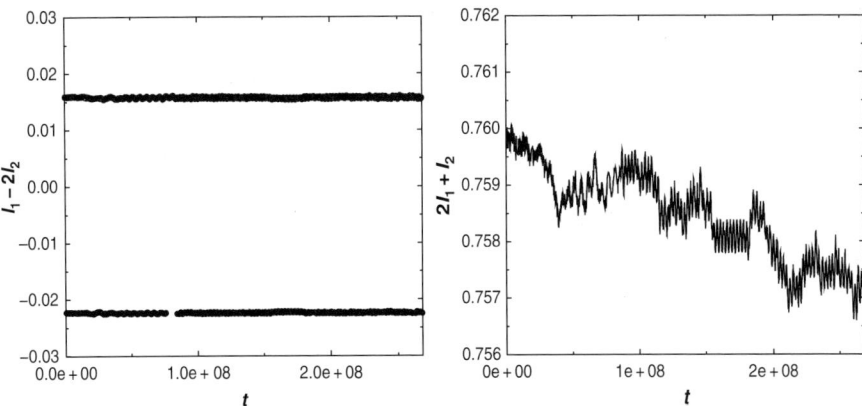

**Fig. 2.9.** Evolution of the quantities $I_1(t) - 2I_2(t)$ (*top*) and of $2I_1(t) + I_2(t)$ corresponding to the intersections of one orbit with the double section $|\phi_1| + |\phi_2| \leq 0.05$, $\phi_3 = 0$, for $\varepsilon = 0.003$. The initial conditions are $I_1(0) = 0.307$, $I_2(0) = 0.145$, $I_3(0) = 0$, $\phi_1(0) = \phi_2(0) = \phi_3(0) = 0$. There is no drift transversal to the resonance (**left**) while a slow diffusion with a velocity of about $10^{-11}$ appears clearly along the resonance (**right**). After [28]

is of the order of $10^{-8}$. At the critical value we expect that the region of the diffusive orbits almost replaces the region of invariant tori. We remark, that even in this critical situation we still appreciate the diffusion along the resonance on a time $t = 10^5$, with a speed of $4 \times 10^{-7}$, while for an higher value of $\varepsilon = 0.04$ let us emphasize that the set of orbits considered already explores the regions between the resonances in a time $t = 5 \times 10^4$.

## Measure of the Diffusion Coefficient

We have computed the quantity $S(t)$ as in (2.3) for the euclidean distance taken along the resonant line $I_1 = 2I_2$ and we have repeated the computation for different values of the perturbing parameter $\varepsilon$. The estimates of $D$ versus $1/\varepsilon$ are reported in Fig. 2.10 in a logarithmic scale. Clearly, data are not well fitted with a linear regression, which would correspond to a power law $D(\varepsilon) = C(1/\varepsilon)^m$. Indeed, if we define three different sets of data, the first containing the values of $D$ for $1/\varepsilon \leq 55$, the second for $62 \leq 1/\varepsilon \leq 250$ and the third for $1/\varepsilon \geq 330$, and we perform local regression for each set, we find the three different slopes $m_1 = -4.5$, $m_2 = -6.9$ and $m_3 = -8.8$. This is sufficient to exclude a global power law and the changes of slope are in agreement with the expected exponential decrease of $D$, although an exponential fit of the form $D(\varepsilon) = C' \exp(-\kappa/\varepsilon)^\alpha$ needs a larger interval of measure in $\varepsilon$ and this is out of our computational possibilities (actually the last point of Fig. 2.10 has been obtained with 4 months of CPU time on a fast workstation).

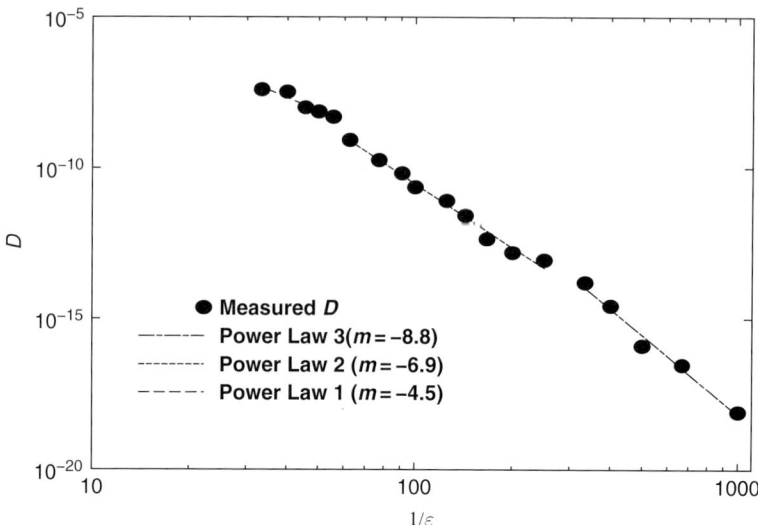

**Fig. 2.10.** Measure of the diffusion coefficient as a function of $1/\varepsilon$. The change of the slope of the three power law fits is in agreement with the expected exponential decrease of D. After [28]

### 2.7.3 Global Arnold's Diffusion

We have considered up to here an Hamiltonian system but the case of quasi–integrable symplectic maps is analogous and, as usual with maps, we have the possibility of exploring the dynamics on much longer integrations time. We considered the quasi–integrable symplectic map of (2.17) studied in [17] for $\alpha = -1$ and $\varepsilon = 0.6$. We have chosen 20 initial conditions near $(I_1, I_2) = (1.71, 0.81)$, i.e. far from resonance crossing and we have computed numerically the map up to $10^{11}$ iterations.

The results are reported in 2.11: on the FLI chart of the action plane $(I_1, I_2)$ we plotted as black dots all points of the orbits which have returned after some time on the FLI section, i.e. having $|\phi_1| \leq 0.005$, $|\phi_2| \leq 0.005$. Figure 2.11a shows only the location of initial conditions (inside the circle), and Fig. 2.11b shows the result after the intermediate times $2 \times 10^9$. Figure 2.11c,d show the result after much longer times. To properly display such long term evolutions we needed to use a zoomed out map of the action plane.

In both cases the orbits filled a macroscopic region of the action plane whose structure is clearly that of the Arnold web. The orbits have moved along the single resonances, and avoided the center of the main resonance crossings, in agreement with the theoretical results of [35] which predict longer stability times for motions in these regions. The larger resonances (which correspond to the smallest orders $|k|$) are practically all visited, while this is not the case for the thinest ones (which correspond to the highest orders $|k|$). This

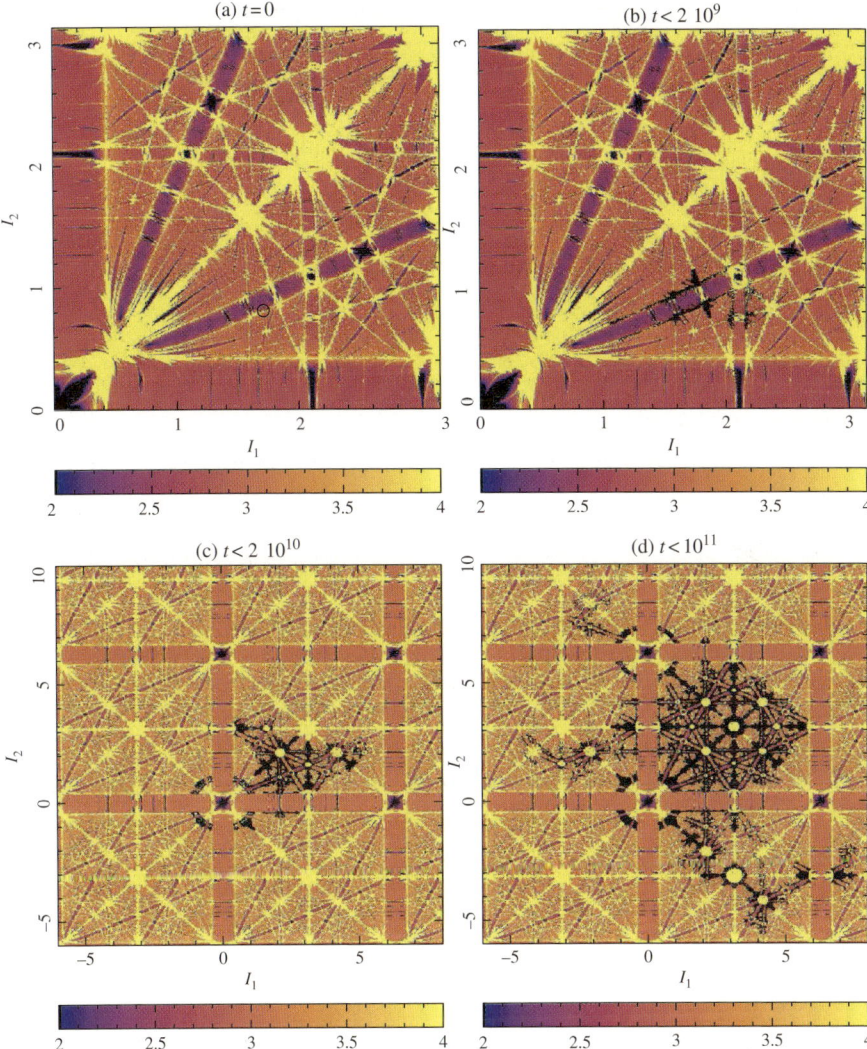

**Fig. 2.11.** The four panels correspond to the FLI map of the action plane $(I_1, I_2)$ for the map 2.17, with initial condition on the section $S$, with different magnifications. The light gray region corresponds to the chaotic part of the Arnold web. Moreover, on panel (**a**) we mark with a circle the location of the 20 initial conditions; on panels (**b,c,d**) we mark with a black dot all points of the 20 orbits which have returned after some time on the section $S$. We consider $2 \times 10^9$ iterations for panel (**b**); $2 \times 10^{10}$ iterations for panel (**c**) and $10^{11}$ iterations for panel (**d**). After [17]

is in agreement with the theoretical results of [34], which predict that the speed of diffusion on each resonance becomes smaller for resonances of high order. Therefore, the possibility of visiting all possible resonances is necessarily limited by finite computational time.

On average, the drift behaves as a diffusion process. In fact, in the case of the map, the average evolution of the squared distance of $(I_1, I_2)$ from the initial datum increases almost linearly with time (Fig. 2.12), so that we can measure a diffusion coefficient $D \sim 1.7 \times 10^{-10}$. The same qualitative behavior appears when dealing with Hamiltonian (2.16). The results [17] are more striking with the mapping for obvious computational reasons. We remark that this diffusion coefficient characterizes the global diffusion process, while diffusion coefficients measured in [28] (and reported in the previous section) characterized local diffusion along a specific resonance.

The described diffusion phenomenon is very different from Chirikov diffusion, where the overlapping of resonances allows diffusion in macroscopic regions of phase space in relatively short time scales and without apparent peculiar topological properties of the stochastic region. For comparison with the Chirikov diffusion regime, in Fig. 2.13 we plot the evolution of 20 orbits for a high value of $\varepsilon$ such that the system is in the Chirikov regime.

It is also interesting to remark that the Largest Characteristic Indicator (whose limit for $t$ going to infinity is the largest Lyapunov exponent) computed on two orbits with very close initial conditions (taken at a distance $d = 10^{-4}$)

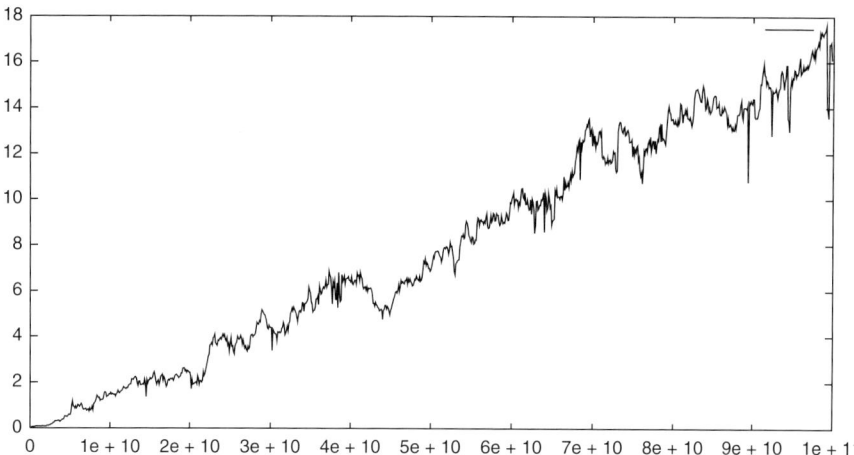

**Fig. 2.12.** Average evolution of the squared distance of $(I_1, I_2)$ from the initial datum for the map, measured for the points on the section $S$. The total computation time $t = 10^{11}$ iterations has been divided in $10^3$ intervals. For each initial condition, and for each interval $[(n-1)10^8, n10^8]$, we have computed the average of the squared distance of $(I_1, I_2)$ from the initial datum, taking into account only points that in the interval $[(n-1)10^8, n10^8]$ are on the section $S$. Then, we averaged over all particles. After [17]

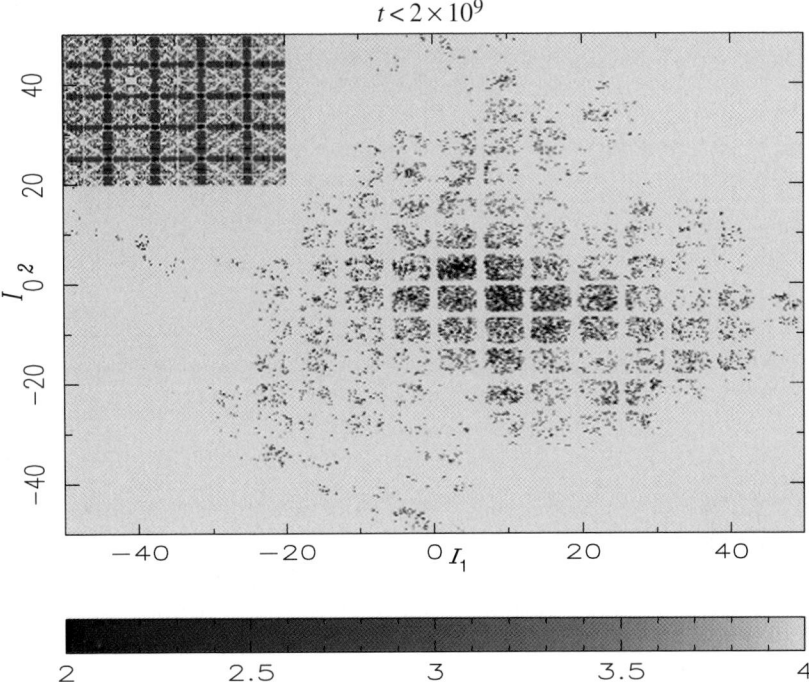

**Fig. 2.13.** Evolution on section $S$ (*black dots*) of 20 orbits for the mapping (2.2) on a time $t < 2 \times 10^9$ iterations for $\varepsilon = 1.6$. This figure as to be compared with Fig. 2.11b–d showing clearly the fundamental differences between the diffusion of Arnold's type and that of Chirikov type. In the left upper corner we plotted, for reference the FLI map. The basic texture of the FLI chart shows that the phase space is almost completely chaotic apart for the regular resonant strips corresponding to $I_1 = 0$ and $I_2 = 0$ and their $2\pi$ periodic repetitions. After [17]

in the Nekhoroshev regime are different (Fig. 2.14, [12]) and also they do not show clear convergence up to $t = 10^{11}$ iterations. Of course convergence can occurs on a much larger integration time. The same quantity, when computed on a single stochastic region, as it is in the Chirikov regime, converges to the largest Lyapunov Exponent.

## 2.8 Diffusion in Non-Convex Systems

We have numerically shown that for systems that satisfy the hypotheses of Nekhoroshev theorem the actions remain close to their initial value up to very long times, that grow faster than a power law of the perturbation norm, possibly exponentially as an inverse power of the perturbation norm. In this section we review the results of [18] concerning the possibility of diffusion for systems that satisfy conditions (i), (ii) and (iii) of Sect. 2.3, but whose $h$ is

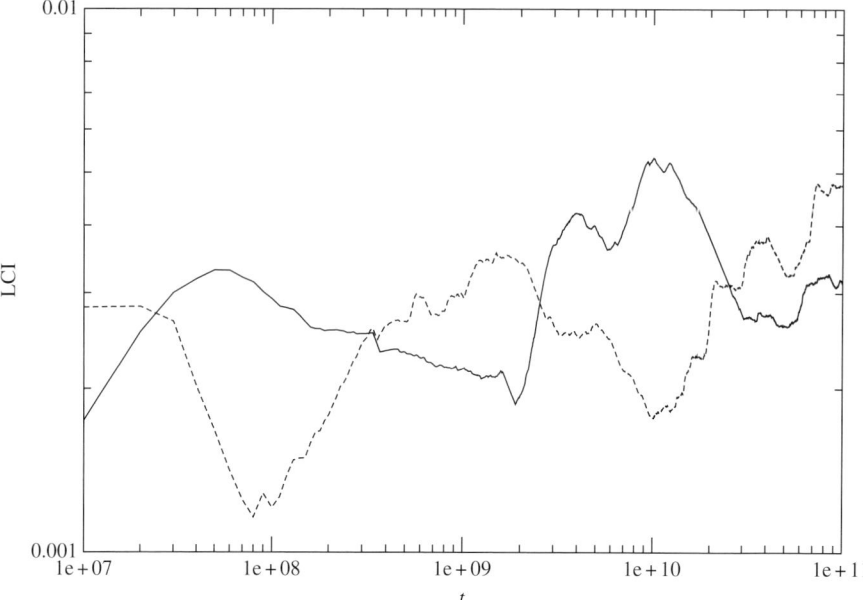

**Fig. 2.14.** Evolution with time of the Largest Characteristic Indicator of two orbits diffusing on the Arnold web and having close initial conditions (initial distance $\sim 10^{-4}$). After [12]

not steep (P–steep for maps, see Chap. 1 for definitions), so that in principle instability is possible already on times of order $1/\varepsilon$.

### 2.8.1 Non-Convexity and Diffusion

To illustrate the mechanism producing this fast diffusion, consider a generic perturbation of the non–convex function $h = \frac{I_1^2}{2} - \frac{I_2^2}{2}$, such as

$$H = \frac{I_1^2}{2} - \frac{I_2^2}{2} + \varepsilon f(\varphi_1, \varphi_2) \ .$$

This system is quasi–integrable with non–degenerate integrable approximation $h$, and therefore the KAM theorem applies to it. However, $h$ is not iso-energetically non–degenerate on the lines $I_1 = \pm I_2$, and therefore action diffusion can occur only near these lines (the systems has $n = 2$), that we call escape lines.

The escape lines correspond also to the resonances: $\dot\varphi_1 \pm \dot\varphi_2 = 0$, and therefore, near the line $I_1 = I_2$ (for simplicity we choose one escape line), by usual normal form construction the Hamiltonian is conjugate by means of a near–to–identity canonical transformation to the resonant normal form

$$\tilde{H} = H_0 + \varepsilon \exp - \left(\frac{\varepsilon_0}{\varepsilon}\right)^b r(I, \varphi)$$

with $H_0$ of the form

$$H_0 = \frac{I_1^2}{2} - \frac{I_2^2}{2} + \varepsilon u(I, \varphi_1 + \varphi_2) \ .$$

The dynamics of the normal form $H_0$ is such that the actions can move only on the line parallel to the vector: $(1,1)$, which is also parallel to the resonance related to the harmonic $\varphi_1 + \varphi_2$. Therefore, with suitable perturbations, actions with initial conditions in the resonance $I_1 = I_2$ can move indefinitely at a speed of order $\varepsilon$ without leaving the resonance (Fig. 2.15).

Following [18], we consider more generic quadratic integrable Hamiltonians with two degrees of freedom, i.e. functions $h$ of the form

$$h = \frac{1}{2} AI \cdot I \ , \tag{2.19}$$

where $A$ is a two-dimensional symmetric square matrix. The previous argument should provide that a condition which is sufficient to prevent the fast

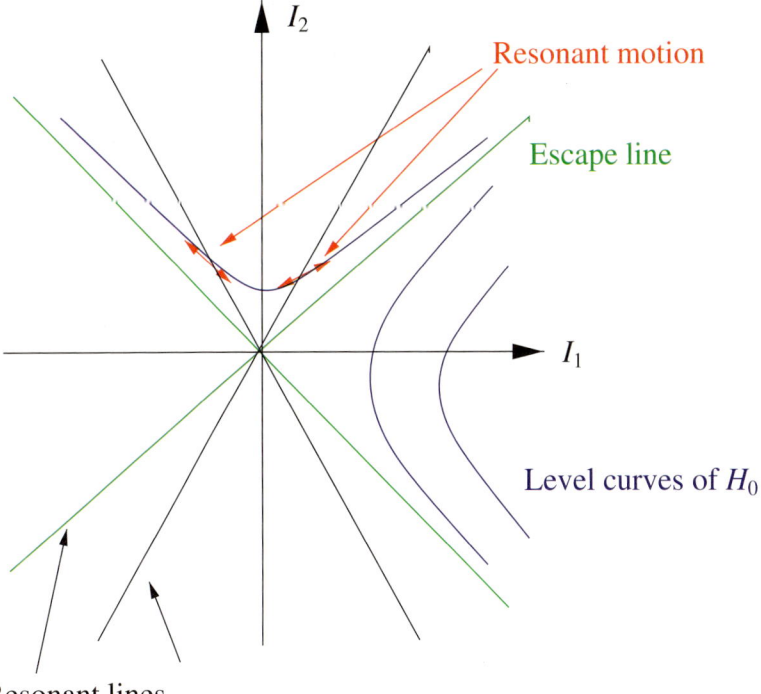

**Fig. 2.15.** Sketch of diffusion and resonant lines in a two-dimensional quadratic non-convex Hamiltonian

diffusion along a given resonance is that the line of fast drift is not contained in the resonance. To be definite, for any $k \in \mathbb{Z}^2 \backslash 0$, the resonance $k \cdot \dot{\varphi} = 0$ is defined by the equation

$$k \cdot AI = 0 \ ,$$

while the line of fast drift, in the action plane, is parallel to the vector $k$. Therefore, a fast diffusion should be possible only if this line is contained in the resonance, which happens only if $Ak \cdot k = 0$. In the rest of the paper we will call "fast diffusion" any diffusion of orbits occurring for systems satisfying ((i), (ii) and (iii) as provided in Sect. 2.3) taking place on resonances of the Arnold web characterized by the fact that a space of fast drift is contained in the resonance.

Fast diffusion in the resonance $k \cdot AI = 0$ is prevented if $k$ satisfies[1]:

$$Ak \cdot k \neq 0 \ .$$

Convex hamiltonians satisfy this condition for any vector $u \in \mathbb{R}^2 \backslash 0$, and therefore also for any integer vector $k \in \mathbb{Z}^2 \backslash 0$.

Following an idea already given in Morbidelli and Guzzo ([31], see caption of Fig. 10) and developed in [18], all the quadratic hamiltonians: $h = \frac{1}{2} AI \cdot I$ with $A$ non–convex, but satisfying:

$$(Ak \cdot k = 0 \ \text{and} \ k \in \mathbb{Z}^2) \ \Rightarrow \ k = 0 \tag{2.20}$$

are not compatible with fast diffusion.[2] For example, the function

$$h = \frac{1}{2}(I_1^2 - \alpha I_2^2) \ , \tag{2.21}$$

with $\alpha > 0$, is non–convex (nor steep), but is compatible with fast diffusion only if $\sqrt{\alpha} \in \mathbb{Q}$, i.e. for $\alpha$ of the form

$$\alpha = \frac{n_1^2}{n_2^2} \tag{2.22}$$

with $n_1, n_2 \in \mathbb{N}$.

We will call "rationally convex" a function $h(I_1, I_2)$ such that its hessian matrix satisfies condition (2.20) at any point of its domain. Function (2.21) is not convex if $\alpha \geq 0$, but is rationally convex if $\sqrt{\alpha} \in \mathbb{R}/\mathbb{Q}$. For $n > 2$ the condition of rational convexity complicates, because of the presence of higher dimensional fast drift planes; the reader can find in [18] the precise definition.

---

[1] More precisely in order to prove exponential stability one should require some algebraic condition such as a Diophantine–like condition: $|k \cdot Ak| \geq \gamma/|k|^\tau$ for any $k \in \mathbb{Z}^2 \backslash 0$.

[2] Again, to prevent diffusion on exponentially long times one should require an algebraic condition, such as $|k \cdot Ak| \geq \gamma/|k|^\tau$ for any $k \in \mathbb{Z}^2 \backslash 0$.

## 2.8.2 Detection of Fast Diffusion

In [18] we have studied numerically the effective impact of the rational convexity on diffusion, and we have compared it with the Arnold diffusion exploring as model problem the map of (2.17) for (many) different values of the parameter $\alpha > 0$. We have considered separately the cases $\sqrt{\alpha} \in \mathbb{R}\backslash\mathbb{Q}$ (rationally convex) and $\sqrt{\alpha} \in \mathbb{Q}$. Repeating the argument given for the hamiltonian case, and recalling that the line of fast drift is parallel to $(k_1, k_2)$, the map (2.17) can have fast diffusion only if $\alpha$ satisfies condition (2.22). In this case there exists a family of resonances that potentially supports fast diffusion, precisely all resonances related to integer vectors $(k_1, k_2, k_3)$ with $(k_1, k_2)$ satisfying

$$\frac{k_1}{k_2} = \pm\frac{n_1}{n_2} = \pm\sqrt{\alpha},$$

and therefore all resonant lines with equation

$$\alpha I_2 = \pm\sqrt{\alpha}I_1 + 2\pi\frac{k_3}{k_2}. \qquad (2.23)$$

This family of resonances constitutes a web, that we call "fast web", and is a subset of the Arnold web.

Because the fast web is dense in the action plane, the phenomenon of fast diffusion can bring the orbits near any point of phase space. In fact diffusing orbits can in principle change several resonances in short times of order $1/\varepsilon$. As a consequence, the orbits can indeed diffuse in the fast web, rather than simply drifting along a line like the simpler $n = 2$ Hamiltonian case. This justifies the use of the name "fast diffusion" for this phenomenon.

Although the steepness hypothesis is no more valid, it still exists a part of the Arnold web that can support action diffusion with a speed that is at most exponentially slow, as in the usual Arnold diffusion. This part is the complement of the "fast web".

The size of the harmonics related to a given resonance is another relevant factor in determining the rate of diffusion. It is related to the order $|k| = |k_1| + |k_2|$. We remark that the fast web contains resonances with minimum order $n_1 + n_2$, that gives a lower bound to the speed of fast diffusion determined by $n_1, n_2$.

As we did for the convex case we integrate chaotic orbits selected through the computation of the FLI and we represent them on the FLI chart considering only those points of the orbits which intersect the section of the phase space: $S = \{(I_1, I_2) \in \mathbb{R}^2, \phi_i = 0, i = 1, 2\}$. As usual the FLI charts are obtained for initial conditions that belong to the section $S$. Figure 2.16 shows the successive intersections of a set of 100 orbits with section $S$ up to a time $t = 5 \times 10^8$ for $\alpha = 4$ and $\varepsilon = 0.1$. The initial conditions have been chosen along $I_2 = I_1/2$ with $0 < I_1 < \pi$. The initial angles are equal to zero.

Figure 2.16 shows a phenomenon of diffusion, involving a macroscopic region of the phase space. Such global diffusion occurs mainly on the "fast

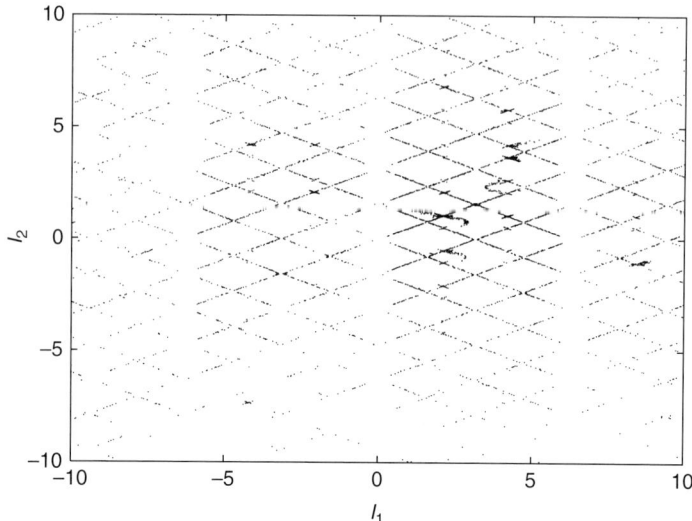

**Fig. 2.16.** Diffusion for $\varepsilon = 0.1$ of a set of 100 chaotic orbits for $\alpha = 4$. The points are the intersections of the orbits with the section $S$ up to a time $t < 5 \times 10^8$ iterations. It appears clearly that diffusion occurs mainly on the on the "fast" web, i.e. the web formed by the set of lines: $I_2 = \pm \frac{1}{\sqrt{\alpha}} I_1 + \frac{2\pi k}{\alpha}$, with $k \in Z$. After [18]

web". We remark that the initial conditions were chosen on the fast web, and a question may arise as to what happens when choosing initial conditions along another resonance.

In fact the Arnold diffusion which is present also in the non–convex case could play an important role. For the non-convex case with $\alpha = 4$, we have selected a set of 100 chaotic initial conditions on the resonance $I_2 = I_1/\alpha$ (that is not in the fast web) and represented their orbits on the FLI chart. Let us remark that the order of this resonance is $|k| = 2$, while the order of the resonances of the fast web satisfies $|k| \geq n_1 + n_2 = 3$.

We have observed that for $\varepsilon = 0.1$ the Arnold diffusion on a time of about $5 \times 10^7$ iterations moves some of the orbits through the Arnold web (Fig. 2.17, top) giving them the possibility of reaching the "fast web" and then rapidly diffusing (Fig. 2.17, bottom). Qualitatively we have observed the same pattern of Fig. 2.16 up to $t = 5 \times 10^8$ iterations.

### 2.8.3 Measure of the Diffusion Coefficient

As we did for the convex case, we have computed the quantity $S(t)$ of (2.4) and measured the diffusion coefficient $D$ as the slope of the regression line of $S(t)$. We have studied the dependence of the diffusion coefficient on the parameter $\varepsilon$ for the rationally convex case with $\alpha_{\rm rc} = ((\sqrt{5}-1)/2)^2$ and, for comparison, for the convex case with $\alpha_{\rm c} = -((\sqrt{5}-1)/2)^2$.

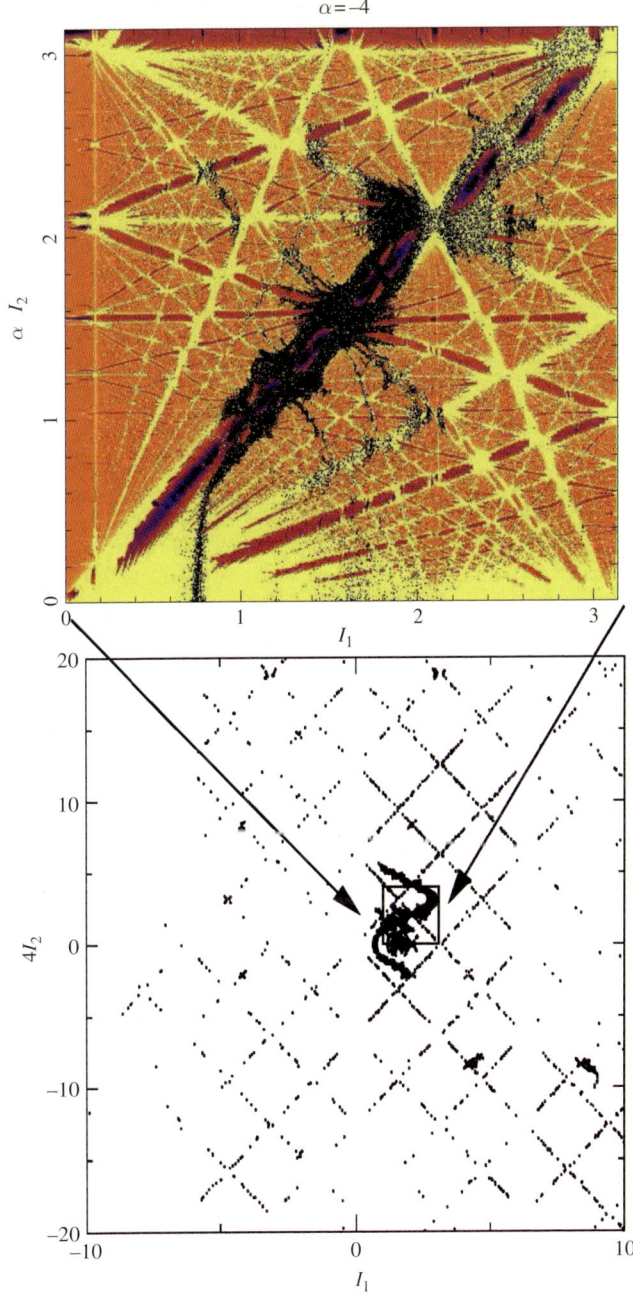

**Fig. 2.17.** Evolution of 100 chaotic orbits with initial conditions in the (1,1) resonance. (**top**) The orbits intersecting section $S$ are plotted as a *black dot* on the FLI chart. Up to $5 \times 10^7$ iterations, Arnold's diffusion moves at least some of the orbits along the resonant line up to the crossing with the "fast web". (**bottom**) Once the fast web reached diffusion becomes a rapid phenomenon occurring mainly on the "fast web". Points intersecting section $S$ up to $t = 5 \times 10^8$ iterations are plotted

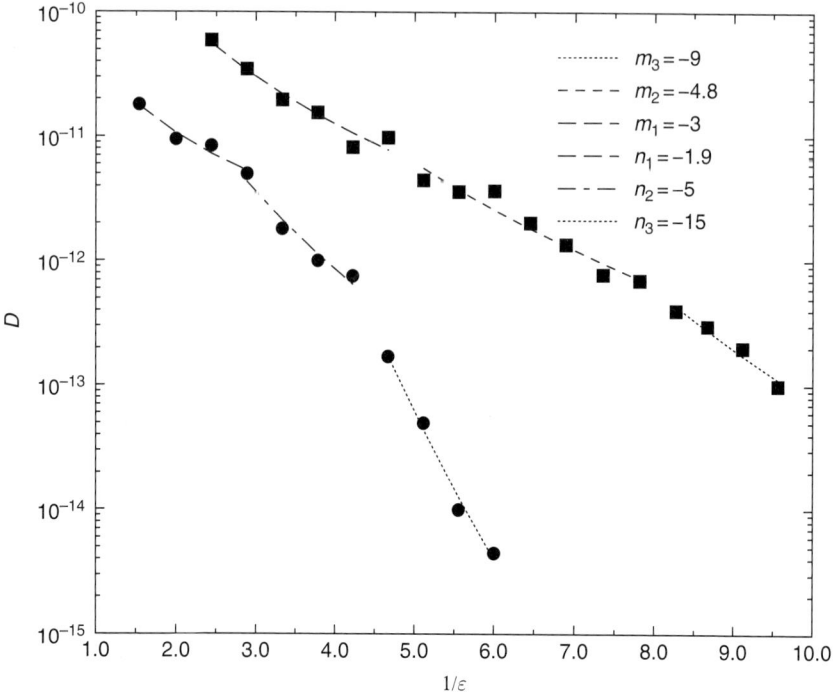

**Fig. 2.18.** Measure of the diffusion coefficient as a function of $1/\varepsilon$ for respectively the rationally convex case with $\alpha_{rc}$ (*square*) and the convex case with $\alpha_c = -\alpha_{rc}$. In both cases data are well fitted by three power laws with slopes $m_i$, $i = 1,..3$ for $\alpha_{rc}$ and $n_i$, $i = 1,..3$ for $\alpha_c$. After [18]

We have integrated, up to $t = 10^9$ iterations, a set of 100 chaotic orbits with initial conditions chosen on the low order resonance $I_2 = I_1/\alpha$ and we have repeated the computation for different values of $\varepsilon$.

The estimates of $D$ versus $1/\varepsilon$ are reported in Fig. 2.18 in a logarithmic scale. Clearly, data are not well fitted with a linear regression, which would correspond to a power law $D(\varepsilon) = C(1/\varepsilon)^m$. Indeed, if we define three different sets of data, and we perform local regressions for each set, we find for $\alpha_c$, the three different slopes $n_1 = -1.9$, $n_2 = -5$ and $n_3 = -15$. This is sufficient to exclude a global power law and the changes of slope are compatible with the expected exponential decrease of $D$. For the non-convex case with $\alpha_{rc}$, we have also found three different slopes $m_1 = -3$, $m_2 = -4.8$ and $m_3 = -9$. Such changes of slopes are in favor of an exponential decrease of $D$ although, in agreement with the theory, slower than in the convex case.

For the non–convex and non-rational convex value $\alpha_{nc} = 4$ we have repeated the experience considering two sets of initial conditions: set (A) on the resonance $I_2 = \frac{1}{\sqrt{\alpha}}I_1$ (that is in the fast web) and set (B) on the resonance

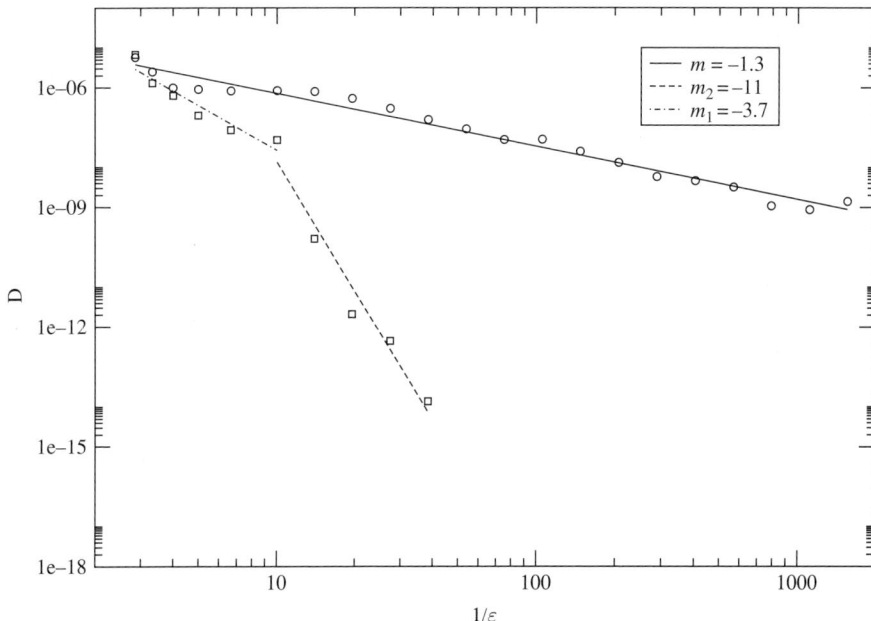

**Fig. 2.19.** Measure of the diffusion coefficient $D$ as a function of $1/\varepsilon$ for orbits of data set (A) (*circle*) and of data set (B) (*square*). For data set (A) the function $D(\varepsilon)$ is fitted by a power law $D(\varepsilon) = (1/\varepsilon)^m$ with $m = -1.3$. For data set (B) the diffusion coefficient is fitted by a power law with $m_1 = -3.7$ up to $\varepsilon = 10$ (global diffusion), then a change of slope is observed with $m_2 = -11$ (local Arnold diffusion). After [18]

$I_2 = \frac{1}{\alpha} I_1$ (that is not in the fast web). Results on diffusion are presented in Fig. 2.19. For the three larger values of $\varepsilon$ we have observed Chirikov diffusion. Its speed is independent on the set (A) or (B) of initial conditions. For lower values of $\varepsilon$, when we have no resonance overlapping, we still observe a global diffusion but of different nature. Such a diffusion occurs mainly along the resonances of the "fast web" as explained in the previous section (Fig. 2.16).

We have observed the global diffusion on the "fast web" for all the values of $\varepsilon$ considered for set (A) and the diffusion coefficient is well fitted by a power law (with slope $m = -1.3$).

For set (B) we observe the same kind of global fast diffusion only for $3 < 1/\varepsilon < 10$. We can fit the points corresponding to global diffusion (both Chirikov and fast) with a power law of slope $m_1 = -3.7$ (Fig. 2.19). For lower values of $\varepsilon$ only local Arnold diffusion along the resonance $I_2 = \frac{1}{\alpha} I_1$ is observed up to $t = 5 \times 10^8$ iterations. The corresponding diffusion coefficient suddenly drops being fitted by a power law of slope ($m_2 = -11$). For values of $\varepsilon$ lower than 0.03 we did not observe any more diffusion up to $t = 5 \times 10^8$. Let us remark that the change of slope of set (B) is not of the same kind than the one presented in Fig. 2.18. Here, it corresponds to the change from global (Chirikov and fast) diffusion to the local Arnold diffusion while in the previous experiment we had always the same phenomenon of local diffusion.

### 2.8.4 About the Influence of $\alpha$ on the Speed of Diffusion

In order to study the influence of $1/\sqrt{\alpha}$ on the speed of diffusion in the non-convex case we have measured the diffusion coefficient $D$ for 300 values of $\alpha$ with $0.5 < 1/\sqrt{\alpha} < 1$. For each value of $\alpha$ a set of 100 chaotic initial conditions has been selected along the resonance $I_2 = I_1/\sqrt{\alpha}$ with $0 < I_1 < \pi$ and the orbits are computed on $10^8$ iterations. The perturbing parameter is $\varepsilon = 0.1$.

Figure 2.20 shows the variation of the logarithm of D as a function of $1/\sqrt{\alpha}$. We observe a drop of four orders of magnitude in D when passing from $1/\sqrt{\alpha} = 0.5$ to $1/\sqrt{\alpha} \simeq 0.55$. The same occurs when going from $1/\sqrt{\alpha} = 1$ to $1/\sqrt{\alpha} \simeq 0.9$. Surprisingly, we do not remark the effect of the low order rational $1/\sqrt{\alpha} = 2/3$. The values of the diffusion coefficient around $2/3$ are quasi-constant around the value $D \simeq 10^{-10}$.

When exploring in detail the case $\alpha = 9/4$ we have observed [18] that the speed of Arnold diffusion on a low order resonance can be of the same order

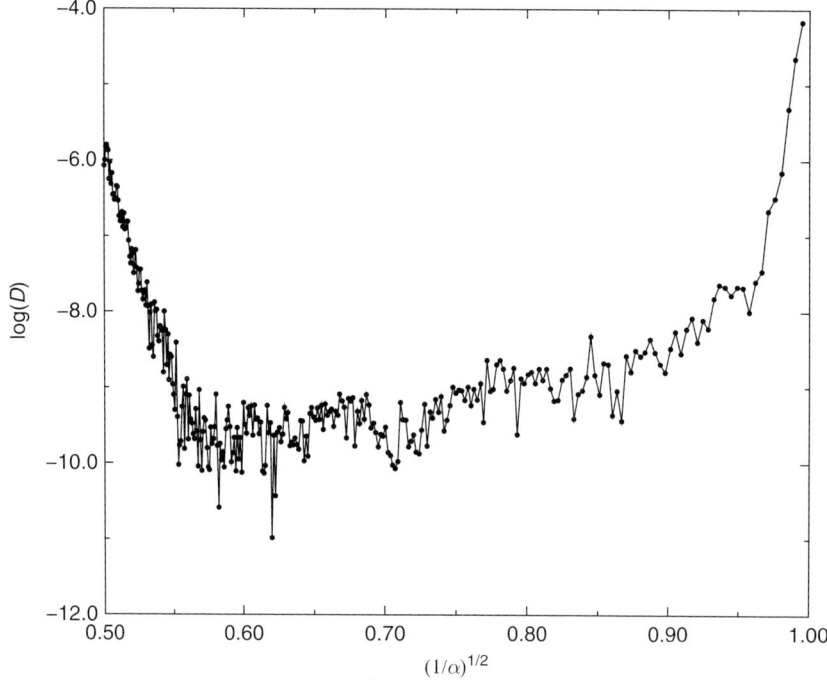

**Fig. 2.20.** Variation of $\log D$ as a function of $1/\sqrt{\alpha}$ for 300 values of $\alpha$ with $0.5 < 1/\sqrt{\alpha} < 1$. For each value of $\alpha$ a set of 100 chaotic initial conditions taken along the fast drift line $I_2 = I_1/\sqrt{\alpha}$ with $0 < I_1 < \pi$ has been integrated up to $t = 10^8$ iterations. The perturbing parameter is $\varepsilon = 0.1$. After [18]

of magnitude, or even greater than the speed of the diffusion along the fast drift line when such a line coincides with a higher order resonance. Let us recall that the order of the fast drift resonance is $|k| = 5$ while the resonance $I_2 = I_1/\alpha$ has order $|k| = 2$. Such a result is a little surprising since we are used to think to Arnold diffusion as a very slow phenomenon. Actually, when $\varepsilon$ approaches the critical value of the transition to the Chirikov regime, as it is for $\varepsilon = 0.18$, and when the fast drift line is of moderately high order, Arnold's diffusion turns out to be competitive with the fast diffusion.

When decreasing $\varepsilon$ the speed of Arnold diffusion on orbits of data set 2 decreases faster than the speed of diffusion of orbits of data set 1. A measure of the diffusion coefficient as a function of $\varepsilon$ for data set 1 and 2 allowed us [18] to quantify the effect of fast and of Arnold diffusion. In particular, it appeared that for $\varepsilon = 0.1$ the speed of Arnold diffusion was of the same order than the speed of fast diffusion. This is the reason why we did not found any difference between the diffusion coefficient computed for $1/\sqrt{\alpha} = 2/3$ and for their irrational neighboring values. In order to observe the influence of the rational value $1/\sqrt{\alpha} = 2/3$ on diffusion it is necessary to decrease $\varepsilon$. We have therefore repeated the computation of $D$ as a function of $1/\sqrt{\alpha}$ for 100 values in a neighborhood of $1/\sqrt{\alpha} = 2/3$ for $\varepsilon = 0.05$. Figure 2.21 shows the

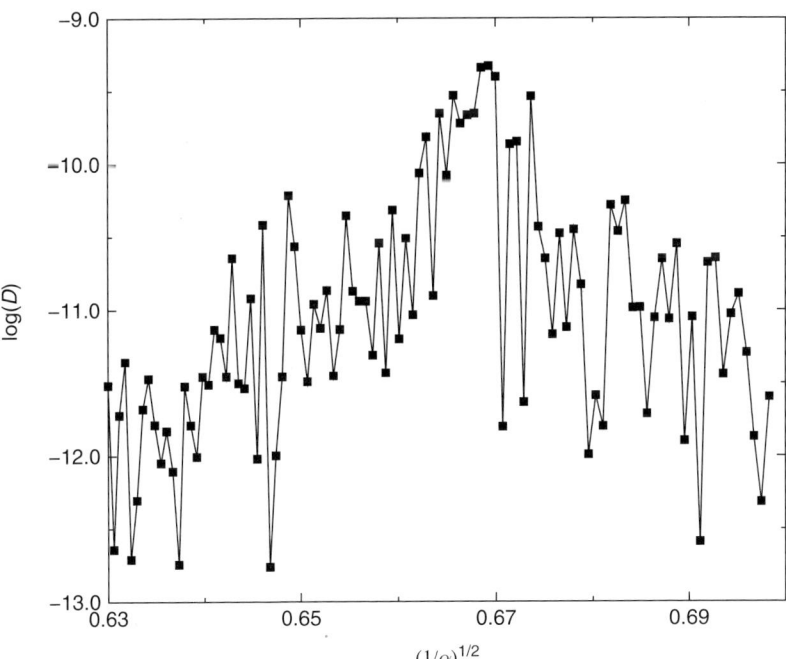

**Fig. 2.21.** Variation of $\log D$ as a function of $1/\sqrt{\alpha}$ for 100 values of $\alpha$ with $0.63 < 1/\sqrt{\alpha} < 0.7$. For each value of $\alpha$ a set of 100 chaotic initial conditions taken along the fast drift line $I_2 = I_1/\sqrt{\alpha}$ with $0 < I_1 < \pi$ has been integrated up to $t = 10^8$ iterations. The perturbing parameter is $\varepsilon = 0.05$ After [18]

emergence of the rational value of $1/\sqrt{\alpha} = 2/3$ characterized by a diffusion coefficient 2.5 order of magnitude higher than for $1/\sqrt{\alpha} \simeq 2/3 \pm 0.03$.

## 2.9 Conclusion

Since the pioneer work of Arnold [3] the numerical detection of Arnold's diffusion has been a delicate and sometimes not clearly defined challenge. In this chapter we recall the Arnold's model and try to detect diffusion directly on it. Motivated by the difficulties (too long CPU time required) and by the fact that it is interesting to generalize this kind of slow diffusion process to a larger class of systems we have studied diffusion on a generic quasi-integrable hamiltonian and on a symplectic map. We have reviewed in this chapter our results [16, 17, 18, 28] about this subject.

Thanks to the model chosen and to results on the transition to the Chirikov regime we were able to select values of the perturbing parameter such that the diffusion along resonant lines could be observed. Using a sensitive tool, the Fast Lyapunov Indicator, for detecting the geography of resonances we have been able to choose and to follow orbits which exhibit diffusive behavior along resonant lines.

We have shown that the dependence of the diffusion coefficient on the perturbing parameter does not follow a power law, and in the explored range is in agreement with the exponential decay predicted by Nekhoroshev theory.

One of the main difficulties in extending the Arnold's mechanism of diffusion to more generic systems is the so-called "large gap" problem, which is related to low order resonance crossings. Therefore, we found it interesting to explore the possibility for orbits of a generic system to globally diffuse, i.e. to explore macroscopic regions of the space, in a model problem in which the "gap problem" is present. The results obtained concerning the global diffusion show that the study of the long–term evolution in quasi–integrable systems must take into account that, at small perturbations, in a subset of phase space of peculiar structure there is an important phenomenon of diffusion of orbits. Indeed, it concerns various problems going from the old question of stability of the Solar System [22, 23] to the modern burden of the confinement of particles in accelerators. Considering the fact that most interesting systems do not satisfy the hypothesis of Nekhoroshev theorem, we have then studied the phenomenon of diffusion in a non convex symplectic map, for different values of the parameter $\alpha$ and of the perturbation parameter $\varepsilon$. We have observed a rapid global diffusion along the resonances forming the "fast web" only when the effect of convexity concerns low order resonances. Otherwise, even if the space of fast drift is contained in a resonance ($\alpha^2 \in \mathbb{Q}$) the diffusion along the fast drift line may not be the most rapid phenomenon. The order of the resonance associated to the fast drift line has to be taken into account. More precisely, for some values of the perturbation parameter, a similar diffusion coefficient for orbits driven by the Arnold mechanism along a low order resonance and for orbits moving along a fast drift line coinciding with an higher order resonance. However, whatever the order of the fast drift resonance, when decreasing $\varepsilon$ the speed of diffusion decreases slower on the fast drift direction than on other lines.

Finally, when the space of fast drift is not contained in a resonance ($\sqrt{\alpha} \in \mathbb{R}/\mathbb{Q}$) we recovered the diffusion properties of the convex case, i.e. we measured a diffusion coefficient decreasing with $1/\varepsilon$ faster than a power law and in agreement with the expected exponential decay.

# References

1. Arnold, V. I., 1963, Small denominators and problems of stability of motion in classical and celestial mechanics. *Russ. Math. Surveys* **18**, 85–191.
2. Arnold, V. I., 1963, Proof of a theorem by A.N. kolmogorov on the invariance of quasi-periodic motions under small perturbations of the hamiltonian. *Russ. Math. Surv.* **18**(9).
3. Arnold, V. I., 1964, Instability of dynamical systems with several degrees of freedom. *Sov. Math. Dokl.* **6**, 581–585.
4. Benettin, G., Fasso, F., and Guzzo, M., 2004, Long-term stability of proper rotations of the Euler perturbed rigid body. *Commun. Math. Phys.* **250**, 133–160.
5. Chirikov, B. V., 1979. *Phys. Reports* **52**, 265.
6. Chirikov, B. V., Ford, J., and Vivaldi, F., 1979, Some numerical studies of Arnold diffusion in a simple model. In Month, M. and Herrera, J. C. (eds), *Nonlinear Dynamics and the Beam-Beam interaction*, American Institute of Physics.
7. Delshams, A., De La Llave, R., and Seara, T. M., 2003, A geometric mechanism for diffusion in hamiltonian systems overcoming the large gap problem: Announcement of results. *ERA Am. Math. Soc.* **9**, 125–134.
8. Efthymiopoulos, C., Voglis, N., and Contopoulos, G., 1998, Diffusion and transient spectra in a 4-dimensional symplectic mapping. In Benest D. and Froeschlé, C. (eds), *Analysis and Modeling of Discrete Dynamical Systems, Advances in Discrete mathematics and Applications*, Gordon and Breach Science Publishers.
9. Froeschle, C., 1971, On the number of isolating integrals in systems with three degrees of freedom. *Astrophys. Space Sci.* **15**, 110.
10. Froeschle, C., 1972, Numerical study of a four-dimensional mapping. *A&A* **16**, 172.
11. Froeschle, C. and Scheidecker, J. P., 1973, On the disappearance of isolating integrals in systems with more than two degrees of freedom. *Astrophys. Space Sci.* **25**, 373–386.
12. Froeschle, C., Guzzo, M., and Lega, E., 2005, Local and global diffusion along resonant lines in discrete quasi–integrable dynamical systems. *Celest. Mech. Dyn. Astron.* **92**, 243–255.
13. Froeschle, C., Lega, E., and Gonczi, R., 1997, Fast Lyapunov indicators. Application to asteroidal motion. *Celest. Mech. Dyn. Astron.* **67**, 41–62.

14. Giordano, C. M. and Cincotta, P. M., 2004, Chaotic diffusion of orbits in systems with divided phase-space. *A&A* **423**, 745–753.
15. Guzzo, M. and Benettin, G., 2001. *DCDS B*.
16. Guzzo, M., Lega E., and Froeschle, C., 2002, On the numerical detection of the effective stability of chaotic motions in quasi–integrable systems. *Physica D* **163**, 1–25.
17. Guzzo, M., Lega, E., and Froeschle, C., 2005, First numerical evidence of global Arnold diffusion in quasi–integrable systems. *DCDS B* **5**, 687–698.
18. Guzzo, M., Lega, E., and Froeschle, C., 2006, Diffusion and stability in perturbed non-convex integrable systems. *Nonlinearity* **19**, 1049–1067.
19. Guzzo, M. and Morbidelli, A., 1997, Construction of a Nekhoroshev like result for the asteroid belt dynamical system. *Cel. Mech. Dyn. Astron.* **66**, 255–292.
20. Guzzo, M., 1999, Nekhoroshev stability of quasi–integrable degenerate Hamiltonian systems. *Regular Chaotic Dyn.* **4**(2).
21. Guzzo, M., 2004, A direct proof of the Nekhoroshev theorem for nearly integrable sysmplectic maps. *Annales Henry Poincare* **5**, 1013–1039.
22. Guzzo, M., 2005, The web of three-planet resonances in the outer Solar System. *Icarus* **174**, 273–284.
23. Guzzo, M., 2006, The web of three-planet resonances in the outer Solar System II: A source of orbital instability for Uranus and Neptune. *Icarus* **181**, 475–485.
24. Kolmogorov, A. N., 1954, On the conservation of conditionally periodic motions under small perturbation of the Hamiltonian. *Dokl. Akad. Nauk. SSSR* **98**, 524.
25. Kuksin, S. B., 1993, On the inclusion of an almost integrable analytic symplectomorphism into a hamiltonian flow. *Russ. J. Math. Phys.* **1**, 191–207.
26. Kuksin, S. B. and Poschel, J., 1994, On the inclusion of analytic symplectic maps in analytic hamiltonian flows and its applications. *Nonlin. Differen. Equat. Appl.* **12**, 96–116.
27. Laskar, J., 1993, Frequency analysis for multi-dimensional systems. Global dynamics and diffusion. *Physica D* **67**, 257–281.
28. Lega, E., Guzzo, M., and Froeschle, C., 2003, Detection of Arnold diffusion in hamiltonian systems. *Physica D* **182**, 179–187.
29. Lichtenberg, A. J. and Aswani, A. M., 1998, Arnold diffusion in many weakly coupled mappings. *Physical Review E* **57**, 5325–5331.
30. Lichtenberg, A. J. and Lieberman, M. A., 1983. *Regular and Chaotic Dynamics*, Springer-Verlag, New York.
31. Morbidelli, A. and Guzzo, M., 1997, The Nekhoroshev theorem and the asteroid belt dynamical system. *Celest. Mech. Dyn. Astron.* **65**, 107–136.
32. Moser, J., 1958, On invariant curves of area-preserving maps of an annulus. *Common Pure Appl. Math.* **11**, 81–114.
33. Morbidelli, A., 2002. *Modern Celestial Mechanics Advances in Astronomy and Astrophysics*, Taylor and Francis, London.
34. Morbidelli, A. and Giorgilli, A., 1995, On a connection between KAM and Nekhoroshev's theorems. *Physica D* **86**, 514–516.
35. Nekhoroshev, N. N., 1977, Exponential estimates of the stability time of near-integrable hamiltonian systems. *Russ. Math. Surveys* **32**, 1–65.

36. Tennyson, J. L., Lieberman, M. A., and Lichtenberg, A. J., 1979, Diffusion in near-integrable hamiltonian systems with three degrees of freedom. In Month, M. and Herrera, J. C. (eds), *Nonlinear Dynamics and the Beam-Beam interaction*, American Institute of Physics.
37. Wood, B. P., Lichtenberg, A. J., and Lieberman, M. A., 1990, Arnold diffusion in weakly coupled standard map. *Phys. Rev. A* **42**, 5885–5893.

# 3
# Weakly Dissipative Systems in Celestial Mechanics

Alessandra Celletti

Dipartimento di Matematica, Università di Roma Tor Vergata,
Via della Ricerca Scientifica 1, I-00133 Roma (Italy)
celletti@mat.uniroma2.it

**Abstract.** We investigate the dynamics associated to nearly integrable dissipative systems, with particular reference to some models of Celestial Mechanics which can be described in a weakly dissipative framework. We start by studying some paradigmatic models provided by the dissipative standard maps in two- and four-dimensions. The dynamical investigation is performed applying frequency analysis and computing the differential fast Lyapunov indicators. After recalling a few properties of adiabatic invariants, we provide some examples of nearly integrable dissipative systems borrowed from Celestial Mechanics, and precisely the spin-orbit coupling and the three-body problem. We conclude with a discussion on the existence of periodic orbits in dissipative autonomous and non-autonomous systems.

## 3.1 Introduction

Celestial Mechanics provides a plethora of physical examples that are described by nearly integrable dissipative dynamical systems. For instance, the celebrated three-body problem is known to be non-integrable, though in many applications it can be considered *close* to an integrable system; however, the conservative setting is not always sufficient to describe the dynamics: accurate investigations of the motion of the celestial objects often require to take into account dissipative effects, like the solar wind, the Yarkowsky effect or the radiation pressure. Nevertheless in many situations the dissipative effects are much less effective than the conservative contribution: for this reason we can speak of a nearly integrable weakly dissipative three-body problem. Another example with similar features is the spin-orbit problem, concerning the motion of a rotating ellipsoidal satellite revolving on a Keplerian orbit around a central body. In this case the conservative setting is described by a nearly integrable problem, which is ruled by a perturbing parameter representing the equatorial oblateness of the satellite. The internal non-rigidity of the satellite provokes a tidal torque, whose effect is typically much smaller than the conservative part.

In order to approach the analysis of the dissipative nearly integrable systems, we start by investigating a simple discrete model known as the *dissipative standard map* (see [3, 4, 6, 8, 20, 29, 32]). Its dynamics is studied through frequency analysis [21, 22] and by means of a quantity called the differential fast Lyapunov indicator as introduced in [8]. We remark that these approaches can be easily adapted to higher dimensional mappings as well as to continuous systems. By means of these techniques we analyze the occurrence of periodic attractors and of invariant curve attractors as the characteristic parameters of the system are varied. The results obtained for the standard map allow an easier approach to continuous systems; indeed after reviewing some results on the adiabatic invariants for a dissipative pendulum, we start by exploring some paradigms of nearly integrable dissipative systems borrowed from Celestial Mechanics. In particular, we focus on the spin-orbit interaction for which we present some explicit expressions of dissipative forces known as MacDonald's and Darwin's torques. In this context we discuss the occurrence of capture into resonance, which depends on the specific form of the dissipation (see, e.g., [10, 12, 15, 16]).

We also provide a short discussion of the restricted planar, circular, three-body problem and related sources of dissipation (see, e.g., [1, 2, 25, 31]). We conclude by mentioning some results about the existence of periodic orbits in (dissipative) autonomous and non-autonomous systems (compare with [9, 27, 28, 30]).

## 3.2 The Dissipative Standard Map

A simple model problem which inherits many interesting features of nearly integrable dissipative systems is given by the so-called *generalized dissipative standard map*, which is described by the equations

$$\begin{cases} y' = by + c + \frac{\varepsilon}{2\pi} s(2\pi x) \\ x' = x + y' , \end{cases} \tag{3.1}$$

where $y \in \mathbf{R}$, $x \in [0, 1)$, $b \in \mathbf{R}_+$, $c \in \mathbf{R}$, $\varepsilon \in \mathbf{R}_+$ and $s(2\pi x)$ is a periodic function. In the case of the classical standard map one defines $s(2\pi x) = \sin(2\pi x)$. The quantity $\varepsilon$ is referred to as the *perturbing* parameter and it measures the nonlinearity of the system. The parameter $c$ is called the *drift* parameter and it is zero in the conservative setting. Finally, $b$ is named the *dissipative* parameter, since the Jacobian of the mapping is equal to $b$. Indeed, for $b = 1$ one reduces to the conservative case, $0 < b < 1$ refers to the (strictly) dissipative case, while for $b = 0$ one obtains the one-dimensional sine-circle-map given by $x' = x + c + \varepsilon s(2\pi x)$.

In the conservative case the dynamics is ruled by the rotation number $\omega \equiv \lim_{j \to \infty} \frac{x_j - x_0}{j}$; indeed, if $\omega$ is rational the corresponding dynamics is periodic, while if $\omega$ is irrational, the corresponding trajectory describes for $\varepsilon$

sufficiently small an invariant curve on which a quasi-periodic motion takes place.

In the dissipative setting it is useful to introduce the quantity

$$\alpha \equiv \frac{c}{1-b}$$

and we immediately recognize that for $\varepsilon = 0$ the trajectory $\{y = \alpha\} \times T^1$ is invariant. Notice that $c = \alpha(1-b) = 0$ for $b = 1$ (i.e., in the conservative case).

In the following we shall specify the function $s(2\pi x)$ by taking $s_1(2\pi x) = \sin(2\pi x)$ or $s_{1,3}(2\pi x) = \sin(2\pi x) + \frac{1}{3}\sin(2\pi x \cdot 3)$. Moreover, we shall take $\alpha$ as the golden mean, $\alpha = \frac{\sqrt{5}-1}{2}$, or as a rational number.

By iterating one of the above mappings, different kinds of attractors appear: invariant curves, strange attractors and periodic orbits, characterized by different values of the largest Lyapunov exponent. Figure 3.1 reports the dynamics of the mapping (3.1) with $s(2\pi x) = \sin(2\pi x)$ and $\alpha = 0.2$, $\varepsilon = 0.8$, for different values of the dissipative parameter; a transient of 10,000 iterations is preliminary performed to get closer to the attractor. For $b = 0.1$ one observes an invariant curve attractor, while a piecewise attractor appears for $b = 0.2718$ and a periodic orbit attractor (denoted with crosses) is evident for $b = 0.28$.

The fate of the trajectories of the dissipative mapping is rather intriguing. Indeed, orbits might wander the phase space running in zigzags through tori and chaotic separatrixes or it may happen that the motion is permanently captured into a resonance. An example is shown in Fig. 3.2 which reports the evolution of the dynamics associated to the mapping (3.1) with $s(x) = s_1(x)$

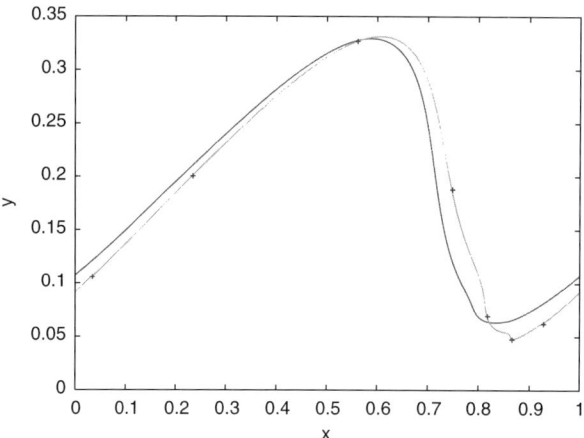

**Fig. 3.1.** Attractors of (3.1) with $s(2\pi x) = \sin(2\pi x)$, $\alpha = 0.2$, $\varepsilon = 0.8$, while $b$ takes the values 0.1, 0.2718, 0.28.

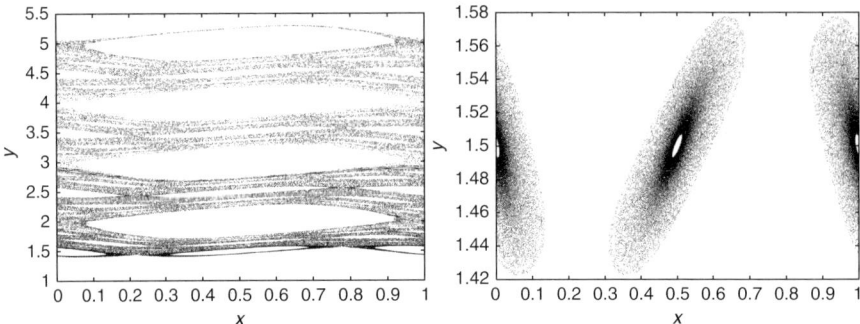

**Fig. 3.2.** Capture into resonance for the mapping (3.1) with $s(2\pi x) = s_1(2\pi x)$, $\alpha = \frac{\sqrt{5}-1}{2}$, $\varepsilon = 0.8$, $b = 1 - 0.00001$, $(y_0, x_0) = (5, 0)$. **Left panel**: the first $10^5$ iterates; **right panel**: some more $5 \times 10^5$ iterations

and for $\alpha = \frac{\sqrt{5}-1}{2}$. The perturbing parameter is set to $\varepsilon = 0.8$, while the dissipation is fairly weak, being $b = 1 - 0.00001$. Taking the initial values $(y_0, x_0) = (5, 0)$, the left panel of Fig. 3.2 shows that the evolution of $10^5$ iterates escapes many resonances before being captured by the 3:2 resonance (at approximately $y_0 = 1.5$). Starting from the last of the previous $10^5$ iterates, we perform some more $5 \times 10^5$ iterations (see the right panel of Fig. 3.2) which manifest a spiralling toward the equilibrium point, after jumping across some secondary resonances. Further iterations would lead to end-up on the center of the resonance.

## 3.3 Techniques for the Numerical Investigation of the Dynamics

In order to analyze the dynamics of the dissipative standard map we implement two complementary numerical techniques, which are based on the frequency analysis (see, e.g., [21, 22, 23]) and on the computation of the so-called fast Lyapunov indicators (see, e.g., [13, 14, 17]), modified in order to work in the dissipative case. We refer the reader to [8] for a complete description and implementation of these techniques.

### 3.3.1 Frequency Analysis

Frequency analysis relies on the computation of the frequency of motion, which is determined by applying the following algorithm. For a given conservative 2-dimensional mapping $M$, let us denote by $P_n = M^n(P_0)$ the $n$th iterate of the point $P_0$ which we assume to belong to an invariant curve with frequency $w$. Over a sample of $N$ points $(P_1, ..., P_N)$ we denote by $P_{n_1}$ the nearest neighbor to $P_0$ and we define the integer $p_1$ through the expression

$n_1\omega = p_1 + \varepsilon_1$, where $\varepsilon_1$ is a small quantity. Since $p_1$ counts the number of revolutions performed around the invariant curve, the quantity $\omega$ can be approximated by the ratio $p_1/n_1$. Increasing $N$, one gets a sequence of better approximations $p_k/n_k$ converging to $\omega$ up to small errors $\varepsilon_k$. Particular care must be taken when applying this method to the dissipative case, since the starting point must be close to the attractor; to this end, a preliminary set of iterations, typically $10^4$, is performed before defining the starting point $P_0$.

In order to investigate the effect of the joined variation of the dissipative and of the perturbing parameters, we use frequency analysis by drawing the curve $\omega = \omega(b)$ for different values of $\varepsilon$ (see Fig. 3.3). This approach allows the recognition of the different kinds of attractors: indeed, invariant curves are characterized by a monotone variation of the frequency curve, periodic orbits show a marked plateau, while strange attractors exhibit an irregular behavior of the function $\omega = \omega(b)$. From experiments on different mappings and different choices of $\alpha$, we notice that invariant curves typically occur more frequently for small values of $\varepsilon$, while periodic and strange attractors appear more often as $\varepsilon$ gets larger. With reference to Fig. 3.3, we also remark that Fig. 3.3a is ruled by the irrational choice of $\alpha$, while in Fig. 3.3b there is a dominant periodic attractor with period $\frac{1}{3}$ whose basin of attraction increases

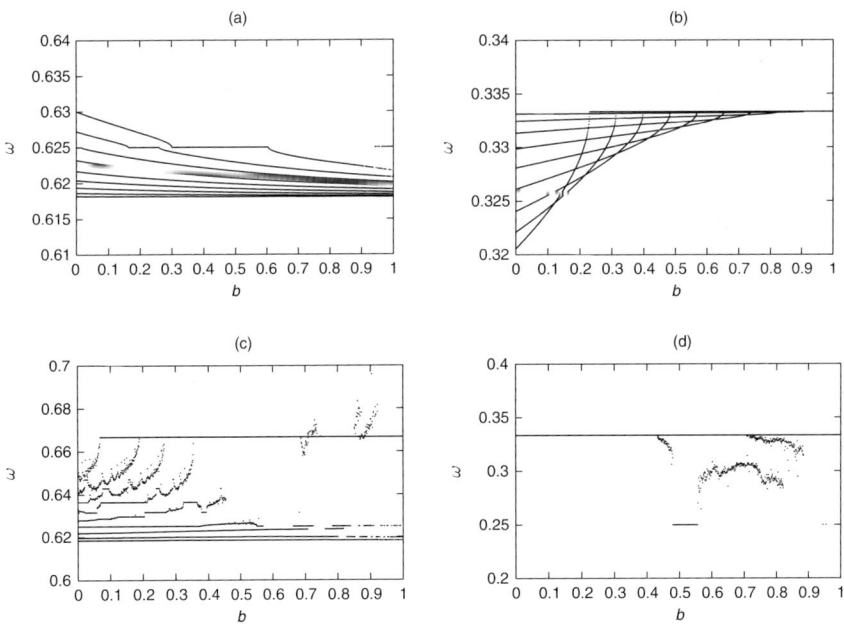

**Fig. 3.3.** Frequency analysis showing $\omega = \omega(b)$ for nine different values of $\varepsilon$ from $\varepsilon = 0.1$ (*lower curve*) to $\varepsilon = 0.9$ (*upper curve*). (**a**) Mapping $s_1(2\pi x)$ with $\alpha = \frac{\sqrt{5}-1}{2}$. (**b**) Mapping $s_1(2\pi x)$ with $\alpha = \frac{1}{3}$. (**c**) Mapping $s_{1,3}(2\pi x)$ with $\alpha = \frac{\sqrt{5}-1}{2}$. (**d**) Mapping $s_{1,3}(2\pi x)$ with $\alpha = \frac{1}{3}$. After [8]

as the parameter $\varepsilon$ gets larger. The remaining panels refer to the two-frequency map $s_{1,3}(2\pi x)$; in Fig. 3.3c the irrational choice of $\alpha$ is compensated by the selection of the harmonics 1 and 3 which appear in the mapping $s_{1,3}(2\pi x)$, while in Fig. 3.3d both the choice of $s_{1,3}(2\pi x)$ and of $\alpha$ induce most of the orbits to be attracted by the periodic attractor with frequency $\frac{1}{3}$.

### 3.3.2 Differential Fast Lyapunov Indicator

A global analysis of conservative systems can be performed through the Fast Lyapunov Indicators (hereafter, FLI) which are defined as follows. Let $\tilde{M}$ be the lift of the mapping; let us denote by $z(0) \equiv (y(0), x(0))$ the initial condition and let $v(0) \equiv (v_y(0), v_x(0))$ be an initial vector with unitary norm. For a fixed time $T > 0$, define the FLI as the quantity

$$\overline{\text{FLI}}(z(0), v(0), T) \equiv \sup_{0 < t \leq T} \log \|v(t)\| ,$$

where $v(t)$ is the solution of the differential system

$$\begin{cases} z(t+1) = \tilde{M}(z(t)) \\ v(t+1) = \frac{\partial \tilde{M}}{\partial z}(z(t))v(t) \end{cases}$$

with initial data $z(0)$, $v(0)$. We stress that in the unperturbed case ($\varepsilon = 0$), the largest Lyapunov exponent of an invariant curve is zero, while in the dissipative case the corresponding FLI can take any value within the range $[\log(|v_x(0)|), +\infty)$. By continuity the same problem holds for $\varepsilon \neq 0$; as a consequence, in the dissipative setting the FLI might not be adequate to differentiate between an invariant curve attractor and a strange attractor. Henceforth we defined in [8] the quantity

$$\text{DFLI}_0(z(0), v(0), t) \equiv F(z(0), v(0), 2t) - F(z(0), v(0), t) ,$$

where $F(z(0), v(0), t) = F(t) \equiv \log \|v(t)\|$. We remark that $\text{DFLI}_0$ is zero for curve attractors, negative for periodic orbits and positive for chaotic attractors, in agreement with the value of the corresponding largest Lyapunov exponent. Finally, in order to kill the oscillations of the norm of the vector $v$ a supremum has been introduced, which corresponds to adopt the following definition of *differential* FLI:

$$\text{DFLI}(T) = G_{2T}(F(t)) - G_T(F(t)) , \qquad (3.2)$$

where

$$\begin{cases} G_\tau(F(t)) = \sup_{0 \leq t \leq \tau} F(t) & \text{if } F(\tau) \geq 0 \\ G_\tau(F(t)) = \inf_{0 \leq t \leq \tau} F(t) & \text{if } F(\tau) < 0 . \end{cases}$$

The DFLI provides a complementary investigation to frequency analysis; to represent it in an effective way, we used a color scale which helps to discriminate among the different attractors. As performed in [8] we computed grids of 500 × 500 initial values of $b$ and $\varepsilon$ regularly spaced in the interval $[0.01:1]$; the initial conditions were set to $y_0 = 5$ and $x_0 = 0$, while $T = 10^3$ (3.2), after a transient of $10^4$ iterations. Then, the color classification is performed on the following basis: invariant curve attractors are denoted by grey and their DFLI values are close to zero; strange attractors are labeled by light grey and their DFLI values are positive; periodic orbit attractors are denoted by dark grey to black with a negative DFLI.

As an example, we consider the two-frequency mapping $s_{1,3}(2\pi x)$ with $\alpha = \frac{\sqrt{5}-1}{2}$; the results are presented in Fig. 3.4. The left panel shows the chart of parameters $b$ versus $\varepsilon$: scanning in the $\varepsilon$-direction we find invariant attractors up to $\varepsilon \simeq 0.36$ and periodic attractors around $\varepsilon \simeq 0.4$; for $\varepsilon > 0.4$ a wide zone filled by periodic attractors is surrounded by two regions of strange attractors. The right panel provides the DFLI chart in the plane $b$ versus the initial condition $y$. We remark that for a fixed $b$, the basin of attraction is typically unique, with the exception of the parameter region $0.65 < b < 0.9$, where different initial conditions can be attracted either by a periodic orbit or by a strange attractor.

An important issue, especially from the point of view of physical applications, is the occurrence of periodic orbits and precisely the dependence of a given periodic attractor upon the choice of $\alpha$ and of the mapping $s(x)$. Numerical experiments (see [8]) show that a $q$-periodic orbit is highly likely whenever $\alpha = \frac{p}{q}$ or $s(2\pi x) = \sin(2\pi x \cdot q)$. We remark also that periodic orbit attractors with small period occur more frequently and that new periodic orbits arise for increasing $b$.

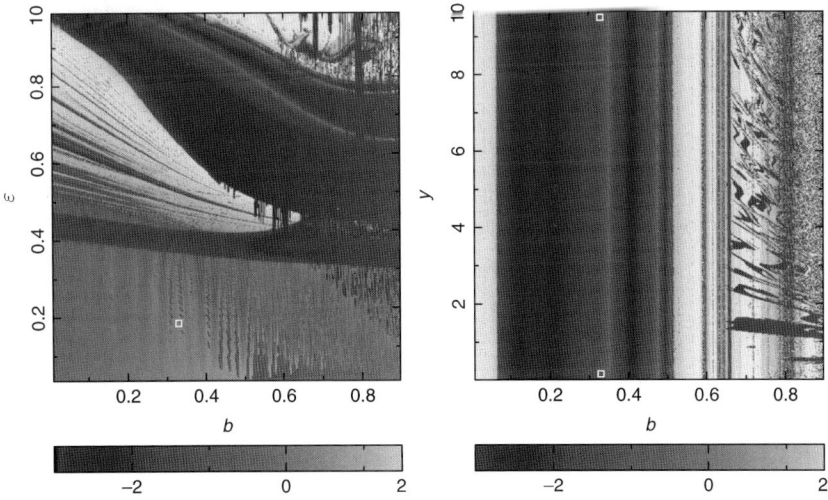

**Fig. 3.4.** Map $s_{1,3}(2\pi x)$ with $\alpha = \frac{\sqrt{5}-1}{2}$. **Left panel**: DFLI chart $b$ versus $\varepsilon$; **right panel**: DFLI chart $b$ versus $y$. After [8]

The applications concerning Celestial Mechanics that we shall consider in the following sections are typically characterized by a small value of the dissipation $b$ when compared to the perturbing parameter $\varepsilon$. In view of such investigations we concentrate on the *weakly dissipative regime*, where $b$ varies in the interval $[0.9, 1]$. To be concrete, let us consider the mapping $s(2\pi x) = \sin(3 \cdot 2\pi x)$ with $\alpha = \frac{1}{2}$ and let us count the number of occurrences of a periodic orbit attractor of period $q$ as $\varepsilon$ varies. This result is presented in Fig. 3.5a using a semi-log scale: the rotation number is computed taking 100 initial conditions, say $x_0 = 0$ and $y_0$ in the interval $[0, 10]$ and 1000 values of $b$ in $[0.901, 0.999]$, while $\varepsilon$ takes the discrete values $0.1, 0.2,..., 0.9$.

This experiment shows that there is a competition between the frequency $q = 3$ (equal to the leading harmonic of $s(2\pi x)$) and the frequency $q = 2$ (as a consequence of the choice $\alpha = \frac{1}{2}$). The occurrence of periodic orbits with period 3 increases as $\varepsilon$ gets larger; on the other hand, the occurrence of the frequency $q = 2$ increases as $\varepsilon$ gets smaller, which means that $\alpha$ is dominant for low values of $\varepsilon$. This example contributes to explain the roles of $\alpha$ and $s(2\pi x)$ in the weakly dissipative regime. In a similar way we interpret the results for the case of the mapping $s(2\pi x) = \frac{\sin(2\pi x)}{\cos(2\pi x)+1.4}$ which admits a full Fourier spectrum (Fig. 3.5b). We remark that the weakly dissipative solution can be analyzed also perturbatively by introducing the small quantity $\beta \equiv 1 - b$. Indeed, let us develop the solution in powers of $\beta$ as $y = y^{(0)} + \beta y^{(1)} + \beta^2 y^{(2)} + ...$, $x = x^{(0)} + \beta x^{(1)} + \beta^2 x^{(2)} + ...$; inserting such equations in the definition of the mapping, one easily gets recursive relations on the quantities $y^{(k)}$, $x^{(k)}$. The investigation of these series expansion might provide information about the solution in the weakly dissipative regime.

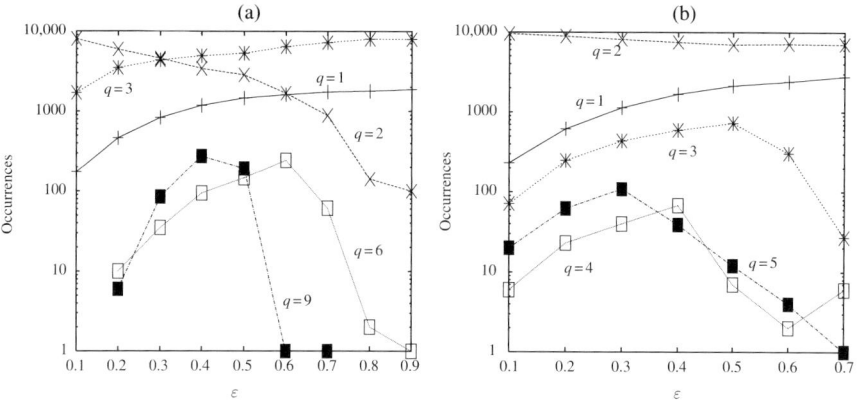

**Fig. 3.5.** Occurrence of periodic attractors versus $\varepsilon$. (**a**) Mapping $s(2\pi x) = \sin(3 \cdot 2\pi x)$ with $\alpha = \frac{1}{2}$; (**b**) mapping $s(2\pi x) = \frac{\sin(2\pi x)}{\cos(2\pi x)+1.4}$ with $\alpha = \frac{1}{2}$. After [8]

### 3.3.3 The Four-Dimensional Standard Mapping

The results presented for the two-dimensional mapping can be easily generalized to higher dimensional maps as well as to continuous systems. For example, let us consider the dissipative four-dimensional standard map described by the equations

$$y' = by + c_1 + \varepsilon \left[ \sin(x) + \gamma \sin(x - t) \right]$$
$$x' = x + y'$$
$$z' = bz + c_2 + \varepsilon \left[ \sin(t) - \gamma \sin(x - t) \right]$$
$$t' = t + z' \,,$$

where $y, z \in \mathbf{R}$, $x, t \in [0, 2\pi)$ and $c_1$, $c_2$ are real constants. The mapping depends also on three parameters: $b \in \mathbf{R}_+$ is the *dissipative* parameter, $\varepsilon \in \mathbf{R}_+$ is the *perturbing* parameter, $\gamma \in \mathbf{R}_+$ is the *coupling* parameter. Indeed, for $\gamma = 0$ we obtain two uncoupled two-dimensional standard mappings; we also remark that for $\varepsilon = 0$ we obtain two uncoupled mappings which admit rotational invariant circles with frequencies $\alpha_1 \equiv \frac{c_1}{1-b}$ and $\alpha_2 \equiv \frac{c_2}{1-b}$. Let $\omega = (\omega_1, \omega_2)$ be the frequency vector. With reference to Fig. 3.6 we select $\omega = \left(\frac{1}{s}, s - 1\right) = (0.754877..., 0.324717...)$, $s$ being the root of the third order polynomial $s^3 - s - 1 = 0$ (i.e., the smallest *Pisot–Vijayaraghavan* number of third degree; see, e.g., [7]). Figure 3.6 shows the two main frequencies as a function of the perturbing parameter in the case $b = 0.7$, $\gamma = 0.8$. A regular behavior is observed for values of $\varepsilon \leq 0.5$, followed by a chaotic motion manifested by an irregular variation of the frequency curves.

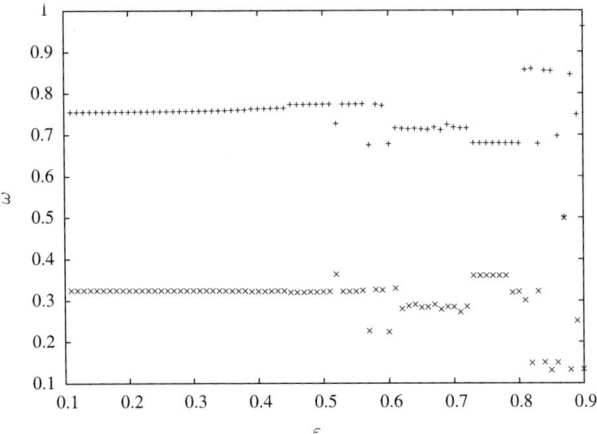

**Fig. 3.6.** Frequency analysis of the four-dimensional standard map with $\omega = (0.754877..., 0.324717...)$, $b = 0.7$, $\gamma = 0.8$ and initial conditions $y = 1$, $x = 0$, $z = 0.7$, $t = 0$

## 3.4 Adiabatic Invariants of the Pendulum

Let us consider a pendulum equation to which we add a small linear dissipative force, say
$$\ddot{x} + \alpha \sin x + \beta \dot{x} - \gamma = 0 ,$$
for $x \in [0, 2\pi]$. We can write the above equation also as
$$\dot{y} = -\alpha \sin x - \beta y$$
$$\dot{x} = y - \frac{\gamma}{\beta} ;$$
we remark that the choice of this example is motivated by the fact that it is very close to the spin-orbit equation described in the following section.

According to [18] the *adiabatic invariant* $Y \equiv \frac{1}{2\pi} \oint y \, dx$ slowly changes for a small variation of the dissipation factor $\beta$ according to $Y(t) = e^{-\beta t} Y(0)$. The phase-space area enclosed by a guiding trajectory is provided by the formula
$$\Gamma \equiv \oint \dot{x} \, dx = 2\pi Y - \frac{\gamma}{\beta} \oint dx .$$

As shown in [18], in case of positive circulation the spin slows down when approaching the resonance; in the librational regime the trajectory tends to the exact resonance; for negative circulation there are two possible behaviors: if $8\sqrt{\alpha} > 2\pi \frac{\gamma}{\beta}$ the guiding trajectory tends to the resonance, while if $8\sqrt{\alpha} < 2\pi \frac{\gamma}{\beta}$ the motion can evolve toward an invariant curve attractor.

To provide a concrete example, let us follow the trajectory with initial conditions $x = 0$, $y = -0.2$; the set of parameters $(\alpha, \beta, \gamma) = (0.0061, 0.01, 0.001)$ satisfies the condition $8\sqrt{\alpha} < 2\pi \frac{\gamma}{\beta}$ and the corresponding dynamics is attracted by an invariant curve (see Fig. 3.7, left panel); on the contrary, such condition is not fulfilled by $(\alpha, \beta, \gamma) = (0.0063, 0.01, 0.001)$ and consistently we find that the corresponding trajectory is attracted by a resonance as shown in Fig. 3.7, right panel.

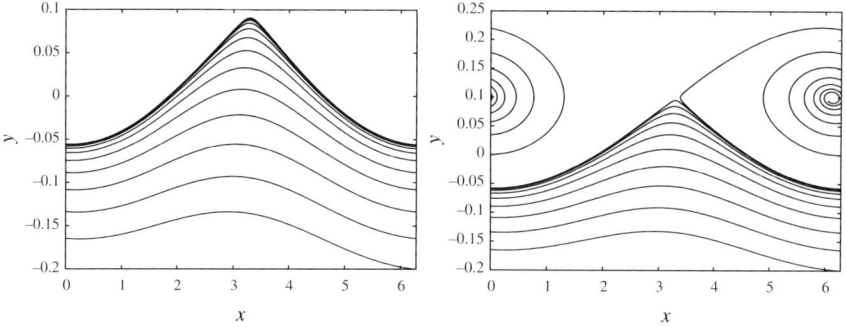

**Fig. 3.7.** Evolution of the dissipative pendulum in the phase space; **left panel**: attraction to an invariant curve; **right panel**: approach to a resonance

## 3.5 A Paradigm from Celestial Mechanics: The Spin-Orbit Model

### 3.5.1 The Conservative Model

A simple interesting physical problem which gathers together many features of nearly integrable weakly dissipative systems is provided by the spin-orbit coupling in Celestial Mechanics. Let us start with the description of the conservative model. We immediately remark that under suitable assumptions the equation of motion describing such model is very similar to the pendulum equation already met in the context of adiabatic invariants. More precisely, the model is the following: we consider a triaxial satellite $S$ orbiting around a central planet $P$ and rotating at the same time about an internal spin-axis. We denote by $T_{\text{rev}}$ and $T_{\text{rot}}$ the periods of revolution and rotation of $S$. The Solar System provides many examples of the so-called *spin-orbit resonances*, which are characterized by peculiar relationships between the revolution and rotation periods, according to the following

**Definition.** A spin-orbit resonance of order $p:q$ (with $p,q \in \mathbf{Z}_+$, $q \neq 0$) occurs whenever
$$\frac{T_{\text{rev}}}{T_{\text{rot}}} = \frac{p}{q} \ .$$

In order to write the equations of motion, we make the following hypotheses:

(i) the satellite moves on a Keplerian orbit around the planet;
(ii) the spin-axis coincides with the smallest physical axis of the ellipsoid;
(iii) the spin-axis is perpendicular to the orbital plane.

Under these assumptions the equation of motion can be written as follows. Let $a$, $r$, $f$ be the semimajor axis, the instantaneous orbital radius and the true anomaly of the satellite; let $A < B < C$ be its principal moments of inertia and let $x$ be the angle between the longest axis of the satellite and the pericentre line. Then, the motion is described by the equation

$$\ddot{x} + \frac{3}{2} \frac{B-A}{C} \left(\frac{a}{r}\right)^3 \sin(2x - 2f) = 0 \ . \tag{3.3}$$

Notice that the quantities $r$ and $f$ are known Keplerian functions of the time; setting $\varepsilon \equiv \frac{3}{2} \frac{B-A}{C}$, (3.3) can be expanded in Fourier series as (see, e.g. [5])

$$\ddot{x} + \varepsilon \sum_{m=-\infty, m\neq 0}^{\infty} W(m,e) \sin(2x - mt) = 0 \ , \tag{3.4}$$

for some coefficients $W(m,e)$ which decay as powers of the eccentricity being proportional to $e^{|m-2|}$.

We remark that the parameter $\varepsilon$ represents the equatorial oblateness of the satellite; when $\varepsilon = 0$ one has equatorial symmetry and the equation of motion is trivially integrable. Moreover, the dynamical system is integrable also in the case of circular orbit, since the radius $r$ coincides with the semimajor axis and the true anomaly becomes a linear function of the time.

### 3.5.2 The Dissipative Model

In writing (3.3) (equivalently (3.4)) we have neglected many contributions, like the gravitational attraction due to other celestial bodies or any kind of dissipative forces. Among the dissipative terms, the strongest contribution is due to the internal non-rigidity of the satellite. This tidal torque may assume different mathematical formulations; among the others we quote the classical MacDonald's ( [24]) and Darwin's ( [11]) torques which are reviewed in the following subsections (see [16, 26]). Let us summarize by saying that MacDonald expression assumes a phase lag depending linearly on the angular velocity, while Darwin's formulation Fourier decomposes the tidal potential, assigning to each component a constant amplitude. For more elaborated formulations of the tidal torque involving the internal structure of the satellite we refer to [19].

#### MacDonald's Torque

Let $\delta$ be the angle formed by the direction to the planet and the direction to the maximum of the tidal bulge. Let us denote the versor to $P$ by $\hat{\underline{r}}$ and the versor to the tidal maximum by $\hat{\underline{r}}_T$, i.e. the sub-planet position on $S$ a short time, say $\Delta t$, in the past; then we have

$$\hat{\underline{r}}_T = \hat{\underline{r}}(t - \Delta t) \simeq \hat{\underline{r}} - \frac{d\hat{\underline{r}}}{dt} \Delta t , \qquad (3.5)$$

where the derivative is computed in the body-frame. Using the relations $\cos\delta = \hat{\underline{r}} \cdot \hat{\underline{r}}_T$, $\sin\delta = \hat{\underline{r}}_T \wedge \hat{\underline{r}}$, we obtain that MacDonald's torque takes the expression (see, e.g., [16])

$$\underline{T} = \frac{3k_2 G m_P^2 R^5}{2r^6} \sin(2\delta)$$

$$= \frac{3k_2 G m_P^2 R^5}{r^6} (\hat{\underline{r}} \cdot \hat{\underline{r}}_T)(\hat{\underline{r}}_T \wedge \hat{\underline{r}}) ,$$

where $k_2$ is the so-called Love's number, $G$ is the gravitational constant, $m_P$ is the mass of the planet, $R$ is the satellite's mean radius. Taking into account (3.5) and the relation $\hat{\underline{r}} \cdot \hat{\underline{r}}_T = 1$, we obtain

$$\underline{T} = \frac{3k_2 G m_P^2 R^5 \Delta t}{r^6} \left( \hat{\underline{r}} \wedge \frac{d}{dt}\hat{\underline{r}} \right) .$$

Let $(\underline{e}_x, \underline{e}_y, \underline{e}_z)$ be the versors of the reference orbital plane; denote by $\Omega$ the longitude of the ascending node, $I$ is the obliquity, while $\psi_m$ is the angle between the ascending node and the body axis of minimum moment of inertia. Then, one obtains the relation ( [26])

$$\hat{\underline{r}} \wedge \frac{d}{dt}\hat{\underline{r}} = \dot{\psi}_m \sin I \cos(f - \Omega)(-\sin f \underline{e}_x + \cos f \underline{e}_y)$$
$$+ (\dot{f} - \dot{\Omega} - \dot{\psi}_m \cos I)\underline{e}_z .$$

Taking only the component along the $z$-axis, one obtains that the average of the tidal torque over the orbital period is given by

$$\langle \underline{T} \rangle = \frac{3k_2 G m_{\rm P}^2 R^5 \Delta t}{a^6} \left( nN(e) - L(e)\dot{\psi}_m \cos I \right) \underline{e}_z ,$$

where $n$ is the mean motion and $N(e)$, $L(e)$ are related to the following averages over short-period terms:

$$\left\langle \frac{a^6}{r^6} \dot{f} \right\rangle \equiv nN(e) = n \left( 1 + \frac{15}{2}e^2 + \frac{45}{8}e^4 + \frac{5}{16}e^6 \right) \frac{1}{(1-e^2)^6}$$

$$\left\langle \frac{a^6}{r^6} \right\rangle \equiv L(e) = \left( 1 + 3e^2 + \frac{3}{8}e^4 \right) \frac{1}{(1-e^2)^{9/2}} .$$

According to assumption $(iii)$ of the previous section we can set $I = 0$, so that $\psi_m$ coincides with $x$. Finally, the equation of motion (3.3) under the effect of the MacDonald's torque is given by

$$\ddot{x} + \frac{3}{2}\frac{B-A}{C}\left(\frac{a}{r}\right)^3 \sin(2x - 2f) = -K\left[L(e)\dot{x} - N(e)\right] ,$$

where we have used $\omega \Delta t = \frac{1}{Q}$, $Q$ being the so-called quality factor ( [26]), and where we have introduced a dissipation constant $K$ depending on the physical and orbital characteristics of the satellite:

$$K \equiv 3n \frac{k_2}{\xi Q} \left(\frac{R}{a}\right)^3 \frac{m_{\rm P}}{m_{\rm S}} ,$$

where $m_{\rm S}$ is the mass of the satellite and $\xi$ is a structure constant such that $C = \xi m R^2$. For the cases of the Moon and Mercury the explicit values of such constants are given in Table 3.1; it results that the dissipation constant amounts to $K = 6.43162 \times 10^{-7}$ yr$^{-1}$ for the Moon and $K = 8.4687 \times 10^{-7}$ yr$^{-1}$ for Mercury.

**Darwin's Torque**

In the case of Darwin's torque we just provide the explicit expression which is related to the Fourier expansion (3.4) as

**Table 3.1.** Physical and orbital data of the Moon and Mercury

|  | Moon | Mercury |
|---|---|---|
| $k_2$ | 0.02 | 0.4 |
| $Q$ | 150 | 50 |
| $m_S$ | $7.35 \times 10^{22}$ kg | $3.302 \times 10^{23}$ kg |
| $m_P$ | $5.972 \times 10^{24}$ kg | $1.99 \times 10^{30}$ kg |
| $R$ | 1737.5 km | 2440 km |
| $\xi$ | 0.392 | 0.333 |
| $a$ | $3.844 \times 10^5$ km | $5.79093 \times 10^7$ km |
| $e$ | 0.0554 | 0.2056 |
| $n$ | 84.002 yr$^{-1}$ | 26.0879 yr$^{-1}$ |
| $K$ | $6.43162 \times 10^{-7}$ yr$^{-1}$ | $8.4687 \times 10^{-7}$ yr$^{-1}$ |

$$\ddot{x} + \frac{3}{2}\frac{B-A}{C}\left(\frac{a}{r}\right)^3 \sin(2x - 2f) = -K\Big(W(-2,e)^2\mathrm{sgn}(x+1)$$
$$+ W(-1,e)^2\mathrm{sgn}(x+\frac{1}{2}) + W(1,e)^2\mathrm{sgn}(x-\frac{1}{2}) + W(2,e)^2\mathrm{sgn}(x-1)$$
$$+ W(3,e)^2\mathrm{sgn}(x-\frac{3}{2}) + W(4,e)^2\mathrm{sgn}(x-2) + W(5,e)^2\mathrm{sgn}(x-\frac{5}{2})$$
$$+ W(6,e)^2\mathrm{sgn}(x-3)\Big),$$

where the coefficients $W_k$ take the form

$$W(-1,e) = \frac{e^4}{24} \qquad W(-2,e) = \frac{e^3}{48}$$

$$W(1,e) = -\frac{e}{2} + \frac{e^3}{16} \qquad W(2,e) = 1 - \frac{5e^2}{2} + \frac{13e^4}{16}$$

$$W(3,e) = \frac{7e}{2} - \frac{123e^3}{16} \qquad W(4,e) = \frac{17e^2}{2} - \frac{115e^4}{6}$$

$$W(5,e) = \frac{845e^3}{48} \qquad W(6,e) = \frac{533e^4}{16}.$$

### 3.5.3 Capture into Resonance

Bearing in mind the discussions about the capture into resonance for the dissipative standard map, we proceed to illustrate some classical results (see [16, 26]) about resonance capture in the spin-orbit problem. We will see that such event strongly depends on the form of the dissipation. Using (3.4) let us write the dissipative spin-orbit equation as

$$\ddot{x} + \frac{3}{2}\frac{B-A}{C}\sum_{m=-\infty, m\neq 0}^{\infty} W(m,e)\sin(2x - mt) = T . \qquad (3.6)$$

Let us introduce the $p$-resonant angle $\gamma \equiv x - pt$; after averaging over one orbital period one gets

$$C\ddot{\gamma} + \frac{3}{2}(B-A)W(p,e)\sin 2\gamma = \langle T \rangle \ . \tag{3.7}$$

**Constant Torque**

Let us consider the case $\langle T \rangle =$ const; a first integral associated to (3.7) is trivially obtained as

$$\frac{1}{2}C\dot{\gamma}^2 - \frac{3}{4}(B-A)W(p,e)\cos 2\gamma = \langle T \rangle \gamma + E_0 \equiv E \ . \tag{3.8}$$

We plot in Fig. 3.8 the behavior of $\frac{1}{2}\dot{\gamma}^2$ versus $\gamma$ as derived from (3.8) (see [16]). We denote by $\gamma_{\max}$ the point at which $\dot{\gamma} = 0$. Assuming an initial positive $\dot{\gamma}$, we proceed along the curve until we reach $\gamma_{\max}$; at this moment the motion reverses sign, thus escaping from the resonance.

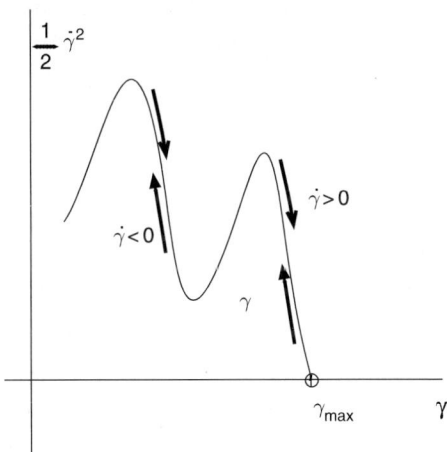

**Fig. 3.8.** Case $\langle T \rangle =$ const

**MacDonald's Case**

Let us now assume that $\langle T \rangle = -K_1(\dot{\gamma} + V)$ for some constants $K_1$ and $V$; then we have:

$$\frac{d}{dt}\left[\frac{1}{2}C\dot{\gamma}^2 - \frac{3}{4}(B-A)W(p,e)\cos 2\gamma\right] = -K_1(\dot{\gamma}^2 + V\dot{\gamma}) = \frac{dE}{dt} \ ,$$

where the behavior of $\frac{1}{2}\dot{\gamma}^2$ is provided in Fig. 3.9. Let us denote by $\Delta E$ the difference of $\frac{\dot{\gamma}^2}{2}$ between two successive minima, say $\gamma_1$ and $\gamma_2$; let $\delta E$ be

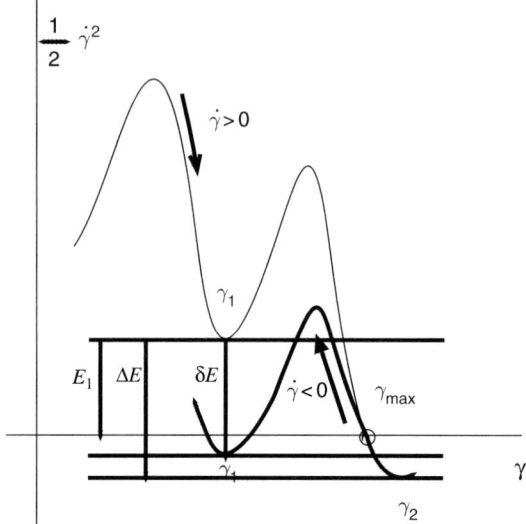

**Fig. 3.9.** MacDonald's case: trapping in libration

the difference between the two minima with the same ordinate $\gamma_1$. Let $E_1$ be the ordinate of the highest minimum at $\Gamma_1$; being $|\delta E| > E_1$ there exists a second zero of $\frac{\dot{\gamma}^2}{2}$ near $\gamma_1$, so that $\dot{\gamma}$ reverses sign again and $\gamma$ is trapped in libration between $\gamma_1$ and $\gamma_2$.

On the other hand, if $|\delta E| < E_1$ the planet escapes from the resonance and continues to despin (see Fig. 3.10).

Following [16] we can compute the probability of capture $P$ as follows. Assuming that the values of $E_1$ are distributed with uniform probability in

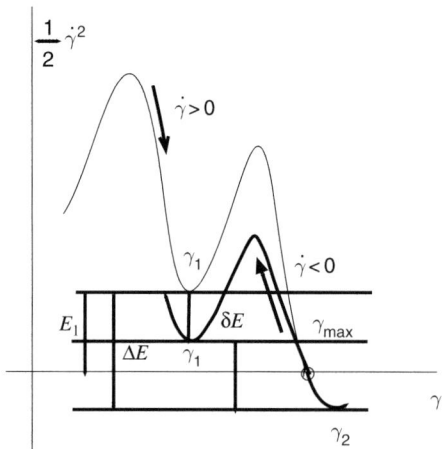

**Fig. 3.10.** MacDonald's case: escape from resonance

$[0, \Delta E]$ we define $P$ as $P \equiv \frac{\delta E}{\Delta E}$; then we obtain

$$P \simeq \frac{2}{1 + \frac{\pi V}{\int_{\gamma_1}^{\gamma_2} \dot\gamma d\gamma}}.$$

**Darwin's Case**

We assume that $\langle T \rangle = -W - Z \, \text{sgn}(\dot\gamma)$, for some constants $W$ and $Z$. Denote by $\delta E'$ the difference of $\frac{1}{2}\dot\gamma^2$ between $\gamma_2$ and the second minimum at $\gamma_1$; then, one can easily show that (see [16])

$$\Delta E = -(W + Z)\pi, \qquad \delta E = -2\pi Z,$$

so that the probability of capture can be written as

$$P = \frac{2Z}{W + Z},$$

which turns out to be independent on $\frac{B-A}{C}$.

## 3.6 The Restricted, Planar, Circular, Three-Body Problem

Another basic example of a dissipative nearly integrable system in Celestial Mechanics is represented by gravitationally interacting bodies, subject to a dissipative force. Focussing our attention to the restricted, planar, circular, three-body problem, we can derive the equations of motion in a synodic reference frame as follows. We investigate the motion of a massless body $S$ under the influence of two primaries $P_1$, $P_2$ with masses $\mu$ and 1-$\mu$; we assume that the motion of the primaries is circular around their common barycenter and that all bodies move on the same plane. If $(x, y, p_x, p_y)$ denote the coordinates of the minor body in the synodic frame, the equations of motion under a linear dissipation read as

$$\begin{aligned}
\dot x &= y + p_x \\
\dot y &= -x + p_y \\
\dot p_x &= p_y - \frac{1-\mu}{r_1^3}(x + \mu) - \frac{\mu}{r_2^3}(x - 1 + \mu) - Kp_x \\
\dot p_y &= -p_x - \frac{1-\mu}{r_1^3}y - \frac{\mu}{r_2^3}y - Kp_y,
\end{aligned} \quad (3.9)$$

where $r_1 \equiv \sqrt{(x+\mu)^2 + y^2}$, $r_2 \equiv \sqrt{(x-1+\mu)^2 + y^2}$. The effect of the dissipation is simulated by adding the terms $(-Kp_x, -Kp_y)$ to the equations for $\dot p_x$ and $\dot p_y$. Let us now look at the survival of periodic orbits under the effect

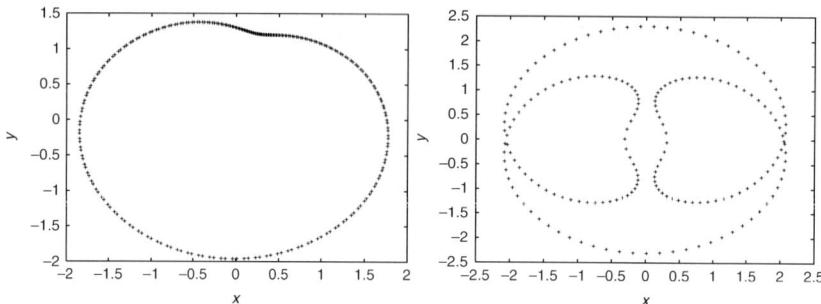

**Fig. 3.11. Left panel**: periodic orbit of period 2 for $K = 0$, corresponding to a semimajor axis equal to 1.58 and eccentricity equal to 0.22. **Right panel**: periodic orbit of period 3 for $K = 0$, corresponding to a semimajor axis equal to 1.31 and eccentricity equal to 0.76

of the dissipation. By Newton's method we determine the periodic orbit in the conservative case; then we slowly increase the dissipation parameter and we compute the periodic orbit through a continuation method. This procedure might fail or work in different situations as shown in Fig. 3.11: the left panel provides a periodic orbit in the conservative setting with period twice the basic period of the primaries; however, such orbit immediately disappears as $K \neq 0$. On the other hand the periodic orbit shown in the right panel of Fig. 3.11 has period 3 times the basic period of the primaries and can be continued in the dissipative context up to $K = 10^{-4}$.

We conclude by mentioning that many other dissipative forces can be considered in the framework of the three-body problem, such as the solar wind (which is caused by charged particles originating from the upper atmosphere of the Sun), the Yarkowsky effect (consisting in the anisotropic emission of thermal photons due to the rotation of the celestial body), the radiation pressure (caused by electromagnetic radiation). This last force is a component of the Poynting–Robertson drag exerted by solar radiation on dust grains. Other dissipative forces widely studied in the literature are the Stokes and Epstein drags, which affect the orbital evolution of a dust grain in a gas planetary nebula, and are respectively valid for low and high Reynold numbers.

## 3.7 Periodic Orbits for Non-Autonomous and Autonomous Systems

A remarkable discussion of the existence of periodic orbits for non-autonomous and autonomous systems can be found in [27] (see also [9, 28, 30]). Indeed, the arguments we are going to present apply to the two examples we have discussed so far, namely the non-autonomous spin-orbit problem and the autonomous restricted, planar, circular, three-body problem. In particular, we want to show that

(i) for the non-autonomous spin-orbit problem described by (3.6) one can find periodic orbits with period equal (or multiple) to that of the conservative problem;
(ii) for the restricted, planar, circular, three-body problem one can apply in the conservative case ((3.9) with $K = 0$) the theory for autonomous systems, which allows to find periodic solutions with the same period of the case in which the perturbing parameter $\mu$ is set to zero;
(iii) for the dissipative restricted, planar, circular, three-body problem, one can use the autonomous theory to find periodic orbits with period close (but not exactly equal) to that obtained for $K = 0$.

We report in the Appendix the perturbative computation up to the second order of periodic orbits in an autonomous dissipative case.

### 3.7.1 Non-Autonomous Systems

Consider the differential equation

$$\dot{x} = f(x, t; \gamma) ,$$

where $x = (x_1, ..., x_n)$ and $f$ is a $T$-periodic function depending on a real parameter $\gamma$. Assume that for $\gamma = 0$ we know a $T$-periodic orbit described by the equations

$$x(t) = \varphi(t) \quad \text{with} \quad \varphi(T) = \varphi(0) .$$

Then, if $\gamma$ is sufficiently small, one can prove the existence of a periodic solution with period $T$. In fact, assume that the initial data of the periodic orbit for $\gamma \neq 0$ are

$$x(0) = \varphi(0) + \beta ,$$

where $\beta = (\beta_1, ..., \beta_n)$. After one period one has

$$x(T) = \varphi(0) + \beta + \psi ,$$

where $\psi = (\psi_1, ..., \psi_n)$ are holomorphic functions in $\beta_1, ..., \beta_n, \gamma$. In order to prove the existence of a periodic solution, the following $n$ equations in the unknowns $(\beta_1, ..., \beta_n)$ must be satisfied:

$$\psi = (\psi_1, ..., \psi_n) = 0 .$$

Applying the implicit function theorem, if $\gamma$ is sufficiently small and if the Jacobian of $\psi$ with respect to $\beta$ satisfies

$$\left(\frac{\partial \psi}{\partial \beta}\right)\Big|_{\gamma=\beta=0} \neq 0 ,$$

then there exists $\beta = \beta(\gamma)$ such that $\beta(0) = 0$ and there exists a $T$-periodic orbit provided $\gamma$ is sufficiently small.

### 3.7.2 Autonomous Systems

Consider the autonomous system

$$\dot{x} = f(x; \gamma) ,$$

where $x = (x_1, ..., x_n)$ and $f$ depends on the real parameter $\gamma$. Assume that for $\gamma = 0$ we know a $T$-periodic orbit, given by the equations

$$x(t) = \varphi(t) \qquad \text{with} \qquad \varphi(T) = \varphi(0) .$$

We immediately remark that if there is one periodic orbit, then there exists an infinity, since if $x(t) = \varphi(t)$ is periodic, also $x(t) = \varphi(t+h)$ for any real $h$ is periodic, being the system autonomous.

Let $\gamma \neq 0$ and look for a solution such that

$$x(0) = \varphi(0) + \beta , \qquad x(T+\tau) = \varphi(0) + \beta + \psi ,$$

where $\psi = (\psi_1, ..., \psi_n)$ are holomorphic in $\beta_1, ..., \beta_n, \gamma, \tau$.

Such solution is $(T+\tau)$-periodic if

$$\psi = (\psi_1, ..., \psi_n) = 0 ,$$

which represents $n$ equations in the $n+1$ unknown quantities $\beta_1, ..., \beta_n, \tau$.

Since there are several choices of the initial conditions which lead to the same orbit (indeed, taking any other point of the orbit we change only the epoch and not the orbit), we can arbitrarily set $\beta_n = 0$. Therefore we have $n+1$ equations $\psi = (\psi_1, ..., \psi_n) = 0$, $\beta_n = 0$ in the $n+1$ unknowns $\beta_1, ..., \beta_n, \tau$. By the implicit function theorem, if the jacobian

$$\begin{pmatrix} \frac{\partial \psi_1}{\partial \beta_1} & \cdots & \frac{\partial \psi_1}{\partial \beta_{n-1}} & \frac{\partial \psi_1}{\partial \tau} \\ \vdots & & & \\ \frac{\partial \psi_n}{\partial \beta_1} & \cdots & \frac{\partial \psi_n}{\partial \beta_{n-1}} & \frac{\partial \psi_n}{\partial \tau} \end{pmatrix}_{\gamma=\beta=\tau=0} \neq 0 ,$$

and if $\gamma$ is sufficiently small, there exists a periodic solution with period $T + \tau$.

### 3.7.3 Autonomous Systems with Integrals

In the autonomous case assume there exists an integral

$$G(x) = C = const .$$

Then the equations

$$\psi_1 = ... = \psi_n = \beta_n = 0 \tag{3.10}$$

are not anymore distinct and one can replace the above equations with

$$\psi_1 = \ldots = \psi_n = \beta_n = 0 , \qquad G = C + \lambda\gamma , \qquad (3.11)$$

where $\lambda$ is a generic constant; alternatively, one can replace (3.10) with the equations

$$\psi_1 = \ldots = \psi_n = \beta_n = 0 , \qquad \tau = 0 . \qquad (3.12)$$

Equations (3.11) imply that the energy level is changed, while (3.12) imply that if there exists an integral $G(x) = C$, one can find a $T$-periodic solution for $\gamma$ small.

## Appendix: Second-Order Computation of Periodic Orbits in the Autonomous Dissipative Case

Consider the differential equations

$$\dot{x} = f(x, \gamma) , \qquad (3.13)$$

where $x = (x_1, \ldots, x_n)$ and $f$ depends also on a real small parameter $\gamma$. We want to look for a periodic solution with period $T_\gamma$, such that

$$x(T_\gamma) = x(0) .$$

To this end we expand the solution $x$ and the period $T_\gamma$ in series of $\gamma$ as

$$x = x_0 + \gamma x_1 + \gamma^2 x_2 + \cdots$$
$$T_\gamma = T_0 + \gamma T_1 + \gamma^2 T_2 + \cdots \qquad (3.14)$$

Let us denote by $f_\gamma = \frac{\partial f}{\partial \gamma}$, $f' = \frac{\partial f}{\partial x}$, $f'' = \frac{\partial^2 f}{\partial x^2}$; inserting (3.14) in (3.13) and equating same orders of $\gamma$ up to the second order we get

$$\dot{x}_0 = f(x_0, 0)$$
$$\dot{x}_1 = f'(x_0, 0)x_1 + f_\gamma(x_0, 0)$$
$$\dot{x}_2 = f'(x_0, 0)x_2 + \frac{1}{2}f''(x_0, 0)x_1^2 .$$

Suppose now that for $\gamma = 0$ we know a $T_0$-periodic orbit, described by $x_0 = x_0(t)$ with initial data $x_0(0)$ and periodicity conditions $x_0(T_0) = x_0(0)$. Then, for $\gamma \neq 0$ we look for a $T_\gamma$-periodic orbit described by $x = x(t)$, whose initial data are displaced with respect to the conservative case as

$$x(0) = x_0(0) + \beta ; \qquad (3.15)$$

moreover we require that the following periodicity conditions are satisfied:

$$x(T_\gamma) = x_0(0) + \beta . \qquad (3.16)$$

Develop $\beta$ in powers of $\gamma$ as $\beta = \gamma\beta_1 + \gamma^2\beta_2 + \cdots$; from (3.15) one has

$$x_0(0) + \gamma x_1(0) + \gamma^2 x_2(0) + \ldots = x_0(0) + \gamma\beta_1 + \gamma^2\beta_2 + \cdots$$

Comparing same orders of $\gamma$, one obtains $\beta_j = x_j(0)$, so that the $\beta_j$s are the corrections at order $j$ to the initial data. Using (3.16) and recalling that $x_0(T_0) = x_0(0)$, one obtains

$$x_0(T_0) + \gamma\dot{x}_0(T_0)(T_1 + \gamma T_2) + \frac{1}{2}\ddot{x}_0(T_0)\gamma^2 T_1^2 + x_1(T_0)\gamma + \dot{x}_1(T_0)T_1\gamma^2$$
$$+ x_2(T_0)\gamma^2 + \cdots = x_0(0) + \gamma\beta_1 + \gamma^2\beta_2 + \cdots$$

Equating same orders of $\gamma$ up to the order 2, one gets

$$\dot{x}_0 = f(x_0, 0)$$
$$x_0(T_0) = x_0(0)$$
$$\dot{x}_1 = f'(x_0, 0)x_1 + f_\gamma(x_0, 0)$$
$$\beta_1 = x_1(T_0) + \dot{x}_0(T_0)T_1 \;;$$
$$\dot{x}_2 = f'(x_0, 0)x_2 + \frac{1}{2}f''(x_0, 0)x_1^2$$
$$\beta_2 = x_2(T_0) + \dot{x}_0(T_0)T_2 + \frac{1}{2}\ddot{x}_0(T_0)T_1^2 + \dot{x}_1(T_0)T_1 \;.$$

Let us analyze the last equation; recalling that $\beta_2 = (\beta_2^{(1)}, ..., \beta_2^{(n)})$, we have $n$ equations in the $n+1$ unknowns $\beta_2^{(1)}, ..., \beta_2^{(n)}, T_2$. To eliminate the ambiguity, we can set $\beta_2^{(n)} = 0$ and solve the equations with respect to the remaining unknowns.

# References

1. Beaugé, C., and Ferraz–Mello, S., 1994, Capture in exterior mean–motion resonances due to Poynting-Robertson drag. *Icarus* **110**, 239–260.
2. Beaugé, C., and Ferraz–Mello, S., 1993, Resonance trapping in the primordial solar nebula: The case of a Stokes drag dissipation. *Icarus* **103**, 301–318.
3. Bohr, T., Bak P., and Jensen, M. H., 1984, Transition to chaos by interaction of resonances in dissipative systems. II. Josephson junctions, charge-density waves, and standard maps. *Phys. Rev. A* **30**(4), 1970–1981.
4. Broer, H. W., Simó C., and Tatjer, J. C., 1998, Towards global models near homoclinic tangencies of dissipative diffeomorphisms. *Nonlinearity* **11**, 667–770.
5. Celletti, A., 1990, Analysis of resonances in the spin-orbit problem in Celestial Mechanics: The synchronous resonance (Part I). *J. Appl. Math. Phys. (ZAMP)* **41**, 174–204.
6. Celletti, A., Della Penna, G., and Froeschlé, C., 1998, Analytical approximation of the solution of the dissipative standard map. *Int. J. Bif. Chaos* **8**(12), 2471–2479.

7. Celletti, A., Falcolini, C., and Locatelli, U., 2004, On the break-down threshold of invariant tori in four dimensional maps. *Regular Chaotic Dyn* **9**(3), 227–253.
8. Celletti, A., Froeschlé, C., and Lega, E., 2006, Dissipative and weakly–dissipative regimes in nearly integrable mappings. *Discrete Cont. Dyn. Sys.—Series A* **16**(4), 757–781.
9. Coddington, E. A., and Levinson, N., 1995, Theory of Ordinary Differential Equations, McGrawHill, New York.
10. Correia, A. C. M. and Laskar, J., 2004, Mercury's capture into the 3/2 spin–orbit resonance as a result of its chaotic dynamics. *Nature* **429**, 848–850.
11. Darwin, G., 1908, Tidal friction and cosmogony, Scientific papers, Cambridge University Press **2**.
12. D'Hoedt, S., and Lemaitre, A., 2004, The Spin-Orbit Resonant Rotation of Mercury: A Two Degree of Freedom Hamiltonian Model. *Celest. Mech. Dyn. Astron.* **89**(3), 267–283.
13. Froeschlé, C., Guzzo M., and Lega, E., 2000, Graphical evolution of the Arnold's web: From order to chaos. *Science* **289**(5487), 2108–2110.
14. Froeschlé, C., Lega E., and Gonczi, R., 1997, Fast Lyapunov indicators. Application to asteroidal motion. *Celest. Mech. Dyn. Astron.* **67**, 41–62.
15. Goldreich, P., 1966, Final spin states of planets and satellites. *Astron. J.* **71**(1), 1–7.
16. Goldreich, P., and Peale, S., 1966, Spin–orbit coupling in the solar system. *Astron. J.* **71**(6), 425–438.
17. Guzzo, M., Lega E., and Froeschlé, C., 2002, On the numerical detection of the stability of chaotic motions in quasi–integrable systems. *Physica D* **163**, 1–25.
18. Henrard, J., 1993, The adiabatic invariant in classical mechanics. *Dyn. Rep.* **2**, new series, 117–235.
19. Hussmann, H., and Spohn, T., 2004, Thermal orbital evolution of Io and Europa. *Icarus* **171**, 391–410.
20. Kim, S. Y. and Lee, D. S., 1992, Transition to chaos in a dissipative standardlike map. *Phys. Rev. A* **45**(8), 5480–5487.
21. Laskar, J., 1993, Frequency analysis for multi-dimensional systems. Global dynamics and diffusion. *Physica D* **67**, 257–281.
22. Laskar, J., Froeschlé C., and Celletti, A., 1992, The measure of chaos by the numerical analysis of the fundamental frequencies. Application to the standard mapping. *Physica D* **56**, 253–269.
23. Lega, E., and Froeschlé, C., 1996, Numerical investigations of the structure around an invariant KAM torus using the frequency map analysis. *Physica D* **95**, 97–106.
24. MacDonald, G. J. F., 1964, Tidal friction. *Rev. Geophys.* **2**, 467–541.
25. Marzari, F., and Weidenschilling, S. J., 2002, Mean Motion Resonances, Gas Drag, and Supersonic Planetesimals in the Solar Nebula. *Cel. Mech. Dyn. Astron.* **82**(3), 225–242.
26. Peale, S. J., 2005, The free precession and libration of Mercury. *Icarus* **178**, 4–18.
27. Poincarè, H., 1892, Les Methodes Nouvelles de la Mechanique Celeste. Gauthier Villars, Paris.
28. Siegel, C., L., and Moser, J. K., 1971, Lectures on Celestial Mechanics. Springer-Verlag, Berlin.

29. Schmidt, G., and Wang, B. W., 1985, Dissipative standard map. *Phys. Rev. A* **32**(5), 2994–2999.
30. Szebehely, V., 1967, Theory of orbits, Academic Press, New York and London.
31. Weidenschilling, S. J., and Jackson, A. A., 1993, Orbital resonances and Poynting-Robertson drag. *Icarus* **104**(2), 244–254.
32. Wenzel, W., Biham, O., and Jayaprakash, C., 1991, Periodic orbits in the dissipative standard map. *Phys. Rev. A* **43**(12), 6550–6557.

# 4

# Connectance and Stability of Dynamical Systems

Marina Cosentino[1], Dimitri Laveder[2], Elena Lega[2] and Claude Froeschlé[2]

[1] Dipartimento di Matematica, Università di Pisa, Via Buonarroti 2, 56127, Pisa, Italy
cosentino@dm.unipi.it
[2] Observatoire de la Côte d'Azur, B.P. 4229, 06304 Nice Cedex 4, France
laveder@obs-nice.fr

**Abstract.** We study the linear stability of dynamical systems as a function of the number of degrees of freedom and of the fraction of direct couplings, i.e. connectance, among them. It is commonly accepted within the community of dynamical systems that the stability decreases when increasing the number of degrees of freedom or when increasing the connectance. Conversely, ecologists have long sought models such that stability increases with the complexity. After an historical overlook, we show in this paper that both results can be obtained and that they are strictly related to the sign and magnitude of the matrix elements characterizing the linearization of the dynamical system about an equilibrium point. More precisely, systems characterized by negative off-diagonal coefficients are shown to display an increased stability at large connectance where the matrix elements fall in suitable ranges. The analytic study of a very specific system and the numerical investigation of more generic cases allow us to give new criteria to establish when stability can increase with the connectance.

## 4.1 Introduction

In the last 30 years many studies have been devoted to the investigation of the relationships between the properties of dynamical systems and their number of degrees of freedom. In the 1970s the problem of looking for the transition from ordered to chaotic motion when increasing the number of interacting particles has been studied both analytically and numerically, leading to the conclusion that the measure of the chaotic domain with respect to the whole phase space volume increases very rapidly with the number of degrees of freedom [1, 2, 4]. In particular, Benettin and et al. [1] took as a model problem an isolated one-dimensional self-gravitating system, consisting of $n$ plane-parallel sheets with uniform density, and they found that the amount of chaoticity increases with the number of sheets. This model has nevertheless the property of being maximally connected, i.e. each particle interacts directly with each other. The question arised then whether changing the fraction of direct couplings could have an effect on the degree of chaoticity of the system.

In the same years Froeschlé [5] turned his attention to this issue, by analyzing the behavior of a $n$-dimensional symplectic mapping in which the degrees of freedom are coupled by an interaction matrix. The interactions (which act as perturbations) are switched on or off by setting to 0 or 1 the elements of the upper or lower diagonals of this matrix. The connectance $n_c$ is then defined as the number of non-zero upper and lower diagonals, so to quantify the number of direct interactions. The emergency of chaos was checked by measuring the size of the non-chaotic zone about a given elliptic equilibrium point, and it was found that it depends not only on the number of degrees of freedom but also on the connectance. Moreover, for $n$ large enough the amount of chaoticity depends only on the connectance and above a critical value of $n_c$ the system becomes completely chaotic. While giving some insight on the influence of "connectedness" on the dynamics of a system, this work was not generalized because of the difficulties in giving an unambiguous definition of connectance in the nonlinear case.

The situation is much more clear in linear systems, where the influence of the reciprocal interactions was considered already by Gardner and Ashby in 1970 [8]. They considered the linearization of a dynamical system about an equilibrium point, reducing their study to a system of ordinary differential equations

$$\dot{\mathbf{x}} = M\mathbf{x} \qquad (4.1)$$

where $M$ is a real $n \times n$ matrix, and defined the *connectance* $C$ as the fraction of non-zero off-diagonal elements $m_{i,j}$ of the matrix $M$. The connectance varies then in the interval $[0, 1]$: if $C = 0$, $M$ is a diagonal matrix, while if $C = 1$, all the elements of the matrix are different from zero.

A matrix element $m_{i,j} \neq 0$ denotes a direct interaction between the variables $i$ and $j$, where if $m_{i,j} = 0$ there is no direct interaction. The non-zero elements of the matrix were chosen as random uniform variables belonging to some interval. In particular, the diagonal terms were taken in the interval $[-1, -0.1]$, so that the individual variables $i$ were assumed to be intrinsically stable. The off-diagonal ones belonged to $[-1, +1]$. According to the definition that the system is stable when all the eigenvalues of $M$ have negative real part, they computed the eigenvalues of a set of 1000 matrices and tested them for stability, when changing the connectance. They defined the *probability of stability* $P(n, C)$ as the fraction of stable systems among all the $N$ randomly chosen ones. $P(n, C)$ was computed for $n = 4, 7$ and 10 and in general it depends both on the connectance and on the dimension of the system. For a given $n$, $P(n, C)$ decreases monotonically with $C$ and when increasing $n$ the probability $P(n, C)$ drastically drops to 0 even for low values of $C$. In other words the larger is $n$, the greater is the rate of decrease. Gardner and Ashby argued that there should exist a critical value $C_{\text{crit}}$ for large $n$, so that $P(n, C < C_{\text{crit}}) \sim 1$ and $P(n, C > C_{\text{crit}}) \sim 0$. A sudden change of behavior should occur in the vicinity of $C_{\text{crit}}$.

In this chapter we reconsider the experiment of Gardner and Ashby, taking randomly constructed linear systems and varying the intervals from which diagonal and off-diagonal terms are drawn. We then show that in some cases systems showing an increase of stability with connectance can exist, and we give general conditions for their realization.

The chapter is organized as follows. In Sects 4.2 and 4.3 we give an historical overlook about the connectance idea in ecology and economics, pointing out how in ecology the problem of the relationship stability-connectance has long been debated. In Sect. 4.4 we recall the main definitions, and in Sect. 4.5 we discuss the relevance of the quasi-diagonal dominance criterion as a guide to enforce stability, pointing out its relationship with a more general result known as Geršgorin theorem [9]. In Sect. 4.6 we analyze the role of the connectance and of the mean intensity of the interactions on the stability of competitive systems (defined as such when the off-diagonal matrix elements are negative), pointing out the conditions to increase stability at high connectance. In Sect. 4.7 we discuss how the size of the system couples with the connectance and with the interaction strength in influencing stability. Conclusions are provided in Sect. 4.8.

## 4.2 Connectance in Ecology: A Historical Perspective

Since Gardner and Ashby's work the concept of connectance in linear systems has been applied to many contexts, in particular ecology. For the ecologists community the connectance, intended as the number of links among living species, gives in fact some measure of complexity of an ecological system. More specifically, linear systems are used in ecology to modelize the dynamical behavior of a so-called *food web*. Loosely speaking, a food web is defined as the multitude of relationships between species in a biotic community. This concept is translated to mathematical language as follows. Let $i$ and $j$ denote two biological species and $x_i$ and $x_j$ the number of organisms belonging to species $i$ and $j$. In a system with $n$ species, the set of all possible relationships among them is modeled at the lowest order by a $n \times n$ Jacobian matrix of interaction coefficients $m_{i,j}$ that describes the impact of each species $i$ on the growth of each species $j$. The connectance associated to this matrix is understood as "complexity", while stability of the food web is identified with the stability of the linear system in Gardner and Ashby's sense. In food webs the type of relationship between the species $i$ and $j$ can be classified according the signs of the entries of the matrix: if $m_{i,j} > 0$, $i$ increases owing to the existence of $j$, whereas if $m_{i,j} < 0$, $j$ effects $i$ negatively. There are four possible relationships:

$$m_{i,j} < 0, \, m_{j,i} < 0 \to \text{negative feedback}$$
$$m_{i,j} > 0, \, m_{j,i} < 0 \to \text{predator/prey}$$
$$m_{i,j} > 0, \, m_{j,i} > 0 \to \text{cooperation}$$
$$m_{i,j} < 0, \, m_{j,i} > 0 \to \text{prey/predator.}$$

In 1972 May [11] reconsidered Gardner and Ashby's results in a food web context. By analytic considerations, he extended Gardner and Ashby's work to an infinitely large number of variables, agreeing with the fact that randomly generated food webs decrease in stability as they increase in complexity.

In 1974 Daniels and Mackay [3] repeated the same numerical calculations of Gardner and Ashby, but they found a slightly different result. They confirmed the decrease of stability when increasing the connectance but the transition from stability to instability is not so sharp and there doesn't exist a critical value $C_{\text{crit}}$. Much later, Martens [10] repeated the same experiment and found results compatible with Daniels and Mackay's ones.

Apart from this discrepancy, the result of May and followers was paradoxical to many ecologists who experience the opposite pattern in nature. It seems in fact that in real food webs complexity helps to increase stability rather than to diminish it. A way out of this paradox was to reconsider the dynamics of real food webs, by taking more complex networks, characterized by structured interactions and dynamical characteristics that would allow complex communities to persist. This is translated into choosing the matrix elements in more complicated ways, not simply as random numbers in a given interval. May himself in 1999 [12] commented on the actual state of food webs theory, affirming that real ecosystems are not randomly constructed and therefore his old results do not apply to real-world networks, which are much more structured.

As an example of a recent work which couples modern ecological considerations to mathematical modeling, in 2000 Wilmers, Sinha and Brede [16] reconsidered the dynamics of interactions in food webs. They examined the stability properties of differents food web configurations and added hierarchical structures to them. They supposed that in nature there are much more prey and predator relationships, which is translated into taking off-diagonal symmetric elements in the Jacobian matrix with opposite sign. Adding this constraint, they made the same experiments as Gardner and Ashby. For a given connectance, they found a larger probability of stability, but they never found an increase of stability with the connectance, so the problem was not really solved. This is the fundamental point, and until now in ecology no simple linear model has been found allowing an increasing of the stability with the connectance.

## 4.3 Connectance in Economics

Another field of application of the relationship between connectance and stability in linear systems is economics, with regards to the stability of equilibria of interdependent markets. A work in this direction was done by Froeschlé and Longhi in 1987 [6] and then deepened by Froeschlé, Lega and Lohinger [7]. They studied the stability-connectance relation in an economic system of $n$ goods characterized by a vector of prices $\boldsymbol{p} = (p_1, \ldots, p_n)$. The time evolution of prices is regulated by the market law of supply and demand. They defined the excess demand $E_i$ as the difference between the global demand and the global supply for the good $i$ obtained for a given vector of prices: $E_i = E_i(\boldsymbol{p})$. Then, the evolution of prices with time is defined by the set of equations:

$$\dot{p}_i = g_i[E_i(\boldsymbol{p})] \ \forall i = 1,..,n \qquad (4.2)$$

where $g_i$ is a regular increasing function which is 0 at the origin. The vector $\boldsymbol{p}^* = (p_i^*, \ldots, p_n^*)$ such that $E_i(\boldsymbol{p}^*) = 0 \ \forall i$ is said to be the vector of prices in general equilibrium. By linearizing the set of (4.2) we get

$$\dot{\boldsymbol{P}} = M(\boldsymbol{p}^*)\boldsymbol{P} \qquad (4.3)$$

where $\boldsymbol{P} = (\boldsymbol{p} - \boldsymbol{p}^*)$ and $M$ is the Jacobian matrix of $\boldsymbol{g}$, i.e. the matrix which defines the relations between the $n$ independent markets. The authors simplified then the problem taking $g_i[E_i(\boldsymbol{p}^*)] = b_i E_i(\boldsymbol{p}^*)$, with $b_i > 0$ ($\forall i$), which for the elements of $M$ gives $m_{i,j} = b_i \partial E_i(\boldsymbol{p}^*)/\partial p_j$. The diagonal elements $m_{i,i}$ represent the influence exerted by a variation of the price of the good $i$ on its global demand, while the off-diagonal terms $m_{i,j}$ represent the influence exerted by a variation in the price of the good $j$ on the global demand for the good $i$. To fix the terminology, if $m_{i,j} > 0$ the good $j$ is said to be a *substitute* of the good $i$, while if $m_{i,j} < 0$ the good $j$ is said to be *complementary* to the good $i$.

Froeschlé and coworkers [6] studied then the stability of the linear system $\dot{\mathbf{x}} = M\mathbf{x}$ as a function of the connectance, a problem perfectly analogous to the one analyzed in the food webs context. In particular, they studied the effects of substitutability/complementarity on the stability, i.e. the effect of varying the fraction of off-diagonal elements with positive/negative signs in the matrix $M$. They not only confirmed the results of Mackay and Daniels, but they also found maximal stability when 50% of the goods are substitutable. By choosing the off-diagonal terms as random variables in a given interval, this coincides with taking this interval symmetric with respect to 0.

This result seemed in contradiction with the general equilibrium theory, for which total substitutability is considered a sufficient condition for stability. In fact there is no contradiction, because general theory of equilibria supposes tacitly the so-called Walras law, which implies that for every line of the matrix $M$, the sum of their elements is smaller than zero. Then, if we want to have any chance to satisfy the Walras law, we have to impose this restriction on the matrix.

Froeschlé and Longhi remarked that coupling this condition with the total substitutability implies that the matrix $M$ satisfies the quasi-diagonal dominance criterion for stability [13], which states the following : all the eigenvalues of a square matrix $M$ are negative if the diagonal terms are negative and there exists a set of positive numbers $w_i > 0, i = 1\ldots n$ such that

$$w_i|m_{i,i}| > \sum_{j \neq i} w_j|m_{i,j}| \ \forall i = 1,\ldots,n \ . \qquad (4.4)$$

Equation (4.4) implies that the relevant parameters for stability are the diagonal terms with respect to the sum of the modules of the off-diagonal terms, i.e the relative size of the intervals from which the diagonal and off-diagonal terms are drawn. With the restriction imposed by Froeschlé and Longhi, the matrix $M$ satisfies exactly this condition with all the weights equal to 1 and then the equilibrium is stable.

Since this work, in the economics community it become common to try to adjust the parameters of an economic system to achieve the quasi-diagonal dominance condition in order to stabilize equilibria. In Sect. 4.5, we will show that this procedure is not optimal, because this criterion for stability is far from being sharp. Moreover, we will point out the relationship with this condition and a more general series of results known as Geršgorin (or Geršgorin-like) theorems.

## 4.4 Linear Stability and Connectance: Basic Definitions

We briefly recall the main definitions and issues related to the connectance problem. Given a $n$-dimensional dynamical system $\dot{\mathbf{x}} = f(\mathbf{x})$, we focus our attention to the study of its linear stability about an equilibrium point. Linearization leads to a system of ordinary differential equations

$$\dot{\mathbf{x}} = M\mathbf{x} \tag{4.5}$$

where $M$ is a real $n \times n$ matrix. We define then the *connectance* $C$ about the equilibrium point as the fraction of non-zero off-diagonal elements $m_{i,j}$ of the matrix $M$. This definition is exactly the same as Gardner and Ashby's [8], presented in Sect. 4.1. The elements of the matrix $M$ are evidently related in sign and order of magnitude to the coupling coefficients among degrees of freedom in the general system, the details depending on the specific model. We make the assumption that these coefficients are chosen in such a way that the diagonal elements of $M$ result negative, so to guarantee stability when $C = 0$.

We repeat then the classical experiment, i.e. we calculate the probability of stability over a given set of $N$ matrices as a function of $C$. In all the following experiments we take $N = 1000$ if not otherwise specified, and 101 values of the connectance between 0 and 1. For each matrix we compute the eigenvalues in order to say if the system is stable or not, considering as usual the system as stable if all its eigenvalues have negative real part. We define the *probability of stability* $\alpha$ as the fraction of stable systems among all the $N$ randomly chosen ones. The non-zero elements of the matrices satisfy the relations:

$$\begin{cases} m_{i,i} = u[a_1, a_2] & \forall i, i = 1, \ldots, n \\ m_{i,j} = u[b_1, b_2] & \forall i, j \; j \neq i, i, j = 1, \ldots, n \end{cases} \tag{4.6}$$

where $u[x_1, x_2]$ denotes a random uniform variable belonging to the interval $U = [x_1, x_2]$. We denote $U_D$ and $U_{OD}$ the intervals from which we draw the diagonal and off-diagonal terms respectively.

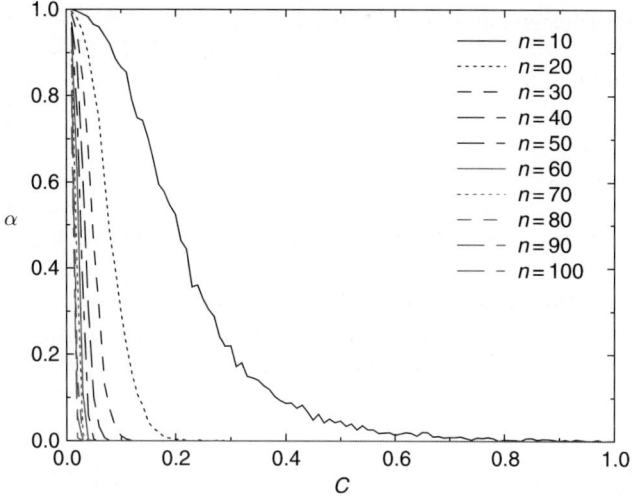

**Fig. 4.1.** Probability of stability $\alpha$ as a function of the connectance for different dimensions $n$ of the system and diagonal interval $U_D = [-1, -0.1]$, off-diagonal interval $U_{OD} = [-1, 1]$

To reproduce the historical results, we take $U_D = [-1, -0.1]$ and $U_{OD} = [-1, 1]$ and ten values of the matrix dimension $n = 10, 20, 30, \ldots 100$. We found in general that $\alpha$ depends both on the connectance and on the dimension of the system. For $C = 0$ the resulting diagonal matrix is evidently stable. For a given $n$, $\alpha$ decreases monotonically with $C$ and when increasing $n$ the probability $\alpha$ drastically drops to 0 even for low values of $C$ (Fig. 4.1). This result confirms previous findings [3, 6]. We found the same discrepancy with Gardner and Ashby's experiment as explained in Sect. 4.1.

## 4.5 The Diagonal Dominance Criterion and its Applicability

For some applications it is interesting to predict a priori the probability of stability $\alpha$ for a given equilibrium, or even to optimize the coupling coefficients among degrees of freedom in order to maximize the stability for a given connectance. In fact, there exist some theorems which can be used to relate the dependence of $\alpha$ on the matrix parameters [15], but which are of difficult application when $n$ is large. The only one which can be easily used is the Geršgorin theorem [9] which states the following: given an $n \times n$ matrix $M = [m_{i,j}]$, define the $i$th disk in the complex plane

$$G_i = \{z \in \mathcal{C} : |z - m_{i,i}| \leq r_i\}, \quad r_i = \sum_{j \neq i} |m_{i,j}| \tag{4.7}$$

then the set of the eigenvalues of $M$ is contained in the union of these disks.

A more refined version of this theorem allows the substitution of $r_i$ by a weighted radius $r_i^w = \sum_{j \neq i} w_j |m_{i,j}|/w_i$, given $n$ strictly positive real numbers $w_i$. From this condition it can be readily derived the sufficient criterion for stability known as quasi-diagonal dominance theorem [13] introduced in Sect. 4.3.

It turns out that when the off-diagonal terms are all positive, the quasi-diagonal dominance criterion becomes a necessary and sufficient condition for stability [14], while nothing can be said if the sign of the off-diagonal terms is not defined.

As explained in Sect. 4.3 for some applications such as economics [6] a classical strategy to maximize the likelihood of equilibria stability consists in forcing the system to satisfy the quasi-diagonal dominance, usually considering the weights $w_i$ equal to one. The main goal of this section is to discuss the limitations of this approach and in general the practical difficulties in using the previous theorems to predict the probability of equilibria for intermediate connectance.

First of all we numerically tried to apply the criterion on a set of off-diagonal positive definite matrices of size $n = 50$ with $U_{OD} = [0, 0.1]$ and $U_D = [-1, -0.1]$ by using $10^p$ randomly distributed weights $w_i \in (0, 1)$, for $p = 3, 4, \ldots, 8$. In Fig. 4.2 we plot the fraction of stable matrices $\gamma$ as a function of $C$ according to the criterion (4.4) compared to the fraction of stable matrices $\alpha$. It is evident that when increasing $p$ the curves $\gamma(C)$ converge extremely slowly to the real stability curve $\alpha(C)$, so that the criterion is in practice inapplicable whenever the weights are unknown. The result does not get better when the $w_i$ belong to a larger interval (not shown), nor taking simply all the $w_i$ equal to one (Fig. 4.2, dark dashed line).

A second aspect concerns the sharpness of the quasi-diagonal dominance criterion, which does not discriminate among the signs of the matrix elements, when $U_{OD}$ contains negative values as well. In Fig. 4.3 we calculated $\alpha(C)$ for $n = 100$, $U_D = [-1, -0.1]$ and three off-diagonal intervals having the same distribution for the module of their elements: $U_{OD} = [-0.1, 0]$, $U_{OD} = [-0.1, 0.1]$ and $U_{OD} = [0, 0.1]$. We remark that in the first two cases the fraction of stable matrices largely exceeds what predicted by (4.4). In other words, when the matrix elements are not positive the sharpness of the criterion (4.4) in checking for stability is very small, making it an inefficient guide to enforce stability of equilibria.

On the other hand, for $U_{OD} = [-0.1, 0]$ $\alpha$ shows a non-monotonic dependence on $C$, becoming larger when the matrix is fully connected. We remark that this behavior is not possible when the off-diagonal matrix elements are drawn from a positive interval $U_{OD} = [0, b_2]$, because in this case matrices with a size $n$ and a connectance $C$ will have the same stability properties if the parameter $nCb_2$ is kept constant, the right-hand side of (4.4) scaling statistically as $nCb_2$. This implies that the probability of stability can just decrease when increasing $C$, $n$ or the amplitude of $U_{OD}$. The counter-intuitive behavior seen for $U_{OD} = [-0.1, 0]$ will be termed "stability recovery phenomenon"

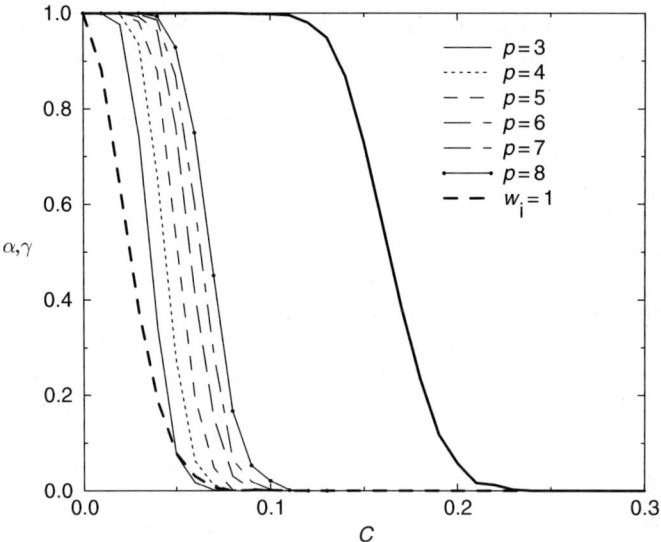

**Fig. 4.2.** *Light lines*: Probability of stability $\gamma$ as a function of the connectance according to the quasi-diagonal dominance criterion for $10^p$ weights $w_i \in (0,1)$ and different values of $p$. *Dark dashed line*: same with all the weights equal to 1. *Dark solid line*: probability of stability $\alpha$ from direct eigenvalues computation

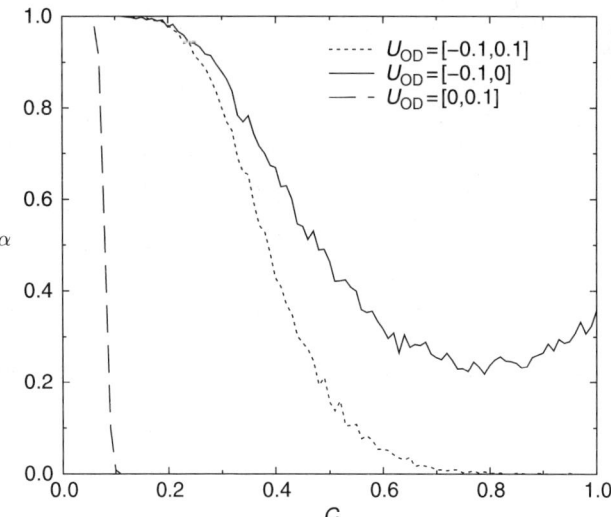

**Fig. 4.3.** Probability of stability $\alpha$ as a function of the connectance for $n = 100$, diagonal interval $U_{\rm D} = [-1, -0.1]$ and off-diagonal intervals $U_{\rm OD} = [-0.1, 0.1]$ (*dotted line*), $U_{\rm OD} = [-0.1, 0]$ (*solid line*), $U_{\rm OD} = [0, 0.1]$ (*dashed line*)

in the following. As this is an example of the behavior long sought by the ecology community, its appearance and the conditions for its realization will be extensively studied in the following sections.

## 4.6 The Stability Recovery Phenomenon: Some Simple Criteria

First of all we want to investigate the behavior of the function $\alpha(C)$ for $C$ close to 1. We start from the simple fully connected case in which the matrix $M$ is given by $m_{i,i} = a\,\forall i$ and $m_{i,j} = b\,\forall i, j \neq i$. This matrix has $n-1$ eigenvalues equal to $a-b$, while the $n$th one equals $a+(n-1)b$. This can be rapidly shown by taking a matrix $B$ with all elements equal to $b$. $B$ has at least one eigenvalue equal to 0 (because $\det B = 0$) with multiplicity $n-1$ (because $\text{rank}(B) = 1$, so $\ker(B) = n-1$). The other eigenvalue equals then $Tr(B) = nb$. Noticing that adding to a matrix the same value $d$ to its diagonal, the eigenvalues are augmented by the same quantity $d$, one recovers the eigenvalues for $M$ by adding $a-b$ to the diagonal of $B$.

A necessary and sufficient condition for the stability is then $|a| > |b|$ if $a, b < 0$, and $|a| > (n-1)|b|$ if $a < 0, b > 0$. If all the matrix elements are negative, the stability condition is independent of the dimension $n$ and much less restrictive than the quasi-diagonal dominance condition, while it obviously coincides with this latter for $a < 0$ and $b > 0$, for $w_i = 1$ in this specific case.

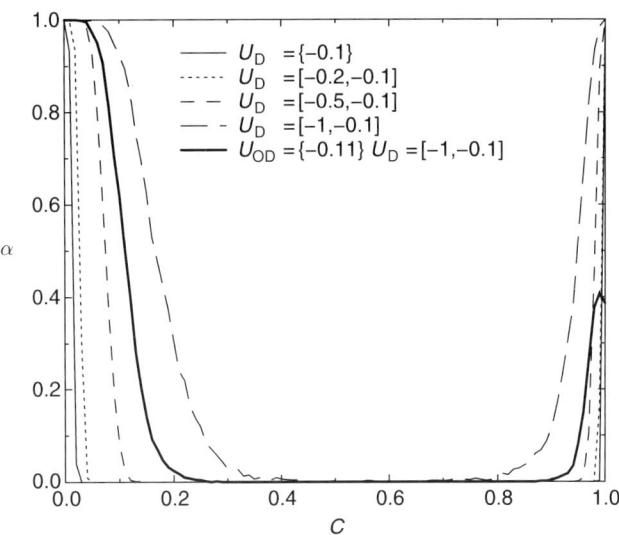

**Fig. 4.4.** Probability of stability $\alpha(C)$ for $n = 100$. *Light lines*: $U_{\text{OD}} = \{-0.09\}$ and four diagonal intervals $U_{\text{D}} = [a_1, -0.1]$ with $a_1 = -0.1, -0.2, -0.5, -1$ (from *solid* to *long dashed lines*). *Dark solid line*: $U_{\text{OD}} = \{-0.11\}$ and $U_{\text{D}} = [-1, -0.1]$

We address now numerically the more general case of $a, b < 0$ taken randomly in a given interval and $C$ between 0 and 1. We first take the same off-diagonal values $b = -0.09$ and use different diagonal intervals $U_D = [a_1, -0.1]$ with $a_1 = -0.1, -0.2, -0.5, -1$. The stability curves $\alpha(C)$ for $n = 100$ are plotted in Fig. 4.4 (light lines). In this case characterized by diagonal elements drawn from a non-zero amplitude interval $U_D = [a_1, a_2]$ such that $a_2 < b$, $\alpha$ is still 1 for $C = 1$ and raises suddenly from 0 when $C$ approaches 1. Moreover, extending the possible diagonal values to more negative numbers widens the range of $C$ for which $\alpha$ is sensibly different from 0. Conversely, when $a_2 > b$ this result does not hold anymore, the probability of stability for $C = 1$ being about 0.98 for $b = -0.101$ (not shown) and only about 0.37 for $b = -0.11$ (Fig. 4.4, dark curve). On this grounds, we concentrate in the following on

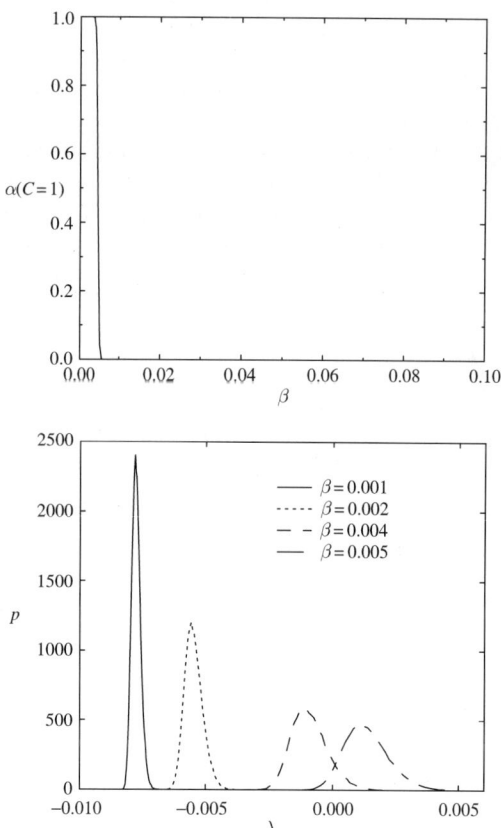

**Fig. 4.5.** Diagonal elements all equal to $a = -0.1$, off-diagonal interval $U_{OD} = [-0.09, -0.09 + \beta]$, $n = 100$. **Top**: Probability of stability $\alpha$ for $C = 1$ as a function of $\beta$. **Bottom**: Probability density $p(\lambda_{\max})$ for the distribution of the real part of the largest eigenvalue for $\beta = 0.001, 0.002, 0.004, 0.005$

the case of intervals $U_D = [a_1, a_2]$ and $U_{OD} = [b_1, b_2]$ with $a_2 < b_1$, for which $\alpha(1)$ is likely to be maximal.

We investigate at first the effect of widening the off-diagonal interval, by keeping the same diagonal elements equal to $a = -0.1$. We analyze first the case $C = 1$. For $n = 100$, we take $U_{OD} = [-0.09, -0.09 + \beta]$ with $0 < \beta \le 0.09$. The stability is rapidly lost when increasing $\beta$, such that already for $\beta = 0.02$ the fraction of stable matrices at $C=1$ is 0 (see Fig. 4.5 (top)). To shade some light on this behavior we study in the same cases the distribution of the largest eigenvalue $\lambda_{\max}$, which determines the stability and is equal to $\lambda_{\max,0} = -0.01$ when $\beta = 0$. For a better statistics we use only for this experience a set of

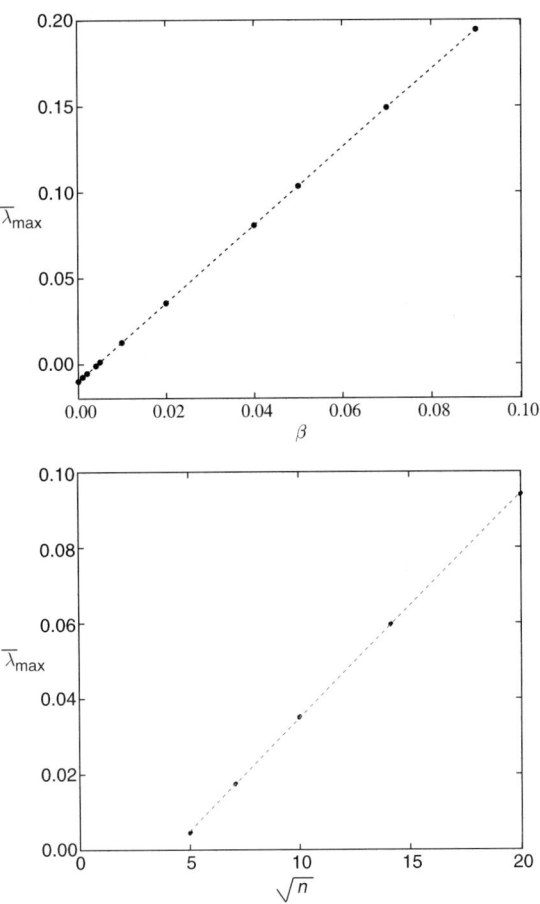

**Fig. 4.6.** Diagonal elements all equal to $a = -0.1$, off-diagonal interval $U_{OD} = [-0.09, -0.09 + \beta]$. **Top**: for $n = 100$, values of $\bar{\lambda}_{\max}$ as a function of $\beta$ and corresponding linear fit. **Bottom**: values of $\bar{\lambda}_{\max}$ as a function of $\sqrt{n}$ for $\beta = 0.02$ and corresponding linear fit

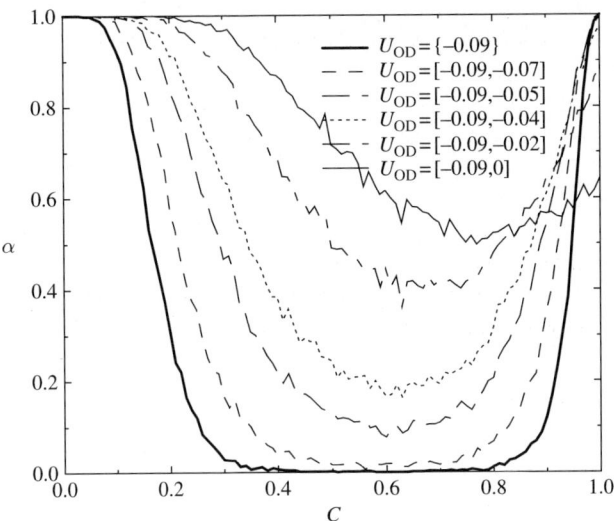

**Fig. 4.7.** Probability of stability $\alpha(C)$ for $n = 100$, $U_D = [-1, -0.1]$ and six off-diagonal intervals $U_{OD} = [-0.09, -0.09 + \beta]$. From the *dark solid line* to the *light solid line* (**bottom to top**), $\beta = 0, 0.02, 0.04, 0.05, 0.07, 0.09$

$10^5$ matrices. A few histograms are plotted in Fig. 4.5 (bottom) for different values of $\beta$. The distribution of $\lambda_{\max}$ has a bell-shape with a higher tail on the right side, and its mean value $\bar{\lambda}_{\max}$ is consistent with a law of the kind:

$$\bar{\lambda}_{\max} - \lambda_{\max,0} = \kappa(\mu + 0.09) = \kappa\beta/2 \tag{4.8}$$

where $\mu = \beta/2 - 0.09$ is the mean value of the off-diagonal terms and $\kappa$ is a proportionality factor depending on the matrix size $n$. This experimental law is plotted in Fig. 4.6 (top). The coefficient $\kappa$ respects an approximate scaling $\kappa \propto \sqrt{n}$, as shown in Fig. 4.6 (bottom) where $\bar{\lambda}_{\max} - \lambda_{\max,0}$ is plotted against $\sqrt{n}$ keeping $\beta$ constant ($U_{OD} = [-0.09, -0.07]$).

Widening the off-diagonal interval corresponds then to destroy the stability for $C = 1$, making impossible a stability recovery at high connectance. One can try to recover stability by taking lower values for the diagonal terms.

On this grounds we now go back to a diagonal interval $U_D = [-1, -0.1]$, considering the whole dependence on $C$. The probability of stability $\alpha(C)$ is plotted in Fig. 4.7 for $U_{OD} = [-0.09, -0.09+\beta]$ and $\beta = 0, 0.02, 0.04, 0.05, 0.07, 0.09$. Taking a more negative diagonal a partial stability is recovered for $C = 1$, while the fast decreasing of $\alpha$ at low $C$ persists. The net effect is a minimum of stability for intermediate values of $C$.

We conclude that even in non-trivial situations in which the diagonal and the off-diagonal coupling coefficients are distributed in a given interval, one can increase the probability of stability at high connectance by choosing non-overlapping intervals $U_D$, $U_{OD}$, such that $U_D$ be wide enough and $U_{OD}$

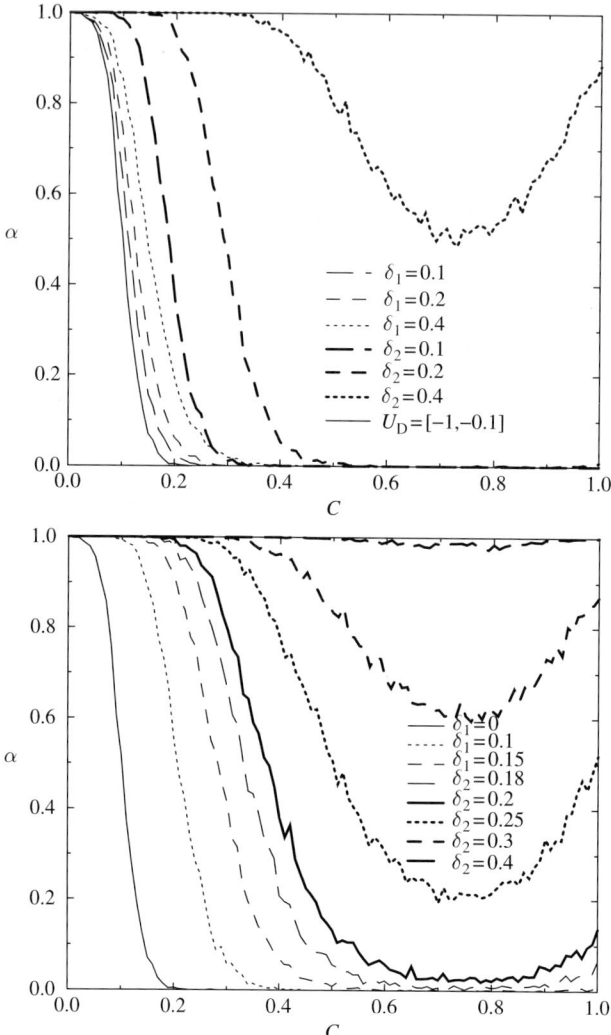

**Fig. 4.8.** Probability of stability $\alpha(C)$ for $n = 100$ and off-diagonal interval $U_{\text{OD}} = [-0.2, 0]$. **Top**: $U_{\text{D}} = [-1, -0.1]$ (*light solid line*); $U_{\text{D}} = [-1 - \delta_1, -0.1]$ and $\delta_1 = 0.1, 0.2, 0.4$ (*light* respectively *long-dashed*, *dashed* and *dotted lines*); $U_{\text{D}} = [-1, -0.1 - \delta_2]$ and $\delta_2 = 0.1, 0.2, 0.4$ (*dark* respectively *long-dashed*, *dashed* and *dotted lines*). **Bottom** :$U_{\text{D}} = [-1 - \delta, -0.1 - \delta]$ and $\delta = 0, 0.1, 0.15, 0.18, 0.2, 0.25, 0.3, 0.4$ (from *light solid* to *dark long-dashed lines*)

be narrow enough, both containing only negative terms. This rule of thumb will be confirmed in the remainder of this section.

To stress the importance of the non-superposition criterion between diagonal and off-diagonal intervals as a way to recover stability at high $C$, we perform two series of experiences with $U_{OD} = [-0.2, 0]$. We first take $U_D = [-1-\delta_1, -0.1]$ varying $\delta_1$ from 0 to 0.4 to widen the diagonal interval but keeping a partial superposition with $U_{OD}$. We take next $U_D = [-1, -0.1-\delta_2]$, thus making narrower the diagonal interval but progressively non-superposed with $U_{OD}$. The results are shown in Fig. 4.8 (top) for $n = 100$. While increasing $\delta_1$ alone has almost no effect on the probability of stability, increasing $\delta_2$ alone leads quickly the system to recover stability at large $C$.

On the other hand, it is necessary to precise that the non-superposition criterion is only probably necessary but evidently non sufficient to increase the stability at large $C$, being related in a complicated way to the size of the intervals themselves. This can be put in evidence by taking still $n = 100$, $U_{OD} = [-0.2, 0]$ and by simply shifting $U_D$ towards the left as $U_D = [-1-\delta, -0.1-\delta]$ with $0 \leq \delta \leq 0.4$. For $\delta = 0.1$ the intervals do not superpose anymore, but the stability recovery is not yet present, appearing for $\delta = 0.18$ and becoming more and more relevant when the separation between the intervals is increased (Fig. 4.8 (bottom)).

When allowing some off-diagonal terms to be also positive, the increase of stability at large $C$ is quickly lost, recovering the more trivial result of $\alpha$ monotonically decreasing with $C$. This can be seen in Fig. 4.9, in which $\alpha(C)$ is plotted for $n = 100$, $U_D = [-1, -0.1]$ and $U_{OD} = [-0.09, b_2]$ with $b_2 = 0, 0.02, 0.045, 0.07, 0.09$, corresponding to a probability of having a fraction between 0% and 50% of non-competitive interactions. As conspicuous in Fig. 4.9, already one tenth of positive off-diagonal terms is sufficient to destroy the stability recovery at large $C$.

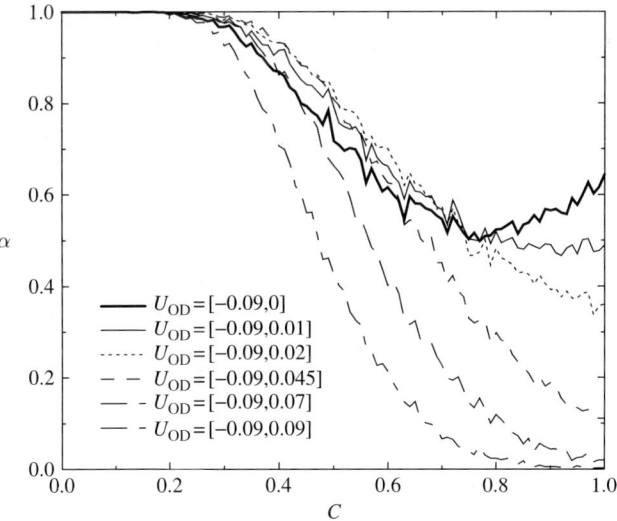

**Fig. 4.9.** Probability of stability $\alpha(C)$ for $n = 100$, $U_D = [-1, -0.1]$ and six off-diagonal intervals $U_{OD} = [-0.09, b_2]$. From the *dark solid line* to the *light dash-dotted line* (**top to bottom**), $b_2 = 0, 0.01, 0.02, 0.045, 0.07, 0.09$

## 4.7 The Influence of the Matrix Size

The problem can be made even more complicated by taking into account the influence of the matrix size $n$. As an example, we plot the stability curves $\alpha(C)$ for $U_D = [-1, -0.1]$ and varying both $n$ and $U_{OD}$. We take $U_{OD} = [-0.09, 0]$ for $n = 50, 100, 200$ and $U_{OD} = [-0.05, 0]$ for $n = 200$. When $U_{OD} = [-0.09, 0]$ one can see that for an intermediate value of $n$ ($n = 50$ or $100$) the stability increases at large $C$, while if $n$ is too large ($n = 200$) $\alpha$ decreases monotonically with $C$ (Fig. 4.10, light lines). Nevertheless, almost total stability everywhere can be recovered when the off-diagonal interval is reduced to $U_{OD} = [-0.05, 0]$ (Fig. 4.10, dark line).

To investigate the dependence of stability on both $n$ and the size of the off-diagonal interval, we calculate $\alpha(C)$ for $U_D = [-1, -0.1]$, taking different intervals $U_{OD} = [b_1, 0]$ ($b_1 < 0$) and different values of $n$ but keeping constant the product $|b_1|\sqrt{n} = \sqrt{2/3}$. The values of the dimension are $n = 150, 200, 400, 600$. The stability curves plotted in Fig. 4.11 do not superpose but are comparable, the stability slightly increasing when $n$ is made larger. We deduce that a situation of overall stability, for $U_D$ and $U_{OD}$ non-overlapping, can be maintained and even strengthened for larger and larger systems if the coupling coefficients are decreased as $1/\sqrt{n}$. This result contrasts with the case of positive off-diagonal terms, in which the stability curve $\alpha(C)$ does not varies when $|b_1|n$ is kept constant.

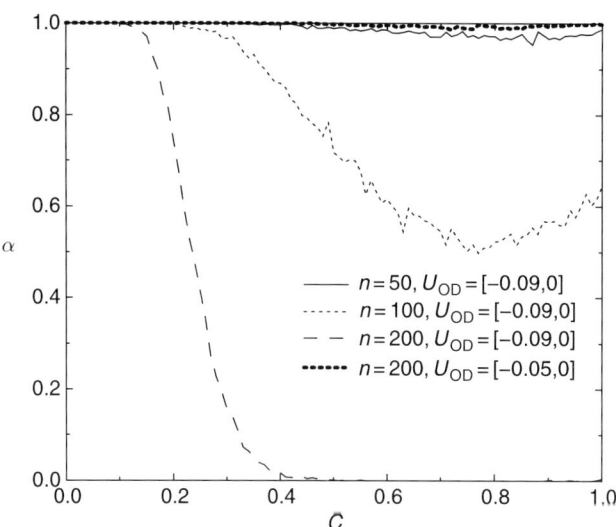

**Fig. 4.10.** Probability of stability $\alpha(C)$ for $U_D = [-1, -0.1]$. *Light lines*: $U_{OD} = [-0.09, 0]$, $n = 50, 100, 200$ (respectively *solid, dotted* and *dashed line*). *Dark dash-dotted line*: $U_{OD} = [-0.05, 0]$, $n = 200$

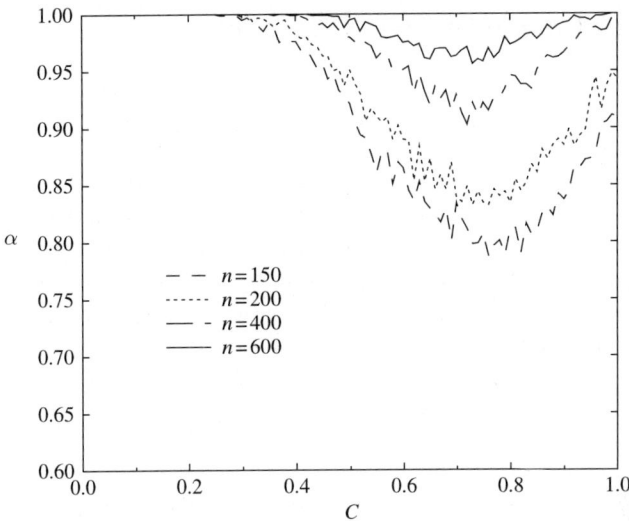

**Fig. 4.11.** Probability of stability $\alpha(C)$ for $U_D = [-1, -0.1]$, $U_{OD} = [b_1, 0]$ and dimensions $n$ such that $|b_1|\sqrt{n} = \sqrt{2/3}$. Dashed line: $n = 150$, $b_1 \approx 0.067$; dotted line: $n = 200$, $b_1 \approx 0.057$; dash-dotted line: $n = 400$, $b_1 \approx 0.041$; solid line: $n = 600$, $b_1 \approx 0.033$

It is interesting to remark that for symmetric off-diagonal intervals $U_{OD} = [-b, b]$ (the diagonal elements being all equal) May [11] conjectured that the relevant parameter for stability was the product $b\sqrt{nC}$. In our case of non-symmetric intervals, we recover an approximate stability parameter $b\sqrt{n}$, but the connectance cannot enters in such a parameter with a monotonic function, which would prevent the stability recovery phenomenon at large $C$ to appear.

## 4.8 Conclusions

The behavior of a dynamical system is influenced both on the number of the degrees of freedom and on their mutual coupling (i.e. connectance), in a way which is far from being understood. In this paper we discussed this topic from an historical point of view and gave some new results to help to clarify this issue, by limiting ourselves to discuss linear stability about an equilibrium point.

We first presented some established results about connectance in its main areas of applications, i.e. ecology and economics. Taking as a starting point a well-known paradigmatic result [3, 8], we have then shown that stability profiles characterized by a non-monotonic dependence on the connectance can be recovered, this behavior being strictly related to the sign and magnitude of the coefficients of the matrix $M$ obtained by linearization of the original model.

We discussed the applicability of the well-known sufficient condition of stability given by the quasi-diagonal dominance criterion, used in literature as a guide to build stable connected systems. That condition is shown to be far from optimal even when the system is only partially competitive (i.e. when some of the off-diagonal coefficients of the matrix $M$ are negative).

By concentrating on this latter case, we looked for different criteria of stability, analyzing in particular the possibility of restabilization at moderate or high connectance. We started our analysis from a specific fully connected system in which the matrix $M$ is characterized by only two values (one for the diagonal terms and one for the off-diagonal ones), for which there exist an analytical necessary and sufficient condition of stability. For competitive systems, this condition is independent of the dimension of the matrix and less restrictive than the quasi-diagonal dominance theorem. We have then investigated numerically the effect of the connectance on stability in the more general case in which the elements of the matrix $M$ are drawn from some intervals. We have obtained this way a new operational criterion to increase the probability of stability at high connectance. Concerning the size of the system we have shown that, for the competitive case, the stability rescales approximately with the square root of the dimension, in agreement with a conjecture concerning symmetric off-diagonal intervals [11].

The question on the influence of the connectance on the dynamics of a fully nonlinear systems is much more complicated. Froeschlé [5] numerically demonstrated a positive relationship between the connectance and the size of the chaotic domain in the phase space of a coupled mapping. Along this line, the extension of the present work to some specific nonlinear models is currently under study.

# References

1. Benettin, G., Froeschlé, C., and Scheidecker J. P., 1979. *Phys. Rev A* **61**, 2454.
2. Casartelli, M., Diana, E., Galgani, L., and Scotti A., 1976. *Phys. Rev A* **13**, 1921.
3. Daniels, J. and Mackay A. L., 1974. *Nature* **251**, 49.
4. Froeschlé, C. and Scheidecker J. P., 1975, *Phys. Rev A* **12**, 2137.
5. Froeschlé C., 1978. *Phys. Rev A* **18**, 277.
6. Froeschlé, C. and Longhi A., 1987. *Econ. Appli* **XL 1**, 49.
7. Froeschlé, C., Lega, E., and Lohinger E., 2000. *Nonlinear Phenom. Complex Sys.* **3:3**, 247.
8. Gardner, M. R. and Ashby W. R., 1970. *Nature* **228**, 784.
9. Geršgorin S., 1931. *Isv. Akad Nauk USSR Ser. Mat.* **7**, 749.
10. Martens B., 1987. *Ecol. Model.* **35**, 157.
11. May R. M., 1972. *Nature* **238**, 413.
12. May R. M., 1999. *Phyl. Trans. R. Soc. Lond. B* **354**, 1951.
13. Mckenzie L., 1959. Matrices with dominant diagonals and economic theory. In Arrow, K. J., Karlin, L. S., and Suppes L. P. (eds), *Mathematical Methods in the Social Sciences*, Stanford University Press, Stanford 1959, pp. 47–62.

14. Šiljak D. D., 1978. *Large-scale Dynamic Systems. Stability and Structure*, North Holland, New York, p. 230.
15. Varga R., 2004. *Geršgorin and His Circles*, Springer-Verlag, Berlin.
16. Wilmers, C., Sinha, S., and Brede M., 2000. *Evolving Stable Ecological Networks*, working paper, Complex Systems Summer School, Santa Fe.

# 5

# Chaotic Diffusion of Asteroids

Kleomenis Tsiganis

Department of Physics, Aristotle University of Thessaloniki,
GR 54124 Thessaloniki, Greece
tsiganis@astro.auth.gr

**Abstract.** A large fraction (> 30%) of the numbered main-belt asteroids follow chaotic orbits, mainly associated to *high-order mean motion resonances* (MMRs) either of *two-body* (asteroid–Jupiter) or *three-body* type (asteroid–Jupiter–Saturn or Mars). These resonances form a dense network of thin chaotic layers throughout the asteroid belt, where small-amplitude variations in the proper elements of asteroids accumulate slowly over time. This effect is commonly referred to as *chaotic diffusion*. In this chapter we review recent results on specific asteroid groups, whose evolution is governed by an interplay between chaotic diffusion and *radial migration*, induced by subttle non-gravitational effects. In particular, we show how chaotic diffusion leads to the slow dispersion of the Trojan swarms and how long-lived resonant populations of individualy unstable asteroids can be formed and sustained in steady state. Furthermore, we show how simple models of chaotic diffusion can be used to estimate the age of asteroid families. In the second part of this chapter, we review different analytic approaches to the problem of chaotic diffusion in MMRs. The domain of validity of each model is discussed, and a comparison between analytical and numerical results is made. We show that, in the absense of 'fast' transport routes, related to secular phenomena, chaotic diffusion in the asteroid belt has a characteristic time-scale of $\sim 1$ Gy, i.e. the time needed for an asteroid to diffuse away from the main belt is of the same order as the age of the solar system.

## 5.1 Introduction

The solution of the Kepler problem (Sun–planet, or Sun–asteroid) suggests that bounded motion around the Sun is described by elliptical orbits, the Sun being located at one of the focii. The size, shape and orientation in space of the orbits can be described by a set of *elliptic orbital elements*: the semi-major axis $a$, eccentricity $e$, inclination of the orbital plane $i$ with respect to some reference plane, longitude of the ascending node $\Omega$ and longitude of the pericenter $\varpi$. The position of the object on the ellipse at any given time is given by the value of its mean longitude $\lambda$.[1] Out of the six elements defined above only $\lambda$ varies with time, while the other five are integrals of motion. However, the Kepler problem does not describe reality with the desired accuracy. The gravitational interaction of the planets forces a slow *precession* of their orbits; $(a, e, i, \Omega, \varpi, \_)$ as well as the *mean motion* $n$ of the body ($n = \dot{\lambda}$) change slowly with time (for a detailed description of elliptic motion and perturbation theory the reader is referred to the book of Murray and Dermott [37]. Similarly, asteroid orbits are forced to precess, under the gravitational influence of the planets.

Asteroids can develop chaotic motion, as a result of *resonant* perturbations, exerted by the major planets. *Secular resonances* (SRs) occur when one of the precession frequencies of the asteroid (related to the motion of $\varpi$ or $\Omega$) becomes equal to a linear combination of the *proper modes* ($g_i, s_i$) of the planetary system ($i = 1, \ldots, N$, $N$ being the number of planets) with rational coefficients. The amplitude of the forced terms decreases as the module of the coefficients of the linear combination increases—the *order* of the resonance. Thus, the strongest resonances are the first-order or *linear secular resonances*, for which $\dot{\varpi} = g_k$ or $\dot{\Omega} = s_k$. On the other hand, *mean motion resonances* (MMRs) occur when the mean motion of the asteroid is equal to a linear combination of the mean motions of the perturbing planets, with rational coefficients. The main MMRs for asteroids are low-order MMRs with Jupiter; the strongest perturber. For a review on resonances in asteroid motion the reader is referred to the book of Morbidelli [35].

Secular and mean-motion resonances of low order are known to lead to *fast chaotic transport* of asteroid orbits on million year time-scales [12, 61, 62]. Orbital evolution is characterized by large-amplitude eccentricity variations, which eventualy place the asteroid on a planet-crossing orbit. It is thus not surprising that low-order resonances are associated with the borders as well as the famous *Kirkwood gaps* of the asteroid distribution (see Fig. 5.1; the reader is referred to the review by Moons [33]). Apart from these main resonances, which host almost no asteroids (see Sect. 5.3 for the small resonant populations in the 2:1 and 7:3 MMR), there is an infinity of MMRs crossing the densely populated asteroid belt. As was already shown by Nesvorný and Morbidelli [43] and Murray et al. [39], a large number of asteroids are moving inside (or, very close to) medium/high-order MMRs with Jupiter as well as inside *three-body MMRs* of the form $p\,n + q\,n_J + r\,n_{S/M} \approx 0$, where $n_J$ is the mean motion of Jupiter, $n_{S/M}$ is the mean motion of Saturn/Mars and $(p, q, r)$ are integers. Note that each of these resonances is in fact a *multiplet* of nearly resonant harmonics, which are closely spaced in mean motion (and, therefore, in semi-major axis). This is a direct consequence of the fact that the integrable approximation (Kepler problem) is *degenerate*. Hence, even if different

---

[1] Other sets of elliptical elements can be defined. Throughout this chapter we use the above defined set, because the corresponding angular variables ($\lambda, \varpi, \Omega$) are a convenient set of canonical coordinates.

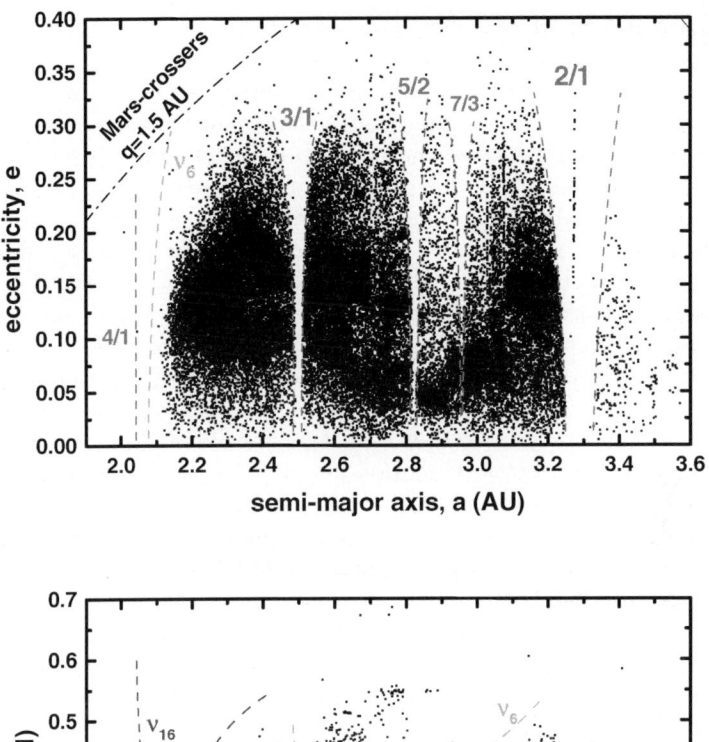

**Fig. 5.1.** Distribution of main-belt asteroids on the $(a, e)$—**top plane**—and $(a, \sin i)$—**bottom plane**. The *dashed lines* denote the approximate location and width of low-order SRs and MMRs, associated with the borders of the main belt and the famous Kirkwood gaps. Data taken from the AstDys web-site

MMRs are far apart, resonance overlap (i.e. chaos) can develop *inside* a MMR multiplet.

Secular resonances may also appear inside the libration zones of low-order MMRs [32]. In the context of the three-body problem, they appear in the form of resonant periodic orbits where $\tilde{\omega} = 0$ (see [14]; called *corotation resonances* by Henrard and Caranicolas [16]).

The co-existence of secular and mean motion resonances leads to a particular partitioning of the phase-space, such that low-$e$ regions are connected to high-$e$ regions through homoclinic orbits (in the integrable approximation). As noted by Wisdom [62], in his perturbative treatment of motion in the 3:1 MMR, this is a highly efficient chaos-generating mechanism, which results to intermittent jumps in eccentricity. In fact, as numerical studies show, this mechanism of fast chaotic transport is the dominant one in low-order MMRs. On the contrary, as shown in Tsiganis et al. [52, 53, 54], SRs do not appear in the small-$e$ regions of medium/high-order MMRs. As a consequence, asteroids in high-order MMRs do not suffer strong variations in $e$ (or $i$) on time-scales of order $10^8$ year.

It can be shown (see also Sect. 5.5) that medium/high-order MMRs are indeed associated with chaotic motion [38, 45]. More importantly, numerical simulations show that this network of MMRs produces a slow macroscopic change of the *proper elements* of asteroids ( [34]; proper elements are 'averaged' elements viewed as quasi-integrals of motion. For definitions and computation methods see Milani and Knežević [27] and Knežević et al. [21]). This effect is known as *chaotic diffusion*. Understanding this process is of great importance for celestial mechanics. Note that, according to the AstDyS catalogue (http://hamilton.dm.unipi.it/cgi-bin/astdys/astibo) more than 30% of the numbered asteroids have chaotic orbits, characterized by a Lyapunov time[2] of only $T_L \leq 10^5$ y (compare with the age of the belt, $\sim 4 \times 10^9$ y). These asteroids are therefore subject to chaotic diffusion (but with different characteristic time-scales, see Sect. 5.5).

From the point of view of pure celestial mechanics, this is a particularly interesting problem, related to the study of chaos in degenerate Hamiltonian systems. From the point of view of solar system dynamics, chaotic diffusion is an important element of the long-term dynamics of the asteroid belt and as such it can be used to 'decode' the present-day asteroid distribution and trace the "initial conditions" of the proto-planetary system. Chaotic diffusion is also considered as one of the main reasons of slow deformation of asteroid families and other asteroid groups (see next sections) and it may be playing an important role in the replenishment of the unstable populations of Mars-crossers and Near Earth Asteroids (NEAs). Thus, unveiling the *microscopic mechanisms* of chaotic diffusion and estimating the corresponding *macroscopic quantities* (i.e. escape rates) is not only a challenging problem for contemporary celestial mechanics, but also a problem of profound importance for solar system science.

---

[2] The Lyapunov time is the inverse of the maximal Lyapunov characteristic exponent, $\gamma$, the mean rate of exponential divergence of initially nearby orbits. An orbit with $\gamma = 0$ is regular (quasi-periodic), while an orbit with $\gamma > 0$ is chaotic.

Besides gravitational perturbations, small asteroids (with diamaters smaller than ∼20 km) are also subject to a subtle non-conservative force, known as the *Yarkovsky effect* [2, 9]. This effect refers to a recoil force that acts on spinning asteroids, who absorb and re-emitt solar radiation in different directions. The tangential component of the force results in *radial migration*, i.e. secular variations of the semi-major axis. The combination of chaotic diffusion and radial migration can lead to a highly complex motion for small asteroids, since they may be forced to cross several high-order MMRs, during their journey inside the asteroid belt [59]. As we will see in the following, the combined effect of chaotic diffusion and radial migration can explain the existence of certain resonant asteroid groups, which are composed of asteroids on highly unstable orbits.

This chapter is divided into two parts. In the first part (Sect. 5.2–5.4), we describe three different asteroid problems, where the footprints of chaotic diffusion are clearly seen. In particular, we first describe the motion of Jupiter Trojans. This large group of asteroids has been found to suffer a slow depletion, which we show to be the direct result of chaotic diffusion. In Sect. 5.3, we study the origin of the group of 7:3-resonant asteroids, all of which follow highly unstable trajectories. We show that this group exists because of a dynamical equilibrium between two competing processes: depletion of the resonance by chaotic transport, on one hand, and replenishment of the resonance by radial migration of neighboring asteroids, on the other. In Sect. 5.4, we show how a kinetic description of chaotic diffusion can be used, in order to compute the age of young, dynamically active, asteroid families, such as the *Veritas* family. The second part of the chapter focuses on analytical models (Sect. 5.5) and numerical studies (Sect. 5.6) of chaotic diffusion in two-body MMRs of different order. We show that analytical estimates of the basic macroscopic quantities can be derived using standard perturbation theory, in the framework of the three-body problem. Moreover, we show that these quantities, although derived on the basis of a simplified dynamical system, are in good agreement with the results of numerical experiments. The basic time-scale of chaotic diffusion is found to be $\tau \sim \mu^3 \sim 10^9$ y ($\mu \approx 10^{-3}$ is the mass of Jupiter in units of the solar mass), i.e. it is comparable to the age of the asteroid belt. Finally, we show that the microscopic mechanism that drives chaotic diffusion is the interaction between the two degenerate degrees of freedom (i.e. $e$ and $i$); in 2D models there is practically no eccentricity diffusion in medium/high-order MMRs. The conclusions of this chapter are discussed in Sect. 5.7.

## 5.2 Slow Dispersion of the Trojan Swarms

Jupiter Trojans are asteroids in 1:1 resonance with Jupiter (i.e. they have approximately the same value of semi-major axis as Jupiter), who lead or trail the planet by ∼ 60° in longitude. Their eccentricities are generally smaller than those of main-belt asteroids. On the other hand, they have a nearly 'flat' inclination distribution, with values going up to ∼ 40°. Up to recently it was assumed that the observed Trojans were formed in situ at the same time as Jupiter. However, it is not easy to explain their current orbital distribution, as representing the evolution of an initially dynamically 'cold' population [25]. An alternative scenario about the origin of the Trojans has been recently suggested by Morbidelli et al. [36]. According to this work, the observed Trojans were initially on circumsolar orbits in the inner part of the proto-Kuiper belt, and were trapped in 1:1 resonance with Jupiter, during the last stages of *planetary migration*. The advantage of this scenario is that it explains simultaneously (i) the total mass, (ii) the orbital distribution, and (iii) the observed spectra of the Trojans. Moreover, it fits well into a more general evolution model for the outer solar system, which explains the orbital configuration of the outer planets [57] and the origin of the Late Heavy Bombardment of the Moon ∼ 3.9 Gy ago [13], as the result of planetary migration.

Stable Trojan orbits exist in the framework of the circular restricted three-body problem. However, perturbations related to Jupiter's eccentricity or the presence of additional planets lead to chaotic motion. The long-term effects of these perturbations are usually studied by numerical integration of the equations of motion. Milani [29, 30] integrated the orbits of 174 Trojans for 1–5 My, showing for the first time that some Trojans follow chaotic orbits. Later, Levison et al. [22] showed for the first time that the Trojan swarms are in fact slowly dispersing in time, as Trojans residing close to the borders of the swarms have dynamical lifetimes smaller than $10^9$ y. However, the presence of Trojans in these dynamically unstable regions of the coorbital zone was not explained.

Tsiganis et al. [56] studied the long-term stability of the numbered Trojans as well as a large population of fictitious Trojans, covering the relevant range of orbital parameters. They were able to determine (numerically) an *effective stability* region, i.e. a region in proper elements space where Trojans are stable for times longer than 1 Gy. As shown in Fig. 5.2, a significant percentage of Trojans follow chaotic orbits (black dots) and many of them are located outside the 1 Gy stability region. Integrating the chaotic Trojans for a time corresponding to 4.5 Gy, Tsiganis et al. [56] found that ∼15% of the Trojans are unstable over the age of the solar system, the median escape time being 1 Gy. Another ∼20% of the Trojans were found to follow chaotic orbits but not escaping within 4.5 Gy.

Dynamicaly unstable Trojans are being transported from the stable to the chaotic region of the coorbital zone. There are basically three transport mechanisms that one could think of: (i) chaotic diffusion due to high-order resonances, (ii) collisions, and (iii) Yarkovsky-induced migration. Note that the latter two mechanisms are size dependent, i.e. small bodies would be transported much more efficiently than large ones. Thus, the size distribution of unstable Trojans should differ significantly from that of stable Trojans, if transport was indeed related to a size-dependent mechanism. Tsiganis et al. [56] combined observations with their numerical integrations and derived the

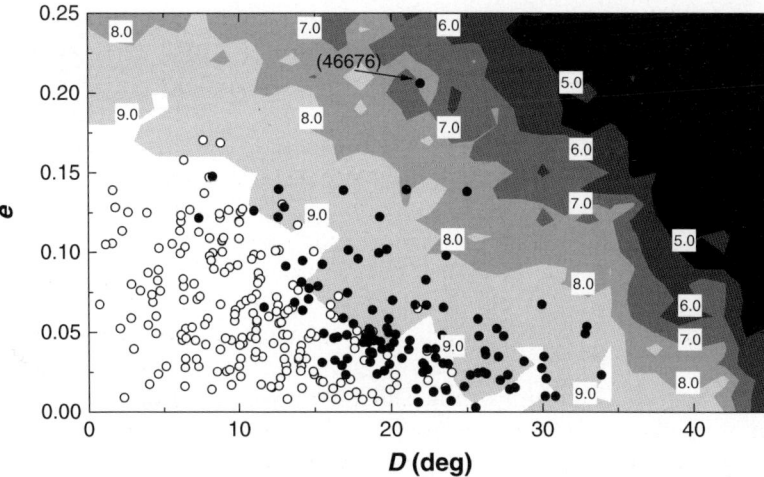

**Fig. 5.2.** Map of escape times of fictitious Jupiter Trojans with $i = 10°$. The coordinates are the proper value of the half-width of libration, $D$ around the equilibrium $|\lambda - \lambda_J| = 60°$ and the proper eccentricity, $e$, of the orbit. The logarithm of the escape time (in years) is color-coded in gray scale, according to the labels shown in the plot. The *white region* is the effective stability region of $T_E > 1$ Gy. The numbered Trojans with $5° \leq i \leq 15°$ are superimposed. The *open circles* are Trojans following regular orbits, while *filled circles* are Trojans following chaotic orbits. Nearly half of the latter lie outside the stability curve. Adapted from Tsiganis et al. [56]

two size distributions (Fig. 5.3). Clearly, the distributions are nearly identical, a result which implies that transport from the stable to the chaotic region does not depend on the size of the body. Thus, chaotic diffusion is responsible for large-scale Trojan transport.

The source of chaotic diffusion in the Trojan region has been revealed and studied by Robutel et al. ( [49]; see also Robutel & Gabern [50]), who demonstrated the existence of a complex network of high-order resonances of different types inside the libration zone of the 1:1 resonance. As shown by Robutel et al. [49], these resonances can indeed give rise to slow diffusion in the space of proper frequencies, which has appreciable effects on giga years time scales. A detailed study of diffusion inside this resonance complex and analytic or numerical estimates of the corresponding diffusion rates are yet to be completed.

## 5.3 Resonant Populations of Unstable Asteroids

Resonant asteroid groups are of particular interest because they contain important information about the early evolutionary phases of the solar system (see previous section for the Trojans). However, they also represent valu-

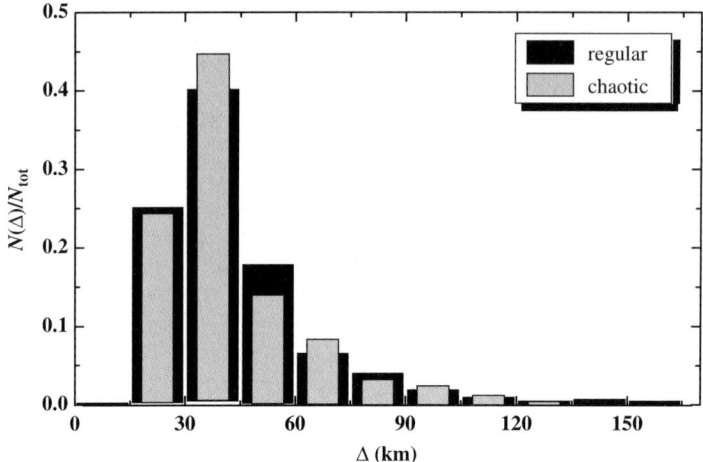

**Fig. 5.3.** The size distribution of regular (*black*) and chaotic (*grey*) numbered Trojans. The effective diameter, $\Delta$, is computed from the value of the absolute magnitude, $H$, assuming a mean albedo of 5%. The two distributions are nearly identical. Adapted from Tsiganis et al. [56]

able case-studies for dynamical models and for understanding the complex interplay between gravitational perturbations and non-gravitational forces. We can distinguish between groups mostly composed of stable asteroids, such as the Trojans in the 1:1 MMR and the Hildas in the 3:2 MMR with Jupiter [11, 42], and groups primarily composed of unstable objects, such as those in the 2:1 and 7:3 MMRs.

The 7:3 group offers a unique example of a long-lived group of individualy unstable asteroids. All asteroids in this group have dynamical lifetimes shorter than 172 My, the median escape time being only 19.3 My. Obviously these objects cannot be dynamically primordial and must have been placed in the 7:3 MMR recently. Tsiganis et al. [53] suggested that the population of resonant objects is sustained in steady state by the continuous flux of bodies from the adjacent asteroid families of *Koronis* and *Eos*. As shown in Fig. 5.4, the resonant population has a roughly bimodal inclination distribution, the two peaks being at 2° and 10° degrees, i.e. the mean inclination of the Koronis and Eos families, respectively. These observations suggest that the low-$i$ group originates from the Koronis family, while the high-$i$ group from the Eos family. As both family-forming events probably took place more than 1 Gy ago [1, 24], the most likely mechanism for transporting family memebrs into the 7:3 MMR is the Yarkovsky effect.

Figure 5.5 shows the distribution of Lyapunov times in the vicinity of the 7:3 MMR, for the two different values of $i$. There is no sizeable region of stable librations and all trajectories are chaotic, at this resolution. There is however a clear increase of $T_L$ towards the center of the resonance and for $e \sim 0.2$,

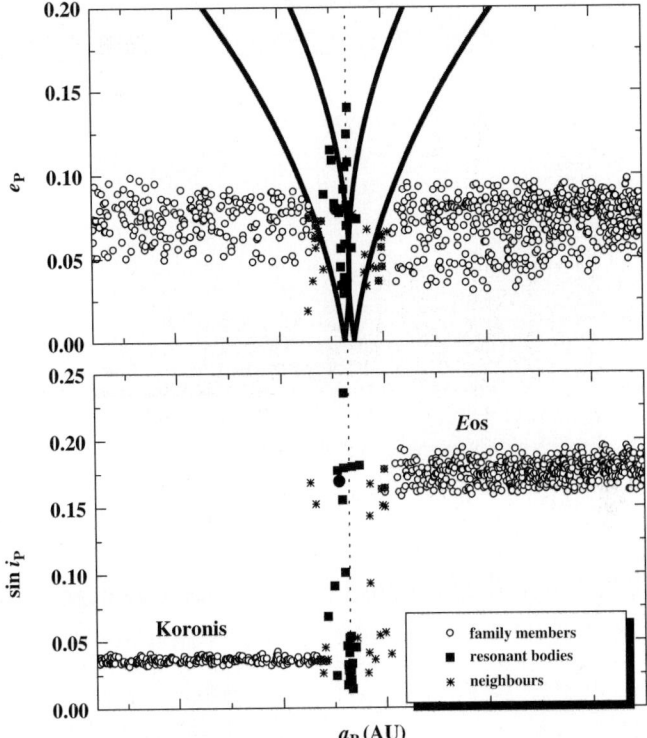

**Fig. 5.4.** The distribution of bodies in the 7:3 MMR with Jupiter (*black dots*), in the $(a, e)$ and $(a, \sin i)$ planes. The members of the *Koronis* and *Eos* families are also shown. The '*V*'-*shaped curves* represent the maximal and minimal width of the 7:3 MMR, in the modulated pendulum approximation (see Sect. 5.5). The *lower panel* suggests that the resonant bodies are a continuation of the two families inside the 7:3 MMR. Adapted from Tsiganis et al. [55]

where secular resonances are excited [32]. Thus, we expect the dynamics of resonant bodies to be different for small ($e \sim 0.05$) and moderate ($e \sim 0.2$) eccentricities. Moreover, since no region of regular motion seems to separate the two regions, we can expect that bodies initially at $e \sim 0.05$ (like most of the observed ones) can slowly increase their eccentricity by chaotic diffusion, eventually reaching the 'fast transport' region.

The above considerations are confirmed by numerical integrations. The evolution of a typical 'Koronis-like' (i.e. with $i(0) = 2°$) 7:3-resonant particle is shown in Fig. 5.6. Both $e$ and $i$ change slowly due to chaotic diffusion, up to $t \sim 80$ My, whence the particle reaches $e \sim 0.2$. At that point the motion changes character abruptly, the body being trapped in a SR, which forces large amplitude oscillations in $e$. The eccentricity grows beyond 0.6 and the particle suffers close encounters with Jupiter, which eventually remove it from the belt.

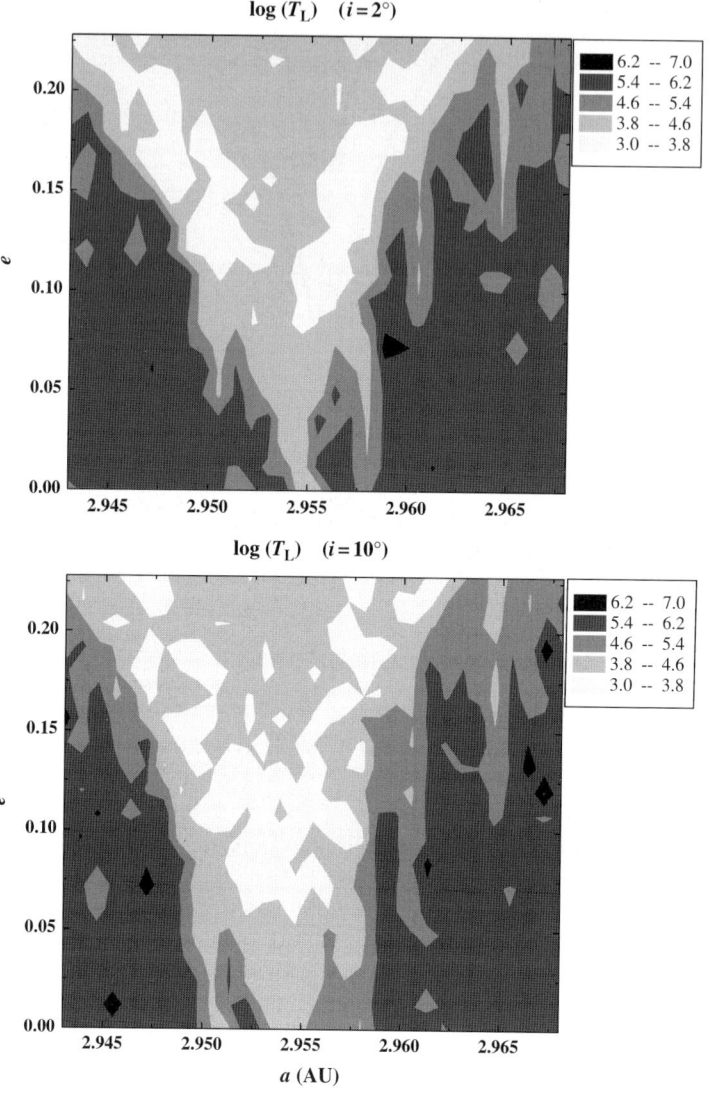

**Fig. 5.5.** Map of the logarithm of the Lyapunov time in the 7:3 MMR, for $i = 2°$ (**top**) and $i = 10°$ (**bottom**). The corresponding scale is shown on the right of each panel. No sizeable region of stable librations is visible, at this resolution. The strongly chaotic region at $e \sim 0.1$ is larger for $i = 10°$. Note that, in both plots, $T_L$ increases as we move towards the center of the resonance and for $e > 0.15$. Adapted from Tsiganis et al. [55]

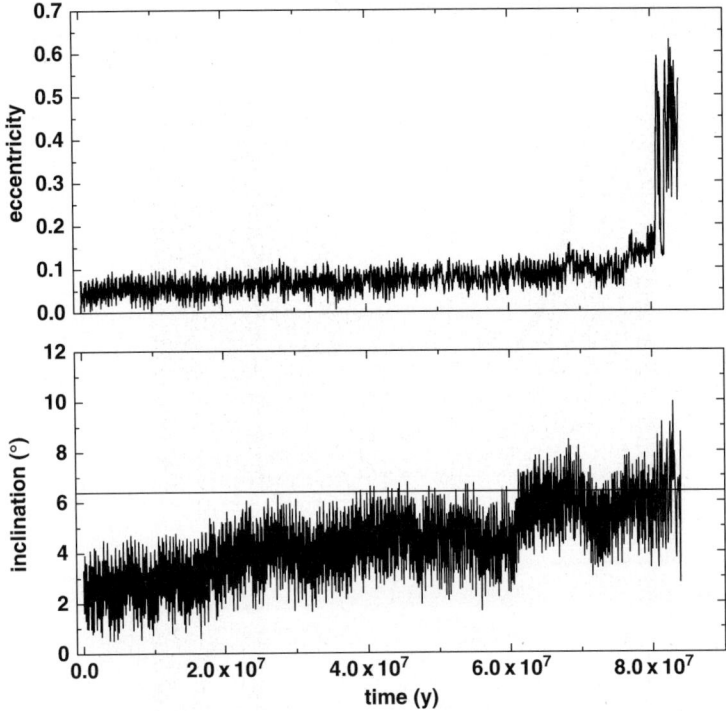

**Fig. 5.6.** Evolution of a typical Koronis-like ($i(0) = 2°$) resonant particle. Initially, both $e$ and $i$ change slowly due to chaotic diffusion. When the orbit reaches $e \sim 0.2$, the body enters into a secular resonance, which forces large-amplitude oscillations in $e$. The object soon approaches Jupiter and is extracted from the resonance. Adapted from Tsiganis et al. [55]

Note that, before it escapes, the particle spends ~20 My at an intermediate inclination of ~6.5°. This explains the existence of a few 7:3 objects with $I \approx 6°$, which should be considered as 'evolved' Koronis-like particles.

So far we have seen that 7:3 asteroids exhibit two types of chaotic transport during their dynamical lifetime: slow chaotic diffusion from small to moderate eccentricities ($e \sim 0.2$), followed by 'fast', intermittent, eccentricity transport to $e \sim 0.6$. As we will see in Sect. 5.5, 'fast' transport can also be described as diffusion, but of a suitably defined action of the problem (the adiabatic invariant) and not $e$. To complete the puzzle of the dynamical processes involved in the dynamics of this group, we return to the problem of its origin. Tsiganis et al. [55] suggested that the resonant population is kept in steady state by the leakage of bodies from the neighboring families. For this statement to hold, the flux of bodies into the resonance should match the escaping flux. Tsiganis et al. [55] showed that this is indeed the case. They first computed the escape rate for both groups, by integrating a large number of fictitious objects

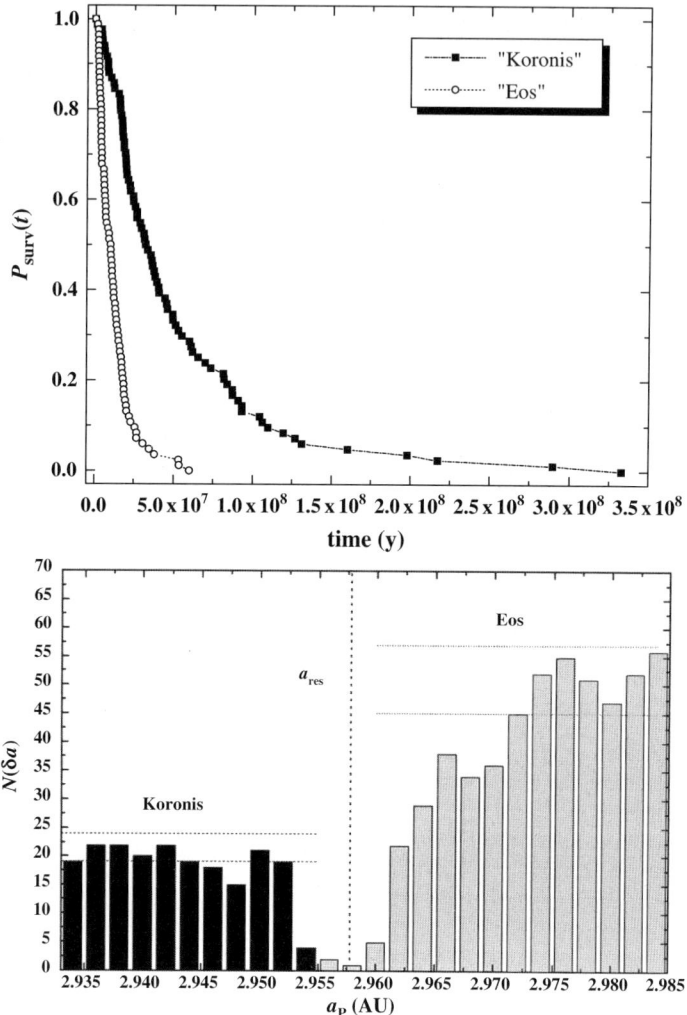

**Fig. 5.7. Top**: Survival probability of fictitious Koronis-like and Eos-like 7:3 resonant objects. Both populations decay exponentially, but with different rates. The median lifetime is 52.8 My for the $i = 2°$ and 12.4 My for the $i = 10°$ group. **Bottom**: Number density of the observed Koronis and Eos family members in semi-major axis, binned to $\delta a = 2 \cdot 10^{-3}$ AU. The *horizontal lines* mark our estimates for the number densities required to sustain the two 7:3-resonant groups in steady state, assuming that the resonant bodies are former family members that drifted towards the 7:3 MMR due to the Yarkovsky effect. The agreement between our estimates and observations is remarkable. Adapted from Tsiganis et al. [55]

at $i(0) = 2°$ and $i(0) = 10°$ (Fig. 5.7—top panel). Then, assuming the mean drift rate due to the Yarkovsky effect to be $\dot{a}_Y = \pm 2.7 \times 10^{-4}$ AU/My [9] they computed the number density in semi-major axis of Koronis and Eos family members, that is required to match the two fluxes. Their estimates match the observed number densities of the two families remarkably well (see Fig. 5.7—bottom panel).

Brož et al. [2] studied the origin of the population of asteroids in the 2:1 MMR (the known *Hecuba gap*). They assumed again that asteroids from the neighborhood of the 2:1 MMR are constantly driven into the resonance by the Yarkovsky effect. Moreover, by using a more refined methodology, they were able to study the time evolution of the size distribution of the trapped particles. They showed that only part of the 2:1 population (the unstable bodies) can be explained by the suggested mechanism; neither the *Zhongguos* nor the *Griquas* fall in this category. As Brož et al. [2] suggested, the existence of these quasi-stable groups could be related either to recent collisional break-ups or to trapping during the late stages of planetary migration (like the Trojans in Morbidelli et al. [36]). Evidently, this subject deserves further study.

## 5.4 Chaotic Chronology of Asteroid Families

Asteroid families [18] are clusters of asteroids in the space of proper elements. They are considered to be the fragments of larger asteroids, disrupted during catastrophic collisions in the past. Their study can provide important information about the composition and mechanical properties of main-belt asteroids [48]. Moreover, family-forming events are nowadays considered to be among the main sources of interplanetary dust particles (IDPs—see Wyatt [63] and references therein). Young asteroid families in particular, corresponding to recent break-ups, are associated with the prominent IRAS dust bands [47]. Thus, the chronology of formation of asteroid families can be used to search for possible correlations with the temporal variations of the flux of IDPs on Earth [6, 10].

The Yarkovsky-induced radial migration of family members can be used to date asteroid families. As time progresses, families form a characteristic 'V'-shaped distribution in the $(H, a)$ plane, since small asteroids drift faster than large ones ($H$ is a function of the body's diameter, $D$). The age of a family can then be approximately found, by fitting the 'V'-shaped borderline with the theoretically expected $(H, \Delta a)$ curve, for given values of $\dot{a}_Y$ [59]. For a family much younger than $\sim$100 My, the Yarkovsky effect would not have had enough time to produce a robust 'V'-shaped signature. However, if the family is located in a quasi-regular phase-space region it can still be dated by dynamical methods, such as the one recently proposed by Nesvorný et al. [46]. This method consists in integrating backwards in time the equations of motion for the family members, until the orientation angles of their perturbed elliptical orbits converge to the values they had at the time of break-up. Using

this method Nesvorný et al. [46, 47] calculated the ages of the *Karin* family
($\sim$5.8 My) and the *Veritas* family ($\sim$8.3 My).

However, the Veritas family is not located in a dynamically quiet region of
the asteroid belt. The distribution of the Veritas family members in the $(a, e)$
plane is shown in Fig. 5.8. Nearly half of the bodies follow chaotic orbits—
those in the regions 'A' and 'B'—and present a large spread in $e$, compared
to the equivelocity curve that surrounds the regular part of the family.[3] As
was shown by Milani & Farinella [31], (490) Veritas, the largest asteroid of
the family, exhibits strongly chaotic behavior (region 'A'). The resonance responsible
for this chaos is the $(5, -2, -2)$ three-body MMR, as later found by
Nesvorný & Morbidelli [43]. Milani and Farinella [31], introducing the concept
of *chaotic chronology*, suggested that the age of the family must be shorter
than 50 My, since this is the time interval needed for (490) Veritas to 'escape'
from his own family. We note that the results of Farley et al. [10] strongly
suggest a single break-up of a $\sim$150 km asteroid $\sim$8.3 My ago (i.e. the age
found by Nesvorný et al. [47]). It is evident that, in order to obtain more

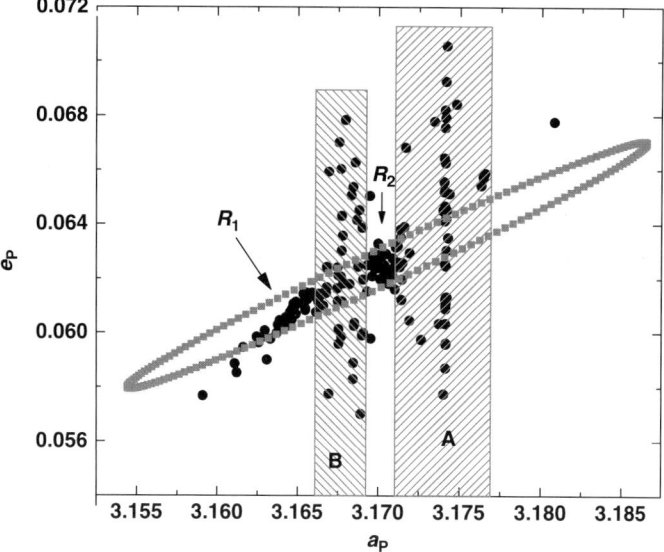

**Fig. 5.8.** Distribution of the Veritas family members in $(a, e)$. The *shaded regions*
mark the two chaotic groups, 'A' and 'B'. Two regular parts of the family, '$R_1$' and
'$R_2$' are also found. The equivelocity curve that fits the regular parts of the family is
indicative of the initial ejection velocity field. The extent in eccentricity of the two
chaotic groups is clearly much larger than that suggested by the equivelocity ellipse

---

[3] The equivelocity curve marks the maximum spread of the fragments from a given
location (break-up location) and for a given maximum relative velocity, assuming
an isotropic field of ejection velocities.

accurate results for the age of the chaotic component of the family, a more refined chaotic chronology method is needed.

Tsiganis et al. [58] studied the long-term behavior of the Veritas family members. They showed that the two chaotic groups have a very different long-term behavior, while being characterized by comparable values of the Lyapunov time; $10^4$ y for group 'A', as opposed to $3 \times 10^4$ y for group 'B'. In particular, group 'B' asteroids show a typical *stable-chaos* behavior [28], as no macroscopic diffusion in proper elements is observed for a 100 My time interval. On the contrary, significant diffusion is observed for the asteroids of group 'A'. As shown in Fig. 5.9, the mean squared displacement in both actions ($I_1 \sim e_P{}^2$ and $I_2 \sim i_P{}^2$) is a linear function of time.

This result suggests that, on average, their motion resembles normal diffusion (see also Sect. 5.5). The slope of each regression line gives the value of the respective *diffusion coefficient*, $\mathcal{D}_i$ ($i = 1, 2$).

Since motion in group 'A' resembles normal diffusion, we can use a simple random-walk model to simulate the evolution of a fictitious population of group-'A' family members. More specificaly, we assume an initial population, whose extent in $e$ is compatible with the equivelocity ellipse, shown in Fig. 5.8.[4] Then, at every time step $\delta t$, each of our random walkers takes a step in $I_1$ and $I_2$, which is randomly chosen from a 2D Gaussian distribution, the projections of the standard deviation in each action being equal to $\sigma_i = \sqrt{\mathcal{D}_i \, \delta t}$. Using this model we evolve our ensemble of random

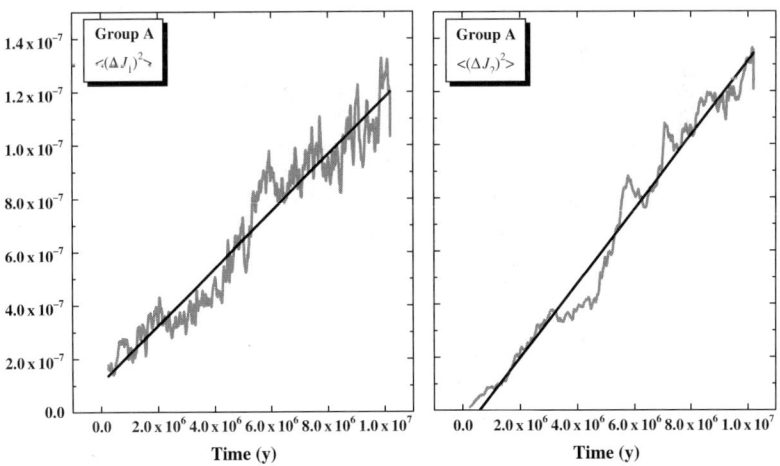

**Fig. 5.9.** Mean squared displacement in $I_1$ (**left**) and $I_2$ (**right**) as a function of time, for the members of group 'A'. The evolution is characterized by small fluctuations superimposed on a linear trend. The *straight lines* on both panel are the results of a linear least-squares fit. Their slopes are equal to the values of the corresponding diffusion coefficients

---

[4] A similar calculation was done for the distribution in $i$.

walkers and estimate the time needed to cover the region occupied by the real asteroids. The result of this experiment is shown in Fig. 5.10. Our random-walkers take ~10 My to cover the same region of orbital elements space as the observed asteroids. Thus, the age of group 'A' must be ~10 My. The above calculation can be refined, by performing a large number of trials (runs) of this 2D model, in which the diffusion coefficients and the width of the diffusion region are varied within the standard errors of their values. Thus, a distribution of ages is derived, whose mean and standard deviation are $8.7 \pm 1.7$ My. Hence, the age of the chaotic component of the Veritas family is statistically the same as the one found by Nesvorný et al. [47], for the regular part of the family.

This modified method of chaotic chronology is based on a statistical approach to chaotic diffusion, rather than the study of individual bodies. This approach drastically reduces the error in estimating the age of a diffusive group of asteroids and can be used to date asteroid groups, located in dynamicaly unstable regions of the belt.

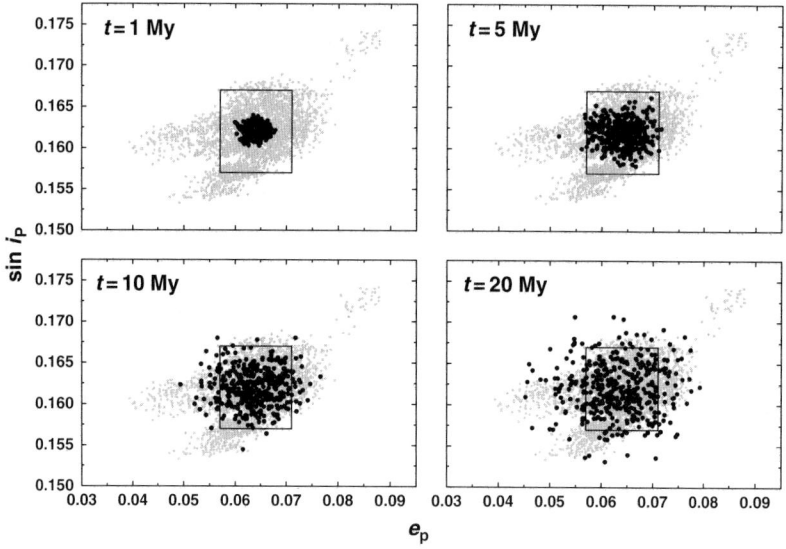

**Fig. 5.10.** Evolution of an ensemble of fictitious asteroids (*black dots*), according to the 2D random-walk model, described in the text. The four panels represent four snapshots during the evolution. The 'box' shows the area covered by the real group-'A' asteroids. The light gray dots denote the extent of the diffusion region, covered by the real asteroids when their orbits are propagated from present to 10 My into the future. As shown in the figure, the random-walkers cover the 'box' within ~10 My. Moreover, they take twice as much time ($t = 20$ My) to cover the same region as the future images of the real family members. This result suggests that the age of group 'A' is close to 10 My

## 5.5 Analytical Models of Chaotic Diffusion in MMRs

Chaotic motion is by definition non-predictable, on time-scales much longer than the Lyapunov time. Thus, analytical theories of chaotic diffusion aim at predicting the long-term behavior of *ensembles* of asteroids, rather than individual orbits, under the assumption that chaotic motion resembles a *random-walk* with certain kinetic properties. This implies that the time evolution of an ensemble of orbits obeys a certain *transport equation*, with coefficients whose functional form depends on the properties of the dynamical system under study. Thus, analytical theories seek to derive reliable estimates of the relevant macroscopic parameters: (i) the decorelation time of orbits, (ii) the size and shape of the chaotic region and (iii) the transport coefficients.

### 5.5.1 Transport in Action Space

For Hamiltonian systems, the study of chaotic diffusion is simplified by considering the transport of ensembles of chaotic trajectories in the *space of actions*. The idea behind this approach is that the perturbation induces small-amplitude variations in the actions of the integrable part of the system, while the angles (or phases) are quickly randomized over the $[0, 2\pi]$ interval. Thus, for a chaotic orbit, the evolution of the actions is on average much slower than that of the angles. Then, ignoring (i.e. 'averaging') the angles, the time evolution of the *distribution function* $f(I,t)$ of the actions of an ensemble of orbits is governed by the Fokker–Planck equation [23]. For Hamiltonian systems with $n > 1$ degrees of freedom, this equation reduces to the known *diffusion equation*

$$\partial_t f = \nabla_I \left( \mathbf{D} \, \nabla_I f \right) \tag{5.1}$$

where **D** is the diffusion tensor (or diffusion coefficient, in 1D), whose elements are given by

$$D_{i,j} = \frac{\langle \Delta I_i \, \Delta I_j \rangle}{2 \, T_c} \tag{5.2}$$

and '$\nabla$' denotes differentiation with respect to the actions. The diagonal elements of D are equal to the mean squared displacement in each action, divided by $2T_c$. The time interval $T_c$ is the *phase-decorrelation* time, i.e. the time after which linear correlations in the angles decay. In principle, $T_c$ has to be short, compared to the evolution time scale of the actions. For a chaotic Hamiltonian system, $T_c$ is usually taken to be equal to the Lyapunov time, $T_L$, although this approximation is not always valid [3]. In the simple case of 1D diffusion (e.g. for a 2D chaotic map) with a constant coefficient, the diffusion equation reads

$$\partial_t f = \mathcal{D}\, \partial_I^2 f \tag{5.3}$$

where

$$\mathcal{D} = \frac{\langle(\Delta I)^2\rangle}{2\, T_\mathrm{L}} \tag{5.4}$$

For $f(I, t=0) = \delta(I_0)$, the solution of the 1D diffusion equation (assuming the boundaries at infinity) is a Gaussian distribution in $I$, whose dispersion grows linearly with time, according to Fick's law

$$\sigma^2(I) = \mathcal{D}\, t \tag{5.5}$$

For $n$-dimensional systems ($n > 1$), a first approximation is to assume that the actions evolve independently of each other. This reduces the problem to solving $n$ independent 1D diffusion equations. However, in 'real' dynamical systems, the problem is usually more complicated than that. First, the actions do not evolve independently of each other. Second, for small perturbations, the diffusion coefficients are not constant in phase space, but functions of the actions [3, 23]. Finally, we frequently have to deal with problems where stable regions intermingle with chaotic regions, so that the angles retain correlations for long times; for example, consider the *stickiness effect* in 2D maps [7, 8]. Such phenomena question the validity of the random-phase approximation and concequently of the simple transport model presented here. Chaotic transport in a non-uniform phase space can resemble *subdiffusive* or *superdiffusive* random walks [20, 26, 51] characterized by solutions of the form

$$\sigma^2(I) \sim t^\alpha \tag{5.6}$$

with $\alpha \neq 1$. Zaslavsky [64] has shown that these solutions can be obtained by a generalized fractional kinetic equation, which describes transport in a fractal space-time domain.

In celestial mechanics the term *diffusion* is used to describe any 'slow' variation of the proper elements of asteroids. However, as discussed in the previous paragraph, there are certain conditions that have to be met, in order for chaotic motion to resemble diffusion. For example, strictly speaking, *stable chaos* cannot be described as 'very slow' diffusion. The reason is that, as shown in Tsiganis et al. [52, 53], linear correlations in the eccentricity and inclination time series of stable-chaotic asteroids do not decay for times longer than $1000 \times T_\mathrm{L}$. Conversely, one should not rush to describe this motion as sub-diffusion, before studying the evolution of the mean squared displacement of the appropriate actions of the problem.

It is not well understood so far which approximation should be considered for a given problem of celestial mechanics. However, experiments show (see e.g. Sect. 5.4) that the simple diffusion approximation can give reasonably good results, if we restrict ourselves to a 'local' description, i.e. study a given phase-space region with specific characteristics, instead of trying to develop a 'global' theoretical framework.

## 5.5.2 Analytical Models of MMR

In this paragraph, we will develop an analytical model of motion in MMRs, using an appropriate set of action-angle variables. We follow closely the line of work of Holman & Murray [19] and Murray & Holman [38], using slightly different notation and variables. We consider the restricted three-body problem, with Jupiter following a fixed Keplerian ellipse and the asteroid being subject to the gravitational forces of the Sun and Jupiter. The Hamiltonian (per unit mass) describing the motion of the asteroid is

$$\mathcal{H} = \frac{v^2}{2} - \frac{\mathcal{G}M}{r} - \mathcal{G}m'\left(\frac{1}{|\mathbf{r}-\mathbf{r}'|} - \frac{\mathbf{r}\cdot\mathbf{r}'}{r'^3}\right) \tag{5.7}$$

where $(\mathbf{r}, \mathbf{r}')$ denote the position vectors of the asteroid and Jupiter respectively (in the heliocentric reference frame), $v$ denotes the modulus of the asteroid's heliocentric velocity vector, $M$ the mass of the Sun, $m'$ the mass of Jupiter and $\mathcal{G}$ the gravitational constant. We define our system of units such that $\mathcal{G} = a' = M + m' = 1$, which gives $n' = 1$ for Jupiter's mean motion. We also define the ratio $\mu = m'/(M+m')$ and set $\mu_1 = 1 - \mu$. We use the standard expansion of the disturbing function (see Murray and Dermott [37], Chap. 6) in orbital elements, to write the Hamiltonian as

$$\mathcal{H} = -\frac{1}{2a} - \mu\mathcal{R}(a, e, i, \lambda, \varpi, \Omega; \lambda') \tag{5.8}$$

where $(a, e, i, \lambda, \varpi, \Omega)$ are the orbital elements of the asteroid and $\lambda' = n't$ is the mean longitude of Jupiter ($t$ is the time). We define the following set of canonical variables

$$\begin{aligned}
\lambda \,,\, \Lambda &= \sqrt{\mu_1 a} \\
\gamma = -\varpi \,,\, \Gamma &= \sqrt{\mu_1 a}\left(1 - \sqrt{1-e^2}\right) \\
\zeta = -\Omega \,,\, Z &= \sqrt{\mu_1 a}\sqrt{1-e^2}\left(1 - \cos i\right)
\end{aligned} \tag{5.9}$$

where, for small values of $e$ and $i$, $\Gamma \approx \Lambda e^2/2$ and $Z \approx \Lambda i^2/2$. Introducing also $\Lambda'$, the momentum conjugate to $\lambda'$, the autonomous Hamiltonian reads

$$\mathcal{H} = -\frac{\mu_1^2}{2\Lambda^2} + n'\Lambda' - \mu\mathcal{R}(\Lambda, \Gamma, Z, \lambda, \gamma, \zeta; \lambda') \tag{5.10}$$

The disturbing function, $\mathcal{R}$, also depends on the (constant) elements of Jupiter's orbit. We can define the orbital plane of Jupiter to be our plane of reference, so that all terms in $\mathcal{R}$ depending on $i'$ disappear. We can also set $\varpi' = 0$, which is equivalent to measuring the longitude of pericenter of the asteroid relative to Jupiter's apsidal line.

We wish to study motion in a specific MMR interior to the orbit of Jupiter, say $k : (k+q)$, where $(k,q)$ are positive integers and $q$ is the *order* of the resonance. The condition for resonance is

$$k\,n - (k+q)\,n' \approx 0 \qquad (5.11)$$

and its location is approximately given by Kepler's third law

$$a_{\text{res}} = a' \left(\frac{k}{k+q}\right)^{2/3} \mu_1^{1/3} \qquad (5.12)$$

Close to this resonance, all terms in the Fourier expansion of $\mathcal{R}$ that contain the combination $k\lambda - (k+q)\lambda'$ will give rise to resonant perturbations, with a characteristic time-scale much longer than the orbital period. Thus, invoking the *averaging principle* we can eliminate all other terms containing the fast angles and restrict our expansion to *resonant* and *secular* terms (not containing mean longitudes).[5] Taking also into account the fact that the coefficients of the Fourier terms are power series in the small quantities $e$ and $s = \sin(i/2) \sim i/2$, we can limit our expansion by retaining secular terms only up to degree 2 in $(e, s)$. The Hamiltonian then becomes

$$\mathcal{H} = -\frac{\mu_1^2}{2\Lambda^2} + n'\Lambda' - \mu \left[c_1(\Lambda)\,\Gamma + c_2(\Lambda)\,e'\,\sqrt{\Gamma}\,\cos(\gamma) + c_3(\Lambda)\,Z\right]$$
$$-\mu \sum_{k,q,p,r} C_{k,q,p,r}(\Lambda,\Gamma,Z;e')\cos\left[k\,\lambda - (k+q)\,\lambda' - p\,\gamma - r\,\zeta\right] \qquad (5.13)$$

The coefficients $c_i$ of the secular part and $C_{k,q,p,r}$ of the resonant part of $\mathcal{H}$ are implicit functions of $a$. The resonant part of $\mathcal{H}$ is a multiple sum of different harmonics, which satisfy the D'Alembert rules[6] and are exactly resonant at almost the same value of $a$. Thus, in the elliptic problem, an MMR is in fact a *resonant multiplet* [35]. The coefficient $C$ of each harmonic is a power series in $\sqrt{\Gamma}$ (and $\sqrt{Z}$), whose lowest-order term has an exponent equal to the coefficient of $\gamma$ (resp. $\zeta$) in the cosine argument, i.e. $|p|$ (resp. $|r|$). We now define the resonant angle $\psi = k\lambda - (k+q)\lambda'$ and complete the canonical transformation via a generating function of second type

$$F_2 = [k\lambda - (k+q)\lambda']\,\Psi + \lambda'\,\Psi' + \gamma\,\Phi + \zeta\,\Theta \qquad (5.14)$$

which defines the new set of canonical variables

---

[5] Strictly speaking, we have to define first a *resonant angle*, $\psi$, and use the Lie-Hori scheme [23] to transform to a new Hamiltonian, whose principal part is the averaged Hamiltonian (over $\lambda$) and all additional terms containing $\lambda$ have coefficients at most of $O(\mu^2)$ in magnitude.

[6] The symmetries of the gravitational potential dictate that (i) the sum of the coefficients of the angles in each cosine argument must be zero and (ii) $r$ must be even.

$$\psi = k\lambda - (k+q)\lambda', \quad \Lambda = k\Psi$$
$$\psi' = \lambda', \quad \Lambda' = \Psi' - (k+q)\Psi$$
$$\phi = \gamma, \quad \Gamma = \Phi$$
$$\theta = \zeta, \quad Z = \Theta \tag{5.15}$$

and transforms $\mathcal{H}$ into

$$\mathcal{H} = -\frac{\mu_1^2}{2k^2\Psi^2} + n'\Psi' - n'(k+q)\Psi - \mu\left[c_1\Phi + c_2 e'\sqrt{\Phi}\cos(\phi) + c_3\Theta\right]$$
$$-\mu\sum_{k,q,p,r} C_{k,q,p,r}(\Psi,\Phi,\Theta;e')\cos(\psi - p\phi - r\theta) \tag{5.16}$$

As $\psi'$ is ignorable, $\Psi'$ is constant and can be dropped from $\mathcal{H}$. Following Murray & Holman [38], we find the fixed point of the secular part of $\mathcal{H}$, which corresponds to $e$ and $i$ being equal to the so-called *forced elements*. Performing a canonical translation around this fixed point we define the *linear free elements* ($\varpi_f, \Omega_f, e_f, i_f$) and the corresponding canonical variables ($\phi, \theta; \Phi, \Theta$) (through 5.9). The full Hamiltonian in the new variables reads,

$$\mathcal{H} = -\frac{\mu_1^2}{2k^2\Psi^2} - n'(k+q)\Psi - \mu d_1\Phi - \mu d_3\Theta$$
$$-\mu\sum_{k,q,p,r} D_{k,q,p,r}(\Psi,\Phi,\Theta;e')\cos(\psi - p\phi - r\theta) \tag{5.17}$$

where ($d_i$; $D_{k,q,p,r}$) are the new coefficients of $\mathcal{H}$. Finally, we expand $\mathcal{H}(\Psi)$ around the exact resonance location $\Psi_{\text{res}} = k\sqrt{\mu_1 a_{\text{res}}}$, keeping terms up to order 2 in the Keplerian part and only the constant term in the $O(\mu)$ perturbation. The final MMR Hamiltonian is

$$\mathcal{H} = \frac{1}{2}\beta J^2 - \mu d_1\Phi - \mu d_3\Theta$$
$$-\mu\sum_{k,q,p,r} D_{k,q,p,r}(\Phi,\Theta;e')\cos(\psi - p\phi - r\theta) \tag{5.18}$$

where $J = \Psi - \Psi_{\text{res}}$ (the new momentum conjugate to $\psi$) and $\beta = -3k^2/a_{\text{res}}^2$. In the following we will focus on the planar problem, i.e. eliminate all terms depending on $(\theta, \Theta)$

$$\mathcal{H}_{2D} = \frac{1}{2}\beta J^2 - \mu d_1\Phi - \mu\sum_{k,q,p,r} D_{k,q,p,r}(\Phi,e')\cos(\psi - p\phi) \tag{5.19}$$

$\mathcal{H}_{2D}$ can also be written in the following form

$$\mathcal{H}_{2D} = \frac{1}{2}\beta J^2 - \mu d_1\Phi - \mu\tilde{D}(\phi,\Phi)\cos(\psi - \tilde{Q}(\phi,\Phi)) \tag{5.20}$$

with $\tilde{D} = \sqrt{A^2 + B^2}$, $\tilde{Q} = \tan^{-1}(B/A)$, and

$$A = \sum_{k,q,p,r} D_{k,q,p,r} \cos(p\phi) \, , \, B = \sum_{k,q,p,r} D_{k,q,p,r} \sin(p\phi) \quad (5.21)$$

Equation (5.20) is the Hamiltonian of a pendulum in $(\psi, J)$, whose amplitude and frequency are modulated by $(\phi, \Phi)$. The ratio of the pendulum libration frequency to the modulation frequency plays a fundamental role in determining the long-term behavior of chaotic MMR trajectories.

According to (5.19), each $p$-harmonic of the $k : (k+q)$ MMR is exactly resonant when $\dot{\psi} - p\dot{\phi} = 0$. Hence, the distance between adjacent harmonics is approximately equal to $\nu = \dot{\phi} = -\mu d_1$ in frequency space and $\delta J = \nu/\beta$ in action space. The half-width of the $p$-harmonic in $J$ and the frequency of small oscillations around the stable fixed point are given by

$$\Delta J = 2\sqrt{|\mu D_p/\beta|} \, , \, \nu_p = \sqrt{|\mu \beta D_p|} \quad (5.22)$$

As shown by Chirikov [4], large-scale stochastic motion is expected when the sizes of adjacent resonant harmonics are larger than their separation, i.e. $K = |\Delta J/\delta J| > 1$. This stochasticity criterion was used by Wisdom [60] to find the extent of the resonance-overlap region for adjacent MMRs of $q = 1$, in the circular restricted problem. His result, known as the '$\mu^{2/7}$ law' explains the paucity of asteroids in a zone of size $\Delta a = 1.5 \, a' \, \mu^{2/7} \approx 1.1$ AU around Jupiter's orbit.

Holman & Murray [19] derived an analytic formula for the coefficients of the resonant harmonics, $D_p$, for different MMRs. According to their results, most outer-belt MMRs have $K \simeq 1$, i.e. the harmonics overlap marginally. We performed numerical calculations of $D_p$, using the definition and properties of Laplace coefficients and different algebraic manipulators. Our results suggest that $K \gg 1$ in all main-belt MMR of order $q \leq 9$. As we will see in the following, the value of $K$ is important for what concerns the selection of the appropriate diffusion model.

The difference between our case here and the one studied by Wisdom [60], is that we study chaos inside a given MMR, caused by the overlapping of adjacent resonant harmonics that belong to the same MMR multiplet. We note that, as discussed in Wisdom [62], other chaos-generating mechanisms can co-exist and even dominate motion in a given MMR. It is well known that in low-order MMRs chaotic orbits evolve mainly by taking intermittent jumps in $e$, as they follow closely the asymptotic invariant curves of unstable resonant periodic orbits (see also Hadjidemetriou [14]). This is an example of secular dynamics dominating motion inside a MMR. As shown in Tsiganis et al. [54], the families of symmetric resonant periodic orbits of period $k : (k+q)$ of the circular problem are generaly not continued in the elliptic problem, except for small-$q$ resonances. Thus, chaotic transport in high-order MMRs of the elliptic problem is qualitatively different, with respect to low-order MMRs, and slow chaotic diffusion in eccentricity—caused by the overlapping between the MMR harmonics—dominates.

## 5.5.3 Estimates of the Lyapunov Time

Murray and Holman [38] used the Hamiltonian (5.19) to calculate the Lyapunov time of chaotic orbits inside a $k : (k + q)$ MMR. Following the modulated pendulum approximation, they derived a symplectic map for the $(\psi, J)$ motion, calculated the maximum eigenvalue, $\lambda_{\max}$, of its tangent map and computed the Lyapunov exponent as the phase average of $\ln \lambda_{\max}$, along the separatrix of the pendulum.[7] We follow here their approach, again with some small differences, such that their formal solution is retrieved, but the values of $T_L$ are somehow different. Starting from (5.20) and assuming that $\phi$ is not resonant, so that its average motion is dictated by the secular term (i.e. $\langle \phi \rangle \approx \nu t$), we can construct a symplectic map for $(\psi, J)$ with a step equal to $\tau_{\sec} = 2\pi/\nu$, using (5.20) as part of the generating function of an infinitesimal canonical transformation [15]. The resulting map has the well-known form of the *standard map*,

$$J' = J - \tau_{\sec}\mu\tilde{D}_* \sin\chi$$
$$\psi' = \psi + \tau_{\sec}\beta J' \qquad (5.23)$$

where $\chi = \psi - \tilde{Q}$. Since $\Phi$ also changes slowly during a secular cycle, we use $\tilde{D} = \tilde{D}_* = [D(\phi = 0) + D(\phi = \pi)]/2$.[8] The Lyapunov exponent is computed by the formula

$$\gamma_{\max} = \frac{1}{2\pi} \int_0^{2\pi} \ln \lambda_{\max}(K_*) \, d\chi \qquad (5.24)$$

where $K_* = -\mu \tilde{D}_* \beta \tau_{\sec}^2$ and $\lambda_{\max}(K_*)$ is the largest eigenvalue of the linearized verion of (5.23). Murray and Holman [38] calculated the integral in (5.24) and found

$$T_L = \tau_{\sec}/\ln\left[1 + \frac{K_*}{2} + \sqrt{\frac{K_*}{2} + \left(\frac{K_*}{4}\right)^2}\right] \qquad (5.25)$$

and, for $K_* \gg 1$, reduces to

$$T_L \approx \tau_{\sec}/(\ln K_*/2) \sim \tau_{\sec} \qquad (5.26)$$

Thus the phase-decorrelation time for chaotic orbits is of the order of the secular period of $\phi$. On the other hand, for $K_* \simeq 1$

$$T_L \approx \tau_{\sec}/\sqrt{K_*} \approx \tau_{\sec} \sim \tau_p \qquad (5.27)$$

---

[7] According to the study of the standard map by Contopoulos et al. [5], this approximation is valid only in the limit of large perturbations!
[8] Normaly, we should have taken the average coefficient, over a cycle of $\phi$. However, we found that this simple approximation for $\tilde{D}$ actually fits better to our numerical data.

where $\tau_p = 2\pi/\nu_p$ is the period of small-amplitude librations in the $p$-harmonic. Hence, the main result is that $T_L$ is always of the same order as the secular period, the latter becoming equal to the period of libration in MMRs with marginaly overlapping harmonics.

The above result can be understood with the following geometrical scheme. Let us suppose that $K_* \sim 100$ in a given MMR. Then, all harmonics lie on top of each other and (5.20) describes the dynamics of a *slowly* modulated pendulum, whose separatrix expands and contracts at a rate equal to $\nu$. A small 'ball' of initial conditions will evolve coherently until the separatrix sweeps through the region, occupied by the phase points. Since the separatrix sweeps the same region twice in a secular cycle, the decoherence time (or $T_L$) should be $\approx \tau_{\rm sec}/2$. Now consider the case of two marginally overlapping harmonics ($K_* \simeq 1$), i.e. two libration regions whose separatrices touch at some point. A ball of initial conditions in the $p$-harmonics will remain coherent until the phase points reach the crossing point of the two harmonics, i.e. after a time approximately equal to the libration period, $\tau_p \sim \tau_{\rm sec}$.

### 5.5.4 Computing the Size of the Chaotic Region

The extent of the chaotic region in a MMR can be found, in the modulated pendulum approximation. Suppose that the modulation is slow, compared to the libration time scale, i.e. $K_* \gg 1$. Then, we can assume the slow degree of freedom $(\phi, \Phi)$ to be 'frozen', during a cycle of $\psi$. For constant values $(\phi_*, \Phi_*)$, the Hamiltonian (5.20) describes the motion of a simple pendulum. This 'fast' sub-system has two equilibria, an unstable one at $(\psi - \tilde{Q}_*, J) = (0,0)$ and a stable one at $(\psi - \tilde{Q}_*, J) = (\pi, 0)$. The value of $\mathcal{H}$ at the separatrix is then

$$h_{\rm SX} = \mathcal{H}(\phi_*, \Phi_*; 0, 0) = d_1 \Phi_* - \mu \tilde{D}(\phi_*, \Phi_*) \tag{5.28}$$

The global study of the phase-space dynamics of (5.20) can be made by taking surfaces of section for different values of $\mathcal{H} = h$ at $\psi - \tilde{Q}_* = \pi$. In Fig. 5.11 we present two sections, one for the 8:3 and one for the 3:1 MMR. Several orbits were integrated numerically and are shown in each section. The trace of the separatrix of the 'fast' pendulum, i.e. the values of $(\phi_*, \Phi_*)$ for which $h_{\rm SX}(\phi_*, \Phi_*) = h$, is superimposed on each section. As shown in the figure, the chaotic region of the surface of section coincides with the region swept by the pendulum separatrix. In particular, all orbits intersecting the separatrix at some point are chaotic. Hence, the extent of the chaotic region can be found by integrating the set of initial coniditions $\{(\phi_*, \Phi_*) / h = h_{\rm SX}\}$.

As shown in Fig. 5.11, there can be significant differences in the size and shape of the chaotic region, depending on secular dynamics inside the MMR. For example, in the 8:3 MMR (and for the selected value of $h$) the stochastic layer covers a small area of the section, since the secular dynamics of $(\phi, \Phi)$ resemble those of a harmonic oscillator, with the separatrix of the frozen pendulum being almost tangent to its orbits. Thus, eccentricity variations

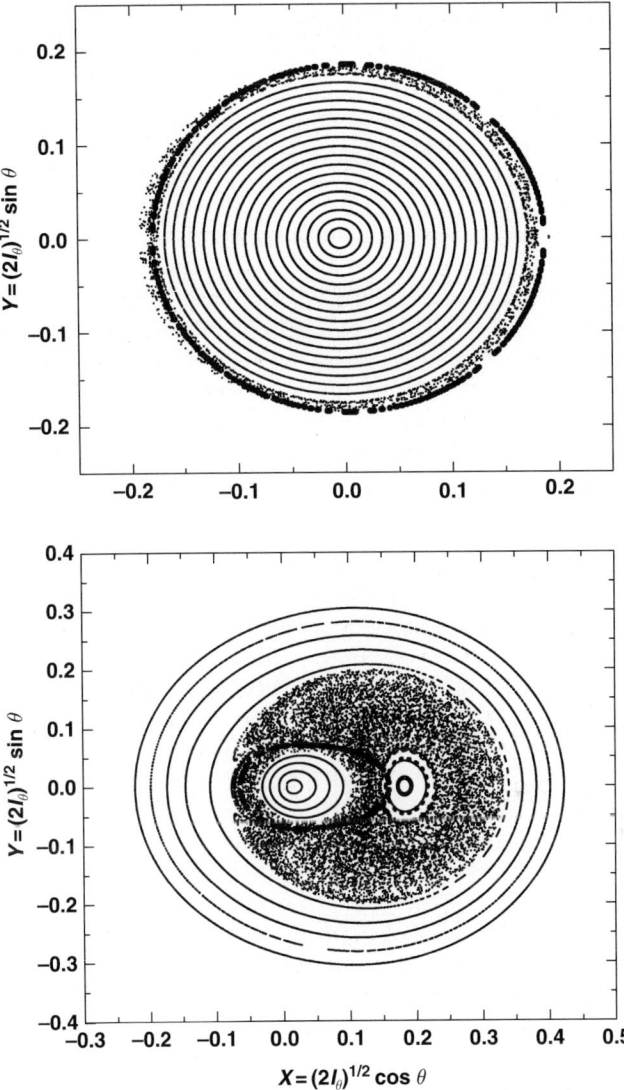

**Fig. 5.11.** Surfaces of section at $\psi - \tilde{Q} = \pi$ for the 8:3 (**top**) and (3:1) MMR. In each panel, the *thick dotted curve* denotes the set of points that belong to the separatrix of the modulated pendulum, i.e. $h = h_{\rm SX}(\phi, \Phi)$. All orbits intersecting this separatrix are chaotic. See text for a discussion on the different dynamics in these two MMRs (the symbols $\theta$ and $I_\theta$ correspond to $\phi$ and $\Phi$, respectively)

are restricted within small values around $e \sim 0.2$. On the other hand, in the 3:1 MMR, two resonant periodic orbits (where $\dot{\phi} = 0$) exist. The homoclinic curve, passing trough the unstable point (at $Y = 0$, $X = -0.1$), transversely intersects the separatrix of the pendulum. As a result, all orbits in the vicinity of the homoclinic orbit are chaotic. Thus, the eccentricity can vary from $e \sim 0.05$ to $e \sim 0.4$, and asteroids can develop Mars-crossing orbits.

### 5.5.5 Estimates of Diffusion Coefficients

Murray & Holman (1997) calculated the diffusion coefficients in $\Psi$ and $\Phi$ in the quasi-linear approximation. According to their model, during a time equal to $T_\mathrm{L}$, the momenta suffer changes $\Delta\Psi$ and $\Delta\Phi$ which are of the order of the size of the strongest MMR harmonic. In other words, orbits starting inside a given harmonic switch randomly from one harmonic to the other, within a time equal to $T_\mathrm{L}$. Using the equations of motion of (5.19), Murray & Holman [38] find[9]

$$\mathcal{D}_\Psi = \frac{1}{4}\mu^2 D_{p_0}{}^2 \tau_{\sec} \sim \mu D_p{}^2$$
$$\mathcal{D}_\Phi = \frac{1}{4}p_0{}^2 \mu^2 D_{p_0}{}^2 \tau_{\sec} \qquad (5.29)$$

where $p_0$ denotes the strongest harmonic and $D_{p_0}$ its coefficient in (5.19). Note that the first of (5.29) can be easily retrieved, using the definition of $\mathcal{D}$ (5.4) and the equations of the standard map (5.23). Given the functional form of $D_p \sim \Phi^{p/2}$, and the fact that $p_0 = q/2$ for $e \approx e'$, the diffusion coefficients take the form $\mathcal{D} \sim \Phi^q$. Solving the 1D Fokker–Plack equation for $\Phi \sim e_\mathrm{f}{}^2$, Murray and Holman [38] find the time of 'removal' from a MMR to be given by

$$T_\mathrm{R} = \frac{\Phi_{\max} \Phi_0}{2\sqrt{\mathcal{D}(\Phi_{\max}) \mathcal{D}(\Phi_0)}}, \qquad (5.30)$$

where an absorbing boundary at $\Phi_{\max}$ is assumed, to account for the effect of close encounters with Jupiter, at large eccentricities. The results of Murray and Holman [38] show a reasonably good agreement with numerical integrations, for several high-order MMRs of the outer asteroid belt. We note however that this approximation may not suited for MMRs of order $q \leq 9$ in general, since numerical calculations suggest that the assumption $K \simeq 1$ is not valid. Moreover, this diffusion model does not take into account the effects of secular dynamics—it assumes that the sole mechanism governing chaotic transport is the marginal overlapping of the MMR harmonics.

---

[9] In their formalism, the diffusion coefficient is defined as $\langle \Delta I^2 \rangle / T_\mathrm{L}$ and a factor $1/2$ (see (5.4)) is embodied in the Fokker–Planck equation. Thus, the formulas given in their paper for $\mathcal{D}_\Psi$ and $\mathcal{D}_\Phi$ have a denominator equal to $1/2$, instead of $1/4$ here.

A different diffusion model, based on the approximation of the slowly modulated pendulum (i.e. $K \gg 1$), has been developed by Neishtadt [40, 41]. The core of this model is the definition of the appropriate action of the problem: the *adiabatic invariant* of the modulated pendulum, $\mathcal{J}$.

Consider the 'frozen' system, defined by (5.20) for fixed values of $(\phi = \phi_*, \Phi = \Phi_*)$. Then, for this pendulum we can define two sets of action-angle variables $(I, w)$, one for librations and one for circulations. Using perturbation theory, Neishtadt [40, 41] calculated an improved adiabatic invariant $\mathcal{J}$, by extending the canonical transformation to the full system and averaging over $w$. Away from the separatrix this new action, $\mathcal{J}$, suffers variations of $O(\mu^2)$ only, while close to the separatrix $\mathcal{J}$ suffers variations of $O(\mu)$. It should be noted that the work of Neishtadt [40] is not restricted to motion in MMRs of the restricted three-body problem, but applies to general Hamiltonian systems with one 'fast' and one 'slow' degree of freedom.

On a surface of section of (5.20) ($\mathcal{H} = h$, $\psi = $ const) each orbit of the averaged system is labeled by the value of $\mathcal{J}$.[10] Neishtadt [40, 41] proved that, upon crossing the separatrix, $\mathcal{J}$ must 'jump' by $\Delta \mathcal{J} \sim \mu$ and that repeated jumps (twice in a secular cycle) occur at random phase and have a Gaussian distribution. Thus, all orbits intersecting the separatrix are chaotic, undergoing diffusion in $\mathcal{J}$. The jumps in $\Phi \sim e^2$ are in fact dictated by the jumps in $\mathcal{J}$ *and* the shape of the chaotic region (i.e. the secular dynamics in the specific MMR). Thus, although the jumps in $e$ may not resemble diffusion, as is the case for the 3:1 MMR, the trajectories are diffusing in $\mathcal{J}$. Hence, when speaking of chaotic diffusion, one has to specify *in which variables*. This result justifies the use of the term 'diffusion' also in the case of 'fast' (intermittent) eccentricity transport, e.g. in the 7:3 MMR that was shown in Sect. 5.3.

Neishtadt [40, 41] applied his method to the Hamiltonian of Wisdom [62] for the 3:1 MMR, which is in fact a particular case of (5.20). He showed that the diffusion coefficient in $\mathcal{J}$ is given by

$$\mathcal{D}(\mathcal{J}) \approx \frac{S^2}{\tau_{\text{sec}}} \sim \mu^3 \qquad (5.31)$$

where $S = 4 d_1 \mu \left(\partial_\phi \tilde{D}\right) / (\beta \tilde{D})$. This result shows that diffusion in $\mathcal{J}$ is indeed slow, as $(1/\mathcal{D}) \sim \mu^{-3}$. However, Neishtadt [41] showed that, due to the intersection of the pendulum's separatrix with the asymptotic curves of the unstable resonant periodic orbit, the mean time between successive 'jumps' in $e$ to Mars-crossing values is $\sim 10^5$ y, i.e. the same as the one found in the numerical integrations of Wisdom [61, 62].

We note that a similar approach can be taken for $K \simeq 1$ MMRs, as shown by the work of Henrard and Sato [17] on the 3:1 inclination resonance of the Miranda-Umbriel system. They derived analytical formulas for computing

---

[10] Wisdom [62] also studied the 3:1 MMR by introducing the action-angle variables of the frozen system, but did not compute the improved invariant.

different quasi-integrals (i.e. invariants) for each resonant harmonic, but did not compute the magnitude of the 'jump' of each invariant, upon separatrix crossings.

Let us now compare the diffusion rates, as given by the models of Neishtadt [40, 41] and Murray & Holman [38]. The theory of Neishtadt [40, 41] best applies to MMRs with $K \gg 1$, for which the diffusion rate is proportional to $\mu^3$ (see (5.31)). On the other hand, the theory of Murray and Holman [38] best applies to MMRs with $K \simeq 1$. Note however that $K \simeq 1$ implies $D_p \sim \mu$, which means that the diffusion coefficient is again proportional to $\mu^3$ (see (5.29)). This observation leads to a general result for chaotic diffusion in MMRs: in the absense of 'fast' chaotic transport routes, related to the secular dynamics, the time scale of chaotic diffusion in all MMRs is of order $\mu^{-3} \sim 1$ Gy. This result explains the ones found in numerical simulations (see also next section). Moreover, it explains the presence of large numbers of chaotic main-belt asteroids in MMRs, since their escape time is comparable to the age of the solar system.

## 5.6 Chaotic Transport in the Main Belt: A Numerical Survey

Let us now present some numerical results on chaotic transport in the main asteroid belt. We will present calculations of the Lyapunov time in the framework of the 2D and 3D elliptic restricted three-body problem and compare our results to the analytical formula presented above (5.26). Furthermore, we will integrate numerically the orbits of a selected set of initial conditions for 1 Gy, in an effort to estimate the removal time from different MMRs. We will discuss similarities and differences between the different models and show that the main mechanism driving macroscopic diffusion is the coupling between the variations in eccentricity and inclination.

The global dynamics in the asteroid belt can be understood by computing dynamical maps, such as the ones presented in Fig. 5.12, where the value of $\log(T_\mathrm{L})$ is given for a portion of the asteroid belt ($2.4 \leq a \leq 3.5$ AU and $0 \leq e \leq 0.42$). These plots were obtained by integrating the variational equations along with the equations of motion for a set of 23,100 particles, in the framework of the three-body problem. The initial conditions were placed on a regular $550 \times 42$ grid in $(a, e)$, with a relative inclination of $i = 0$ (top—2D problem) or $i = 5°$ (bottom—3D problem), with respect to Jupiter's orbital plane. In the context of these simple models, the total volume of the chaotic region for $e < 0.3$ is small, chaotic motion being restricted to thin 'V'-shaped layers around the main MMRs. For $a > 3.4$ AU an extended region of overlapping MMRs appears for $e > 0.2$. It is clear from the figure that there are no significant differences between the two experiments. Thus, for small inclinations, the value of $T_\mathrm{L}$ is already determined by the planar component

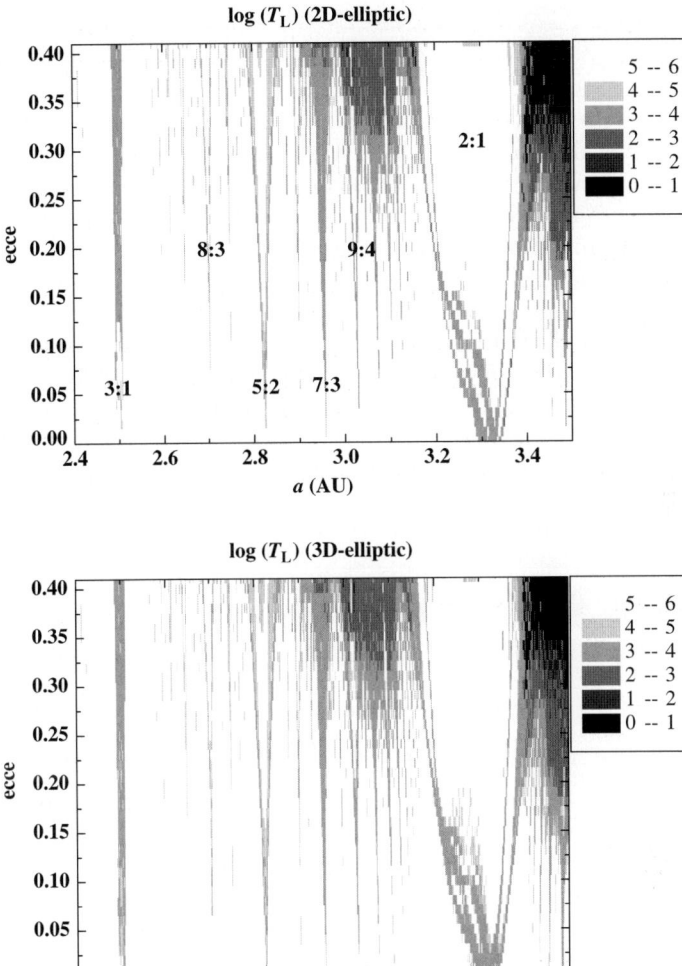

**Fig. 5.12.** Maps of $\log T_L$ for a section of the main asteroid belt. The test particles were placed on a regular $550 \times 42$ grid in $(a, e)$ and were given a relative inclination of $i = 0$ (**top**) or $i = 5°$ (**bottom**), with respect to Jupiter's orbital plane. The values of $\log T_L$ are color-coded, according to the scale shown on the right (*black = chaotic*, *white = regular*). The location of some low-order resonances is noted in the upper panel. The two plots differ very little

of the disturbing function, throughout the main belt. As we will see in the following, this is not the case for the transport time scale.

We now focus on a group of MMRs for a more detailed study. We selected a set of 10 particles for every MMR of order $q \leq 7$ in the inner and central belt ($a \leq 3.4$ AU) and $q \leq 5$ in the outer belt ($a \geq 3.4$ AU), i.e. a total of 22 MMRs. The particles were chosen from a short integration ($\sim 10^5$ y) of $\sim 600$ test particles in the framework of the 2D elliptic three-body problem, by demanding the respective critical argument (i.e. $\psi - \tilde{Q}$) to show irregular transitions between libration and circulation. In this way we were left with a set of 220 initial conditions with $0.05 \leq e \leq 0.1$, all of which leading to chaotic motion. The variational equations were integrated for 2 My, in order to calculate the Lyapunov time. The integration was repeated in the framework of the 3D problem, by giving an initial inclination of $i = 5°$ to each of our test particles.

The results of the above experiment are shown in Fig. 5.13, where the value of $T_L$ is plotted against the initial value of $a$ for all particles. The corresponding $T_L$ value obtained from (5.26) is also given. As seen in this plot, the values of $T_L$ are practically the same for both integrations. This confirms our previous observation that the differences between the 2D and the 3D problem are minimal, for small inclinations.

Moreover, it is clear that, despite the large number of approximations behind the derivation of (5.26), the agreement with the numerical results is

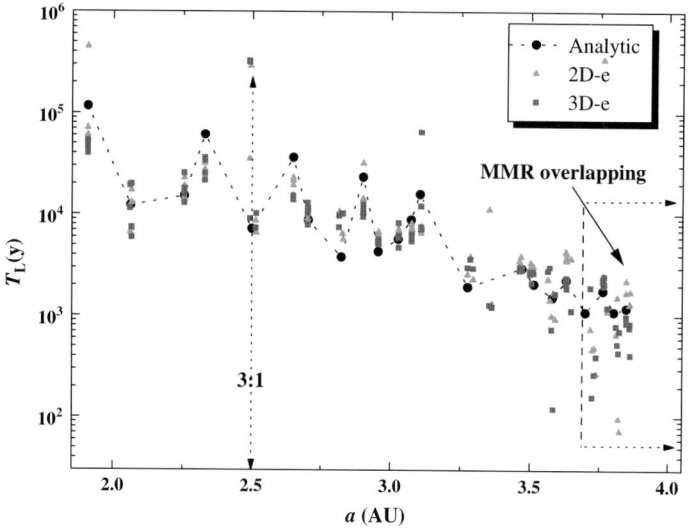

**Fig. 5.13.** Lyapunov time in main belt MMRs. All particles had initially $0.05 \leq e \leq 0.10$. The *triangles* correspond to particles integrated in the framework of the planar problem, while the squares are particles integrated in the 3D problem. The *solid circles* (connected with a *dotted curve*) denote the values obtained from (5.26). The agreement between the analytic and the numerical results is better than a factor of 2 for most MMRs. Some important exceptions to this rule are noted in the region of the 3:1 MMR (see text) and the outer part of the belt ($a \geq 3.6$ AU), where extensive overlapping between adjacent MMRs occurs

better than a factor of two, for most MMRs. A few notable exceptions can be found in the vicinity of the 3:1 MMR and for $a \geq 3.6$ AU. In the first case (3:1 MMR) the particles with $T_{\mathrm{L}} \sim 3 \times 10^5$ y may in fact follow regular dynamics. As for the region with $a \geq 3.6$ AU, we note that overlapping between adjacent MMRs is expected even for small values of $e$. Thus, our single-MMR model is not able to capture the basic dynamics in this phase-space region.

The same set of initial conditions was integrated for a maximum time of 1 Gy, i.e. a time comparable to the lifetime of the solar system. The integration was stopped if a particle reached planet-crossing values of $e$. Four different integrations were performed, using four different models: the 2D and 3D three-body problem, and the 2D and 3D *secularly precessing problem*, in which the orbit of Jupiter was forced to precess, according to the secular solution of the Jupiter–Saturn system. The results of the four long-term integrations are shown in Fig. 5.14, where each point corresponds to the maximum eccentricity reached by the particle during the integration. Only particles that survived for 1 Gy are shown in these plots. The evolution is clearly different for different groups of resonances. We can distinguish between three main groups of MMRs: outer-belt MMRs ($a \geq 3.4$ AU), inner-belt ($a \leq 3.4$ AU) low-order MMRs ($q \leq 3$), and inner-belt high-order MMRs ($q > 4$). The effects of 3D motion and secular precession are different for these three groups.

In the outer asteroid belt, a significant number of particles develops Jupiter-crossing orbits on a variety of time scales, ranging from a few $10^5$ y (for the 7:4 and 5:3 MMRs) to several times $10^8$ y. These numbers agree with the numerical results of Holman & Murray [19] and partly with the analytic estimates of Murray & Holman [38]. It should be noted though that, as shown in Tsiganis et al. [54], fast chaotic transport in the 5:3 and 7:4 MMRs is related to the existence of resonant periodic orbits (i.e. secular dynamics) and not chaotic diffusion à la Murray and Holman [38]. The percentage of escaping particles in the outer belt is 36% in the 2D-elliptic and 51% in the 2D-precessing model. In the 3D models, the numbers become 88% (elliptic) and 85% (precessing). Hence, although secular precession increases the volume of the chaotic phase-space region, the coupling between the two degenerate degrees of freedom (eccentricity and inclination) is the main mechanism that enhances chaotic transport. The observed depletion of the outer asteroid belt can be explained in the context of the 3D-precessing problem. We point out that the percentage of escaping particles in the 2D-elliptic model is unexpectedly high, given the geometrical constraints of this model (see the following discussion). However, it should be noted that most Jupiter-crossing particles were initially located at $a \geq 3.6$ AU, where resonance overlapping between adjacent MMRs [60] leads to chaotic diffusion in semi-major axis.

In the inner belt, the percentage of particles that develop planet-crossing orbits within 1 Gy is only 9% in the 2D-elliptic and 28% in the 2D-precessing model. Moreover, all these planet-crossing particles were initially in low-order MMRs (2:1, 3:1, 4:1, and 5:2; also the 7:3 in the 2D-precessing model), in which fast chaotic transport is related to the local secular dynamics [32, 33].

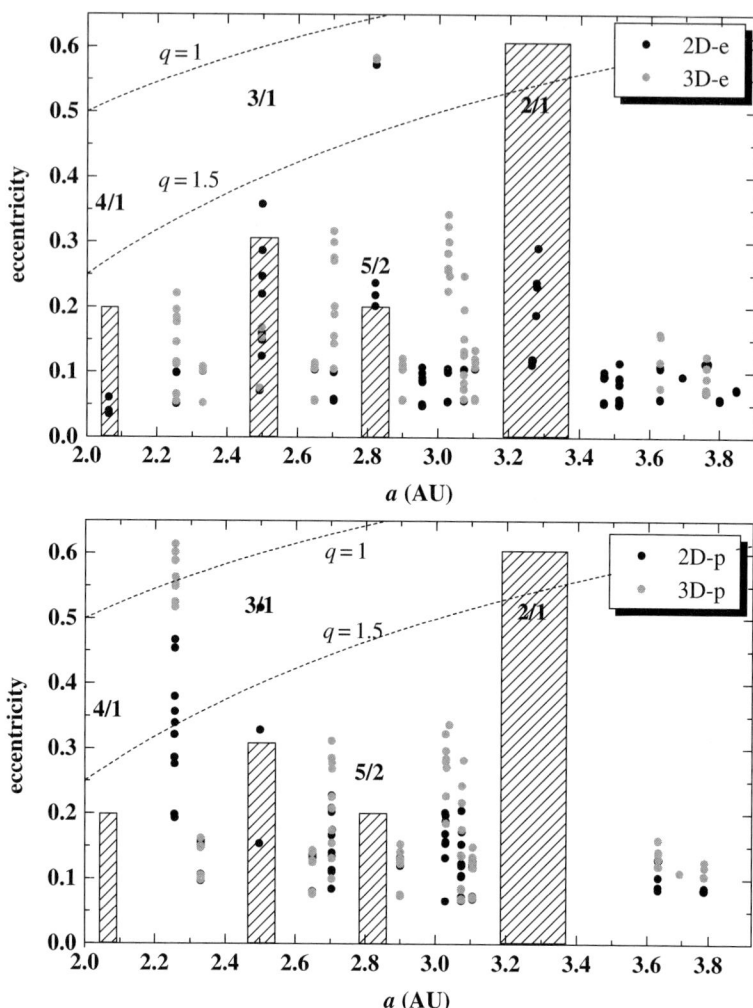

Fig. 5.14. Final eccentricities for the surviving particles in the 1 Gy integrations. The top panel corresponds to integrations in the framework of the 2D and 3D three-body problem, while the lower panel to integrations in the framework of the 2D and 3D secularly precessing problem. The Mars-crossing (pericenter distance $q_p = 1.5$ AU) and Earth-crossing ($q_p = 1$ AU) limits are also shown. Black dots correspond to 2D and grey dots to 3D motion. The shaded regions mark the neighborhoods of the main low-order MMRs. Note that chaotic diffusion is strongly enhanced in 3D models. However, orbits in $q \geq 3$ MMRs of the central belt ($2.5 \leq a \leq 3.2$ AU) do not reach planet crossing eccentricities within the integration time

In all other inner-belt MMRs there is practically no diffusion in the 2D-elliptic problem. When the 2D-precessing model is considered, particles in high-order MMRs can diffuse up to at most $e \sim 0.2$ within 1 Gy.[11] Thus, secular precession (in 2D) leads to a wider diffusion region within high-order MMRs, but still cannot drive particles to planet-crossing eccentricities.

The situation changes when we consider 3D motion, but not dramatically. The percentage of planet-crossing particles increases to 35% in the elliptic and 44% in the precessing problem. Low-order MMRs become almost completely void of asteroids. Enhanced chaotic diffusion is observed also in high-order MMRs, where particles can diffuse up to $e \sim 0.35$. As shown in the upper panel of Fig. 5.14, MMRs of order $q = 5$ show a larger dispersion in $e$ than MMRs of higher order, the dispersion increasing as we move towards higher values of $a$ (i.e. closer to Jupiter). However, the time needed to reach Mars-crossing values of $e$ is evidently longer than 1 Gy, even in the precessing problem. These results agree qualitatively with the analytic results presented in Sect. 5.5. Finally, as in the case of the outer belt, we see the decisive role of the coupled evolution of $e$ and $i$, which is the main mechanism driving diffusion.

The observed differences between 2D and 3D motion in MMRs can be understood in terms of geometry. Consider a MMR that is 'isolated', i.e. no MMRs of similar order exist in its vicinity. Then, 2D motion is well described by (5.20). Since this is a Hamiltonian system with two degrees of freedom, motion is restricted on the 3D energy manifold. Moreover, for $\mathcal{J} = $ const, there is no diffusion, since the value of $\mathcal{J}$ defines a single orbit on the surface of section $\psi =$ const, i.e. a 1D curve in 2D space. Small changes in the adiabatic invariant, induced by separatrix crossing ($\Delta \mathcal{J} \sim \mu$), produce chaotic diffusion which is restricted within thin layers in the 2D surface (see top panel of Fig. 5.11). In the 3D problem, a constant value of $\mathcal{J}$ does not label a single orbit but instead defines a 3D subspace of the 4D surface of section. Let us suppose an orbit, starting with $\mathcal{J} = \mathcal{J}^*$ and $\phi^* = 0$ (i.e. on one axis of the section). As the orbit diffuses, it has to pass arbitrarily close to the initial point. In the 2D problem, if the system would return to $(\mathcal{J}^*, \phi^*)$ (no diffusion), $\Phi$ would necessarily be equal to its initial value (i.e. the system would return on the initial phase point). In the 3D problem, if the orbit would return to $(\mathcal{J}^*, \phi^*)$, the other variables $(\Phi, \theta, \Theta)$ could take any value, as long as they stayed on the 3D subspace of the 4D section, defined by $\mathcal{J} = \mathcal{J}^*$. In other words, in the 3D case, the orbit is 'free' to diffuse in an extended region of $(\Phi, \Theta)$. This is the essential difference bewteen the 2D and 3D problems. Of course, even in the 2D problem, the 'true' (non-averaged) Hamiltonian contains terms that depend on another degree of freedom (i.e. $\lambda$), but their magnitude is of $O(\mu^2)$ only. This explains why particles in our numerical survey do not reach planet-crossing eccentricities, in isolated MMRs of the

---

[11] With the exception of particles starting in the 7:2 MMR, which are strongly affected by the nearby $\nu_6$ SR.

2D-elliptic problem. The far outer belt ($a > 3.6$ AU) is an exception, since the condition of isolation is not met for most MMRs.

As a final note on the importance of coupling between the two degenerate actions, we performed the following numerical experiment. We selected a set of 100 initial conditions, placed in each of the following $q = 5$ resonances: the 7:2, 8:3 and 9:4 MMRs. All particles had initially $0.13 \leq e \leq 0.15$ and $i = 0$ or $5°$. We integrated the variational equations to make sure that all initial conditions lead to chaotic motion. Then, we integrated the equations of motion for 2 My in the framework of the 2D and 3D elliptic problem (no precession), and computed the mean squared displacement in the two degenerate actions as a function of time. On-line digital filtering was used, to eliminate short-periodic fluctuations in the orbital elements and linear free elements were calculated for each object, by taking the maximum values of $h = e \sin(\varpi - \varpi')$ and $p = i \sin \Omega$ over a running window of $\Delta t = 10^5$ y.

The result of the above experiment are shown in Fig. 5.15, where the time evolution of the mean squared displacement in all three actions (two in the 2D problem) is shown (all panels are in the same scale). The mean squared displacement in $a$ is constant with time in all panels, as no particle can leave the resonance region. In the top panels, we see that the mean squared

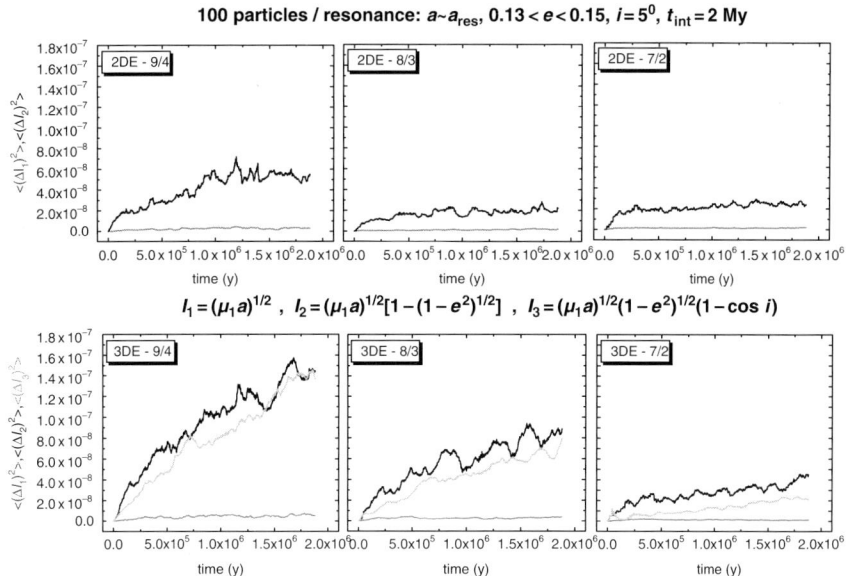

**Fig. 5.15.** Evolution of the mean squared displacement as a function of time, for all three actions. The *top* (resp. bottom) *panels* correspond to integration in the framework of the 2D (resp. 3D) model. All panels are in the same scale. Saturation of $\langle (\Delta I_2)^2 \rangle$ is observed in the 2D problem. Conversely, in the 3D problem, the coupling between $I_2$ and $I_3$ drives diffusion in both actions

displacement in $e$ (actually, $I_2 \sim e^2$) increases with time in the beginning, but soon reaches saturation; the corresponding stochatic layers are filled. On the contrary, in the 3D model (lower panels), macroscopic diffusion is driven by the coupling between $I_2$ and $I_3$. The diffusion rate is similar for $I_2$ and $I_3$ in each of the three MMRs. Finally, as we move away from Jupiter, i.e. towards smaller values of $a$ (from left to right in Fig. 5.15) the diffusion rate continuously decreases by a small factor.

## 5.7 Conclusions—Discussion

In this chapter we reviewed the problem of chaotic transport in the asteroid belt, emphasizing on the role of slow chaotic diffusion. In the first part of the paper (Sect. 2–4) we presented recent results on the origin and evolution of certain groups of asteroids, in different places of the asteroid belt, where several chaos-generating mechanisms interact. In the second part of the chapter we reviewed analytical theories of chaotic diffusion in MMRs and compared analytical and numerical estimates of the Lyapunov and transport time scales. In this last section we summarize our conclusions and discuss some of the open problems of this exciting subject.

The slow dispersion of the Trojan swarms is a direct result of slow chaotic diffusion [56], generated by a variety of high-order resonances [49] which overlap the 1:1 MMR with Jupiter. The flux of Trojans to Jupiter-crossing orbits is estimated to be $\sim 1.5 \times 10^{-6}$ Earth masses per Gy (i.e. $\sim 10\%$ of the total Trojan mass, see [36]). The dynamics of the escaping Trojans and their contribution to the population of Near Earth Asteroids needs to be determined.

The observed group of asteroids in the 7:3 MMR is a typical example of a long-lived population of individualy unstable asteroids. We have shown that escape from the 7:3 MMR proceeds in two steps: 'slow' diffusion at first, leading nearly circular orbits to $e \sim 0.2$, followed by 'fast' chaotic transport (related to secular resonances) that leads to Jupiter-crossing values of $e$. However, although 7:3 asteroids have a median escape time of $\sim 20$ My, their population is kept in steady state, owing to the continuous influx of bodies from neighboring asteroid families, caused by the Yarkovsky effect [55]. This marvelous interplay of chaotic transport and radial migration governs the motion of asteroids smaller than $\sim 20$ km. The study of such resonant groups, combined with asteroid observations and analysis of meteorite samples, can render important information about the composition of the proto-solar nebula and the early evolutionary phases of the solar system.

Chaotic diffusion leads to a slow spread of asteroid family members in the space of proper elements, as shown in Sect. 5.4 for the Veritas family. The main chaos-generating mechanisms in the region of Veritas are three-body MMRs of the Jupiter–Saturn–Asteroid type. Despite the small amplitude of the resonant harmonics [43, 44], chaotic diffusion produces measurable effects, even on time scales of $\sim 10$ My. This is exactly what we can take advantage of and estimate the age of the family. The results of Tsiganis et al. [58] suggest that chaotic motion can be approximated locally by simple diffusion and a reasonably accurate estimate of the family's age can be found, by using a simple random-walk model. Accurate dating of asteroid families is of profound importance in asteroid science. The method described here is particularly suited for young asteroid families, residing in chaotic regions of the main belt.

Analytical models of motion in MMRs can be used to study the statistical properties of chaotic diffusion and estimate the fundamental parameters related to transport in the space of actions. Such a model was presented in Sect. 5.5, where two-body MMRs were studied, in the framework of the elliptic restricted three-body problem. We note that a similar model can be constructed for three-body MMRs, when Saturn's perturbations are also taken into account [43, 44]. In both cases, chaos is generated by the overlapping of the harmonics of the resonant multiplet. A distinct behavior is observed in low-order MMRs, essentially the ones associated with the Kirkwood gaps (2:1, 3:1, 4:1, 5:2 and 7:3 MMRs). Inside these MMRs, secular dynamics are of profound importance for chaotic transport. Secular resonances (or resonant periodic orbits, in the three-body framework) can co-exist with MMRs, providing fast transport routes to planet-crossing eccentricities. On the other hand, in high-order MMRs, secular dynamics do not play an important role, as suggested by numerical results. A systematic study of secular dynamics in high-order MMRs of the main asteroid belt - either of two-body or three-body type - has not been done so far.

The main parameters needed to understand the long-term effects of chaotic diffusion in MMRs are (i) the Lyapunov time, (ii) the size and shape of the chaotic region and (iii) the diffusion coefficients. As shown in Sect. 5, $T_\mathrm{L}$ is of the order of the secular period in all MMRs, i.e. the inverse of the separation between adjacent harmonics of the resonant multiplet. In resonances with $K_* \simeq 1$, this time scale is also approximately equal to the period of small-amplitude librations. However, depending on the value of $K_*$, asteroids spread in action space following different types of 'random walk'.

For $K_* \simeq 1$ resonances, trajectories perform uncorrelated jumps of magnitude $\Delta J \sim \Delta \Phi \sim \sqrt{\mu D_p}$ on time intervals of order $T_\mathrm{L}$, as shown by Murray and Holman [38]. The diffusion coefficient in both actions is of order $\mathcal{D} \sim \mu D_p^2$ and has a power-law dependence on eccentricity, $\mathcal{D} \sim \Phi^q$. For $K_* \gg 1$, $J$ changes by an amount equal to the width of the (modulated) resonance, $\Delta J \sim \sqrt{\mu \tilde{D}}$, on a time-scale which is much shorter than the decoherence time, i.e. $\tau_p \ll T_\mathrm{L} \sim \tau_\mathrm{sec}$. If we average over the short time-scale, the change in $\langle J \rangle$ over a secular period is much smaller and the appropriate action of the problem is the adiabatic invariant $\mathcal{J}$, related to the area of the pendulum (equivalently, one could use the width of libration $J_\mathrm{max}$). As shown by Wisdom [62] and Neishtadt [40, 41], the chaotic region can be defined as the set of orbits whose initial conditions belong to the separatrix of the modulated pendulum, for a given value of the averaged Hamiltonian, i.e. $h = h_\mathrm{SX}$. Neishtadt [40, 41] proved that $\mathcal{J}$ suffers repeated random jumps upon crossing the sweeping separatrix of the modulated pendulum. The diffusion coefficient depends on the sweeping rate, which is a function of the eccentricity (i.e. $\tilde{D}$), and has a magnitude of order $\mathcal{D} \sim \mu^3$. A distinct advantage of Neishtadt's model is that it is equally valid in low-order MMRs, where $\mathcal{J}$ diffuses in spite of the fact that secular dynamics may force $\Phi \sim e^2$ to evolve in a non-diffusive manner.

Interestingly enough, both analytical models of chaotic diffusion suggest that the characteristic diffusion time-scale in all MMRs is of order $\mu^{-3} \sim 1$ Gy, in the absense of secular resonances. As numerical calculations suggest, the theory of Neishtadt [40, 41] is better suited for two-body MMRs of order $q \leq 9$. The same should hold for three-body MMRs with $q = 1, 2$ [43, 44]. As $q$ increases, the size of the resonant harmonics decreases, so that $K_* \to 1$ beyond some (yet unspecified) value $q_*$. However, as shown in Sect. 5.5, $K_* \simeq 1$ implies $D_p \sim \mu$, so that the diffusion coefficient in $\Phi \sim e^2$ takes values of order $\mathcal{D} = \mu D_p^2 \sim \mu^3$. This explains what is seen in numerical integrations of both real and fictitious asteroids (e.g. Sect. 5.6): chaotic diffusion is slow and the time needed for chaotic asteroids to develop planet-crossing orbits is comparable to the lifetime of the solar system. We note though that the transport time-scale can be much shorter, as we approach the edges of the asteroid belt, where the number of observed bodies also falls steeply. In the outer belt ($a \geq 3.4$ AU), overlapping of adjacent MMRs leads to more efficient asteroid transport, because the semi-major axis also varies significantly. A similar effect was observed by Morbidelli and Nesvorný [44] for inner-belt asteroids (with $a \leq 2.4$ AU), where low-order three-body MMRs with Jupiter and Mars are dense in semi-major axis.

Macroscopic diffusion in high-order MMRs is driven by the coupled evolution of the two degenerate degrees of freedom, i.e. the motion of the pericenter and the node. Contrary to what observed for 2D motion, where diffusion is restricted within thin layers of phase space due to geometrical reasons (for isolated MMRs), 3D motion leads to the slow growth of both $e$ and $i$ at similar rates. We note though that, for small initial inclinations, 2D models give good estimates for $T_L$ and $\mathcal{D}$. In two-body MMRs, the coupled evolution of $e$ and $i$ is already determined by the three-body problem. Indeed, as the numerical results of Sect. 6 suggest, the basic time scales of chaotic diffusion ($T_L$ and $1/\mathcal{D}$) change very little when the secular perturbations of Saturn are considered. What seems to change considerably is the size of the chaotic MMR region, as the volume of initial conditions leading to chaotic motion increases. Analytical models of chaotic diffusion can be easily extended in principle to three degrees of freedom, although it has not been done so far.

As discussed in the beginning of Sect. 5.5, numerical results suggest that the diffusion approximation works well, on a 'local' scale. Chaotic transport in the asteroid belt is a complicated phenomenon, related to several 'microscopic' mechanisms. In this respect, the construction of a global analytical model for asteroid transport may not be feasible. Thus, analytical studies would need to be performed on a 'resonance-by-resonance' basis. Such a systematic study of chaotic transport in two-body and three-body MMRs of medium/high order is still missing.

## Acknowledgments

I would like to thank the organizers of the Aussois 2006 workshop, Claude Froeschlé, Elena Lega and Daniel Benest, for inviting me to participate in this exciting meeting and for encouraging me to present this review talk. The last part of this chapter was based on joint work and extensive discussions with Alessandro Morbidelli, during my stay in Nice. I would also like to thank M. Guzzo and A. Neishtadt for many interesting discussions, especially during a visit of myself and A. Morbidelli to the Mathematics Department of the University of Padova.

## References

1. Bottke, W. F., Vokrouhlicky, D., Broz, M., et al., 2001. *Science* **294**, 1693.
2. Brož, M., Vokrouhlicky, D., Roig, F., et al., 2005, *Month. Not. Royal Astron. Soc.* **359**, 1437.
3. Bruhwiler, D. L. and Cary, J. R., 1989. *Physica D* **40**, 265.
4. Chirikov, B. V., 1979, *Phys. Rep.* **52**, 263.
5. Contopoulos, G., Efthymiopoulos, C., and Voglis, N., 1999, The form and significance of dynamical spectra. In Dvorak, R., Haupt, H.F., and Wodnar, K. (eds), *Modern Astrometry and Astrodynamics, Proceedings of the International Conference honoring Heinrich Eichborn Held at Vienna observatory, Austria*, 25–26 May 1998, p. 171.
6. Dermott, S. F., Kehoe, T. J. T., Durda, D. D., et al., 2002, Recent rubble-pile origin of asteroidal solar system dust bands and asteroidal interplanetary dust particles. In Warmbein, B. (ed.), *Proceedings of Asteroids, Comets, Meteors*, ACM, ESA-SP-500, Noordwijk, pp. 319–322.
7. Dvorak, R., Contopoulos, G., Efthymiopoulos, C., et al., 1998. *Plan. Space Sci.* **46**, 1567.
8. Efthymiopoulos, C., Contopoulos, G., Voglis, N., et al., 1997. *J. Phys. A* **30**, 8167.
9. Farinella P. and Vokrouhlicky, D., 1999. *Science* **283**, 1507.
10. Farley, K. A., Vokrouhlicky, D., Bottke, W. F., et al., 2006. *Nature* **439**, 295.
11. Ferraz-Mello, S., Michtchenko, T. A., Nesvorny, D., et al., 1998. *Plan. Space Sci.* **46**, 1425.
12. Gladman, B., Migliorini, F., Morbidelli, A. et al., 1997. *Science* **277**, 197.
13. Gomes, R., Levison, H. F., Tsiganis, K., et al., 2005. *Nature* **435**, 466.
14. Hadjidemetriou, J. D., 1993. *Celest. Mech. Dyn. Astron.* **56**, 563.
15. Hadjidemetriou, J. D., 1993. *Celest. Mech. Dyn. Astron.* **56**, 201.
16. Henrard, J. and Caranicolas, N., 1990, *Celest. Mech. Dyn. Astron.* **47**, 99.
17. Henrard, J. and Sato, M., 1990. *Celest. Mech. Dyn. Astron.* **47**, 391.

18. Hirayama, K., 1918. *Astron. J.* **31**, 185.
19. Holman, M. J. and Murray, N. W., 1996. *Astron. J.* **112**, 1278.
20. Klafter, J. et al.: *Levy Description of Anomalus Diffusion in Dynamical Systems*, Lect. Notes Phys. **450**, 196–215. Springer, Heidelberg (1995).
21. Knezevic, Z., Lemaitre, A., and Milani, A., 2002, The determination of asteroid proper elements. In Bottke, W. F., Cellino, A., Paolicchi, P., and Binzel, R.P. (eds), *Asteroids III*, University of Arizona Press, Tucson, pp. 603–612.
22. Levison, H., Shoemaker, E. M., and Shoemaker, C. S., 1997, *Nature* **385**, 42.
23. Lichtenberg, A. J. and Lieberman, M. A., 1983. *Regular and Stochastic Motion*, Springer-Verlag, New York.
24. Marzari F., Farinella, P., and Davis, D. R., 1999. *Icarus* **142**, 63.
25. Marzari, F., Scholl, H., Murray, C., et al., 2002, Origin and evolution of Trojan asteroids. In Bottke, W. F., Cellino, A., Paolicchi, P., and Binzel, R. P. (eds), *Asteroids III*, University of Arizona Press, Tucson, pp. 725–738.
26. Metzler, R. and Klafter, J., 2000. *Phys. Rep.* **339**, 1.
27. Milani, A. and Knezevic, Z., 1990. *Celest. Mech. Dyn. Astron.* **49**, 347.
28. Milani, A. and Nobili, A. M., 1992. *Nature* **357**, 569.
29. Milani, A., 1993. *Celest. Mech. Dyn. Astron.* **57**, 59.
30. Milani, A., 1993, The Dynamics of the Trojan Asteroids. In Milani, A., Di Martino, M., and Cellino, A. (eds), *IAU Symp. 160: Asteroids, Comets, Meteors*, Kluwer Academic Publishers, Dordrecht, pp. 159–174.
31. Milani, A. and Farinella, P., 1994. *Nature* **370**, 40.
32. Moons, M. and Morbidelli, A., 1995. *Icarus* **114**, 33.
33. Moons, M., 1997. *Celest. Mech. Dyn. Astron.* **65**, 175.
34. Morbidelli, A. and Nesvorny, D., 1999. *Icarus* **139**, 295.
35. Morbidelli, A., 2002. *Modern Celestial Mechanics: Aspects of Solar System Dynamics*, Taylor & Francis, London.
36. Morbidelli, A., Levison, H. F., Tsiganis, K., et al., 2005. *Nature* **435**, 462.
37. Murray, C. D. and Dermott, S. F., 2002. *Solar System Dynamics*, Cambridge University Press, Cambridge, UK.
38. Murray, N. and Holman, M., 1997. *Astron. J.* **114**, 1246.
39. Murray, N., Holman, M., and Potter, M., 1998. *Astron. J.* **116**, 2583.
40. Neishtadt, A. I., 1987. *Prikl. Mat. Mekh.* **51**, 750.
41. Neishtadt, A. I., 1987. *Sov. Phys. Dokl.* **32**, 47.
42. Nesvorny, D. and Ferraz-Mello, S., 1997. *Icarus* **130**, 247.
43. Nesvorny, D. and Morbidelli, A., 1998. *Astron. J.* **116**, 3029.
44. Nesvorny, D. and Morbidelli, A., 1999. *Celest. Mech. Dyn. Astron.* **71**, 243.
45. Nesvorny, D., Ferraz-Mello, S., Holman, M., et al., 2002, Regular and chaotic dynamics in mean-motion resonances: Implications for the structure and evolution of the asteroid belt. In Bottke, W. F., Cellino, A., Paolicchi, P., and Binzel, R. P. (eds), *Asteroids III*, University of Arizona Press, Tucson, pp. 379–394.
46. Nesvorny, D., Bottke, W. F., Dones, L., et al., 2002. *Nature* **417**, 720.
47. Nesvorny, D., Bottke, W. F., Levison, H. F., et al., 2003. *Astrophys. J.* **591**, 486.
48. Paolicchi, P., Dell'Oro, A., Cellino, A., et al., 2002, Fitting the mass distributions of Koronis family: New ideas and related physical constraints. In Warmbein, B. (ed.), *Proceedings of Asteroids, Comets, Meteors*, ACM, sESA-SP-500, Noordwijk, pp. 525–528.
49. Robutel, P., Gabern, F., and Jorba, A., 2005. *Celest. Mech. Dyn. Astron.* **92**, 53.

50. Robutel, P. and Gabern, F., 2006. *Month. Not. Royal Astron. Soc.* **372**, 1463.
51. Shlesinger, M. F., Zaslavsky, G. M., and Klafter, J., 1993. *Nature* **363**, 31.
52. Tsiganis, K., Varvoglis, H., and Hadjidemetriou, J. D., 2000. *Icarus* **146**, 240.
53. Tsiganis, K., Varvoglis, H., and Hadjidemetriou, J. D., 2002. *Icarus* **155**, 454 (2002a).
54. Tsiganis, K., Varvoglis, H., and Hadjidemetriou, J. D., 2002. *Icarus* **159**, 284 (2002b).
55. Tsiganis, K., Varvoglis, H., and Morbidelli, A., 2003. *Icarus* **166**, 131.
56. Tsiganis, K., Varvoglis, H., and Dvorak, R., 2005. *Celest. Mech. Dyn. Astron.* **92**, 71.
57. Tsiganis, K., Gomes, R., Morbidelli, A., et al., 2005. *Nature* **435**, 459.
58. Tsiganis, K., Kneževi, Z., and Varvoglis, H., 2007. *Icarus*, **186**, 484.
59. Vokrouhlicky, D., Broz, M., Bottke, W. F., et al., 2005, Non-gravitational perturbations and evolution of the asteroid main belt. In Kneževiæ, Z. and Milani, A. (eds), *Dynamics of Populations of Planetary Systems*, Proceedings of IAU Colloquium No. 197,Cambridge University Press, pp. 145–156.
60. Wisdom, J., 1980. *Astron. J.* **85**, 1122.
61. Wisdom, J., 1983. *Icarus* **56**, 51.
62. Wisdom, J., 1985. *Icarus* **63**, 272.
63. Wyatt, M. C., 2005, The origin and evolution of dust belts. In Kneževiæ, Z. and Milani, A. (eds), *Dynamics of Populations of Planetary Systems*, Proceedings of IAU Colloquium No. 197, Cambridge University Press, pp. 383–392.
64. Zaslavsky, G. M., 1994. *Physica D* **76**, 110.

# 6

# An Overview of the Rotations of Planets in the Solar System

Jean Souchay

Observatoire de Paris/DANOF, URA 1125 du CNRS,
61 Avenue de l'Observatoire, F-75014, Paris
Jean.Souchay@obspm.fr

**Abstract.** This chapter can be divided into three sections. The first one presents two alternative theoretical ways for calculating the rotation of any celestial body, insisting on the difference between the free and force rotation modes. The first one, constructed by Woolard [109] is starting from the theorem of the conservation of the angular momentum, whereas the second one, elaborated by Kinosita [49, 51], is using canonical equations obtained from Hamiltonian theory. The second section is devoted to the rotation of the Earth, which is nowadays determined with a very good accuracy from very up-to-date techniques (VLBI, Lunar Laser ranging, GPS etc...). We insist in the very accurate modelization of the EOP (Earth Orientation Parameters) which leads to a detailed understanding of the interior of our planet as well as geophysical effects on its rotation. Therefore Earth's rotation study stands as an ideal starting point to modelize at best the rotation of the other planets, and this is the goal of the third section, where we explain sequentially the specificities of each of them. One of our purposes is to recapitulate past and recent studies leading to a better knowledge of the rotational history and of the rotational models of each planet.

## 6.1 Introduction

The term "rotation" applied to a celestial body is spontaneously understood for the specialist of astronomy as well as for the common people, as the global motion of this body "around itself". At first eye, after experiencing the rotation of a sphere, for instance a tennis ball, this concept seems easily transferable to the Earth or any of the eight planets. Nevertheless the subject becomes far more complicated when we want to modelize this rotation with a high accuracy, or when we want to explain the changes undergone by this rotation back from the origin of the solar system until now.

Moreover the eight bodies of the solar system named as "planets", as well as the huge population of asteroids present a very large variety of rotational modes. In order to characterize them it is necessary to elaborate a theoretical framework from which the rotation of the body can be computed as given some macroscopic characteristics of the body (mass, nominal value of the angular speed of rotation, moments of inertia, obliquity etc...). The first attempts were done by Newton in the Philosophiae Naturalis Mathematica Principa in 1687, who in particular explained in some extensive manner the origin of the phenomena of precession which is chracaterized by a slow motion of the Earth pole of rotation (and of figure!) in space, discovered by Hipparque at the IInd. Century B.C.

But Newton's theory contained a lot of approximations and errors which were mentioned by D'Alembert in his introduction of his treatise entitled "De la Nutation de l'Axe de la Terre dans le Systeème Newtonien" published in 1749. In this work the author achieved what we can consider as a masterpiece i.e. the first complete and precise theory of the rotation of the Earth, including some fundamental results as an exact determination of the Earth to Moon mass ratio, and the precise evaluation of the main nutation, which is a small oscillation of the axis of roation of the Earth to be added to the precession, discovered observationnaly by Bradley only one year before the edition of the treatise. To have an extended explanation of D'Alembert's treatise, the reader can refer to a recent book by Chapront and Souchay [21] in the frame of the edition of D'Alembert complete works.

In the third section we explain with details the particularities of the rotation of each planet. A general tendency consists in admitting that at the birth of the solar system all the planets were rotating fastly in the prograde direction, with an axis roughly perpendicular to the orbital plane. Presently most of the planets do not obey to these simple rules: Mercury is locked in a 3/2 resonance, which means that its period of rotation is equal to 2/3 its period of revolution around the Sun (88d). Venus has a very slow retrograd rotation (243d) which clearly looks as the testimony of large and dominant interaction with its heavy atmosphere. The Earth is gradually slowing down and its present obliquity ($23°27'$) is large enough to offer significant seasons, which is also the case of the next planet Mars. Uranus presents a very strange rotational feature: its axis of rotation is nearly lying on its orbital plane, so that at a given epoch a whole hemisphere is exposed at sunlight whereas the other one is in the dark. Only the three remaining big and gazeous planets Jupiter, Saturn and Neptune seem to have kept their initial angular momentum and to rotate according to primitive rules. In the two last decades, thanks to the dramatic evolution of numerical integration, it was possible to extrapolate in the past with more or less success the rotation modes of some of these planets, starting from their present knowledge. Therefore we can recognize two ways of studies of these rotations: one is to restrict to a very short time interval and trying to modelize with the best accuracy the rotation of the planets, as given the drastically improving precision of the physical and orbital parameters. The other one is to modelize qualitatively the variations of the rotation above for a very long interval of time (i.e. 100 million years or more) by adopting hypothesis concerning the evolution of the fundamental parameters.

## 6.2 Theories of the Motion of Rotation of a Celestial Body

### 6.2.1 Generalities

In the present subsection we will consider two different cases: the simple case of an axisymmetric ellipsoidal rigid body and the more complicated case of a triaxial ellipsoidal rigid body. In the axisymmetric case the body is symmetric with respect to an axis of revolution which is coinciding with its *axis of figure*, generally defined itself as the axis with *largest moment of inertia C*. The condition of symmetry leads obviously to the equality between the two moments of inertia perpendicular to the figure axis, whatever be their choice: $A = B$ with: $A < C$. In the case of a triaxial body the moments of inertia $A$ and $B$ about two axis perpendicular to the figure axis are defined such that $A < B < C$

Let us consider two reference frames. One, $\Re_0$ is supposed to be inertial which means in particular that the equations of motion as they will be preseneted and solved are valid in this reference frame. $\Re_0 = (O, X_0, Y_0, Z_0)$, where O stands for the center of mass of the celestial body, and $X_0, Y_0, Z_0$ is a triad. Then we consider a second reference frame, namely $\Re = (O, X, Y, Z)$ such that $(O, Z)$ is oriented towards the axis of maximum moment of inertia $C$, $(O, X)$ towards a point on the equator of figure supposed to give the origin of longitudes in the body, and $(O, Y)$ completes the triad. We adopt a convention according to which $(O, X)$ is directed towards the axis of minimum moment of inertia $A$.

$\Re$ is rigidly tied to the body, which means that each point at the surface and at the interior of the body has its coordinates $(X, Y, Z)$ remaining fix with respect to the time. The intersection of the surface of the body with a plane through $O$ perpendicular to the axis of figure is the *equator of figure*, crossed by both axes $(O, X)$ and $(O, Y)$.

The body is supposed to rotate with an angular speed $\Omega$, $\boldsymbol{\Omega}$ being the rotation vector of $\Re$ with respect to $\Re_0$. Determining the rotation of the body consists in positioning at any instant each axis of $\Re$ with respect to each axis of $R$. This can be done through a large number of parametrizations, one of the most common one being the famous Euler's parametrization.

The *axis of rotation* of the body is the defined at any instant as the set of points belonging to the body whose the speed with respect to $\Re_0$ is equal to 0. Let us call **A**, **B**, **C** respectively the intersections of $(O, X)$, $(O, Y)$, $(O, Z)$ with the surface of the body.

### 6.2.2 Two Kinds of Parametrization

In order to describe the rotation of any celestial body, as for instance the Earth, two kinds of parametrization are generally proposed: one with the help of the Eulerian set of three angle variables, the other one with the help of six Andoyer angle-action variables. In the following we summarize how theses two parametrization are constructed.

**The Parametrization with Eulerian Angles**

The Eulerian angles can be defined as follows (cf. Fig. 6.1):

- $\theta$, the *obliquity angle* (or simply the *obliquity*) represents the inclination of the equator of figure with respect to the fixed plane $(P_0) = (O, X_0, Y_0)$. It is generally reckoned positive from the fixed $(O, Z_0)$ axis to the moving $(O, Z)$ axis, in the direction for which $\theta < 90°$.
- $\psi$, the *precession angle* is the angle defined in the fixed plane $(O, X_0, Y_0)$, between a reference point $\gamma_0$ on $(P_0)$, and the line $(O, \gamma)$ along which the moving plane of the equator of figure crosses the plane $(P_0)$. For the sake of simplicity, we will choose $\gamma_0$ in coincidence with the axis $(O, X_0)$. $\psi$ is here reckoned positive eastward from $\gamma_0$ to the ascending node of the equator of figure $\gamma$ on $(P_0)$. Then it corresponds to the *longitude of the ascending node* of the equator of figure, reckoned from $\gamma_0$. We can write: $\psi = \gamma_0\gamma$.

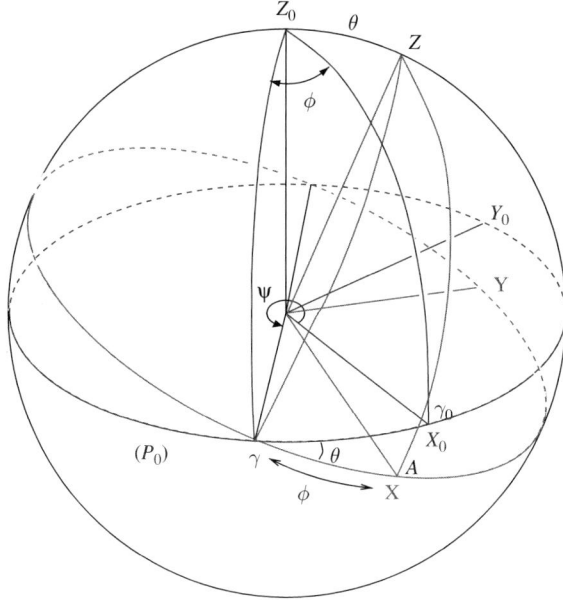

**Fig. 6.1.** Parametrization of the rotation of a solid body with Eulerian variables. $\Re_0 = (O, X_0, Y_0, Z_0)$ is a fixed inertial reference frame and $\Re = (O, X, Y, Z)$ is the body-fixed moving reference frame. The obliquity angle $\theta$ is the angle between the axes $(O, X_0)$ and $(O, X)$. The precession angle $\psi$ enables one to determine the position of the nodal line $(O, \gamma)$ between the body fixed equatortial plane $(O, X, Y)$ and the fixed plane $(P_0) = (O, X_0, Y_0)$. The angle of proper rotation $\phi$ enables one to determine the position of $(O, X)$ with respect to the nodal line

- $\phi$, sometimes called the *proper rotation* angle, is the angle $(O, \gamma), (O, \mathbf{A})$ in the plane of the equator of figure between $(O, X)$ (or $O\mathbf{A}$) and the moving line of nodes $(O, \gamma)$. It is reckoned positive eastward from the descending node $\gamma$ to $\mathbf{A}$.

## The Parametrization with Andoyer Variables

An alternative and other famous way to describe the rotation of the rigid body around its center of mass $O$ consists in using the Andoyer variables [1], fully used by Kinoshita [49, 51] in his theory of the rotation of a rigid body. In this case three fundamental definitions of axes are necessary (cf. Fig. 6.2): in addition to the fixed axis $(O, Z_0)$ and to the moving axis $(O, Z)$ rigidly tied to the body, we define the axis $(O, \mathbf{L})$ directed towards the angular momentum $\mathbf{L}$ of the body. Therefore we can define six action-angle variables as in the following: Let us call $(P_L)$ the plane perpendicular to the angular momentum vector, and $Q$ the ascending node of $(P_L)$ with respect to $(P_0)$. In a similar way, let us call $P$ the ascending node of the equator of figure $(O, X, Y)$ with

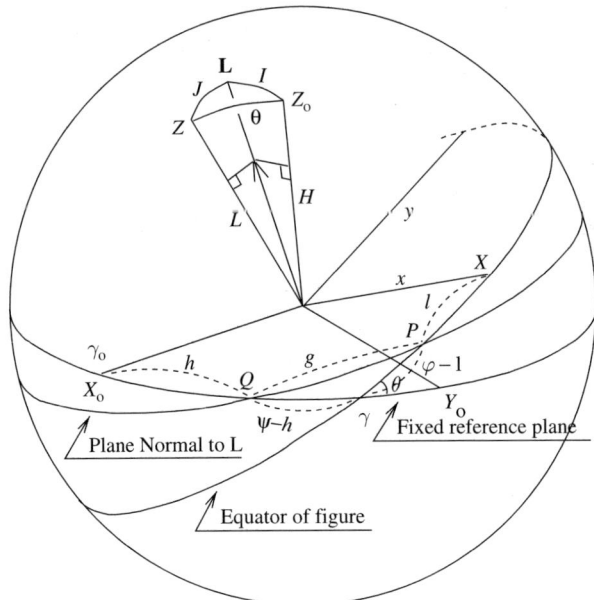

**Fig. 6.2.** Parametrization of the rotation of a solid body with Andoyer variables. The three action variables are $G$, the angular momentum vector, $H$ its projection on the fixed axis $(O, Z_0)$, and $L$ its projection on the body-fixed $(O, Z)$ axis. The three angle variables $l$, $g$ and $h$, precession of the angular momentum equator with respect to the fixed plane $(O, X_0, Y_0)$ enable one to represent the motion of the body-fixed frame $\Re = (O, X, Y, Z)$ with respect to $\Re_0 = (O, X_0, Y_0, Z_0)$.

respect to $(P_L)$. Finally let us recall that $\gamma$ is the ascending node of the equator of figure with respect to the inertial plane $(P_0)$. The Andoyer variables [1] are then defined as in the following.

*Action Variables*

- $G$ is the amplitude of the angular momentum vector $\mathbf{L}$.
- $L$ is the component of $\mathbf{L}$ along the $(O, Z)$ axis.
- $H$ is the component of $\mathbf{L}$ along the $(O, Z_0)$ axis.

Thus we can introduce to angles $I$ and $J$ giving respectively the inclination of the angular momentum vector with respect to the inertial axis $(O, Z_0)$, and with respect to the body-fixed figure axis $(O, Z)$:

$$L = G \cos J$$
$$H = G \cos I$$

*Angle Variables*

- $h$ is the angle measured along the reference plane $(P_0)$ between the fixed point $\gamma_0$ and the node $Q$. Notice that it is equivalent to the precession angle $\psi$ of the above set but for the equator of angular momentum instead of the equator of figure.
- $g$ is the angle along the equator of angular momentum $(P_L)$, between the two nodes $Q$ and $P$.
- $l$ is the angle along the equator of figure between the ascending node $P$ and the axis $(O, X)$ rigidly tied with the body.

### 6.2.3 Euler Kinematical Equations

The rotational motion of the rigid body about its center of mass with respect to the inertial reference frame $\Re_0$ is the combination of three components which enables one to make the transfer between the reference frames $\Re_0$ and $\Re$:

- a rotation at rate $\dot{\psi}$ around $Z_0$
- a rotation at rate $\dot{\theta}$ around the the moving line of nodes of the equator of figure on the fixed plane $(P_0)$
- a rotation at rate $\dot{\phi}$ around the axis of figure.

These three rotations compound into a resultant rotation $\boldsymbol{\omega}$ around the *instantaneous axis of rotation* passing through the center of mass $O$. Its amplitude $\omega$ is the angular speed of rotation around this axis. Notice that $\omega$ must not be confused with the rate of proper rotation $\dot{\phi}$, although theses two quantities are very closed one to the other in the majority of cases, characterized by: $\dot{\psi} \ll \dot{\phi}$ and: $\dot{\theta} \ll \dot{\phi}$.

The axis of rotation is more or less inclined to the axis of figure. Its position with respect to $\Re_0$ can be represented through the intermediary of two

Eulerian parameters $\psi_r$ and $\theta_r$ which are not represented in the figure for the sake of clarity. The position of the axis of rotation with respect to the Earth coordinate system (or reciprocally) $\Re$ is determined at any instant by the rectangular coordinates $\omega_1$, $\omega_2$ and $\omega_3$ of the velocity vector $\boldsymbol{\omega}$ with respect to $\Re$. These coordinates are linked to the derivatives $\dot{\psi}$, $\dot{\theta}$, $\dot{\phi}$ of the angles which determine the position of the body with respect to $\Re_0$, by kinematical equations, which are easily found by resolving each of the vectors $\dot{\psi}$, $\dot{\theta}$, $\dot{\phi}$ into its components along the $\Re$ axes.

$$\omega_1 = -\dot{\theta}\cos\phi - \dot{\psi}\sin\theta\sin\phi \tag{6.1}$$

$$\omega_2 = \dot{\theta}\sin\phi - \dot{\psi}\sin\theta\cos\phi \tag{6.2}$$

$$\omega_3 = \dot{\psi}\cos\theta + \dot{\phi} \tag{6.3}$$

Reciprocally

$$\dot{\psi}\sin\theta = -\omega_1\sin\phi - \omega_2\cos\phi \tag{6.4}$$

$$\dot{\theta} = -\omega_1\cos\phi + \omega_2\sin\phi \tag{6.5}$$

$$\dot{\phi} = \omega_3 - \dot{\psi}\cos\theta \tag{6.6}$$

These classical relations are known as *"Euler's kinematical equations"*.

### 6.2.4 Relations Between Eulerian Variables and Andoyer Variables

These relationships are easily derived from the spherical triangle $(P, Q, \gamma)$ where these three points have already been defined (cf. Fig 6.2). With the notations adopted above, we have [51]:

$$\cos\theta = \cos I \cos J - \sin I \sin J \cos g \tag{6.7}$$

$$\frac{\sin(\psi - h)}{\sin J} = \frac{\sin(\phi - l)}{\sin I} = \frac{\sin g}{\sin\theta} \tag{6.8}$$

For a large majority of celestial bodies the angle $J$ is very small, which means that the axis of angular momentum vector is very close to the axis of figure. In that case we can write, from these two last equations

$$\psi = h + \frac{J}{\sin I}\sin g + O(J^2) \tag{6.9}$$

$$\theta = I + J\cos g + O(J^2) \tag{6.10}$$

$$\phi = l + g - J\cot I \sin g + O(J^2) \tag{6.11}$$

Where $O(J^2)$ stands for a small quantity with order 2 with respect to $J$. Therefore we have simple expressions for the link between the set of Andoyer variables $(l, g, h)$ and the set of Eulerian variables $(\psi, \theta, \phi)$.

## 6.2.5 The Dynamical Equations with Lagrangian Formalism

The rotational motion of the body can be obtained by applying the laws of motion to each of the individual particles that compose the body, including all the internal and external forces that act on each particle, then by summing the equations of motion of all individual particles by suitable integration. This integration is considerably facilitated after the following remarks. At first the internal forces annul one another. At then the the external forces taken as a whole are equivalent to a single force applied at the center of mass $O$. In a similar way, the combination of all the external couples exerted on each individual particle is equivalent to a single couple with a moment resultant of the moments of the individual forces [109]

In the inertial reference frame $\Re_0$ the motion relative to the center of mass $O$ can be expressed through the equation of the angular momentum, according to which the derivative of the angular momentum vector is equal to the resultant of the external forces $\boldsymbol{M}$ exerted on the body:

$$\left(\frac{d\boldsymbol{L}}{dt}\right)_{\Re_0} = \boldsymbol{M} \tag{6.12}$$

When converting these equations to those of the motion of the body with resepect to the body-fixed reference frame $\Re$, we must take into account an additional component due to the fact that this last reference frame is not inertial. We thus obtain

$$\left(\frac{d\boldsymbol{L}}{dt}\right)_{\Re} + \omega \times \boldsymbol{L} = \boldsymbol{M} \tag{6.13}$$

Where the symbol X stands for the vectotial product. Notice that the difference between these two equations come from the fact that in the first case the components of the vectors involved are expressed in $\Re_0$ whereas in the second case they are expressed in $\Re$.

An alternative and more straightforward way to determine the equations of motion consist in applying Lagrange's equations. For this purpose, let us express the kinetic energy of the rotating body in the following classical form:

$$T = \frac{1}{2}(A\omega_1^2 + B\omega_2^2 + C\omega_3^2) \tag{6.14}$$

Then let us choose the Eulerian angles as generalized coordinates $q_i$ and consider the force function (potential) $U$ for the external forces. The Lagrangian function is:

$$L_a = U + T = U + \frac{1}{2}(A\omega_1^2 + B\omega_2^2 + C\omega_3^2) \tag{6.15}$$

The system being conservative, we can write the Lagrangian equations as

$$\frac{d}{dt}\left(\frac{\partial L_a}{\partial \dot{q}_i}\right) - \frac{\partial L_a}{\partial q_i} = 0 \tag{6.16}$$

When applied to the three Eulerian angles $\psi$, $\theta$ and $\phi$ as the generalized coordinates $q_i$, we find, after some developments

$$A\frac{d\omega_1}{dt} + (C - B)\omega_3\omega_2 = \frac{\sin\phi}{\sin\theta}\left(\cos\theta\frac{\partial U}{\partial \phi} - \frac{\partial U}{\partial \psi}\right) - \cos\phi\frac{\partial U}{\partial \theta} \tag{6.17}$$

$$B\frac{d\omega_2}{dt} - (C - A)\omega_3\omega_1 = \frac{\cos\phi}{\sin\theta}\left(\cos\theta\frac{\partial U}{\partial \phi} - \frac{\partial U}{\partial \psi}\right) + \sin\phi\frac{\partial U}{\partial \theta} \tag{6.18}$$

$$C\frac{d\omega_3}{dt} + (B - A)\omega_2\omega_1 = \frac{\partial U}{\partial \phi} \tag{6.19}$$

These equations are generally known as *Euler's dynamical equations*.

### 6.2.6 The Dynamical Equations with Hamiltonian Formalism

Another way to determine the dynamical equations for the rotation of a rigid body was developed in detail by Kinoshita [49, 51] starting from Andoyer variables [1]. Indeed, these variables constitute a canonical set which enables one to apply a perturbation theory based on canonical transformations.

$U$ having the same meaning as in the precedent section, the Hamiltonian $K$ for the rotation of the rigid body can be expressed in the following way:

$$K = T - U = \frac{1}{2}\left(\frac{\sin^2 l}{A} + \frac{\cos^2 l}{B}\right)(G^2 - L^2) + \frac{1}{2}\frac{L^2}{C} - U \tag{6.20}$$

Then the equations of motion are directly coming from the canonical equations:

$$\frac{d}{dt}(L, G, H) = -\frac{\partial K}{\partial(l, g, h)} \tag{6.21}$$

$$\frac{d}{dt}(l, g, h) = \frac{\partial K}{\partial(l, g, h)} \tag{6.22}$$

These equations enable one to determine after suitable integration the expressions of the six angle-action canonical variables once the potential function $U$ has been explicited. In the section dealing with the rotation of the Earth we will present all the details of the theoretical and practical developments starting from these equations, to determine the quantities related to the precession and to the nutation of the figure axis of the body considered as rigid.

### 6.2.7 The Torque-Free Rotational Motion in the Axi-symmetric Case

The *torque-free rotational motion* corresponds to the case for which no external force is exerted on the body: $U = 0$.

The particular *axi-symmetric case* is represented by the condition of equality of the two moments of inertia along the two axes located on the equator of figure: $A = B$. Lagrangian formalism is particularly well suited for this case and (6.17)–(6.19) can then be re-written, after the right-hand members of the first two equations reduce to zero (because $U = 0$):

$$\frac{d\omega_1}{dt} + \frac{(C-A)}{A}\omega_3\omega_2 = 0 \qquad (6.23)$$

$$\frac{d\omega_2}{dt} - \frac{(C-A)}{A}\omega_3\omega_1 = 0 \qquad (6.24)$$

$$\frac{d\omega_3}{dt} = 0 \qquad (6.25)$$

The third equation is equivalent to

$$\omega_3 = cte = \omega_0 \qquad (6.26)$$

When derivating once more the two first equations we get directly the trivial second-order differential equations:

$$\frac{d^2\omega_1}{dt^2} + \left(\frac{C-A}{A}\right)^2 \omega_1 = 0 \qquad \frac{d^2\omega_2}{dt^2} + \left(\frac{C-A}{A}\right)^2 \omega_2 = 0 \qquad (6.27)$$

By putting

$$\sigma = \frac{C-A}{A}\omega_0 \qquad (6.28)$$

We can write

$$\omega_1 = \nu_0 \cos(\sigma t - \kappa_0) \qquad \omega_2 = \nu_0 \sin(\sigma t - \kappa_0) \qquad (6.29)$$

In where which $\nu_0$ and $\kappa_0$ are constants of integration.

Remark that $(\omega_1, \omega_2)$ represent the bi-dimensional motion of the projection of the axis of rotation on a plane perpendicular to the axis of figure, generally called the *polar motion*.

The above equations show that this motion in the axi-symmetric case is circular ($\nu = cte$) and described uniformly with respect to the time. It is classically called the *Eulerian motion*. Its corresponding period $T_{\text{euler}}$ is given by

$$T_{\text{euler}} = \frac{2\pi}{\sigma} = \frac{2\pi}{\omega_0} \times \left(\frac{C-A}{C}\right) = T_0 \times \left(\frac{C-A}{C}\right) \qquad (6.30)$$

# 6 An Overview of the Rotations of Planets in the Solar System

Notice that when $\omega_1 \ll \omega_0$ and $\omega_2 \ll \omega_0$, then $T_0 = 2\pi/\omega_0$ is a good approximation of the period of rotation and in that case we can remark that the period of the Eulerian motion corresponds to the period of rotation multiplied by the coefficient $(C - A)/C$ representing the dynamical flattening of the planet.

By using the Eulerian kinematical equations (6.4)–(6.6) we can also obtain the equations of the Earth relative to the fixed coordinate frame $\mathfrak{R}_0$:

$$\dot\psi \sin\theta = -\omega_1 \sin\phi - \omega_2 \cos\phi = -\nu \sin(\phi + \sigma t - \kappa) \quad (6.31)$$

$$\dot\theta = -\omega_1 \cos\phi + \omega_2 \sin\phi = -\nu \cos(\phi + \sigma t - \kappa) \quad (6.32)$$

$$\dot\phi = \omega_3 - \dot\psi \cos\theta \quad (6.33)$$

## 6.2.8 The Torque-Free Rotational Motion in the Triaxial Case: Kinoshita's Theory

The general case for the free rotational motion of the body can be called *the trixial case*, which means that the three moments of inertia of the body are different from each other: $A < B < C$. This case, when treated in a rigorous manner, is much more complicated than the axi-symmetric case. A general and complete treatment of this case has been successfully done by Kinoshita [53], by using a Hamiltonian formalism. It involves elliptic functions and in the following we present the successive steps of the resolution according to Kinoshita's work.

The Hamiltonian for the torque-free motion ($U = 0$) with the use of 6Andoyer variables, according to (6.20), is

$$F = \frac{1}{2}(\sin^2 l/A + \cos^2 l/B)(G^2 - L^2) + \frac{1}{2}L^2/C \quad (6.34)$$

Notice that $g$, $h$ and $H$ are not present in the expression of the Hamiltonian. Therefore according to the canonical equations, this means that the amplitude of the angular momentum $G$, its projection $H$ on the fixed axis $(O, Z_0)$ as well as the precession angle $h$ are constant. Thus we obtain the logical result that the angular momentum vector is constant. The solution of the above equations can be found in standard textbooks of celestial mechanics (see for instance Whittaker [106]). Moreover the analytical solution with action-angle variables was done by Saadov [85] and Kinoshita [53], whereas Deprit [28] used the isoenergetic curves to discuss the features of the torque-free motion. Kinoshita [49] gave the expressions of the coordinates of the angular momentum vector $(L_x, L_y, L_z)$ with respect to the body fixed frame $\mathfrak{R}$, in the case of the short-axis mode (the body is rotating around the axis with maximum moment of inertia $C$ corresponding to the shortest axis):

$$L_x = G \sin J \sin l = G \sin j \, cnu \quad (6.35)$$

$$L_y = G \sin J \cos l = G\sqrt{(1+e)/(1-e)} \sin jsnu \tag{6.36}$$

$$L_x = G \cos J \sin l = G \cos j dnu \tag{6.37}$$

The variable $g$ is expressed as a function of the elliptic integral of third kind $\Pi$

$$g = \frac{G}{C}t + \tilde{g}_0 - G\left(\frac{1}{C} - \frac{1}{A}\right)\frac{\pi}{2Kn_{\tilde{l}}}\Pi(\phi, -2e/(1-e), k) \tag{6.38}$$

The parameter $e$ depends on the moments of inertia of the body in the following manner:

$$e = \frac{1}{2}\left(\frac{1}{B} - \frac{1}{A}\right)D \tag{6.39}$$

With

$$\frac{1}{D} = \frac{1}{C} - \frac{1}{2}\left(\frac{1}{A} + \frac{1}{B}\right) \tag{6.40}$$

The other definitions are

$$u = \frac{2K}{\pi}\left(\tilde{l} - \frac{\pi}{2}\right) \qquad \Phi = amu \tag{6.41}$$

Then the angle $g$ can be expanded in the following series:

$$g = \tilde{g} - 2\sum_{m=1}^{\infty}\frac{(-q)^m}{m(1-q^{2m})}\sinh(m\pi c/K)\sin 2m\tilde{l} \tag{6.42}$$

In this last equation, $c = F(\kappa, k')$ is the incomplete elliptic integral of the first kind with

$$\kappa = \frac{\pi}{2} - j \qquad k' = \sqrt{(1-k^2)} \tag{6.43}$$

$q$ is the Jacobi's nome. In the above equations the variables $\tilde{l}$ and $\tilde{g}$ are linear functions of the time:

$$\tilde{l} = n_{\tilde{l}}t + \frac{\pi}{2} \qquad \tilde{g} = n_{\tilde{g}}t + \tilde{g}_0 \tag{6.44}$$

$K$ is the complete elliptic integral of the first kind with modulus $k^2$, where $k^2$ is given by

$$k^2 = \frac{2e}{1-e}\tan^2 j \tag{6.45}$$

The mean motions of $\tilde{l}$ and $\tilde{g}$ are

$$n_{\tilde{l}} = \frac{\pi G}{2KD}\sqrt{1-e^2}\cos j \qquad n_{\tilde{g}} = \frac{1}{2}(1/A + 1/B)G - n_{\tilde{l}}\Lambda_0 + \frac{G}{D}\cos^2 \tilde{J} \tag{6.46}$$

where $\Lambda_0$ has the following expression:

$$\Lambda_0 = \frac{2}{\pi}[EF(\kappa, k') + KE(\kappa, k') - KF(\kappa, k')] \quad (6.47)$$

$E$ is the complete elliptic integral of the second kind, $F(\kappa, k')$ the incomplete elliptic integral of the first kind, $E(\kappa, k')$ the incomplete elliptic integral of the second kind. $\Lambda_0$ is the Heuman's Lambda function (see Byrd and Friedmann [16]). $\bar{J}$ is such that

$$\cos \bar{J} = \sqrt{1 - (1+e)\sin^2 j} = \cos j \sqrt{1 - e\tan^2 j} \quad (6.48)$$

From (6.35) we remark that $J = j$ when choosing $t_0 = 0$.
The epoch of time $t_0$ being chosen arbitrarily, so that we consider here that $\tilde{l} = \frac{\pi}{2}$ and $\tilde{g}_0 = 0$ at $t_0 = 0$, from (6.44).
According to (6.35) we have the relationships between $J$ and its minimum value $j$ which is a constant of integration of the problem:

$$\sin j \leq \sin J \leq \sqrt{(1+e)/(1-e)} \sin j \quad (6.49)$$

Two rotational modes can then be considered: the *short-axis mode*, which is represented by a rotation around the shortest axis $(O, Z)$, for which the moment of inertia $C$ is maximum, and the *long-axis mode* which is represented by a rotation around the longest axis, for which the moment of inertia A is minimum. It can be shown (for instance from isoenegetic diagramms) that the short-axis mode is secularly stable, whereas the long-axis one is secularly unstable. Therefore in the following, for the sake of simplicity, we will only consider the short-axis mode, which represents the overwhelming majority of rotational cases observed in the solar system.

The two kinds of rotation are separated by a circle called *separatrix* which can be determined by its angle $j_s$ from the body-fixed $(O, Z)$ axis:

$$\sin j_s = \sqrt{(1-e)/(1+e)} \quad (6.50)$$

Then it is easy to check from the two last equations that $j \leq j_s$.
As given the constants of integration of the problem $G$ (amplitude of the angular momentum) and $j$ (angle between the axis of figure and the axis of angular momentum at $t_0 = 0$) the solutions for the torque-free motion are obtained from (6.38) to (6.46). The function $F$ which represents in (6.34) the energy of the system can be expressed as a simple function of $G$ and $j$:

$$F = \frac{1}{2}\left(\frac{\sin^2 j}{A} + \frac{\cos^2 j}{C}\right) G^2 \quad (6.51)$$

The relation between the "true" value of the angular variable $l$ and its "mean value" $\tilde{l}$ can be deduced from the set of equations above:

$$\tan l = \sqrt{(1-e)/(1+e)} \tan\left[am\frac{2K}{\pi}\left(\tilde{l} - \frac{\pi}{2}\right) + \frac{\pi}{2}\right] \quad (6.52)$$

In Sect. 6.2.4 we have shown that Eulerian angles can be calculated easily from Andoyer variables starting from geometrical relations (6.7) and (6.8). Notice that in the case of the torque-free motion, these relations become still more simpler when the quasi inertial reference frame $\Re_0$ is chosen such that the axis $(O, Z_0)$ is coinciding with the angular momentum axis (which is fixed in space): in that case $\phi = g + h$, $\theta = J$ and $\psi = l$. Notice that in the presence of external torques this simplification is no more available, for the axis of angular momentum does not remain fixed in space.

In conclusion we remark that the solutions found for the variables involved in the torque-free motion are far from being straightforward, because they involve elliptic integrals, as for instance the elliptic integral of third kind in (6.38), whose the numerical determination is cumbersome. Nevertheless in some specific cases Kinoshita [49] transformed the solutions in the form of analytical expansions which give values as close as we wish to the exact ones, and which are suitable for hand calculations. These transformations are given in the following.

### 6.2.9 Solutions of the Torque-Free Rotational Motion for Small e Value

A large sample of celestial bodies with known sizes have a small $e$ parameter, which means that the relative difference between the moments of inertia with respect to the two axes perpendicular to the figure axis are much smaller than the realtive difference between one of them and the moment of inertia with respect to the figure axis:

$$\frac{B-A}{C} \ll \frac{C-A}{C} \tag{6.53}$$

This is particularly the case of the major planets, from Mercury to Neptune, and of the majority of large asteroids (Pluto, Ceres etc...) in the solar system. In that case, Kinoshita [49] showed that the solutions for the parameters $l, g$, $L$ and $J$ representing the orbital motion of the body can be expressed as analytical series, given by Kinoshita [49, 53]:

$$l = \tilde{l} - \frac{1}{4}(\alpha^2 + 1)e\sin 2\tilde{l} + \frac{1}{64}(\alpha^4 + 6\alpha^2 + 1)\sin 4\tilde{l} + O(e^3) \tag{6.54}$$

$$g = \tilde{g} + \frac{1}{2}\alpha e \sin 2\tilde{l} - \frac{1}{16}\alpha(\alpha^2 + 1)e^2 \sin 4\tilde{l} + O(e^3) \tag{6.55}$$

$$L = G\cos\bar{J}\left[1 - \frac{1}{8}(\alpha^4 - 1)e^2 - \frac{1}{2}(\alpha^2 - 1)e\cos 2\tilde{l} + \frac{1}{16}(\alpha^2 - 1)^2 e^2 \cos 4\tilde{l} + O(e^3)\right] \tag{6.56}$$

In these equations $\alpha$ is given by

$$\alpha = \frac{1}{\cos \bar{J}} = G/\tilde{L} \tag{6.57}$$

where $\tilde{L}$ is the action variable canonically conjugate to $\tilde{l}$ and one of the integral constants. $\tilde{J}$ is related to $j$ by

$$\cos\tilde{J} = \cos j\left[1 - \frac{1}{2}e\tan^2 j - \left(\frac{1}{4}\tan^2 j + \frac{3}{16}\tan^4 j\right)e^2\right] + O(e^3) \quad (6.58)$$

Which gives

$$\tilde{J} = j + \frac{1}{2}e\tan j + e^2\tan j\left(\frac{1}{8} + \frac{3}{16}\tan^2 j\right) + O(e^3) \quad (6.59)$$

And the mean motions $n_{\tilde{l}}$ and $n_{\tilde{g}}$ given by (6.46) can also be transformed to the simplified expressions

$$n_{\tilde{l}} = \frac{G}{D}\cos\tilde{J}\left[1 - \frac{1}{8}(\alpha^4 + 3)e^2\right] + O(e^4) \quad (6.60)$$

and:

$$n_{\tilde{g}} = \frac{1}{2}\left(\frac{1}{A} + \frac{1}{B}\right)G + \frac{G}{4D}(\alpha^2 + 1)e^2 + O(e^4) \quad (6.61)$$

Notice that only three integration constants ($G$, $j$ and $\tilde{g}_0$) are needed to solve the above equations at a given order. Once the value of $\tilde{J}$ is given by (6.59) it can be introduced in (6.60) and (6.61) to deduce the mean motions $n_{\tilde{l}}$ and $n_{\tilde{g}}$. Then the other equations give directly the values of $l$, $g$ and $J$ at any given date.

### 6.2.10 Solutions of the Torque-Free Rotational Motion for a Large $e$ Value

On the contrary of planets and large asteroids, the majority of the small bodies of the solar system (asteroids, comets) are characterized by a very irregular shape, and consequently by very different values of their principal moments of inertia $A$, $B$ and $C$. Thus from the definition (6.39) of the parameter $e$ giving the triaxility it is easy to understand that for these small bodies the value of $e$ is relatively very large. Kinoshita [53] calculated the $e$ value for ten bodies with well known aspect and found values between $e = 0.311$ (for the asteroid Telesto) and $e = 0.971$ (for Halley's comet).

In this case also expansions are available [53] but in terms of $j$ instead of $e$, because $j$ has a small value, according to the equation (6.50). By using the substitution

$$\tan l^\star = \sqrt{\frac{(1-e)}{(1+e)}}\tan\tilde{l} \quad (6.62)$$

we find

$$l = l^\star - \frac{1}{4}e\sqrt{\frac{(1+e)}{(1-e)}}j^2\frac{\sin 2\tilde{l}}{1 + e\cos 2\tilde{l}} + O(j^4) \quad (6.63)$$

$$g = \tilde{g} + \frac{G}{Dn_{\tilde{l}}}\left(-\sqrt{1-e^2}(l^* - \tilde{l})\right.$$
$$\left.+\frac{1}{4}(1+e)j^2\left[\frac{e\sin 2\tilde{l}}{1+e\cos 2\tilde{l}} + \frac{2}{\sqrt{1-e^2}}(l^* - \tilde{l})\right]\right) + O(j^4) \qquad (6.64)$$

$$L = G\left(1 - \frac{1}{2}j^2 - \frac{e}{1-e}j^2\cos^2\tilde{l}\right) + O(j^4) \qquad (6.65)$$

and

$$J = j\sqrt{1 + \frac{2e}{1-e}\cos^2\tilde{l}} + O(j^3) \qquad (6.66)$$

The mean motions are given by

$$n_{\tilde{l}} = \frac{G}{D}\sqrt{1-e^2}\left[1 - \frac{1}{2(1-e)}j^2\right] + O(j^4) \qquad (6.67)$$

and

$$n_{\tilde{g}} = \frac{1}{2}\left(\frac{1}{A} + \frac{1}{B}\right)G + \frac{G}{D}(1 - \sqrt{1-e^2}) \times \left[1 + \frac{1}{2}\sqrt{(1+e)/(1-e)}j^2\right] + O(j^4) \qquad (6.68)$$

Moreover it is possible to express the angle $\tilde{J}$ as a function of the constant of integration $j$:

$$\cos\tilde{J} = \cos j\left[1 - \frac{1}{2}(\sqrt{\frac{1+e}{1-e}} - 1)\tan^2 j\right] + O(j^4) \qquad (6.69)$$

or:

$$\tilde{J} = (\frac{1+e}{1-e})^{1/4} + O(j^3) \qquad (6.70)$$

The procedure to get the solutions of the motion are similar to the last section: $G$ and $j$ being integration constants, (6.67) and (6.68) enable one to evaluate the mean motions of $\tilde{l}$ and $\tilde{g}$. Then we obtain $l$, $g$ and $J$ at any date from (6.63) to (6.66).

### 6.2.11 The Forced Rotational Motion of the Figure Axis in the Axi-Symmetric Case

Here we consider the presence of external forces, which means that the potential function $U$ is not equal to zero. In that case it is possible to deduce the forced motion starting from the solutions for the torque free motion by the

## 6 An Overview of the Rotations of Planets in the Solar System

intermediary of the well known method of the variation of the parameters. In this section we summarize the method given by Woolard [109].

Let us take the solutions for the torque free motions given by (6.29). The method consists in replacing the constant terms $\nu_0$ by a function $\nu$

$$w_1 = \nu \cos(\sigma t - \kappa_0) = f \cos \sigma t + g \sin \sigma t \tag{6.71}$$

$$w_2 = \nu \sin(\sigma t - \kappa_0) = f \sin \sigma t - g \cos \sigma t \tag{6.72}$$

$$w_3 = cte \tag{6.73}$$

with

$$f = \nu \cos \kappa_0 \qquad g = \nu \sin \kappa_0 \tag{6.74}$$

Then it is easy to give the expressions of $f$ and $g$ after substituing (6.71) and (6.72) inside (6.17) and (6.18).

$$f = f_0 - \int \left[ \frac{\sin(\phi + \sigma t)}{A \sin \theta} \frac{\partial U}{\partial \psi} + \frac{\cos(\phi + \sigma t)}{A} \frac{\partial U}{\partial \theta} \right] dt = h \cos \kappa \tag{6.75}$$

$$g = g_0 + \int \left[ \frac{\cos(\phi + \sigma t)}{A \sin \theta} \frac{\partial U}{\partial \psi} - \frac{\sin(\phi + \sigma t)}{A} \frac{\partial U}{\partial \theta} \right] dt = h \sin \kappa \tag{6.76}$$

Then we apply (6.31), (6.32) and (6.33) to get the differential equations at the first order involving the eulerian variables:

$$\sin \theta \frac{d\psi}{dt} = -f \sin(\phi + \sigma t) + g \cos(\phi + \sigma t) = -h \sin(\phi + \sigma t - \kappa) \tag{6.77}$$

$$\frac{d\theta}{dt} = -f \cos(\phi + \sigma t) - g \sin(\phi + \sigma t) = -h \cos(\phi + \sigma t - \kappa) \tag{6.78}$$

$$\frac{d(\phi + \sigma t)}{dt} = \frac{C}{A} w_3 - \cos \theta \frac{d\psi}{dt} \tag{6.79}$$

Therefore after substitution of $f$ and $g$, given by (6.75) and (6.76), in (6.77) and (6.78) respectively, it is possible to determine by quadratures the solutions respectively for the sets $(w_1, w_2, w_3)$ and $(\theta, \phi, \psi)$. In particular after differenciating both members of the equations with respect to the time $t$ we get

$$\frac{d}{dt}\left(\sin \theta \frac{d\psi}{dt}\right) = (\dot{\phi} + \sigma)\frac{d\theta}{dt} + \frac{1}{A \sin \theta} \frac{\partial U}{\partial \psi} \tag{6.80}$$

$$\frac{d}{dt}\left(\frac{d\theta}{dt}\right) = -(\dot{\phi} + \sigma)\sin \theta \frac{d\psi}{dt} + \frac{1}{A} \frac{\partial U}{\partial \theta} \tag{6.81}$$

Substituing to $\dot\phi+\sigma$ its value obtained in (6.79) in the two equations above leads to the final expressions for the derivatives of the two Eulerian angles $\theta$ and $\psi$. Therefore it is possible to get the solutions of the rotational motion, after double integration.

## 6.3 Rotation of the Earth

For obvious reasons the Earth is by far the planet from which we have the most accuarte knowledge concerning the rotation. Thanks to modern techniques as VLBI (Very Long Baseline Interferometry), SLR (Satellite Laser Ranging), LLR (Lunar Laser Ranging) GPS (Global Positioning System) and DORIS (Orbitography by Radiopositioning Integrated on Satellite) the accuracy of Earth's rotation measurements has considerably improved in the two last decades, leading to the use of such units as the "µas" (microarcsecond) for the determination of the axis of figure in space (nutation, precession) the "mm" (millimeter) for the positioning of the axis of rotation with respect to an Earth's frame, and the "µs" for the variations of the length of day.

The observational data concerning the rotation of the Earth is collected by the IERS (International Earth Rotation and Reference Systems Service) which provides also publications and general informations on the Earth's orientation, the International Celestial Reference Frame (ICRF) and the International Terrestrial Reference Frame (ITRF). The reader can consult the IERS website to get further informations (http://www.iers.org).

The measurements of the Earth's rotation are under the form of time series of the so-called EOP (Earth Orientation Parameters) which are divided into three categories:

- $[\delta\psi, \delta\varepsilon]$ represent the O–C (observed–calculated) off-sets of the nutation in longitude $\Delta\psi$ and in obliquity $\Delta\varepsilon$ with respect to a reference model of precession-nutation. Recall that these two last parameters are those enabling one to determine the positioning of the axis of figure of the Earth in space.
- $[x, y]$ are the two parameters giving the position of the axis of rotation of the Earth with respect to a conventional point located near the pole of figure axis. This two dimensional parametrization is done perpendicularly to this axis.
- $UT1$ is the parameter representing the true rotation of the Earth around its rotational axis. Therefore if the rotational speed of rotation were constant, $UT1$ should vary linearly with respect to absolute time. Thus $\Delta UT1$ reflects the variations of the speed of rotation of the Earth with respect to a fixed value. Sometimes instead of $\Delta UT1$ specialists use $\Delta l.o.d$ (length of day) as the parameter to represent the irregularities of the rotational speed.

In the following we summarize the present state of knowledge concerning the EOP.

## 6.3.1 Earth's Precession

The Earth's precession is generally expressed by the conventional $p_A$ angle which represents the motion of the node of the mean equator of the Earth with respect to the moving mean ecliptic plane, i.e. the equinox. $p_A$ is called the *general precession in longitude*. For a very long time the astronomical community used a conventional value of $p_A$ adopted at the IAU 1976 General Assembly [62]. As shown by Williams [107] various observational techniques in the 1990s converged towards a correction to this conventional value : lunar laser ranging [108], VLBI [43], a combination of these two techniques [22], and also systematic proper motions in star catalogues [70]. The correction was close to $-0".3$/cy, so that the new value of $p_A$ is $p_A = 5028".7700$/cy instead of $p_A = 5029".0966$/cy. For recent developments about the improved modelizations of the precession model we can refer to a recent paper by Capitaine [17].

As shown by Williams [107] and by Souchay et al. [93] many contributions are included in the theoretical determination of $p_A$: these are second-order coupling effects, the influence of the gravitational harmonic $J_4$ of the Earth, the direct planetary influence on the Earth's flattening, the relativistic geodetic precession, the important effect due to the ecliptic motion (called planetary precession). The related amplitudes are gathered in Table 6.1.

The value of the precession $p_A$ is itself very important for the scaling of the coefficients of nutation, whose the value is directly dependent on the dynamical ellipticity of the Earth $H_d$, defined by

$$H_d = \left( \frac{2C - (A+B)}{2A} \right) \quad (6.82)$$

**Table 6.1.** Contributions to the general precession in longitude $p_A$ according to Souchay et al. [93]

| Contribution | Value ("/cy) |
|---|---|
| **First order** | |
| Lunisolar $p'_A$ | 5040.6445 |
| **Second order** | |
| Coupling effect | −0.3080 |
| r $J_4$ effect | 0.0026 |
| Direct planetary effect | 0.0321 |
| Geodetic precession | −1.9194 |
| Ecliptic motion (plan. precession) | −9.6865 |
| Total | −11.8745 |
| **First order + second order $p_A$** | 5028.7700 |

Indeed we have

$$p'_A = 3 \times H_d \times \left( (\frac{m_M}{m_M + m_E})(\frac{n_M^2}{\omega})M_0 + (\frac{m_S}{m_S + m_E + m_M})(\frac{n_S^2}{\omega})S_0 \right) \cos \varepsilon_A \quad (6.83)$$

Where $p'_A$ is the lunisolar contribution to the general precession in longitude $p_A$ obtained by substracting from $p_A$ all the second-order contributions mentioned above and shown in Table 6.1. $m_S$, $m_E$ and $m_M$ are respectively the masses of the Sun, the Earth and the Moon, $n_M$ and $n_E$ are respectively the mean motions of the Moon and of the Earth–Moon system barycenter, and $\omega$ is the angular speed of rotation of the Earth. $M_0$ and $S_0$ are the constant terms in the computations of the quantities $\frac{1}{2}(a/r)^3(1 - 3\sin^2 \beta)$ from the Moon and for the Sun. $\varepsilon_A$ is the value of the obliquity of the date which is the angle between the plane of the mean ecliptic and the mean equator of the date. As all the parameters of the equation above other than $H_d$ are known with a very good accuracy, a relative correction in $p'_A$ is accompanied by the same relative correction in the determination of $H_d$. Thus a corrected value of $H_d$ using the new determination of $p_A$ and $p'_A$ is [93]: $H_d = 0.0032737548$.

### 6.3.2 The Nutation of the Earth

Accurate knowledge of Earth nutation is of importance for studies in astrometry and geophysics as well as in space navigation. Such modern techniques as VLBI (Very Long Baseline Interferometry), Lunar Laser Ranging (LLR) and Global Positioning System (GPS) enable one to provide the capability to measure the bi-dimensional nutation components ($\Delta\psi$, $\Delta\varepsilon$) with a very high precision, i.e., at the level of the sub-milliarcsecond level for the global amount and of a few 10 µas for individual coefficients. One of the fundamental recent astronomical goals concerning the nutation consisted in constructing an analytical theory matchnig at best the observational data obtained from the various techniques above. A new nutation series called MHB2000 constructed by Mathews et al. [68] was adopted by the IAU General Assembly (2000) at Sydney. It was obtained by direct solution of the linearized equations, with best fit values adopted for various estimated Earth parameters, and leads to a considerably better fit to the nutation data than any of the earlier series based on geophysical theory.

Forced nutation is entirely due to the gravitational torques exerted by the external bodies (the Moon, the Sun but also the planets whose the effects can no more be neglected). In order to built the MHB2000 series, Mathews et al. [68] started from a proxy, so called rigid Earth nutation series representing the action of these torques on an hypothetical rigid Earth having the same physical characteristics (mass, harmonic coefficients of the geopotential etc...) as the real Earth. In order to be in acquaintance with the very accurate observational data, as that obtained from VLBI, the precision to what the amplitudes of the spectral coefficients of the rigid Earth series have been computed has increased by over 2 orders of magnitude in the past decade. Whereas the series of Kinoshita and Souchay [52] were calculated with a 0.01 mas truncation level, the REN-2000 series are tabulated with a 0.1 µas truncation level. This made it possible to estimate from the observed nutation dataset the amplitudes of a drastically larger number of nutation coefficients with uncertainties that are about an order of magnitude smaller than at the beginning of the 1990s.

## The SMART97, RDAN97 and REN-2000 Nutation Series for a Rigid Earth Model

The tables REN-2000 constructed by Souchay et al. [93] were obtained starting from an Hamiltonian theory set previously by Kinoshita [49, 51]. They include a lot of contributions which were neglected or partly considered in previous tables [51, 52]. These are the indirect planetary effects contained in the luni-solar perturbing function [90], as well as the direct planetary contributions [91], together with second-order terms coming from crossed nutation effects and spin-orbit coupling when considering the Earth–Moon system as a whole. Moreover they take into account the influence of the second-order geopotential of the Earth ($J_3, J_4$). Quasi-diurnal and quasi semi-diurnal components coming from the harmonics of degree 2, 3 and 4 [33, 34] of the geopotential were also added to the series. One of the by-products of the series REN-2000 was a complete expression of the general precession in longitude $\psi_A$ with its various contributions [90].

In parallel with these works, two concurrent theories of a rigid Earth nutation were set up: one has been developed by Bretagnon et al. [11, 12] and was called SMART-97. The other one has been constructed by Roosbeek and Dehant [84] and was called RDAN97. In order to validate the quality of the tables REN-2000, Souchay [92] compared the global nutation of the figure axis given by the corresponding tables with that given given by the above authors. He showed that no significant differences exist between these authors, at the level of a few µas, except for the 18,6 y component. The agreement is still better when he compared the global nutation with that obtained from a numerical integration. All these various checks justify that REN-2000, with its coefficients truncated at 0.1 µas, looks as valid theory of nutation for a rigid Earth model at the µas level. Notice that each of the three nutation series above (REN-2000, SMART97, RDAN97) gives the nutation for each of the three following axes: the axis of angular momentum, the axis of rotation and the axis of figure.

## The Computations of the REN-2000 Nutation Series for a Rigid Earth Model

The conventional series REN-2000 for a rigid Earth model [93] have been computed from the same theoretical developments as Kinoshita [51], but with a level of truncation of the coefficients of 0.1 µas instead of 0.1 mas in this last work. In this section we recall the principal steps of the computations of these series.

First of all the disturbing potential due to the disturbing bodies considered i.e. the Moon, the Sun and the planets, is represented by expansions in spherical harmonics, with the origin at the center of the Earth [51, 94]

$$U = U_1 + U_2 \tag{6.84}$$

With

$$U_1 = \frac{GM}{r^3}\left[\frac{2C - A - B}{2}P_2(\sin\delta) + \frac{A - B}{4}P_2^2(\sin\delta)\cos 2\alpha\right] \tag{6.85}$$

$$U_2 = \sum_{n=3}^{\infty} \frac{GMM_E a_E^n}{r^{n+1}}$$

$$\times \left[J_n P_n(\sin\delta) - \sum_{m=1}^{n} P_n^m(\sin\delta) \times (C_{nm}\cos m\alpha + S_{nm}\sin m\alpha)\right] \tag{6.86}$$

Where the perturbing body is considered as a point mass. $\alpha$ and $\delta$ are the geocentric longitude and latitude of the disturbing body with mass $M$, referred to the principal axes of the Earth.

Traditionally the precession and the nutation of the Earth are given by the two angles $\psi + \Delta\psi$ and $\varepsilon + \Delta\varepsilon$ referred to moving mean ecliptic of the date. Therefore the Hamiltonian of the system can be divided in three components: the Hamiltonian for the torque-free motionb $F$, the disturbing potential $U$ as written above, and an additional component $E$ which represents the motion of the ecliptic caused by planetary perturbations.

$$F = \frac{1}{2}\left(\frac{\sin^2 l}{A} + \frac{\cos^2 l}{B}\right)(G^2 - L^2) + \frac{1}{2C}L^2 \tag{6.87}$$

$$E = H(1 - \cos\pi_A)\frac{d\Pi_A}{dt}$$
$$+ G\sin I\left[\sin\pi_A \cos(h - \Pi_A)\frac{d\Pi_A}{dt} - \sin(h - \pi_A)\frac{d\Pi_A}{dt}\right] \tag{6.88}$$

Then the determination of the coefficients of nutation can be done from the dynamical canonical equations (6.21) and (6.22) after using an averaging method derived from the Lie transformations and proposed in 1966 by Hori [47] to separate the secular components (giving the precession) from the short-periodic ones (giving the nutation). In the problem, involving Andoyer variables, the angular variables $l$ and $g$ appearing in the motions of the perturbing body are considered as short-period arguments. To separate the short-period terms, Kinoshita [51] used a determining function $W$ and a new Hamiltonian $K^\star$ such that:

$$W = W_1 + W_2 + \ldots \ldots \qquad K^\star = K_0^\star + K_1^\star + K_2^\star + \ldots \qquad (6.89)$$

with

$$K_0^\star = F + E \qquad K_1^\star = U_1^{\text{sec}} \qquad W_1 = \int U_1^{\text{per}} dt$$

$$K_2^\star = U_2^{\text{sec}} + \frac{1}{2}\int \{U_1 + K_1^\star, W_1\}_{\text{per}} dt \quad W_2 = \int U_2^{\text{per}} dt + \frac{1}{2}\int \{U_1 + K_1^\star, W_1\}_{\text{per}} dt \qquad (6.90)$$

Here the symbols $\{\ldots\}$ stand for the Poisson brackets. Each part $U_i$ of the potential $U$ has been separated into two components, a secular one $U_i^{\text{sec}}$ and a periodic one $U_i^{\text{per}}$. The set of original canonical Andoyer variables $(l, g, h, L, G, H)$ is replaced by a new set of canonical variables $(l^\star, g^\star, h^\star, L^\star, G^\star, H^\star)$, and the nutation quantities represented by the symbol $\Delta$ are given by [51]:

$$\Delta(L, G, H) = (L, G, H) - (L^\star, G^\star, H^\star)$$

$$= -\frac{\partial(W_1 + W_2)}{\partial(l^\star, g^\star, h^\star)} - \frac{1}{2}\frac{\partial W_1}{\partial(l^\star, g^\star, h^\star)} \qquad (6.91)$$

$$\Delta(l, g, h) = (l, g, h) - (l^\star, g^\star, h^\star)$$

$$= \frac{\partial(W_1 + W_2)}{\partial(L^\star, G^\star, H^\star)} + \frac{1}{2}\frac{\partial W_1}{\partial(L^\star, G^\star, H^\star)} \qquad (6.92)$$

From the equations $L = G\cos J$ and $H = G\cos I$ (see Sect. 6.2.2) which enable one to make the link between the angles $I$ and $J$ in one side, and the action variables $L, G$ and $H$ in another side, it is easy to show that

$$\frac{\partial J}{\partial L} = -\frac{1}{G\sin J} \quad \frac{\partial J}{\partial G} = \frac{1}{G}\cot J \quad \frac{\partial I}{\partial H} = -\frac{1}{G\sin I} \quad \frac{\partial I}{\partial G} = \frac{1}{G}\cot I \qquad (6.93)$$

*Nutations of the Angular Momentum Axis*

The nutations in longitude and in obliquity for the angular momentum axis are thus represented respectively by $\Delta h$ and $\Delta I$, such that, at the first-order:

$$\Delta h = -\frac{1}{G\sin I}\frac{\partial W_1}{\partial I} \qquad (6.94)$$

$$\Delta I = \frac{1}{G}\left(-\frac{1}{\sin I}\Delta H + \cot I \Delta G\right) = \frac{1}{G}\left(\frac{1}{\sin I}\frac{\partial W_1}{\partial l} - \cot J\frac{\partial W_1}{\partial g}\right) \qquad (6.95)$$

Where $W_1$ is given by (6.90). Then solving (6.94) and (6.95) is possible once $W_1$ has been developed suitably. More precisely for a given perturbing body (Moon, Sun, planet) with geocentric coordinates $\alpha, \delta, r$:

$$W_1 = \int U_1^{\text{per}} dt = \int \frac{GM}{r^3}\left[\frac{2C-A-B}{2}P_2(\sin\delta) + \frac{A-B}{4}P_2^2(\sin\delta)\cos 2\alpha dt\right] \tag{6.96}$$

The difficulty in the calculation of such an expression is that semi-analytical ephemerides for the Moon [20] and of the Sun [10] give Fourier series of $r$ and of the orbital elements $\lambda$, $\beta$ referred to the ecliptic. Consequently, the coordinates $\delta$ and $\alpha$ of the perturbing body introduced in (6.96) have to be expressed in function of the ecliptic coordinates $\lambda$ and $\beta$. This is possible thanks to the modified Jacobi polynomials as was whown by Kinoshita [49] and Kinosita et al. [50]. Thus

$$\begin{aligned}
P_2(\sin\delta) &= \frac{1}{2}(3\cos^2 J - 1)\left[\frac{1}{2}(3\cos^2 I - 1)P_2^0(\sin\beta)\right.\\
&\quad \left. - \frac{1}{2}\sin 2I P_2^1(\sin\beta)\sin(\lambda - h) - \frac{1}{4}\sin^2 I P_2^2(\sin\beta)\cos(2\lambda - h)\right]\\
&\quad + \sin 2J\left[-\frac{3}{4}\sin 2I P_2^0(\sin\beta)\cos g\right.\\
&\quad - \sum_{\varepsilon=\pm 1}\frac{1}{4}(1+\varepsilon\cos I)(-1+2\varepsilon\cos I)P_2^1(\sin\beta)\sin(\lambda - h - \varepsilon g)\\
&\quad \left. - \sum_{\varepsilon=\pm 1}\frac{1}{8}\varepsilon\sin I(1+\varepsilon\cos I)P_2^2(\sin\beta)\cos(2\lambda - 2h - \varepsilon g t)\right]\\
&\quad + \sin^2 J\left[\frac{3}{4}\sin^2 I P_2^0(\sin\beta)\cos 2g\right.\\
&\quad + \frac{1}{4}\sum_{\varepsilon=\pm 1}\varepsilon\sin I(1+\cos I)P_2^1(\sin\beta)(\sin(\lambda - h - 2\varepsilon g))\\
&\quad \left. - \frac{1}{16}\sum_{\varepsilon=\pm 1}(1+\varepsilon\cos I)^2 P_2^2(\sin\beta)\cos 2(\lambda - h - \varepsilon g)\right] \tag{6.97}
\end{aligned}$$

Thus we can finally express analytically the nutations $\Delta\psi^{\text{AM}}$ and $\Delta\varepsilon^{\text{AM}}$ for the angular momentum axis at the first order after using the equations (6.94) and (6.95). We finally obtain [51]

$$\Delta\psi^{\text{AM}} = -\Delta h = -k\sum \frac{E_\nu}{N_\nu}\sin\Theta_\nu \tag{6.98}$$

$$\Delta\varepsilon^{\text{AM}} = -\Delta I = -\frac{k}{\sin I}\sum \varepsilon \frac{B_\nu}{N_\nu}\cos\Theta_\nu \tag{6.99}$$

Here the constant term $k$ is defined by

$$k = \frac{3GM}{a^3\omega}\frac{2C-A-B}{2C} \tag{6.100}$$

where $a$ being an arbitrary constant term generally chosen close to the mean distance of the perturbing body, $\omega$ representing the angular speed of sidereal rotation of the Earth, $M$ the mass of the perturbing body (Moon, Sun, planet).

The coefficients $B_\nu$ and $E_\nu$ are expressed as a function of coefficients $A_\nu^i$ ($i = 0, 1, 2$) in the following from

$$B_\nu = -\frac{1}{6}(3\cos^2 I - 1)A_\nu^0 - \frac{1}{2}A_\nu^1 - \frac{1}{4}A_\nu^2 \qquad (6.101)$$

$$E_\nu = [A_\nu^0 - \frac{1}{2}A_\nu^2]\cos I - \frac{\cos 2I}{\sin I}A_\nu^1 \qquad (6.102)$$

Where $A_\nu^i$ are directly coming from the computation in a form of Fourier series of the Legendre polynomials. They are given by

$$\frac{1}{2}\left(\frac{a}{r}\right)^3 (1 - 3\sin^2 \beta) = \sum_\nu A_\nu^0 \cos \Theta_\nu \qquad (6.103)$$

$$\left(\frac{a}{r}\right)^3 \sin \beta \cos \beta \sin \lambda = \sum_\nu A_\nu^1 \cos \Theta_\nu \qquad (6.104)$$

$$\left(\frac{a}{r}\right)^3 \cos^2 \beta \sin 2\lambda = \sum_\nu A_\nu^2 \cos \Theta_\nu \qquad (6.105)$$

From (6.98) to (6.102) we remark that at the first order the coefficients of the nutation can be calculated in a straightforward manner once we have the analytical developments at the right-hand side of (6.103), (6.104) and (6.105).

Such developments are possible by using analytical ephemerides as ELP 2000 computed by Chapront-Touzé and Chapront [20] for the Moon and VSOP computed by Bretagnon et al. [10] for the planets. In these ephemerides the geocentric ecliptic cordinates $\lambda$ and $\beta$ are expressed themselves as Fourier series, the argument $\Theta_\nu$ of each term being a linear combination of the Delaunay variables $l,l',F,D,\Omega$, of the longitude of the planets ($\lambda_i$, $i = 1, \ldots, 8$) or a mixture of these two sets of variables (for instance when computing the indirect planetary second-order effects on the Moon's orbital motion).

*Nutations for the Figure Axis*

The nutations for the figure axis are obtained from the nutations for the angular momentum axis by the intermediary of the *Oppolzer terms*, which are coming from the Euler kinematical equations (6.9) and (6.10) (in the present case $\theta$ is represented by the obliquity angle $\varepsilon$).

$$\Delta\psi - \Delta\psi^{A.M.} = -\Delta\left(\frac{J\sin g}{\sin I}\right) \qquad (6.106)$$

$$\Delta\varepsilon - \Delta\varepsilon^{A.M.} = -\Delta\left(J\cos g\right) \qquad (6.107)$$

The corresponding analytical developments lead to the following formula for the differences between the nutations for the figure axis and that for the angular momentum axis:

$$\Delta\psi_{\text{Oppolzer}} = \Delta\psi - \Delta\psi^{AM} \approx -\frac{k}{\sin I}\sum_{\nu}\sum_{\tau=\pm 1}\frac{\tau C_\nu(\tau)}{n_g - \tau N_\nu}\sin\Theta_\nu \qquad (6.108)$$

$$\Delta\varepsilon_{\Delta\varepsilon} - \Delta\varepsilon^{AM} \approx -k\sum_{\nu}\sum_{\tau=\pm 1}\frac{C_\nu(\tau)}{n_g - \tau N_\nu}\cos\Theta_\nu \qquad (6.109)$$

In these two expressions, $N_\nu$ represents the frequency associated with the argument $\theta_\nu$ ($N_\nu = 1/\dot{\theta}_\nu$).

## The Conventional MHB2000 Nutation for a Non-rigid Earth Model

The MHB2000 nutation series for a non rigid Earth model (Mathews et al. [68]) can be considered as the results of numerous efforts to understand various effects dealing with the geophysical response of our planet to the external torque exerted by the Moon, the Sun, and the planets. Taking as a basis the series of nutation for a rigid Earth model REN-2000 [93] the authors modified the dynamical equations developed in previous works [43, 66, 67] by including several contributions: the effects of mantle anelasticity, of ocean tides as well as the electromagnetic coupling produced between the fluid outer core (FOC) and the mantle on the one hand, and the solid inner core (SIC) on the other hand. Moreover the authors introduced non linear terms. Among the new characteristics of these theoretical developments one is the inclusion of complex and frequency dependent parameters in the equations. Another one is the construction of an empirical ocean loading and of current admittance functions for the purpose. At last, the nature of the dependence functions of nutation amplitudes on the basic parameters of the theory is fully discussed and some of these parameters are estimated through a least squares fit to observational data. Thanks to the very nice agreement between the nutation given by MHB2000 and the nutation data obtained through modern techniques as VLBI [42], MHB2000 was adopted as the conventional nutation series by the IAU General Assembly at Sydney in 2000. It replaces advantageously the nutation series of Wahr [104, 105] which was adopted as the IAU 1980 nutation series [87].

As a confirmation we can refer to the comparisons of MHB2000 nutation with 20 years of VLBI observational data [42]: analysis of these comparsions yields to estimates of the coefficients of the nutation series with standard deviations ranging from 5 µas for the terms with periods <400d to 38 µas for the longest period terms. The largest deviations between the VLBI estimates of the amplitudes of the terms in the nutation series are 56±38 µas (associated with the leading 18,6 y nutations.

Notice that a secular change of the obliquity of $-0.252 \pm 0.003$ mas/y was also detected which quite in agreemnt with theoretical estimations by Williams [107] confirmed by Souchay et al. [93]

In Tables 6.2 and 6.3 we present the leading coefficients of the MHB2000 nutations respectively in $\Delta\psi$ and $\Delta\varepsilon$. Recall that $\Delta\psi$ represents the oscillations of the true equinox with respect to the mean equinox along the ecliptic of the date, whereas $\Delta\varepsilon$ represent the oscillations of the obliquity, i.e. the angle between the ecliptic of the date and the true equator. After the leading term of lunar origin with argument $\Omega$ (longitude of the node of the Moon's orbit) the largest one is the semi-annual term of solar origin with argument $2F - 2D + 2\Omega = 2\lambda_E$, where $L_s$ is the mean longitude of the Earth. Only these two terms are larger than 1".

In Tables 6.4 and 6.5 we show, for the leading terms of nutation (respectively $\Delta\psi$ and $\Delta\varepsilon$), the difference between their value as calculated for the rigid Earth nutation model REN-2000 [93] and for the new non rigid earth conventional model MHB2000 [68]. These tables enable one to understand the effect of the non-rigidity which reaches roughly 74 mas (milliarcseconds) for the $\Omega$ component and 22 mas for the $2\lambda_E$ component.

### 6.3.3 The Polar Motion

This section is devoted to what is called the *polar motion* or *polhody* of the Earth: it represents the relative motion of the Earth's axis of rotation with respect to the Earth's axis of figure. We have demonstrated in Sect. 6.2.7

**Table 6.2.** Coefficients of nutation $\Delta\psi$ as given by the non rigid Earth theory MHB-2000 [68]

| Argument | Period | $\Delta\psi$ sine (as) | $\Delta\psi$ cosine (as) |
|---|---|---|---|
| $\Omega$ | 6798.384 | $-17.206416$ | 0.003338 |
| $2\Omega$ | 3399.192 | 0.207455 | $-0.001369$ |
| $l'$ | 365.260 | 0.147587 | 0.001181 |
| $2F - 2D + \Omega$ | 182.621 | $-1.317090$ | $-0.001369$ |
| $l' + 2F - 2D + 2\Omega$ | 121.749 | $-0.051682$ | $-0.000052$ |
| $l$ | 27.555 | 0.071115 | $-0.000087$ |
| $2F + 2\Omega$ | 13.661 | $-0.227641$ | 0.000279 |
| $2F + \Omega$ | 13.633 | $-0.038730$ | 0.000038 |
| $l + 2F + 2\Omega$ | 9.133 | $-0.030146$ | 0.000082 |

**Table 6.3.** Coefficients of nutation $\Delta\varepsilon$ as given by the non-rigid Earth theory MHB-2000 [67]

| Argument | Period | $\Delta\varepsilon$ sine (as) | $\Delta\varepsilon$ cosine (as) |
|---|---|---|---|
| $\Omega$ | 6798.384 | 0.001537 | 9.205233 |
| $2\Omega$ | 3399.192 | −0.000029 | −0.089749 |
| $l'$ | 365.260 | −0.000192 | 0.007387 |
| $2F - 2D + \Omega$ | 182.621 | −0.000458 | 0.573034 |
| $l' + 2F - 2D + 2\Omega$ | 121.749 | −0.000017 | 0.022438 |
| $2F + 2\Omega$ | 13.661 | 0.000137 | 0.097846 |
| $2F + \Omega$ | 13.633 | 0.000032 | 0.020073 |
| $l + 2F + 2\Omega$ | 9.133 | 0.000037 | 0.012902 |

that in the case of a rigid axi-symmetric body ($A = B$) the polar motion, in absence of external torque, is characterized by a circular motion of the axis of figure with respect to the axis of rotation (or reciprocally), called the Eulerian motion. The amplitude of this motion is a free parameter, but the frequency is directly dependent on the dynamical ellipticity of the planet: it is equal to corresponds to $(C - A/C)\omega$ where $\omega$ is the sidereal angular speed of rotation of the Earth. This leads to a period of 305.5 days, when taking for

**Table 6.4.** Coefficients of nutation $\Delta\psi$ with the largest difference between the non-rigid Earth model MHB-2000 [68] and the rigid Earth model REN-2000 [93]

| Argument | Period | $\Delta\psi$ sine (mas) | $\Delta\psi$ cosine (mas) | $\Delta\psi$ Total | Amplitude Ratio |
|---|---|---|---|---|---|
| $\Omega$ | 6798.384 | 74.1760 | 2.9560 | 74.2349 | 0.0043 |
| $2\Omega$ | 3399.192 | −1.5742 | −0.0757 | 1.5760 | 0.0076 |
| $l'$ | 365.260 | 22.0841 | 1.1817 | 22.1157 | 0.1498 |
| $2F - 2D + \Omega$ | 182.621 | −39.6154 | −1.3696 | 39.6391 | 0.0301 |
| $l' + 2F - 2D + 2\Omega$ | 121.749 | −1.7443 | −0.0524 | 1.7451 | 0.0338 |
| $-l + 2D$ | 31.812 | 0.7410 | −0.0168 | 0.7412 | 0.0472 |
| $l$ | 27.555 | 3.3482 | −0.0872 | 3.3493 | 0.0471 |
| $2F + 2\Omega$ | 13.661 | −6.1301 | 0.2796 | 6.1365 | 0.0270 |
| $2F + \Omega$ | 13.633 | −0.8772 | 0.0380 | 0.8780 | 0.0227 |
| $l + 2F + 2\Omega$ | 9.133 | −0.5633 | 0.0816 | 0.5692 | 0.0189 |

**Table 6.5.** Coefficients of nutation $\Delta\varepsilon$ with the largest difference between the non-rigid Earth model MHB-2000 [68] and the rigid Earth model REN-2000 [93]

| Argument | Period | $\Delta\varepsilon$ sine (mas) | $\Delta\varepsilon$ cosine (mas) | $\Delta\varepsilon$ Total | Amplitude Ratio |
|---|---|---|---|---|---|
| $\Omega$ | 6798.384 | 1.5408 | −22.6769 | 22.7292 | 0.0025 |
| $2\Omega$ | 3399.192 | −0.0323 | 0.5877 | 0.5886 | 0.0066 |
| $l'$ | 365.260 | −0.1924 | 7.5220 | 7.5245 | 1.0182 |
| $2F - 2D + \Omega$ | 182.621 | −0.4587 | 19.6988 | 19.7041 | 0.0344 |
| $l' + 2F - 2D + 2\Omega$ | 121.749 | −0.0174 | 0.8187 | 0.8189 | 0.0365 |
| $l$ | 27.555 | 0.0358 | 0.2975 | 0.2996 | 0.4433 |
| $2F + 2\Omega$ | 13.661 | 0.1374 | 2.9250 | 2.9282 | 0.0299 |
| $2F + \Omega$ | 13.633 | 0.0318 | 0.6603 | 0.6611 | 0.0329 |
| $l + 2F + 2\Omega$ | 9.133 | 0.0367 | 0.2928 | 0.2951 | 0.0229 |

instance the value of the dynamical ellipticity $H_d = 0.0032737548$ determined by Souchay et al. [93].

Actually, the real Earth's polar motion is far from following this simple mode, for two main reasons: first, the Earth is not a rigid body, which means that its elastic mantle, its liquid outer core and its solid inner core are considerably increasing the complexity of the motion. Second, the polar motion undergoes excitations coming both from atmospheric and oceanic parts. As a consequence the *Chandler wobble* which is an alternative terminology for the Earth's polar motion, contains variations in a wide range of frequencies from a secular drift to high-frequency variations. These two extreme parts can be easily removed from the data by appropriate smoothing or filtering [110]. The remaining and leading part consists of two major compomnents:

- The Chandler component which corresponds to the Eulerian motion but for a non rigid Earth. It is characterized by a broadened peak in the power spectrum at about 0.84 cpy (cycle per year), which means roughly a 435 d period.
- The annual component which is characterized by a narrow peak centered exactly at 1 cpy.

The two oscillations alternatively combine or oppose each other during a cycle of about six years. Nevertheless, the amplitude of each of the oscillations is not constant, so that the feature of the motion is not the same during different cycles [110]. Therefore long data series and a careful analysis are required to discriminate precisely the two oscillations. On the opposite short time records are also necessary when studying the changes with the time. Numerous studies starting from various approaches were devoted to solve this problem [48, 69, 72, 110]. One of the approaches consists in assuming that the annual component is relatively stable in phase and in amplitude with respect to the Chandler component Therefore it is removed from the data and the Chandler component only remains.

The usual features of the Chandler components can be summarized as follows [110]

- The amplitude undergoes large variations from 0".005 to 0".3 sometimes associated with a beat phenomeneon when considering a suitable time span.
- The phase is subject to important changes at it was the case between 1923 and 1940.
- Both the amplitude and the phase are subject to irregular variations.

The irregular nature of the polar motion appears clearly in Fig. 6.3 which corresponds to two years interval of time, from January 2005 to December 2006.

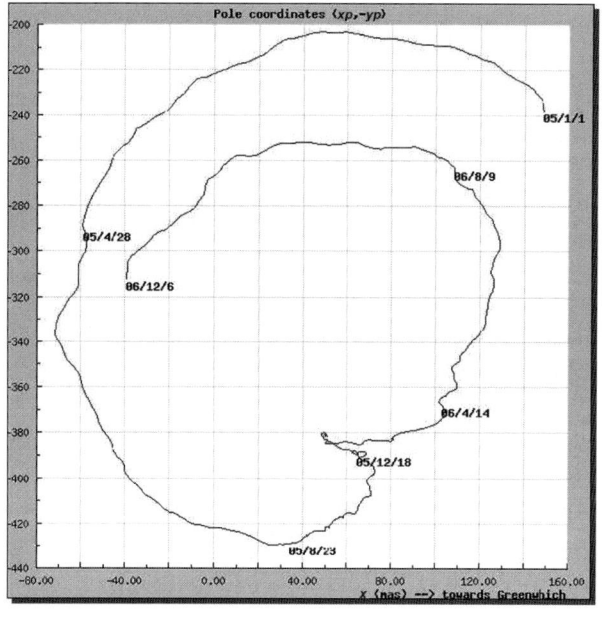

**Fig. 6.3.** The polhody or "polar motion" of the Earth between January 2005 and December 2006: the pole of rotation is describing a curve around a body-fixed conventional axis of figure. Careful analysis of this motion for a long time interval reveals the presence of two leading oscillations, one annual and the other with period around 435 d. This motion is subject to large amplitude variations related to excitation mechnism involving both oceans and the atmosphere of the Earth

## Empirical Models for the Chandler Motion

Different kinds of empirical models of the Chandler component were proposed.

- The "damped model" [14, 97] involves the resonant quality factor $Q$ and a complex excitation function $\chi$ such that $\chi = \chi_1 + \chi_2$ and

$$i\frac{m}{2\pi f_c} + m = \chi \tag{6.110}$$

Where $m = m_1 + im_2$ represents the coordinates of the Celestial Ephemeris Pole (CEP) which is a conventional definition of the pole very close to the axis of rotation. The quality factor $Q$ is given by

$$Q = \pi f_0 \alpha^{-1} = \pi f_0 t_d \tag{6.111}$$

where $t_d$ is the damping time. When adopting this model, the frequency response function is

$$\bar{f} = \frac{1 - \frac{f}{f_0} + \frac{(1}{(2Q)}^2 - if/(2Qf_0)}{(1-f/f_0)^2 + (1+(2Q))^2} \tag{6.112}$$

The two parameters $f_0$ and $Q$ characterizing the damping are then obtained by minimizing a functional, or by means of an appropriate time series analysis [71]. Notice that the value of the quality factor $Q$ of the Chandler wobble is varying very much according to the authors. Its most probable value is about $Q = 60$ but the estimates oscillate between $Q = 25$ and $Q = 1000$.

- The "time-variable model" is defined by the the same kind of equations as above but with a time-dependent Chandler frequency $f_c(t)$ such that

$$i\frac{m}{2\pi f_c(t)} + m = \chi \tag{6.113}$$

With the following expression for the two-dimensional polar motion $m$

$$m = A(t)exp\left(-i2\pi \int_0^t f_c(\tau)d\tau\right) \tag{6.114}$$

Various expressions can be proposed for the function $f_c(\tau)$ as the fluctuation between two values [88], or a relationship with the amplitude [98]. A supplementary difficulty consists in separating the fluctuations of $f(t)$ from those of $\chi(t)$.

- The "multi-components" model of the Chandler motion has been proposed after noticing that its amplitude for a long period of observations (more than one century) showed a non-random behaviour. It is represented by the following equation:

$$p(t) = \sum_i A_i e^{i2\pi f_i t} \qquad (6.115)$$

Thus, following this model, several authors as Dickman [31] and Chao [19] have proposed a set of frequencies with relative periods of about 406, 426, 435 and 452 days. Nevertheless we must be cautious concerning these estimations : for instance Vicente and Wilson [96] when estimating the Chandler periods from a set of various polar motion series showed that the period was significantly dependent on the time series analysed.

Although it has been under investigation for more than a century, the excitation mechanism has remained elusive [41]. It seems that the Chandler wobble is excited by a combination of atmospheric and oceanic processes, with the dominant excitation mechanism being ocean bottom fluctuations [15, 41, 81].

We can conclude that although polar motion has been observed for more than one century, and that its separation into two main components (an annual one and Chandler one) is well known, a the full understanding of its variations and the underlying process of excitation are not fully understood.

## 6.3.4 The Variations of the Length of Day (l.o.d.)

The demonstration of the expected fact that the Earth was subject to variations of its angular speed of rotation, and consequently of the *length of day*, often quoted as "l.o.d." (we adopt this terminology in the following, was obtained at the end of the nineteenth century, when S. Newcomb (1835–1909) observed some significant differences between the coordinates of the Moon given by the tables of Hansen, done in 1857, and those obtained from observations: these differences, which amounted to 8" in 1875, were due to the variations of the rotation of the Earth, which affected the recorded time of observations, when supposing the Earth as rotating at a constant rate.

In 1937 Stoyko showed directly the existence of a seasonal variation of the Earth's rotation, by detecting annual correlated irregularities of various high precision astronomical clocks located at different places of the Earth, which were in fact due to a correspondant irregularity of the Earth's angular speed of rotation. This was confirmed by a group of quarz clocks between 1943 and 1949.

Today we can distinguish three kinds of variations of the l.o.d., and consequently of UT1 (Universal Time 1) which is an angle directly associated with the Earth's rotation.

- A secular trend, which is characterized by a slow increase of the l.o.d. and which is mainly caused by the oceanic tides leading to a friction.

- Irregular variations at a typical decadal time scale, which have been subject to many investigations until it became certain that they are due to internal geophysical processes involving the core, the mantle and also their mutual boundary.
- The seasonal variations discovered by Stoyko and due to a coupling between the Earth and its atmosphere.

## Short-Periodic and Decennal Variations of the Earth's Rotation

In Fig. 6.4 we show the variations of the l.o.d. between 1986 and 2007 with respect to a nominal value of 86,400 s. We can easily observe the annual term with peak to peak amplitude of roughly 2.5 ms, and a decadal variation with about the same amplitude for the interval of time considered.

The atmospheric seasonal effects on Earth's rotation are generally investigated using the atmospheric angular momentum (AAM) approach: it assumes an angular momentum balance between the Earth and the atmosphere. More precisely, the global angular momentum of the [Earth + atmosphere] system is conserved, which means that to any change of the AAM corresponds an identical change of the solid Earth, with opposite sign. This last change

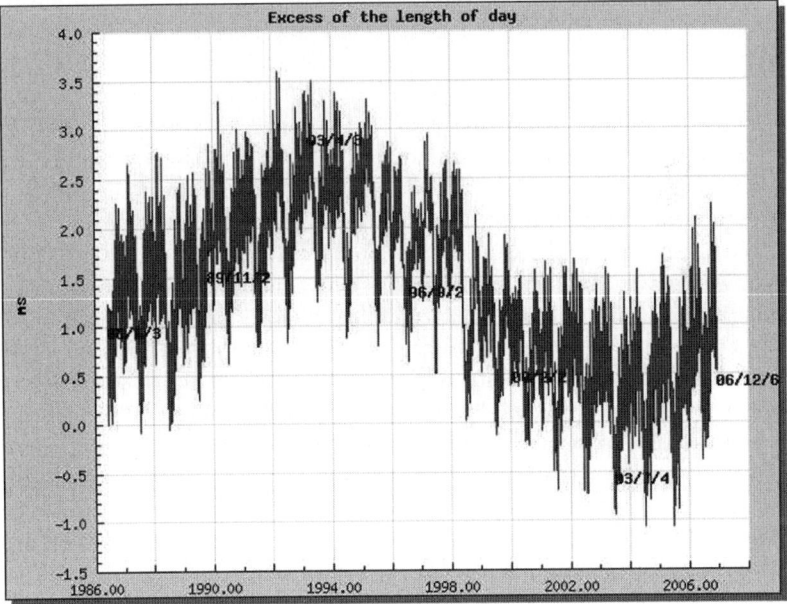

**Fig. 6.4.** Variations in millisecond of time of the length of day (l.o.d.) of the Earth with respect to a nominal value of 86,400 s, between 1986 and 2006. Remark the combination of a seasonal signal with annual frequency, due to a coupling mechanism with the atmosphere, and a decadal non regular signal probably due to friction mechanisms at the interior of the Earth

affects directly the angular speed of rotation of our planet. The AAM values are derived from meteorological data according to a procedure introduced by Munk and Mc. Donald [71] and described in detail by Barnes et al. [4]. An alternative method, called the *torque approach*, has been tested more recently by De Viron et al. [30]. It consists in considering only the Earth as the system and the atmospheric effect as an external torque acting on that system. The way of calculation of the pressure torque has been described by some authors as Dehant et al. [27] or Bizouard and Lambert [6].

The study of atmospheric effects on Earth's rotation has been recentl subject to numerous refined models including the influence of El Nino [65, 114]) and taking into account various meteorological centers as the US National Centers for Environmental Prediction (NCEP), the Japan Meteorological Agency (JMA), the European Centre for Medium Range Weather Forecasts (ECMWF) and the UK Meteorological Office (UKMO). Diurnal and semidiurnal angular momentum budgets of the atmosphere and its consequences for Earth's rotation have also been investigated [64].

### Long-Term Variations of the Earth's Rotation

It is now a well established fact that according to Milankovitch theory [5] ice ages are related to important variations of the orbital parameters of the Earth as well as the Earth's obliquity. Laskar et al. [58] investigated the stability of the Earth's orientation for various values of the initial obliquity. They showed that a chaotic zone exists between 60° and 90° of initial obliquity. On the contrary, in its present state, the Earth avoids this chaotic zone, with small variations of ±1.3° around a mean value of 23.3°. Nevertheless, these authors showed that the Moon had a real stabilizing effect. Indeed, if the Moon were not present, the torque exerted on the earth would be significantly lowered and large variations of the obliquity should result from a chaotic behaviour which might have affected dramatically the climate of our planet. As the solution of the orbital motion of the Earth is chaotic [59, 60] the study of the authors above is based on a qualitative understanding of the behaviour of this solution after analysis of the Fourier spectrum of the parameters involved in the solutions for the rotation. For instance the value of the mean precession speed of the Earth over 18 My (50".4712/y) is very close to the opposite of a small term due to the perturbations of Jupiter and of Saturn. This might lead to a secular resonance accompanied by variations of ±0.5° in obliquity and the presence of ice ages.

It is also well reckoned that tidal motions both of the solid Earth and of the oceans are affecting the Earth's rotation through small frictional effects which dissipate energy and thus slow down this rotation. The basic principles can be explained in a simple geometrical manner [56]. Indeed the total angular momentum of the Earth-Moon system is constant:

$$L_E + L_M = \text{const.} \tag{6.116}$$

where $L_E$ denotes the angular momentum of the Earth's rotation, and $L_M$ the angular momentum of the Moon's orbital motion. This can be written in the following way:

$$\frac{dL_M}{dt} = -\frac{dL_E}{dt} = \bar{W} \qquad (6.117)$$

$\bar{W}$ being a torque exerted between the Earth and the Moon. As can be seen in Fig. 6.5 (top) if we consider a perfectly elastic model of the Earth, the tidal bulge due to the gravitational action of the Moon is directed toward along the line joining the Earth's geocenter to the Moon's geocenter. In the reality, as shown in Fig. 6.5 (bottom), because of the inelastic response of our planet, there is a phase lag $\delta$ between the axis of the bulge and the line joining the geocenters. Thus because of the dissymetry of the actions of the half bulges on the Moon, a tangential force $F_t$ is exerted on the Moon which leads to the torque:

$$\bar{W} = d \times F - t \qquad (6.118)$$

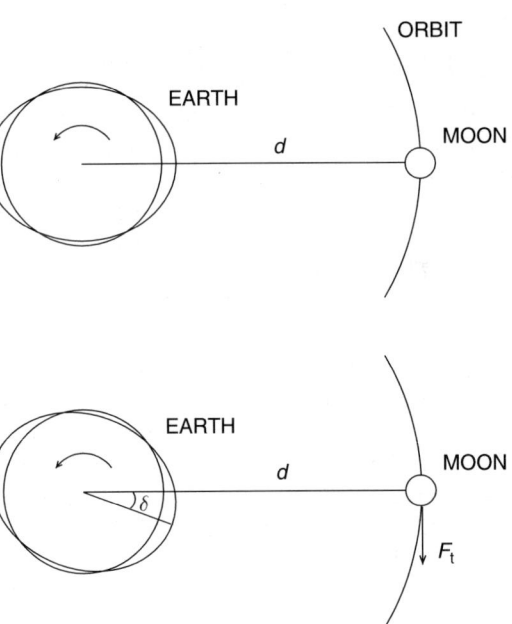

**Fig. 6.5.** Explanation for the secular deceleration of the rotation of the Earth: if the response of the Earth to the differential attraction of the Moon were purely elastic, the tidal bulge corresponding to terrestrial tides or oceanic tides would be oriented along the line joining the Earth and the Moon barycenters (*top* figure). In fact a phase lag $\varepsilon$ exists, which is represented by the angle between this line and the bulge symmetry axis (*bottom* figure). This lag creates a friction characterized by an energetic loss, which at its turn leads to a slowing down of the rotational speed

$d$ being the Earth–Moon distance After several calculations involving the sectorial harmonic of the form $P_{22}(\cos\theta)\cos 2\lambda$ where $\theta$ and $\lambda$ are respectively the co-latitude and the longitude of the Moon it can be shown [56] that $\bar{W}$ can be approximated with:

$$\bar{W} = \frac{3}{2}kGM_{\text{Moon}}^2 \frac{R^5}{d^6}\sin 2\delta \qquad (6.119)$$

Here $k$ is the secular Love number, $R$ the Earth's radius, and $M_{\text{Moon}}$ the Moon's mass. Reducing the angular momentum to its $z$ component $C\omega$ and using (6.117) leads to:

$$\dot{\omega} = -\frac{3}{2}\frac{GM_{\text{Moon}}^2}{C}\frac{R^5}{d^6}\sin 2\delta \qquad (6.120)$$

This equation enables one to explain the secular deceleration of the rotational speed of the Earth, which is accompanied by a secular deceleration of the Moon in its orbital motion. This last phenomenon is not considered in the topic of this paper. Rough estimations [56] lead to the estimation of $\dot{\omega}/\omega \approx -2.5\times 10^{-8}$/cy. This corresponds to a secular increase of the length of day by 2 ms per century [56].

## 6.4 The Rotation of the Planets

In the following we study one by one the rotational chareteristics of each planet.

### 6.4.1 The Rotation of Mercury: A Singular Case of 3/2 Spin-Orbit Resonance

#### History of a Discovery

From the end of the nineteenth century until 1965, all the astronomers were in agreement on the value of the period of rotation of Mercury: they believed it was exactly equal to its period of revolution around the Sun, that is to say 87.96 d. In other terms, as it is the case for the Moon with respect to the Earth, Mercury would have been in a 1/1 spin-orbit resonance, which means that the same face should be oriented towards the Earth. Then to everybody's great suprise a short paper published on 19 June 1965 in *Nature* by Pettengill and Dyce [79] gave the results of direct measurements of the Mercury's rotational speed. The technique consisted in analysing the spectrum of short radar echos reflected at various points of the surface, by taking into account for the Doppler effect. The authors arrived to an outstanding conclusion: the planet shows a direct rotation with a sidereal period of 59±5 days. Moreover the axis of rotation is very close to the axis of the orbit, the obliquity thus having a value close to 0° [24]. A lot of various determinations during three years, noticably by combining old visual and photographic observations lead to the precise value of the period of rotation of 59.65 ±0.01 days [32] which corresponds exactly to two-thirds. the period of revolution of the planet.

At the same time, specialists of celestial mechanics tried to justify this strange commensurability. As soon as at the end of 1965, i.e. the same year as the discovery of the real rotation of the planet, Colombo [25] as well as Liu and O'Keefe [63] explain the motion by the intermediary of a resonance phenomena.

In the following we detail the mechanism leading to the 3/2 spin-orbit resonance, after doing some approximations which do not influence the demonstration. Then we present the main results of theoretical developments done in the last decades concerning the accurate determination of the rotational regime based on both the Cassini laws and the librations.

## Explanations of the 3/2 Spin-Orbit Resonance

### Hypothesis and Notations

Let us consider the center of mass of Mercury $M$ describing an elliptic trajectory in a prograde direction, with excentricity $e$, around the Sun $S$ (Fig. 6.6). Mercury axis of figure is supposed to be coincident with its axis of rotation and perpendicular to the orbital plane, so that we treat the problem in the following reference frame $(S, \boldsymbol{I}, \boldsymbol{J}, \boldsymbol{K})$. $P$ being the perihely, we have the following definitions, with bold notations for the vectors:

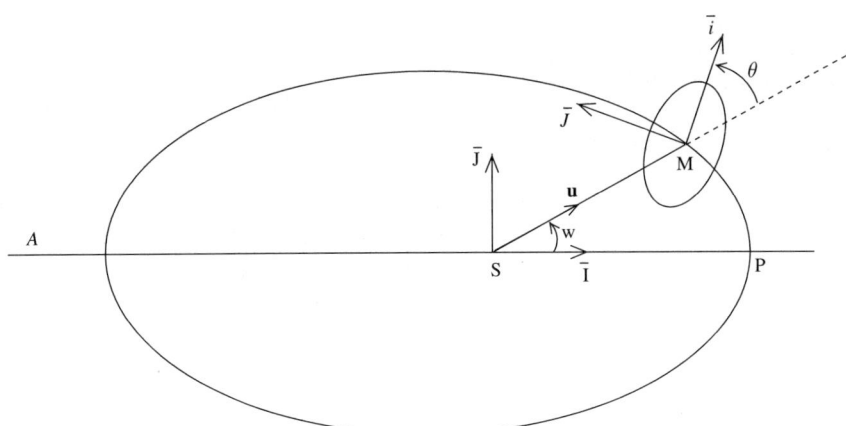

**Fig. 6.6.** Geometrical parametrization of the rotation of Mercury. The axis $(O, \boldsymbol{i})$ is directed toward the axis of minimum moment of inertia $A$. The axis $(O, \boldsymbol{k})$ perpendicular to the plane of the figure is oriented towards the axis of maximum moment of inertia $C$. $\theta$ is the libration angle with respect to the position of equilibrium for which $\bar{p}$ and $\boldsymbol{i}$ are aligned

$$I = \frac{SP}{SP} \qquad (I, J) = \frac{\pi}{2} \qquad \frac{SM}{SM} = p \qquad r = SM \qquad (6.121)$$

$K$ completing the triad. In the Fig. 6.6 we consider at first approximation that the axis of rotation of Mercury, considered as a rigid body, is perpendicular to the plane of figure which corresponds to the orbital plane of the planet. Moreover we define a second reference frame $(M, i, j, k)$, where the three vectors $i, j, k$ are aligned along the three principal moments of inertia $A, B, C$, with $A < B < C$. We admit also the hypothesis that the axis of figure $k \approx K$ which comes from the condition that the axis of figure and the axis of rotation coincide ($SM = ru$). Then we define an angle $\theta$ such that: $\theta = (u, i)$ where $u$ is the unit vector along $SM$ With these hypothesis the position of Mercury around its center of mass is completely done through the determination of $\theta$.

*Calculation of the Moment and Formulation of the Problem*

Each particle $P_i$ of Mercury with mass $m_i$ is undergoing a gravitational force $F_i$ from the Sun given by

$$F_i = -kM_S m_i \frac{SP_i}{SP_i^3} \qquad (6.122)$$

Where $K$ is the gravitational constant and $M_S$ is the mass of the Sun. Then the resulting moment of all these individual forces with respect to Mercury's center of mass is given by

$$\mu = \sum_i MP_i \wedge F_i = -kM_S \sum m_i \frac{MP_i \wedge SP_i}{SP_i^3} \qquad (6.123)$$

Where

$$MP_i \wedge SP_i = MP_i \wedge [SM + MP_i] = MP_i \wedge SM = rMP_i \wedge u \qquad (6.124)$$

Then we can use the classical transformation

$$SP_i^{-3} = [(SM + MM_i)^2]^{-\frac{3}{2}} = [r^2 + 2ruMP_i + MP_i^2]^{-\frac{3}{2}} \qquad (6.125)$$

$$(SP_i)^{-3} = r^{-3}\left[1 + 2\frac{MP_i.u}{r} + \frac{MP_i^2}{r^2}\right]^{-3} \qquad (6.126)$$

Then the distances $MP_i$ being very small in comparison with $r$, we can write

$$SP_i^{-3} \approx \frac{1}{r^3}\left(1 - 3\frac{MP_i.u}{r}\right) \qquad (6.127)$$

Then the moment of the external forces exerted by the Sun can be written in the following way:

$$\mu = 3\frac{kM_S}{r^3} \sum_i m_i(MP_i.u)(MP_i \wedge u) \qquad (6.128)$$

But according to our definitions (cf. Fig. 6.6):

$$\boldsymbol{u} = \cos\theta \boldsymbol{i} - \sin\theta \boldsymbol{j} \tag{6.129}$$

Substituing this expression in the precedent equation leads finally to

$$\boldsymbol{\mu} = -\frac{3kM_S}{r^3}(B-A)\sin\theta\cos\theta \boldsymbol{k} \tag{6.130}$$

Let us call now $\delta$ the angle enabling one to give the position of the body-fixed axis $(O, \boldsymbol{i})$ with respect to the inertial frame (cf. Fig. 6.7).

$$\delta = (O, \boldsymbol{I}), (O, \boldsymbol{J}) = v + \theta \tag{6.131}$$

As the axis of rotation is along $\boldsymbol{k} = \boldsymbol{K}$ the angular momentum can be written as $C(d\delta/dt)\boldsymbol{k}$ and the theorem of the angular momentum gives:

$$C\frac{d^2\delta}{dt^2} + \frac{3K}{r^3}(B-A)\sin\theta\cos\theta = 0 \tag{6.132}$$

With $K = kM_S$ Then determining the rotation of Mercury around its principal axis $\boldsymbol{k}$ consists in solving this equation which involves $\theta$, included explicitely but also implicitely through the varaible $\Phi$. Solving this equation is not straightforward and in the following subsection we present an elegant way of resolution based on Hamiltonian formalism.

*An Equation of Motion Based on Hamiltonian Formalism*

To solve the equations abobe we show in this section an elegant way set up by Liu and O'Keefe [63] and described in detail by Pacault [73] It is based on

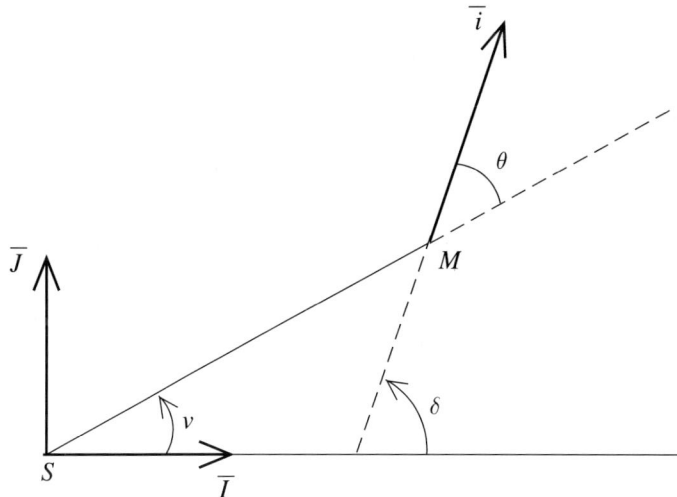

**Fig. 6.7.** Geometrical parametrization of the rotation of Mercury showing the angles involved in the equations

Hamiltonian developments with canonical equations. Let us call $\eta$ the constant representing the triaxiality of the planet Mercury:

$$\eta = 3\frac{B-A}{C} \tag{6.133}$$

Then we replace $\theta$ by its doubled value $x = 2\theta$, so that the precedent equation is transformed in

$$\frac{d^2\delta}{dt^2} + \frac{K\eta}{2r^3}\sin x = 0 \tag{6.134}$$

By taking into account the fact that $\delta = v + \theta = v + x/2$ this leads to

$$\frac{d^2x}{dt^2} + \frac{K\eta}{r^3}\sin x + 2\frac{d^2v}{dt^2} = 0 \tag{6.135}$$

It is easy to show that this last equation is equivalent to the canonical system of equations with two variables $x$ and $y$ [73]

$$\frac{dx}{dt} = \frac{\partial H}{\partial y} \qquad \frac{dy}{dt} = -\frac{\partial H}{\partial x} \tag{6.136}$$

$H$ being a function of $x$, $y$, and also explicitely of the time $t$:

$$H(x,y,t) = -\frac{K\eta}{r^3}\cos x + \frac{y^2}{2} - 2\frac{dv}{dt}y \tag{6.137}$$

By substituing true anomaly $v$ to the time variable $t$ we replace the original Hamiltonian $H$ with $H' = H(x,y,v)$: Then $x$ and $y$ verify

$$\frac{dx}{dt} = \frac{dx}{dv}\cdot\frac{dv}{dt} = \frac{\partial H'}{\partial y} \qquad \frac{dy}{dt} = \frac{dy}{dv}\cdot\frac{dv}{dt} = -\frac{\partial H'}{\partial x} \tag{6.138}$$

Which can still be written as

$$\frac{dx}{dv} = \frac{\partial}{\partial y}\left(H'\frac{dt}{dv}\right) \qquad \frac{dy}{dv} = -\frac{\partial}{\partial x}\left(H'\frac{dt}{dv}\right) \tag{6.139}$$

Then the substitution above with $v$ as new independent variable leads to the new Hamiltonian:

$$H'' = H'\frac{dt}{dv} = H'\frac{r^2}{C} = -\frac{K\eta}{Cr}\cos x + \frac{r^2y^2}{2C} - 2y \tag{6.140}$$

By applying the basic equations for the eulerian motion:

$$r = \frac{p}{1+e\cos v} \qquad C^2 = pK \tag{6.141}$$

with $K = kM_S$. $H''$ can be written as

$$H" = -\frac{CK}{p^2}(1+e\cos v)\cos x + \frac{p^2}{2C}\frac{y^2}{(1+e\cos v)^2} - 2y \tag{6.142}$$

From this new expression of the Hamiltonian, we can deduce the equation, found by Liu and O'Keefe [63]

$$\frac{d^2 x}{dv^2} - \frac{2e\sin v}{1+e\cos v}\left(\frac{dx}{dv}+2\right) + \eta\frac{\sin x}{1+e\cos v} = 0 \tag{6.143}$$

This equation has an interesting form: indeed, it contains two dimensionless parameters which are the orbit excentricity $e$ and the coefficient $\eta$ representing the triaxiality of the planet Mercury. The constant of gravitation $k$ has disappeared, as well as the dimension of the orbit represented by the semi-major axis $a$ or the ellipse parameter $p$. The Hamiltonian represented by the (6.142) can still be replaced by a simplified one:

$$\bar{H} = -\eta(1+e\cos v)\cos x + \frac{y^2}{(1+e\cos v)^2} - 2y \tag{6.144}$$

In conclusion, the problem consists in solving the set of two canonical equations, i.e. a system of two differential equations at the first order:

$$\frac{dx}{dt} = \frac{\partial \bar{H}}{\partial y} = -\frac{y}{(1+e\cos v)^2} - 2 \tag{6.145}$$

$$\frac{dy}{dt} = -\frac{\partial \bar{H}}{\partial x} = -\eta(1+e\cos v)\sin x \tag{6.146}$$

Then it is possible to study the periodic solutions of the Liu and O'Keefe equation or of the two equivalent equations above. A periodic solution, in its extended meaning, characterizes any solution such that

$$x(v+2\pi) = x(v) + 2n\pi \tag{6.147}$$

Then the solution will be called *periodic of order n*. The solution will be called *strictly periodic* in the case $n = 0$.

It is easy to understand the physical signification of periodic solution of order $n$. The angle $\delta$ giving the orientation of the vector $\boldsymbol{i}$ with respect to the inertial direction $\boldsymbol{I}$. We have

$$\delta = v + \theta = v + \frac{x}{2} \tag{6.148}$$

$x$ will be periodic of order $n$ at the condition that

$$\delta(v+2\pi) = v + 2\pi + \frac{x(v+2\pi)}{2} = \delta(v) + (n+2)\pi \tag{6.149}$$

Therefore the angle $\delta$ is periodic with order $(n+2)/2$, which means that the planet is undergoing $(n+2)/2$ rotations during one revolution. $T_{\text{orbit}}$ being the period of the sidereal revolution of the planet and $T_r$ being its period of rotation around its center of mass, we have

$$T_r = \frac{2T_{\text{orbit}}}{n+2} \qquad (6.150)$$

Thus $n = 0$ corresponds to a strict periodic solution (order 0) still called *synchronous rotation* or $1/1$ resonance. This is the case of the Moon around the Sun, and of the Galilean satellites around Jupiter. In the case of Mercury, $n = 1$.

*Existence of Stable Periodic Solutions*

As explained in detail in the report by Pacault [73] it is possible to study the periodic solutions of the equation of Liu and O'Keefe by using several theorems found in H. Poincaré famous *Méthodes nouvelles de la mécanique céleste* [80] The demonstrations are rather long and cannot be included in the scope of this chapter. Here we only recall the methodology. It consists in solving a linear differential system of equations as (6.143) whose the second member depends on the parameter $\eta$. Once the solutions for $\eta = 0$ are found, one important theorem from Poincaré shows that for small values of $\eta$ it is possible to develop analytically $x$ and $y$ as a function of $v$ and $\eta$. Then the calculation of characteristic exponents enables one to study the stability of the periodic solutions. It can be shown [73] that if we consider the excentricity of the orbit $e$ and the order of the periodicity $n$ as fixed, then a periodic solution of order $n$ of the equations exists.

## The Cassini's Laws

Although being close to 0, Mercury's obliquity is in fact locked in a state governed by Cassini's laws. Cassini's laws, which describe the rotation and precession of the Moon, have been generalized to apply to Mercury, as was shown by Colombo [25] Peale [77, 78] and Goldreich and Peale [40] just after the discovery of the planet 3/2 resonance. This means that the spin axis and the orbit normal of Mercury precess around the normal to the proper plane or Laplacian plane and that the three vectors remain coplanar. One of the results of this configuration is that the angle between the spin axis and the orbit normal, generally quoted as the obliquity, remains constant. Moroever the value of this angle is directly determined by the value of the relative ration $(C - A)/C$ of the moments of inertia of the planet.

Peale [75] showed that there are two stable positions of Mercury's spin vector which allow coplanar precession. Nowadays it is not still possible to differenciate between these two positions because they are both within the current errors of the determination of the spin orientation. One of the positions is near the orbit normal, with an obliquity increasing with a decreasing value of $(C - A)/C$ and the other one is near the normal to the proper plane, with an obliquity increasing with $(C - A)/C$.

Notice that a reasonable minimum value of $(C-A)/C$ is that corresponding to hydrostatic equilibrium as explained by Munk and Mc. Donald [71]. The condition is given by

$$\frac{C-A}{C} > \frac{k'R^5\omega^2}{3GC} \qquad (6.151)$$

where $k'$ is the secular Love number, $G$ is the gravitational constant, $R$ is Mercury's mean radius and $\omega$ is the angular rotational velocity of the planet. $k'$ and $C$ being unknown, it is useful to approximate these values by the corresponding ones for the Earth [75], that is to say $k' = 0.96$ and $C = MR^2/3$.

A value of $(C - A)/C \approx 10^{-6}$ is obtained from this last equation which shoul lead to an obliquity close to $30°$ in the second state, whereas radio and optical observations have shown that the planet obliquity is very small (no more than a few degrees) leading to the conclusion that Mercury is in the first state.

**Recent Theoretical Developments**

The interest with respect to Mercury was significantly renewed recently by two upcoming space missions, respectively called MESSENGER (NASA) [89] and BepiColombo(ESA,ISAS) [2]. Up-to-date knowledge on the planet can be found in a review report by Balogh and Giampieri [3]. Some investigations concerning its physical librations were done by Rambaux and Bois [82]. In this chapter a new determination of the mean obliquity (1.6') is proposed which looks very small in comparison with the above theoretical estimations but seems in agreement with Cassini's states. The authors put forward two proper frequencies with associated periods 15.847 y. and 1066 y, and a secondary spin-orbit secular resonance with period 278,898 y.

On the opposite D'Hoedt and Lemaître [29] considered an obliquity of $\approx 7°$ as a basic hypothesis when constructing an analytical theory of Mercury's rotation in the form of a two degrees of freedom Hamiltonian system, averaged over short periods. This theory succeeds in describing the libration abouth the 3/2 spin-orbit resonance by two pairs of action-angle variables, obtained after a succession of canonical changes of variables and simplifications. The authors agreed perfectly with Rombaux and Bois [82] on the values of the two basic frequencies. Such studies coupled with observational results obtained by Bepi Colombo might constraint the physical parameters of the planet as well as its internal model.

The Bepi Colombo joint mission is scheduled for 2013 with an arrival on Mercury in 2019. In preparation to this mission, which will study the surface and the interior of the planet, a lot of theoretical efforts have been made on the topic of the rotation. The most recent ones deal with various aspects: Peale [78] determined the free precession as well as the librations, and also discussed the proximity of Mercury's spin to Cassini state 1 from adiabatic invariance [75]. Yseboodt et al. [112] studied the evolution of Mercury's obliquity. A review paper has just been published on the subject by Lemaître et al. [61]

### 6.4.2 The Rotation of Venus: A Predominant Effect of the Atmosphere

#### A Long History of Measurements

Because of the thickness of the atmosphere covering the surface of the planet, the period of rotation of Venus has remained a long time unknown. A lot of erroneous estimations were done in the past. Several astronomers as Jean-Dominique Cassini, between 1666 and 1667, his son Jacques, and Francesco de Vico, at the beginning of the eighteenth century, proposed a period of roughly $23^h 20^{mn}$. On the contrary Giobvanni Schiaparelli, during a campaign of observations in 1877 and 1878, did not remark any substantial diurnal change of appearance of the planet and concluded that the period should be much longer than that proposed by the above scientits. He proposed in 1890 a synchronous rotation (with the orbital motion) of 224.5 d, much closer to the real value.

In fact the discovery of true period of rotation of Venus was possible only in 1962, thanks to the radar observations lead by the Jet Propulsion Observatory (JPL). To the great surprise of astronomers it was shown by Goldstein [39] and Carpenter [18] that the planet had a very slow retrograd rotation with a period of 243.16 d. Naturally, as it was the case for Mercury, specialists of celestial mechanics rushed into the dynamical interpretation of such a period. Some resonant behaviour with the Earth was proposed: indeed, if the rotation of Venus is 243.16 d retrograd, the axis of Venus which points toward the Earth at one inferior conjunction will point toward the Earth at all subsequent inferior conjunctions. If the moment of inertia of Venus about this axis is minimal, it is possible that the planet is locked into this synodic commensurability with its permanent bulge or longest axis toward the Earth at each inferior conjunction. This means that the gravitational attraction of the Earth, although being very small in comparison with that of the Sun, may have helped the stabilzation of the sidereal rotation of Venus at 243.16 d retrograde.

#### Up-to-Date Explanation of Venus Retrograd Rotational Motion

Certainly at the beginning of its history Venus might have rotated with a standard angular rate, with a period of the order of one day. Then remains the double question: how it has been slowed down so that its period of rotation reached its present value and what made possible the fact that this rotation is retrograd. Already more than 30 years ago Gold and Soter [38] proposed a scenario explaining the unusual rotation of the planet Venus as a result of a state of equilibrium between two tidal effects: solid tides and atmospheric tides.

In the case of solid tides the differential attraction exerted by the Sun is deforming the Venusian globe such that it creates a bulge in the direction of the Sun, as well as on the opposite side. This is exactly the same kind of phenomena that leads to oceanic tides on the Earth because of the attraction of the Moon and of the Sun, and what explained already in Sect. 6.3.4. If Venus was purely elastic the bulge should be permanently orientated toward the Sun, but because of a non-elastic behaviour a phase lag $\delta$ between the direction of the axis of the bulge and the direction planet-Sun (cf. Fig. 6.8). This non-coincidence induces a return torque whose the effect is to reduce the rotational speed of the planet, which remains as long as this angular speed is larger than the mean angular speed of revolution. Thus the final stage in these conditions is the synchronous rotation or 1/1 resonance, as it is the case for the Moon around the Earth.

In the case of Venus, this evolutional scheme is not complete and in fact erroneous, because we must take into account the atmospheric tidal effect due to the extremely dense atmosphere of the planet (leading to a surface pressure approximativley 100 times larger than for the Earth). Naturally the Sun is warming the subsolar point thus leading to a substantial pressure difference between the dark side and the lightened side. This difference of pressure is accompanied by a mass redistribution of the atmosphere. In consequence, as the angular speed of rotation of the planet is larger than the angular speed of revolution, another phase lag exists between the direction of the atmospheric bulge and the direction perpendicular to the planet-Sun direction (cf. Fig. 6.9).

One of the pioneer developments trying to explain the present venusian rotational state was done by Lago and Cazenave [55] who extrapolated the past evolution of the rotation parameters in Eulerian formalism by using a numerical integration method with the hypothesis that only solar tidal torques and core–mantle coupling have been active since formation. They found quite

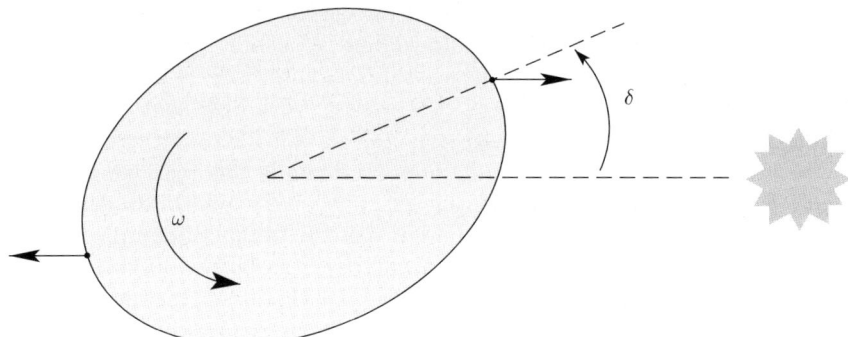

**Fig. 6.8.** Representation of the solid Earth tide exerted by the gravitational action of the Sun on Venus

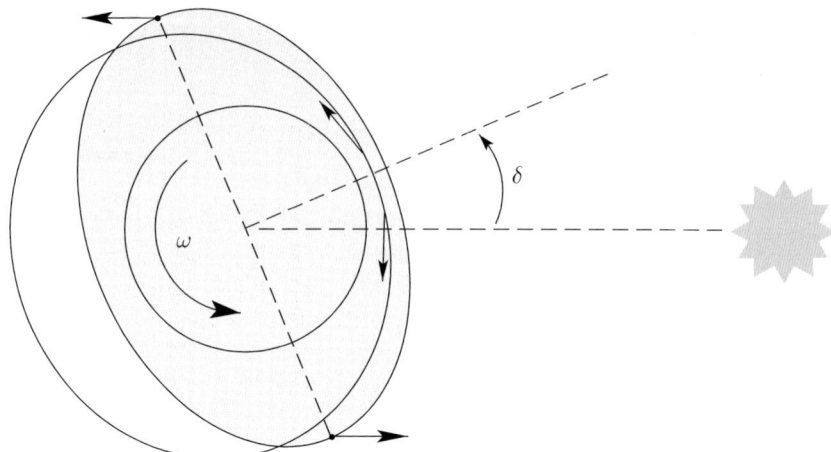

**Fig. 6.9.** Representation of the thermic tide exerted by the warming effect of the Sun on the Venus'atmosphere

conceivable that Venus had originally a rotation similar to the other planets and has evolved in roughly $4.5 \times 10^9$ y from a rapid and direct rotation (around 12 hours) to the present slow retrograd one. The authors proposed an evolutionary explanation of the current rotation of Venus leading place to constraints on the solid body quality factor $Q$. They showed that a $Q \approx 15$ at the annual frequency is required, and a dominant diurnal thermal response is required, whereas a value $Q \approx 40$ can also be allowed.

Recently, Correia and Laskar [26] showed that when the atmospheric torque is large enough, as it is the case for Venus, the synchronous rotation becomes unstable, and two new configurations appear, one prograde and the other retrograd. These same authors investigated what dynamical history a planet can undergo to reach a state of equilibrium as Venus today. They showed starting from a large number of initial conditions that under the effect of the dissipative tidal effects above, but also of the core-mantle coupling the obliquity of Venus naturally tends towards the value of 0° (prograde rotation) or 180° (retrograde rotation), whereas the rotation mode reaches a position of equilibrium between the various tidal forces, with only four possibilities, two ones in a prograde regime with a period of 76.8 days, and two ones in a retrograd one, with a period of 243 days. The fact is that the two retrograd scenarios correspond exactly to the present rotational regime of the planet [26] but they correspond to a totally different history. In one of them Venus had initially a prograd rotation, then it slowed down under the influence of the dissipative processes until it stops, and begins to rotate in the inverse sense, to reach the resent 243.0 days value. In the other one, the planet, after having a near to 0° obliquity undergoes large changes of this last parameter whereas the rotational speed is gradually slowing down. At some moment of the rotational history the axis is turning down which means that the rotaton suddenly goes from a prograde regime to a retrograd one.

### 6.4.3 The Rotation of Mars: Similitudes with the Earth and Determinations

At the end of the seventeenth century the period of rotation has been calculated with a very good precision, for instance by Jean-Dominique Cassini at the Paris observatory. This was possible because of the thinness of the atmosphere (the surface pressure is roughly 1/100 that of the Earth) and the presence of significant details at the surface.

Two fundamental characteristics of Mars rotation are very closed to the Earth: the present obliquity ($25°11'$ instead of $23°27'$) and the rotation rate ($24^h37^{mn}$ instead of $23^h56^{mn}$). Therefore we have to treat a standard rotation without resonance as it is the case for Mercury, without a very slow retrograd proper rotation as it is the case for Venus and without a near zero obliquity as it is the case for these two planets: then the way of computation of the combined precession and nutation motion is quite similar to that used for the Earth. It is even more simple because of the absence of a large perturbation a big satellite, the two Mars' satellites Phobos and Deimos having a very small influence on the displacement of the polar axis of the planet.

As early as the 1970s it was found by Ward [99, 100] and by Christiensen and Balmino [23] that the total variation of Mars obliquity at long time scales attained very large amplitudes, roughly $20°.6$ in comparison with the Earth (less than $2°$). These large oscillations are a consequence of the motion of Mars orbital plane, which is more perturbed than the Earth one, mainly by Jupiter and by the Earth. Nevertheless the long term behaviour of the obliquity depends on the long terml evolution of the orbital plane which is not known with precision over millions of years. Consequently the results are essentially of qualitative value. Ward [101] made a very complete study of the climatic implications of these long term obliquity variations.

In the 1970s Borderies [7] calculated for the first time with a good precision the variations of the rotation of Mars at short and long time scales, resulting from the solar torques, in an absolute reference frame. By using the classical method based on Eulerian angles and Euler's dynamical equations she gave a precise quantitative evaluation of the short period nutations, useful for the computation of trajectories of an orbiter around the planet. This work was done soon after the touchdown of the two Viking landers on Mars. With the use of nine months of Viking data it was possible to improve the knowledge of both the rotation rate and the mean orientation of the spin axis of Mars, referred to its mean orbit [8].

The polar motion on Mars as well as the variations in its speed of rotation has never been detected, in particular because of their smallness and because space missions time scales are too limited in time for this purpose. Nevertheless prospective models of polar motion have been proposed by several authors as Zhang [113], Kubo [54] and Hilton [45], based on various values for the moments of inertia of the core and of the mantle. Comparisons between these models have been made by Gauchez and Souchay [37].

## Mars' Rotation at Short Time Scale

With an increasing number of spatial missions around Mars and also at the surface, the knowledge of the planet orbital and rotational parameters has significantly improved. Precious information was obtained from the Viking–Pathfinder combined data set. Whereas the Viking lander data gave the mean spatial orientation of the pole of rotation of Mars at the midpoint of the experiment, Pathfinder data gave the orientation 20 years later. Thus in 1997 Doppler and range measurements to the Mars Pathfinder lander made using its radio communications system have been combined with similar measurements from the Viking landers to give an improved value of the Mars rotational characteristics, in particular the precession which was set to $\dot{\psi} = -7".576 \pm 0.035/y$ [36]. This nominal value is significantly different from another one obtained by Yoder and Standish [111] that is to say $\dot{\psi} = -7".83 \pm 0.30$ by using Viking data. Notice that these last authors showed that seasonal variations in Mars'rotation were detected at semi-annual and annual frequencies, which are $279 \pm 100$ and $311 \pm 100$ mas of angle respectively, after correcting for a general relativity effect.

The recent Mars Pathfinder mission [36] was also an opportunity to improve the knowledge of Mars' polar moment of inertia $C$ as well as well as Mars interior. Indeed, the precession constant measured above gives a directly the value of the dynamical ellipticity $H_d^{\text{Mars}}$ of the planet, given by

$$H_d^{\text{Mars}} = \frac{C - (A + B)}{2A} \qquad (6.152)$$

with $A < B < C$. Moreover from the detection of Mars gravity field through tracking data we have the following relationship between the physical parameters of the planet:

$$C - (\frac{A + B}{2}) = J_2 M R^2 \qquad (6.153)$$

where $M$ and $R$ are respectively the mass and the radius of the planet. The confrontation of the two last equations leads to the determination of $J_2$ which at its turn gives information in the repartition of mass at the planet interior. More generally, the knowledge of the moment of inertia, combined with Mars parameters as mass, size, shape and low-order gravity provides key information for models of the interior structure [36].

All the new constraints obtained from space missions lead to an improved version of the Martian precession and nutation models in the early 1990s by Hilton, both considering Mars as a rigid body [44] or by including the effects of an elastic mantle and a liquid core [45].

In the first paper, Hilton showed that the nutation in longitude $\Delta\psi$ with amplitudes larger than 0".001 consist of eight solar nutations terms and one nutation each contributed by the motions of the nodes of Phobos and Deimos, and a single nutation from the direct gravitational torque of Jupiter. Concerning the nutation components in obliquity $\Delta\varepsilon$ they consist of six solar terms and one nutation term from the motion of Phobos and Deimos. There is no precession term contributed by the Martian satellites because they are both located the plane of Mars' equator. The amplitudes and the periods of the solar nutation terms are in agreement with nutation calculated previously by Reasenberg and King [83] in 1979.

In the second paper Hilton [45] computed a first-order approximation of the effects of an elastic mantle and a liquid core on the motion of Mars'pole. Among interesting results, the author showed that the effect of Mars' Chandler wobble (Eulerian free nutation) should much less dependent on Mars' structure than the Earth's Chandler wobble depends on the Earth's structure. Moreover the period of the liquid core free-core nutation (FCN) is found to be very sensitive to the mean core radius. For instance if the FCN period is known with an uncertainty of two days, then the mean core radius could be inferred with an uncertainty of 6 km. Unfortunately the amplitude of the FCN might be very small and has not yet been detected.

Bouquillon and Souchay [9] computed with an optimal truncation of 0.1 mas the coefficients of nutation of Mars, including the effects of the two satellites Phobos and Deimos, the direct influence exerted by the planets, as well as the Oppolzer terms which differenciate the nutation for the axis of figure and the axis of angular momentum. Their values concerning the terms due to the solar influence agree very well with other determinations by Borderies [7] and Hilton [44]. The leading nutation term is semi-annual (when considering the Martian year with 687 days), with expression: $\Delta\psi = -1".0969\sin(2M + 2\Lambda_M)$    $\Delta\varepsilon = 0".51589\cos(2M + 2\Lambda_M)$, where $\Lambda_M$ is the angle giving the longitude of Mars perihely. Note that the amplitude of this term is nearly the same as the amplitude of the semi-annual term of nutation of the Earth, due to the influence of the Sun. The largest direct planetary effect on Mars' nutation is due to Jupiter with an amplitude of 0".00019 (for $\Delta\psi$)and a $2\lambda_{Ju}$ argument.

In conclusion the direct gravitational attraction of the satellites and planets on Mars does not contribute significantly to its precession and nutation. Nevertheless, the change in the orbit of Mars arising from perturbations by the planets does cause significant changes in the solar precession and nutation of Mars over very long period of time [9, 44]. This is the object of the next subsection.

**Evolution of Mars' Rotation at Long Time Scale**

About 25 years ago, some authors as Ward [99] have examined a number of processes which have the potential to alter the spin axis precession rate, discussing both the expected magnitude and time scales involved. Of these, Ward showed that the Tharsis uplift should have played a dominant role among other processes, as differenciation, mantle convection etc. One of the fundamental effects of these physical processes with respect to the rotation should be to modify significantly the planet's quadripole moment in the past. Accompanying these changes the consequences of passage through a secular spin-orbit resonance have been investigated analytically [99, 101]. Near resonances, the overall spin axis motion could be largely reproduced by a superposition of two solutions: a linearized response to non resonant orbital terms plus a nonlinear solution that incorporates the effects of resonance [101]. Bouquillon and Souchay [9] calculated the long term variations of Mars' obliquity and precession and showed that Pathfinder recent data [36] could constraint their numerical simulations so that the uncertainty on the obliquity did not exceed $\pm 0.5°$ for 5 My instead of $\pm 4°$ when using Viking data. Their results for Mars'obliquty are shown in Fig. 6.10.

More precisely, numerical integrations [101] show that the obliquity of Mars executes large-scale oscillations as a result of spin axis and orbit plane precessions. These changes undoubtedly had drastic effects on the climatic system of the planet. They are characterized by two dominant periodicities [101]: oscillations driven by the differential spin-axis and orbit precession rates and amplitudes modulations due to changes in the orbital inclination. The typical time scale of the first phenomena is $\approx 10^6$ years, with obliquity variations of $\approx 20°$. A third mechanism is the spin-orbit resonance mentioned before with a typical time scale of $10^7$ y.

Laskar and Robutel [57] made numerical studies of the global stability of the spin-axis orientation of Mars against secular orbital perturbations. By the

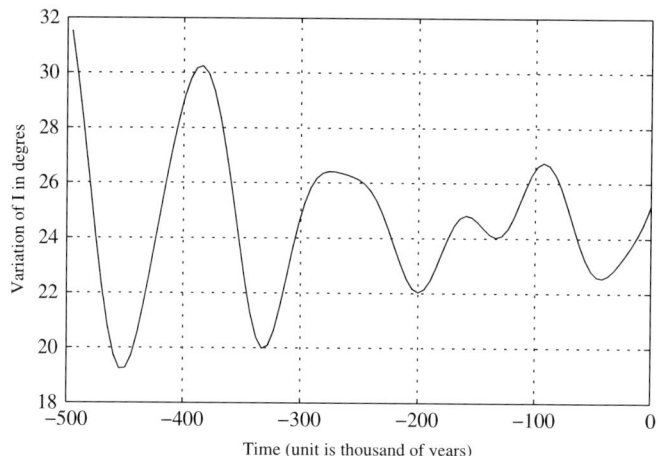

**Fig. 6.10.** Variations of Mars' obliquity from 500 Myr. ago until now, according to Bouquillon and Souchay. Notice that the variations reach $12°$ peak to peak

intermediary of an appropriate frequency analysis of the obliquity for 45 My, they found a large chaotic zone ranging from 0° and 60°.

### 6.4.4 The Rotation of the Giant Planets

In comparison with the telluric planets, the rotation of the giant planets has been poorly investigated. They concern mainly their rate of precession, their obliquity and also some models explaining the evolution of their rotational speed. Recently Ward [102, 103] studied the precession rate of both Jupiter and Saturn starting from the following general formalism giving the precessional constant $\alpha$:

$$\alpha = \frac{3}{2}\frac{n^2}{\omega}\frac{J_2+q}{\lambda+l} = \frac{3}{2}\frac{GM}{\omega a^3}\left(\frac{J_2+q}{\lambda+l}\right) \tag{6.154}$$

where $\omega$ is the spin frequency of the giant planet considered, $a$ its semi major axis, $n$ is the heliocentric mean motion, $J_2$ is the coefficient of the quadrupole moment of its gravity field, and $\lambda$ is the moment of inertia of the giant planet normalized to $MR^2$ where $M$ and $R$ are respectively the mass and the radius of the planet.

The satellite system is considered as a whole, as participating to an additive contribution of the planet quadrupole moment, in the following form:

$$q = \frac{1}{2}\sum_j \frac{(m_j/M)(a_j/R)^2 \sin(\theta - i_j)}{\sin\theta} \tag{6.155}$$

where $q$ is the effective quadrupole moment of the satellite system. $m_j$, $a_j$ and $i_j$ are respectively the mass, the semi-major axis and the inclination of the satellite with index $j$, $\theta$ being the obliquity of the planet. Moreover the angular momentum of the satellite system normalized to $MR^2\omega$ can be expressed as

$$l = \sum_j (m_j/M)(a_j/R)^2 (n_j/\omega) \tag{6.156}$$

In the case of Jupiter the term $\sin(\theta - i_j)/\sin\theta$ at the right-hand side of (????) vanishes for $i_j = 0$ in the case of the four dominant galilean satellites (the other satellites can be neglected). The values of the precessional constant obtained from (6.154) for the Jupiter's precession rate and Saturn's one are respectively $\alpha_J = 0".8306/\text{y}$ [102] and $\alpha_S = 2".741/\text{y}$ [103].

Ward [102] pointed out that it is a remarkable occurence in the solar system that the spin-axis precession of Jupiter ($\approx 4.74(\lambda/0.25)\times 10^5$ y) and Saturn ($\approx 1.75(\lambda/0.22)\times 10^6$ y) are correspondingly very close to the precession periods of the orbit planes of Uranus ($4.33\times 10^5$ y) and Neptune ($1.87\times 10^6$ y).

More precisely, for Jupiter Ward [102] found that the spin-axis precession period of Jupiter is near that of a fundamental Laplace–Lagrangian solar system mode describing the precession of Uranus' orbit plane ($\approx 4.3\times 10^5$ y).

As a result they showed that a portion of the 3.1° obliquity of Jupiter may be forced by a spin-orbit secular resonance with Uranus. In the case for which this scenario is valid, it should bring constraints on Jupiter's moment of inertia independent of the interior models.

Saturn's obliquity, in comparison with Jupiter's one, has a large amplitude of 26.°7. Here also a similarity between the precession period of Saturn's spin axis and the precession period of Neptune's slightly inclined orbit plane may implicate a resonant interaction between these planets which should be responsible for tilting Saturn from a more upright state [103]. A sequence of events could account for the establishment of the resonant state above. For instance initially forming with a small obliquity, Saturn may have entered a resonance during the erosion of the Kuiper belt, which acted on the decreasing the rate of regression of Neptune's orbit plane. Saturn's spin axis may also be librating in the resonance with an amplitude $\psi > 31°$ which adds constraints on Saturn's moments of inertia [103].

The evolution of the obliquities of the giant planets was studied as a whole recently by Brunini [13] who reported numerical simulations of the phenomena already suggested by Tsiganis [95] according to which the orbital architechture of the outer solar system is in agreement with an initial configuration for which it was compact and where Jupiter and Saturn crossed the 2:1 orbital resonance by divergent migration, which lead to close encounters among the giant planets, but with eccentricities and inclinations damped to their current values by interactions with planetisimals. Large obliquities may have been generated during the encounter phase and the observed values could be explained. Notice that a simple analytic argument proves that the change in a spin axis direction relative to an inertial frame during an encounter between the planets is very small and that in comparison the change in obliquity (which is measured with respect to the orbit normal), is due to change in the orbital inclination [46].

Generally no clear indication was obtained on the variations of the rotational angular speed of the giant planets, noticably because of the lack of a solid surface. Indeed there are two ways to determine the rotation rate of a planet without a solid surface: one way consists in measuring the motions of features on the visible part of the atmosphere. This method is really difficult to apply because such motions may not reflect the rotation of the planet itself. The second way consists in tracking the rotational modulation of the radio emission, whose some components, deeply depending on the magnetic field of the planet and then on its interior, are supposed to be directly related to the rotation rate of the planet itself.

The latest results obtained from this last method with the help of the Cassini spacecraft seem to indicate that Saturn' rotation rate has drastically slowed down by 6 minutes since the Voyager 1 and 2 spacecraft flew by the planet in 1980 and 1981 [86]. It is difficult to believe that Saturn's rotation period has really changed, because angular momentum conservation would require a large change in the interior angular velocity or in the inertial moment of the planet to explain the 6 min difference. Future results from the Cassini mission in the next years might fully answer to the question

## 6.5 Conclusion

In this chapter we have given the main theoretical developments devoted to a complete modelization of the rotation of a planet considered as a solid body, starting from two kinds of theoretical frameworks, developed by Woolard [109] and Kinoshita [49, 51]. Then we have explained in detail the current state of knowledge of the rotation of the Earth, which is by far the most accurately known of all the planetary rotations in the solar system. In particular we have shown the remarkable accuracy got for the determination of the Earth Orientation Parameters (EOP), enabling one to give the position of the axis of figure of the Earth in space, and the axis of rotation with respect to a body-fixed reference frame, thanks to the drastic improvements of ground-based techniques as VLBI.

Then we have explained in some details the mechanisms leading to the uncommon rotation of the two inner planets Mercury and Venus, insisting on the causes of their behaviour at short and long time scales. Some future missions as Bepi Colombo for Mercury will certainly bring drastic constraints on the models of rotation and on the interior of these planets. At the exception of the Earth, the planet whose we know at best the rotation is Mars. Thanks to recent space missions, the knowledges concerning the rotation of this telluric planet has considerably improved. We have shown the high accuracy of the determination of Mars' precession constant as well as nutation coefficients.

Nevertheless a lot of future investigations concerning the rotation of the planets in general remain to be done. The incredibly high precision of present day measurements of the rotation of our planet lead to a good amount of interrogations when exploring the O–C (observed–calculated) residuals, noticably concerning the influence of the atmosphere and of the oceans on the detected remaining signals of the EOP. Polar motion with its excitation driven mechanism is particularly concerned, together with UT1 variations.

The present and future Mars' missions will certainly bring new and decisive insights in the knowledge of the red planet, whose the polar motion has not still been detected although its measurement should lead to very interesting informations concerning its interior, in particular when comparing with the polar motion of the Earth.

The lack of informations about the rotation of giant and gazeous planets might be compensated in the future by appropriate measurements and gives a large opportunity for theoretical modelizations to understand better the rotational history of these planets, in particular in the case of Uranus whose the axis is nearly lying in its orbital plane.

## Acknowledgements

I am strongly indebted to Mr. Olivier Becker for his help in making the figures, and to Dr. Anne Lemaître for some informations concerning Mercury.

## References

1. Andoyer, H., 1923. *Cours de Mécanique Céleste, Vol. I*, Gauthier-Villars, Paris, p. 54.
2. Anselin, A. and Scoon, G. E. N., 2001. *Planet. Space Sci.* **49**, 1409.
3. Balogh, A. and Giampieri, G., 2002. *Rep. Prog. Phys.* **65**, 529.
4. Barnes, R. T. H., Hide, R., White, A. A., and Wilson, C. A., 1983. *Proc. R. Soc. Lond. Ser. A* **387**.
5. Berger, A., Imbrie, J., Hays, J., Kukla, G., and Saltzman, B. (eds), 1984. *Milankovitch and Climate Reidel*, Dordrecht.
6. 2002, *Planet. Space Sci.* **50**, 323.
7. Borderies, N., 1980. *A&A* **82**, 129.
8. Borderies, N., Balmino, G., Castel, L., and Moynot, B., 1980. *Earth Moon Planets* **22**, 191.
9. Bouquillon, S. and Souchay, J., 1999. *Astron. Astrophys.* **345**, 282.
10. Bretagnon, P., 1982. *A&A* **114**, 278.
11. Bretagnon, P., Rocher, P., and Simon, J. L., *Astron. Astrophys.* **319**, 305.
12. Bretagnon, P., Francou, G., Rocher, P., and Simon, J. L., *Astron. Astrophys.* **329**, 329.
13. Brunini, A., 2006. *Nature* **440**, 1163B.
14. Brzezinski, A., 1992. *Proc. J. Syst. Réf. Spatio-Temporels*, Paris, **102**.
15. Brzezinski, A. and Nastula, J., 2002. *Adv. Space Res.* **30**(2), 195.
16. Byrd, P. F. and Friedman, P. M., *Handbook of Elliptical Integrals for Engineers and Physicists* Springer-Verlag, Berlin, p. 58.
17. Capitaine, N., 2005. *A&A* **432**, 355.
18. Carpenter, R. L., 1964. *Astron. J.* **69**, 2C.
19. Chao, B. F., 1993. *Geophys. Res. Lett.* **20**, 253.
20. Chapront-Touze, M. and Chapront, J., 1988. *A&A* **190**, 342.
21. Chapront, M. and Souchay, J., 2006, Précession et Nutation 1749–1752, *Les Oeuvres Completes de Jean Le-Rond d'Alembert*, Vol. I, 7, CNRS-EDITIONS, Paris.
22. Charlot, P., Sovers, O. J., Williams, J. G., and Newhall, X. X., 1995. *Astron. J.* **109**, 418.
23. Christiensen, E. J. and Balmino, G., 1979. *J. Geophys. Res.* **84**, 7943.
24. Colombo, G., 1965. *Nature* **208**, 575.
25. Colombo, G., 1966. *AJ* **71**, 891.
26. Correia, A. and Laskar, J., 2001. *Nature* **411**, 767.
27. Dehant, V., Bizouard, C., Hinderer, J., Legros, H., and Lefftz, M., 1996. *Phys. Earth Planet. Inter.* **96**, 25.
28. Deprit, A., 1967. *Am. J. Phys.* **35**, 424.
29. D'Hoedt, S. and Lemaitre, A.: *Spin-Orbit Resonant Rotation of Mercury*, Lect. Notes Phys. **682**. Springer, Heidelberg (2006).

30. De Viron, O., Bizouard, C., Salstein, D., and Dehant, V., 1999. *J. Geophys. Res.* **104**(B3), 4861.
31. Dickman, S. R., *J. Geophys. Res.* **86**, 4904.
32. Dollfus, A., 1968. La Recherche Spatiale.
33. Folgueira, M., Souchay, J., and Kinoshita, H., 1998. *Celest. Mech.* **69**, 373.
34. Folgueira, M., Souchay, J., and Kinoshita, H., 1998. *Celest. Mech.* **70**, 147.
35. Ibach, H. and Luth, H., 1996. *Solid-State Physics*, 2nd Ed., Springer, Berlin, pp. 45–56.
36. Folkner, W. M., Yoder, C. F., Yuan, D. N., Standish, E. M., and Preston, R. A., 1997. *Science* **278**, 1749.
37. Gauchez, D. and Souchay, J., 2000. *Earth Moon Planets* **84**, 33.
38. Gold, T. and Soter, S., 1966. *Nature* **209**, 1117.
39. Goldstein, R. M., 1964. *Astron. J.* **69**, 12G.
40. Goldreich, P. and Peale, S. J., 1968. *Ann. Rev. Astron. Astrophys.* **6**, 287.
41. Gross, R. S., 2000. *Geophys. Res. Lett.* **27**(15), 2329.
42. Herring, T. A., Mathews, P. M., and Buffett, B. A., 2002. *J. Geophys. Res.* **107**(B4), 10.129.
43. Herring, T. A., Buffett, B. A., Mathews, P. M., and Shapiro, I. I., 1991. *J. Geophys. Res.* **96**, 8529.
44. Hilton, J. L. 1991. *Astron. J.* **102**(4), 1510.
45. Hilton, J. L., 1992. *Astron. J.* **103**(2), 619.
46. Hoi Lee, M., Peale, S. J., Pfahl, E., and Ward, W. R., 2006. *DPS Meeting*, Am. Astron. Soc.
47. Hori, G., 1966. *Publ. Astron. Soc. Jpn* **18**, 287.
48. Kimura, H., 1917. *Month. Notices R. Astron. Soc.* **78**, 163.
49. Kinoshita, H., 1972. *Publ. Astron. Soc. Jpn* **24**, 423.
50. Kinoshita, H., Hori, G. and Nakai, H., 1974. *Annals Tokyo Astr. Obs.* **14**, 14.
51. Kinoshita, H., *Celest. Mech.* **15**, 277.
52. Kinoshita, H. and Souchay, J., 1990. *Celest. Mech.* **48**, 187.
53. Kinoshita, H., 1992. *Celest. Mech.* **53**, 365.
54. Kubo, Y., 1991. *Celest. Mech.* **50**, 165.
55. Lago, B. and Cazenave, A.,1979. *Earth Moon Planets* **21**, 127.
56. Lambeck, K., 1980. *The Earth's Variable Rotation*. Cambridge University Press, London.
57. Laskar, J. and Robutel, P., 1993. *Nature* **361**, 608.
58. Laskar, J., Joutel, F., and Robutel, P., 1993. *Nature* **361**, 615.
59. Laskar, J., 1989. *Nature* **338**, 237.
60. Laskar, J., 1993. *Icarus* **361**, 615.
61. Lemaitre, A., D'Hoedt, S., and Rambaux, N., *Celest. Mech.* **95**, 213.
62. Lieske, J. H., Lederle, T., Fricke, W., and Morando, B., 1977. *A&A* **58**, 1.
63. Liu and O'Keefe, 1965. *Science* **150**, 1717.
64. Madden, R. A., Lejenas, H., and Hack, J., 1998. *J. Atm. Sci.* **55**, 2561.
65. Marcus, L., Dickey, J. O., and De Viron, O., 2001. *Geophys. Res. Letters* **28**(18), 3465.
66. Mathews, P. M. and Dehant, V., 1995, *Highlights Astron.* **10**, 243.
67. Mathews, P. M. and Shapiro, I. I. *Ann. Rev. Earth Planet Sci.* **20**, 469.
68. Mathews, P. M., Herring, T. A. and Buffett, B. A., 2002. *J. Geophys. Res.* **107**(B4), 10.129.
69. Melchior, P., 1957. *Latitude Variation, Progress in Physics and Chemistry of the Earth*, Pergamon Press.

70. Miyamoto, M. and Soma, M., 1993. *Astron. J.* **105**, 1.
71. Munk, W. H., Mc Donald, G. J. F., 1960. *The Rotation of the Earth*, Cambridge University Press, Cambridge.
72. Okubo, S., 1982. *Geophys. J. R. Astron. Soc* **71**, 134.
73. Pacault, R., 1969. Synchronisme de la rotation et du mouvement orbital des planets (Mercure) et des satellites, Note Scientifique, de ESRO (ESRO SN-92).
74. Peale, S. J. and Gold, T., 1965. *Nature* **206**, 1240.
75. Peale, S. J., 2005. *Icarus* **178**(1), 4.
76. Peale, S. J., 2006. *Icarus* **338**(2), 338.
77. Peale, S. J., 1969. *AJ* **74**, 483.
78. Peale, S. J., 1972. *Icarus* **17**, 168.
79. Pettengill, G. H. and Dyce, R. B., 1965. *Nature* **206**, 1240.
80. Poincare, H., 1892. *Les éthodes Nouvelles de la Mécanique Céleste*, Paris.
81. Ponte, R. M., Stammer, D., and Marshall, J., 1998. *Nature* **391**, 476.
82. Rambaux, N. and Bois, E., 2004. *A&A* **413**, 381.
83. Reasenberg, R. D., 1977. *J. Geophys. Res.* **82**(2), 369.
84. Roosbeek, F. and Dehant, V. 1998. *Celest. Mech.* **70**, 215.
85. Saadov, Y., 1970. *Prikl. Mat. Mekh.* **34**, 962.
86. Sanchez-Lavega, A., 2005. *Science* **307**, 1223.
87. Seidelmann, P. K., 1980. *Celest. Mech.* **27**, 79.
88. Sekiguchi, N., 1975. *J. Geodetic Soc. Jpn* **21**, 131.
89. Solomon, S. C., et al., 2001. *Planet. Space Sci.* **49**, 1445.
90. Souchay, J. and Kinoshita, H., 1996. *A&A* **312**, 1017.
91. Souchay, J. and Kinoshita, H., 1997. *A&A* **318**, 639.
92. Souchay, J., 1998. *Astron. J.* **116**, 503.
93. Souchay, J., Loysel, B., Kinoshita, H., and Folgueira, M., 1999. *A&A Suppl. Series* **135**, 111.
94. Tisserand, E. W., 1892. Méecanique Céeleste, Gauthier Villars, Paris, Chapt. 16–19.
95. Tsiganis, K., Gomes, R., Morbidelli, A., and Levison, H. F., 2005. *Nature* **435**, 459.
96. Vicente, R. O. and Wilson, C. R., 1997. *J. Geophys. Res.* **102**(B9) 23439.
97. Vondrak, J., 1992. *J. Syst. Réf. Spatio-Temporels, Paris*.
98. Vondrak, J., 1985. *Ann. Geophys.* **3**, 351.
99. Ward, W. R., 1973. *Science* **181**, 260.
100. Ward, W. R., Burns, J. A., and Toon, O. B., 1979. *J. Geophys. Res.* **84**(B1), 243.
101. Ward, W. R. and Rudy, D. J., 1991. *Icarus* **94**, 160.
102. Ward, W. R. and Canup, R. M., 2006. *Astrophys. J.* **640**, L-91.
103. Ward, W. R. and Hamilton, D., 2004. *Astrophys. J.* **128**, 2501.
104. Wahr, J. M., 1981. *Geophys. J. R. Astron. Soc.* **64**, 651.
105. Wahr, J. M., 1981. *Geophys. J. R. Astron. Soc.* **64**, 677.
106. Whittaker, E. T., 1961. *A Treatise on the Analytical Dynamics of Particles and Rigid Bodies*, 4th. Ed., Cambridge University Press, Cambridge, p. 144.
107. William, J. G., 1994. *Astron. J.* **108**, 711.
108. Williams, J. G., Newhall, X. X., and Dickey, J. O., 1991. *A&A* **241**, L9.
109. Woolard, E. W., 1953. *Astron. Papers Am. Ephemeris* **15**, pt. 1.
110. Yatskiv, Y., 2000. *Polar Motion: Historical and Scientific problems*, ASP Conference Series Vol. 208, 383.

111. Yoder, C. F. and Standish, E. M., *J. Geophys. Res.* **102**(E2), 4065.
112. Yseboodt, M. and Margot, J. L., *Icarus* **181**(2), 327.
113. C. Z., 1998. *Earth Moon Planets* **64**, 117.
114. Zhou, Y. H., Zheng, D. W., and Liao, X. H., 2001. *J. Geodesy* **75**, 164.

# On the Stability of Extra-Solar Planetary Systems

Elke Pilat-Lohinger and Rudolf Dvorak

Institute for Astronomy, University of Vienna, Türkenschanzstrasse 17, A-1180 Vienna, Austria
lohinger@astro.univie.ac.at;
dvorak@astro.univie.ac.at

**Abstract.** Since the discovery of the first extra-solar planet near a Sun-like star (i.e., 51Peg b) numerous planets outside the solar system have been discovered during the last century (by October 2006, 168 extra-solar planetary systems are known). For dynamical studies the 20 multi-planet systems and the 21 binary systems that host one or more planets are important to be examined, since gravitational interactions between the celestial bodies restrict the stable planetary motion to certain regions in the phase space of such a system. Besides the investigation of these systems, stability studies of hypothetical terrestrial-like planets in all systems are of special interest.

In this chapter we give a short overview about the discovered extra-solar planetary systems and the different detection methods. Then we discuss the multi-planet systems according to a classification of Ferraz-Mello. For the planetary motion in binary systems we introduce the three types of motion and show the results of general stability studies thereto on the one hand and the application of these studies to real binary systems on the other hand. Furthermore, we discuss the influence of the perturbing star (i.e., secondary) on the planetary motion in brief.

In the last part of this chapter we concentrate on the stability of Earth-like planets moving in the so-called habitable zone. This study is important, especially with regard to current (CoRoT) and forthcoming space missions (like e.g., Kepler, Darwin, TPF) that try to find a planet—similar to our Earth—where complex life could have developed. Since we know that the development of a biosphere and the further evolution to intelligent life-form is a process on a long timescale, the long-term stability of such a planet is an important and necessary condition for the habitability of a planet.

## 7.1 Introduction

The famous discovery of the first planet orbiting the Sun-like star 51 Peg more than one century ago [41] was certainly a milestone of astronomy. Since that time, we know that the planets of our solar system are not the only ones. And we learned that the new systems are quite different compared to our planetary system, since most of the planets are very massive; a significant number of planets move in orbits quite close to the host star (much closer than Mercury, the innermost planet of our solar system) and we are faced with a high eccentricity motion of the planets. But this is probably a bias due to the observation technique. In principal, extra-solar planets can be detected either by direct or indirect methods. Most of the exo-planets have been found by a very sensitive Doppler technique, where perturbations of the star due to an accompanying planet can be observed (1) in the *radial velocity* signal, (2) in the *astrometric position* of the star and (3) in the arrival time of a signal (i.e., *pulsar timing*).

More precisely, the gravitational attraction of an accompanying planet forces the star into a motion around the center of mass which can be measured as the spectrum is shifted to the blue color when the star moves towards the Earth and is shifted to the red when the star moves away (= Doppler shift). From these measurements we are able to determine the planet's mass (or more precisely the minimum mass[1]), its orbital period and its distance to the star with the aid of the Newtonian law. Since the precision of such measurements is about 3 m/s it is easy to detect giant planets (Jupiters) but there is no hope to detect an Earth-size planet at the moment.

With astrometry we measure a wobble in the star's motion which is caused by an accompanying planet. Nowadays the best astrometric measurements can reach a precision of about 100 µarcsec which allows therefore the detection of Jupiter-size planets.

As already mentioned above, dynamical perturbations of a planet are also measurable in the arrival time of a signal, like in the case of pulsars where a precision of a few µsecs is needed for the detection of planets. Even Earth-like planets can be discovered easily by the timing of pulsars.

Other detection methods are:

The *transit*, where a planet produces a drop in the star light when passing in front of the star – this method allows the discovery of (a) Jupiter-size planets for ground-based observations and (b) Earth-size planets from space. In fact, it is presently the only method to detect an exo-earth.

*Gravitational lensing*, where an exo-planet can produce a gravitational amplification of the light of background stars for a certain time depending on its transverse velocity.

*Direct imaging.* A direct observation of planets outside the solar system near a Sun-like star is up to now impossible,[2] even the Hubble Space Telescope would not be able to detect them at the expected distances from their stars, since Sun-like stars are about 1 billion times brighter than planets in the visible light. Even if four objects have been imaged so far, we have to point out that either they are very massive (most probably brown dwarfs)

---

[1] We cannot determine the exact mass with this method, since the inclination of the system with respect to the line of sight is not known.

[2] The announcement of GQ Lupi – the possibly first direct discovered extra-solar planet by Neuhäuser and co-workers in 2004 with the ESO VLT NACO – is still to be proven and is not accepted by all researchers yet.

or they move around very low-massive M-dwarfs so that their detection was easier.

From the observations (until October 2006) we can summarize the following:

*196 planets* have been detected by the *radial velocity method*
*Four planets* by *microlensing*
*Four planets* by *imaging*
*Four planets* by *pulsar timing*

From the 200 extra-solar planetary systems (=EPSs) near Sun-like stars (cf. web site of J. Schneider)[3] that have been discovered during the last decade we can distinguish the following systems:

1. Single-star single-planet systems
2. Single-star multi-planet systems
3. Planets in double-star systems

where the latter two are the most interesting ones from the dynamical point of view, since the stable planetary motion is restricted to certain regions of the phase space of a system due to the gravitational interactions between the celestial bodies. Up to now we have knowledge of about 20 multi-planet systems and about 21 binary systems that host one or more planets. An overview thereto will be given in the next two sections.

Besides the study of these systems, an investigation of fictitious terrestrial-like planets in all known EPSs is of great interest. Even if most of the discovered planets are very massive, first detections of smaller planets (of some Earth masses) raises our hopes to find terrestrial-like planets. And maybe some of them move in the so-called *habitable zone* of a Sun-like star, so that the requirements for the evolution of life, like on our Earth, do exist. However, even if we find such planets, we have to examine their habitability – this is an interdisciplinary venture which also includes astrophysical studies. As we know from intensive studies of our own solar system, the evolution of a biosphere is a process over a very long time, which requires long-term stability of the planetary orbits. This fact underlines the importance of dynamical stability studies for the EPS research, which will be discussed in the following sections.

## 7.2 Multi-planet Systems

The necessity to verify the dynamical stability of multi-planet systems was shown by Ferraz-Mello, when a numerical simulation of the system HD82943 ended in a catastrophe after about 50,000 years, using the orbital parameters given by the observations. Therefore, it is quite evident that the determination

---
[3] http://exoplanet.eu

of the orbital parameters is quite a tricky task. Especially, when only few observations for a system are available, the errors of the data set are relatively high, particularly in the eccentricity. For single-planet systems this is not a problem but when a planetary system hosts more than one planet, the stability of the system has to be studied in detail. In principal we can distinguish four classes of multi-planet systems (according to [25]):

*Class Ia – planets in mean-motion resonance (MMR)*: This class contains planet pairs with large masses and eccentric orbits that are relatively close to each other, where strong gravitational interactions occur. Such systems remain stable if the two planets are in mean-motion resonance (examples are given in Table 7.1). There are many studies that examined the stability of such systems (see e.g. [2, 3, 28, 36, 37], and many others).

As an example we show the planet pair of HD82943, which is in 2:1 MMR and where both initial configurations – i.e., the two peri-astron are aligned or anti-aligned (see Fig. 7.1) – lead to stable solutions (cf. E. Bois, private communication). Another interesting system is the "old" HD160691 one, where two planets were found in 2:1 MMR. A detailed dynamical study of the two planets [11] showed that (1) they have to be in resonance, (2) combined with an apsidal secular resonance, (3) in an anti-aligned configuration and (4) a high eccentricity motion of the outer planet is needed for the stability of this system. Even if the system parameters changed totally, this stability mechanism can be helpful for other similar systems that may be found in the future.

*Class Ib – low-eccentricity near-resonant planet pairs*: In this class, a mean-motion resonance of the planet pair is not needed to guarantee the long-term stability of the system. Therefore, the eccentricities of the planets have to be

**Table 7.1.** Some examples for planet pairs in MMR (class Ia):

| Star mass [$M_{Sun}$] | Planet | Mass$_P$ [$M_{Jup}$] | $a_P$ (AU) | $e_P$ | Period (days) |
|---|---|---|---|---|---|
| GJ 876 | b | 0.597 | 0.13 | 0.218 | 30.38 |
| (0.32) | c | 1.90 | 0.21 | 0.029 | 60.93 |
| 55 Cnc | b | 0.784 | 0.115 | 0.02 | 14.67 |
| (1.03) | c | 0.217 | 0.24 | 0.44 | 43.93 |
| HD82942 | b | 1.7 | 0.75 | 0.39 | 219.5 |
| (1.15) | c | 1.8 | 1.18 | 0.15 | 436.2 |
| HD202206 | b | 17.5 | 0.83 | 0.433 | 256.2 |
| (1.15) | c | 2.41 | 2.44 | 0.284 | 1296.8 |
| HD160691* | b | 1.7 | 1.5 | 0.31 | 638 |
| (1.08) | c | 1.0 | 2.3 | 0.8 | 1300 |

* Note that the parameter set for the planet pair HD160691 is an old one, the configuration of this system has changed during the last year twice.

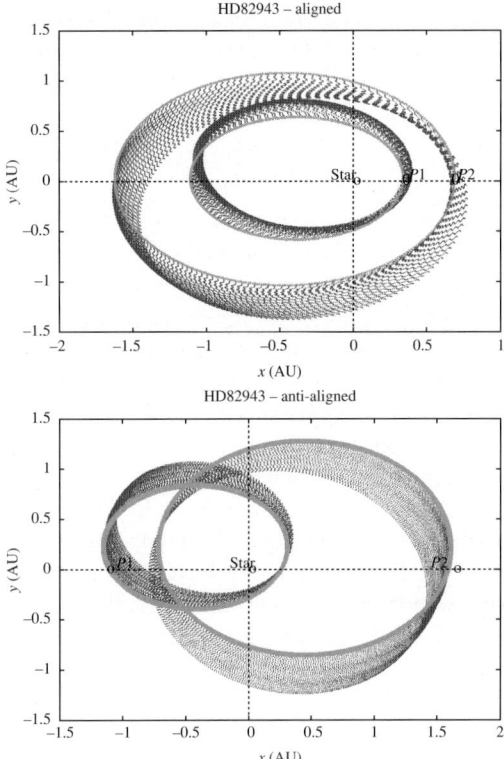

**Fig. 7.1.** Possible orbits of the two planets in HD82943 in the 2:1 MMR where the peri-astron positions are aligned (**upper panel**) or anti-aligned (**lower panel**)

small to exclude a crossing of the orbits. As an example we state the 47Uma system, which is possibly near the 5:2 or 7:3 MMR but the orbital parameters show that a resonance is not needed for the stability of the system (Table 7.2).

*Class II – non-resonant planets with significant secular dynamics*: Planet pairs of this class can have strong gravitational interactions, where long-term variations are ascribed to secular perturbations, large variations of the eccentricities and dynamical effects like the alignment and anti-alignment for the apsidal

**Table 7.2.** An example for a planet pair near a MMR (class Ib)

| Star Mass [$M_{Sun}$] | Planet | Mass$_P$ [$M_{Jup}$] | $a_P$ (AU) | $e_P$ | Period (days) |
|---|---|---|---|---|---|
| 47Uma (1.03) | b | 2.9 | 2.1 | 0.05 | 1079.2 |
|  | c | 1.1 | 4.0 | 0.0 | 2845.0 |

**Table 7.3.** Some examples for planets of class II

| Star Mass [$M_{Sun}$] | Planet | Mass$_P$ [$M_{Jup}$] | $a_P$ (AU) | $e_P$ | Period (days) |
|---|---|---|---|---|---|
| 55 Cnc | e | 0.045 | 0.038 | 0.17 | 2.808 |
| (1.03) | b | 0.784 | 0.115 | 0.02 | 14.67 |
| HD169830 | b | 2.88 | 0.81 | 0.31 | 225.62 |
| (1.4) | c | 4.04 | 3.6 | 0.33 | 2102 |
| HD37124 | b | 0.72 | 0.54 | 0.1 | 153 |
| (0.91) | c | 1.3 | 2.5 | 0.7 | 1595 |

We have to note that the parameters of planet HD37214 c is still uncertain.

lines [43]. For the long-term stability of such a system, it is not necessary that the planets are in MMR (Table 7.3).

*Class III – hierarchical planet pairs*: Roughly speaking this class is for all planet pairs with a large ratio of their orbital periods – $P1/P2 > 10$ (see Table 7.4). Due to the large ratio of periods, the gravitational interaction are not so strong like in class II and the probability of a capture in a MMR is negligible. The weaker interactions lead to stable motion in the numerical simulations, even if the orbits of the planet are not so well determined. For a detailed description of the interesting topic of multi-planet systems we refer the reader to [25].

**Table 7.4.** Some examples for hierarchical planet pairs

| Star Mass [$M_{Sun}$] | Planet | Mass$_P$ [$M_{Jup}$] | $a_P$ (AU) | $e_P$ | Period (days) |
|---|---|---|---|---|---|
| HD168443 | b | 7.7 | 0.29 | 0.529 | 58.116 |
| (1.01) | c | 16.9 | 2.85 | 0.228 | 1739.5 |
| HD74156 | c | 1.86 | 0.294 | 0.636 | 51.643 |
| (1.27) | b | 6.17 | 3.40 | 0.583 | 2025.0 |
| HD38529 | b | 0.78 | 0.129 | 0.29 | 14.309 |
| (1.39) | c | 12.7 | 3.68 | 0.36 | 2174.3 |

## 7.3 Binary Systems

Stability studies of binary systems are very important, since it is well known that more than 60% of the stars build double- or multiple-star systems (at least in the solar neighborhood). Due to this fact, we expect that in the future more and more planets will be found in binary systems. From the dynamical point of view, we distinguish three types of motion in such systems ([14]):

1. *The satellite-type (or S-type) motion*, where the planet moves around one stellar component
2. *The planet-type (or P-type) motion*, where the planet surrounds both stars in a very distant orbit
3. *the libration-type (or L-type) motion*, where the planet moves in the same orbit as the secondary but 60° before or behind; furthermore, they are locked in 1:1 mean motion resonance.

Since the stability of L-type motion is limited to mass-ratios of the binary $< 1/26$, this motion is not so interesting for double stars. Therefore, we will restrict our discussion to the first two types.

Long before the first planet in a binary system has been discovered, astronomers, working in dynamical astronomy, carried out stability studies for the different types of motion (see, e.g. [14, 15, 16, 17, 32, 38, 46, 47, 50] three-body problem (ER3BP)[4] for numerical simulations). Between 1988 and 1998 Benest [4, 5, 6, 7, 8, 9], studied in a series of papers several binaries. The discovery of planets in such systems encouraged other research groups to examine special double-star systems (see e.g., [18, 19, 20, 31, 49]).

Additionally, there are investigations that used the general three-body problem: see e.g., [10, 27, 30] and more recently by [33] or [44].

Following the observations about planets in double stars (see Table 7.5), one can clearly see that they have to move in S-type orbits due to the fact that most of these systems are wide binaries, where the distance between the two stars is more than 100 AU.

### 7.3.1 S-Type Motion

Most of the general stability studies of S-type motion[5] in the planar ERTBP determined the stable region as a function of the binary's eccentricity, where the motion of the planet is initially circular (see, e.g., [50] or [32]). Only the numerical investigation by [46] analyzed also the influence of the planet's eccentricity. The three cited works determined the stable regions of planetary motion in a similar way. The host star about which the planet (which is the massless body) moves is always $m_1$.

*The initial conditions of the binaries are:* a fixed semi-major axis of 1 AU, a variation of the eccentricity between 0 and 0.9 with a step of 0.1 and two starting positions for the second star $m_2$ that are the peri-center and the apo-center.

---

[4] The elliptic restricted three-body problem describes the motion of a massless body in the gravitational field of two massive bodies; the so-called primaries move in elliptic orbits (Keplerian motion) around their center of mass, without being influenced by the massless body.
[5] S-type motion is also called circumstellar motion.

**Table 7.5.** Some of the detected extra-solar planets in double stars (E. Günther from the Tautenburg Observatory, 2006, private communication)

| Star | $a_b$ (AU) | $a_{pl}$ (AU) | $M_{pl} \sin i$ [$M_{Jup}$] |
|---|---|---|---|
| HD40979 | ∼6400 | 0.88 | 3.16 |
| Gl 777 A | ∼3000 | 3.65 | 1.15 |
| HD80606 | ∼1200 | 0.469 | 3.90 |
| 55 Cnc | ∼1065 | 0.11 | 0.84 |
|  |  | 0.24 | 0.21 |
|  |  | 5.9 | 4.05 |
| Ups And | ∼750 | 0.059 | 0.69 |
|  |  | 0.83 | 1.96 |
|  |  | 2.56 | 3.98 |
| 16 Cyg B | ∼700 | 1.66 | 1.64 |
| HD178911 | ∼640 | 0.32 | 6.29 |
| HD219542 | ∼288 | 0.46 | 0.25 |
| tau Boo | ∼240 | 0.05 | 4.09 |
| HD142 | ∼200 | 0.98 | 1.00 |
| HD195019 | ∼150 | 0.14 | 3.55 |
| HD114762 | ∼130 | 0.35 | 11.0 |
| HD19994 | ∼100 | 1.30 | 2.00 |
| γ Cep | ∼18.7 | 2.1 | 1.76 |
| Gl 86 | ∼20 | 0.11 | 4.00 |
| HD41004 AB | 23 | 1.7 | 2.64 |

For the test-planets we used semi-major axes between 0.1 and 0.7 AU with various steps $\delta a$ and four starting positions for each orbit (i.e., mean anomaly $= 0°$ or $90°$ or $180°$ or $270°$). The initial eccentricity was 0 in [50] and [32] and was varied between 0 and 0.5 with a step of 0.1 for all mass-ratios and in some cases up to 0.9 (see [46]).

We determine the orbital behavior by means of the fast Lyapunor indicator (FLI [26] – for details see Sect. 2.4 this volume) which are computed for 1000 periods[6] of the binary. A summary of the results is given in Table 7.6, where one can see the extension of the stable zone for different mass-ratios. The results for $e_{binary} = 0$ are in good agreement with those of the two other studies. In principal we have seen that the reduction of the stable zone due

---

[6] Test computations for a longer time (10,000 or 100,000 periods) did not change the result significantly, therefore we used the shorter computation time for the whole sample.

**Table 7.6.** Stable zone (in dimensionless units) of S-type motion

| Mass-ratio $\mu = m_2/(m_1 + m_2)$ | $e_{binary}$ | Stable zone $e_{planet} = 0$ | $e_{planet} = 0.5$ |
|---|---|---|---|
| 0.1 | 0   | 0.45 | 0.36 |
|     | 0.5 | 0.18 | 0.13 |
| 0.2 | 0   | 0.40 | 0.31 |
|     | 0.5 | 0.16 | 0.12 |
| 0.3 | 0   | 0.37 | 0.28 |
|     | 0.5 | 0.14 | 0.11 |
| 0.4 | 0   | 0.30 | 0.25 |
|     | 0.5 | 0.12 | 0.07 |
| 0.5 | 0   | 0.27 | 0.22 |
|     | 0.5 | 0.12 | 0.07 |
| 0.6 | 0   | 0.23 | 0.21 |
|     | 0.5 | 0.10 | 0.07 |
| 0.7 | 0   | 0.20 | 0.18 |
|     | 0.5 | 0.09 | 0.07 |
| 0.8 | 0   | 0.16 | 0.16 |
|     | 0.5 | 0.09 | 0.05 |
| 0.9 | 0   | 0.13 | 0.12 |
|     | 0.5 | 0.06 | 0.04 |

to an increase of the binary's eccentricity is between 0.07 AU (i.e., for the initially circular motion in a binary with $\mu = 0.9$) and 0.28 AU (i.e., for the initially circular motion in a binary with $\mu = 0.1$). Even if the dependence of the size of the stable region on the eccentricity of the planet is not so strong, it is not negligible, especially if a planet is close to the border of chaotic motion and moves in a highly eccentric orbit.

A presentation of the 3D stability plots of all mass-ratios and a detailed discussion of this work is given in [46]. As an example we show the results for $\mu = 0.2$ (see Fig. 7.2) as an application to the real binary $\gamma$ Cephei that hosts a giant planet of 1.76 Jupiter masses.

**Application to Real Systems**

$\gamma$ Cephei is one of the most interesting double-star systems that host a planet. It is about 11 pc away from our solar system. The binary consists of a K1 IV star (of 1.6 solar masses) – around which the detected planet moves – and a M4 V star (of 0.4 solar masses). Thus the mass-ratio ($m_2/(m_1 + m_2)$) of this system is 0.2. An overview of the size of the stable region for this mass-ratio is given in Fig. 7.2, that shows for each ($e_{binary}, e_{planet}$) pair on the $(x,y)$ plane the respective extension of the stable zone ($z$-axis), which is defined by the semi-major axis of the last stable orbit (corresponding to the largest distance of the planet to its host star, where all computed orbits remained

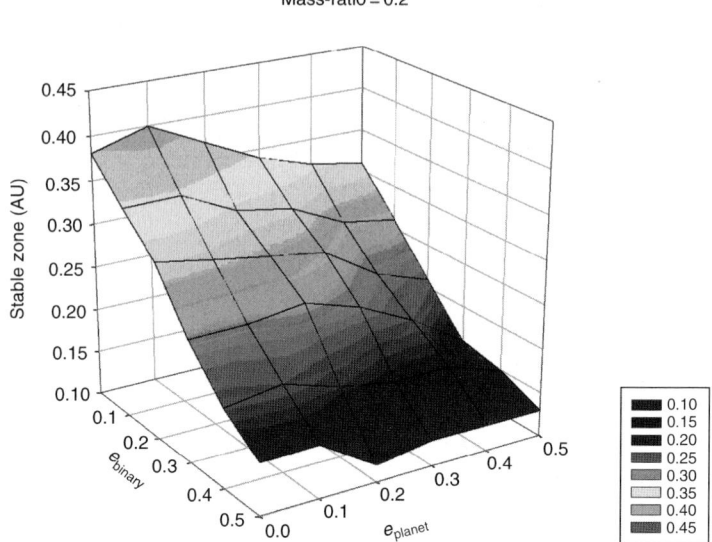

**Fig. 7.2.** The size of the stable zone of S-type motion in a binary with mass-ratio $\mu = 0.2$ depending on the eccentricity of the binary (x-axis) and of the planet (y-axis). It is clearly seen that the variation of $e_{\text{binary}}$ influences the extension of the stable zone stronger than the variation of $e_{\text{planet}}$

stable for 1000 time units).[7] Taking the orbital parameters of this system: semi-major axis ($a_{\text{binary}}$) and eccentricities ($e_{\text{binary}}$) of the binary as well as the eccentricity of the planet ($e_{\text{planet}}$), we get for the old orbital parameter set ($a_{\text{binary}} \sim 22\,\text{AU}$, $e_{\text{binary}} = 0.44$ and $e_{\text{planet}} = 0.209$) an extension of the stable zone up to 3.6 AU and for the new parameter set ($a_{\text{binary}} \sim 23\,\text{AU}$, $e_{\text{binary}} = 0.36$ and $e_{\text{planet}} = 0.12$) up to nearly 4 AU.

The work by Holman and Wiegert [32] is often used to confirm the stability of a detected planet in a binary. However, in Fig. 7.3b we show that one has to be careful, especially in the case of eccentricity motion of the planet, since the study by Holman and Wiegert gives a larger stable zone due to the limitation of circular planetary motion. This is well visible in panel 7.3b, where the two dashed lines show the zone for the stability borders defined by the relation given in [32]. The plotted results of all eccentricities of a planet show that the majority of the stability borders is outside the given zone by Holman and Wiegert. Figure 7.3a shows the results for circular orbits (full line with crosses)and for $e_{\text{planet}} = 0.1$ (dashed line with stars). Furthermore, the dotted line with white squares shows the result obtained by [32]. It is clearly seen that the results for the circular problem ($e_{\text{binary}} = 0$) are the same for the three cases, but for the elliptic problem ($e_{\text{binary}}$ from 0.1 to 0.8)

---

[7] Test computations for $\mu = 0.3, 0.5$ and $0.7$ up to 100,000 time units showed the same qualitative results.

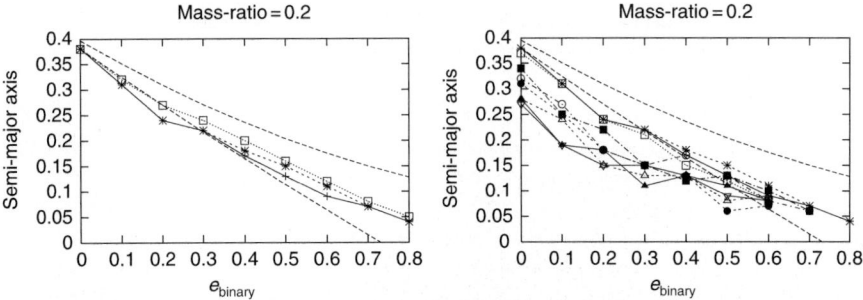

**Fig. 7.3.** A comparison of our results with those of [32]. The area between the two *dashed lines* defines the zone for the stability border according to the relation given in [32]; (**a**) shows our results for $e_{planet} = 0$ (*full line with crosses*) and $e_{planet} = 0.1$ (*dashed line with stars*) and the result of Holman and Wiegert (*dotted line with white squares*). (**b**) The results for all $e_{planet}$ (from 0 to 0.9) in comparison with the theoretical zone for the borderline of stability

the new determined stability borders are closer to the host-star. Moreover, the results for $e_{planet} = 0$ and $0.1$ are the same up to $e_{binary} = 0.3$ and again for $e_{binary} \geq 0.7$; and two cases ($e_{binary} = 0.1$ and $0.2$) are outside the zone determined with the relation of Holman and Wiegert. Therefore, we claim to verify the stability through numerical computations especially if the detected planet is quite close to the border of the stable zone.

The application to the other two "close" binary systems, Gliese 86 and HD 41004 AB shows obviously the stability for the detected planet Gl 86 Ab, since it is a close-in planet with a semi major axis of 0.11 AU. For the detected planet of HD 41004 A we have three parameter sets, where the position of the planet varies between 1.31 and 1.7 AU and the eccentricity of the planet seems to be quite high (between 0.39 and 0.74). Since we have no indication for the binary's eccentricity, we have determined the stable zone for the different parameter sets as a function of $e_{binary}$. Depending on the planet's eccentricity, $e_{binary}$ has to be $<0.6$ in all cases, and even $<0.15$ when $e_{planet} = 0.74$, otherwise the detected planet would not be in the stable region. A detailed study of this system can be found in [49].

**Influence of the Secondary**

To study the influence of the secondary, we examined in two systems ($\gamma$ Cephei and HD 41004 AB) the region between the host star and the detected planet with (left panels of Fig. 7.4) and without (right panels of Fig. 7.4) secondary. A comparison of the two results show significant differences. In the system $\gamma$ Cephei, the presence of the perturbing star (see upper left panel Fig. 7.4) decreases the stable region (i.e., the faint region in the panels) and shows an arc-like chaotic path with a stable island around 1 AU (which corresponds to the 3:1 MMR). The same study for HD 41004 AB does not show a significant

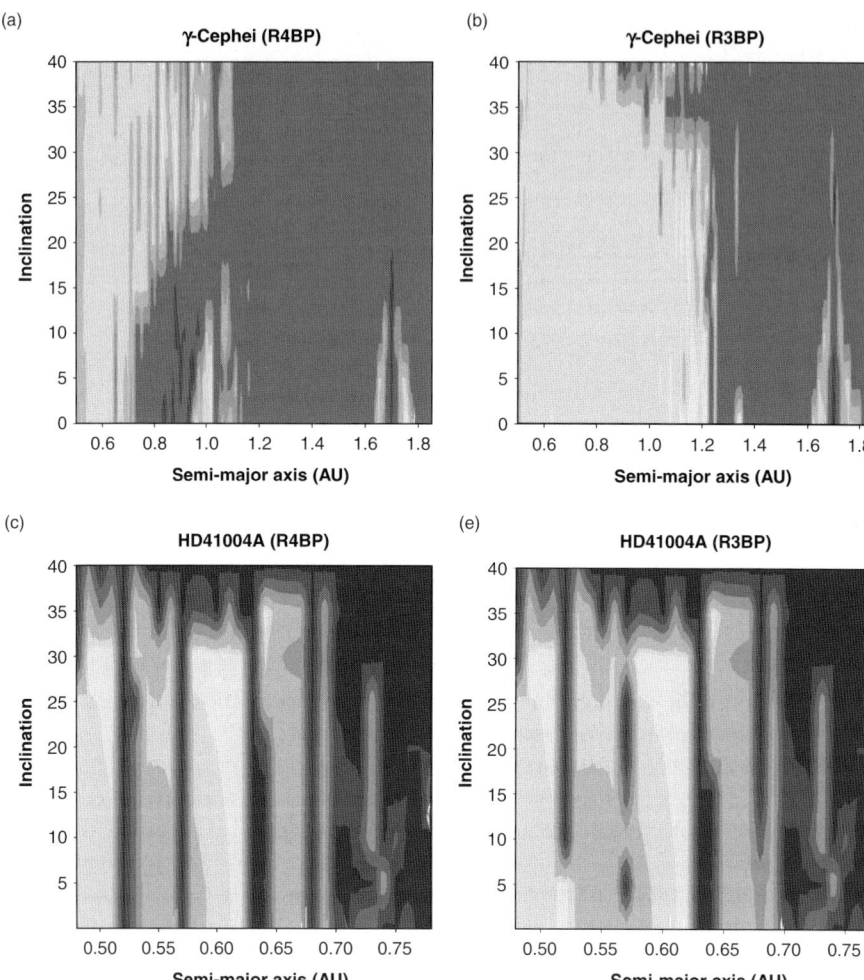

**Fig. 7.4. Upper two panels**: Stability maps for a fictitious planet in the vicinity of $\gamma$ Cephei: **left panel** shows the FLI-result in the restricted four body problem (R4BP) (i.e., $\gamma$ Cephei + secondary + detected planet + fictitious planet) and **right panel** shows the FLI-result in the restricted three body problem (R3BP) (i.e., $\gamma$ Cephei + detected planet + fictitious planet). The *dark region* shows the chaotic zone and the *white area* the stable one. **Lower two panels**: Stability maps (i.e., maximum-e maps) for a fictitious planet in the vicinity of HD 41004 A: **left panel** shows the result in the R4BP (i.e. binary HD 41004 AB + detected planet + fictitious planet) and **right panel** (b) shows the result in the R3BP (i.e., HD 41004 A + detected planet + fictitious planet). The different *gray scales* indicate the zones of different maximum ecentricities: $e < 0.2$ (*white*), $0.2 < e < 0.3$ (*light grey*), ..., $e > 0.8$ (*black*, i.e., unstable region)

difference between the two models for eccentricities of the binary <0.3 (compare the two lower panels in Fig. 7.4).

A first study about this different behavior is given in [48], where a variation of the semi-major axis of the detected giant planet in both systems shows the following: when the giant planet is close enough to the host star (i.e., around 1.3 AU for the two binaries) the region is mainly perturbed by MMRs with respect to the giant planet. A curved chaotic structure appears if the giant planet is shifted toward the secondary, so that a secular perturbation occurs, since the secondary will cause a precession of the perihelion of the giant planet (since we started all massive bodies in the same plane a precession of the ascending node cannot be modeled).

### 7.3.2 P-Type Motion

This type of motion in a binary is for very close double stars, since the planet moves in a wide orbit around both stars. Our stability study of P-type motion in the spatial elliptic restricted three body problem defines the border of the stable zone depending on the eccentricity of the binary and on the inclination of the planetary orbit. The *initial conditions of the binaries* are: a fixed semi-major axis to 1 AU, a variation of the eccentricity between 0 and 0.5 with $\delta e = 0.05$ and two starting positions for the binary at peri-astron and the apo-astron. For the planets we vary the initial distance of the planet to the barycenter of the binary between 1.8 and 4.5 AU with a step of 0.05 AU and the inclination of the planetary orbit to the plane of the binary is increased from 0° to 50° with a step of 2.5°. We use four starting positions (mean anomaly = 0° or 45° or 90° or 135°) and study circular planetary orbits with respect to the barycenter. To determine the stable region we use two numerical methods (a) orbital computations over 50,000 periods of the binary by means of the Lie-series (see [39] and [29]) and (b) the FLI. From the orbital computations we determine an upper critical orbit (UCO) and a lower critical orbit (LCO) (see [16]) which split the phase space into three zones: a stable zone, an unstable one and in between a so-called mixed zone, where both stable and unstable motions can be found for a certain distance to the barycenter.

Figure 7.5 shows the global result of the UCO in the $(e, i)$-plane, which is the border of the stable zone and therefore defined by the distance to the barycenter where all starting positions indicate stable motion. One can see that an increase of the binary's eccentricity shifts the stable zone to larger distances from the barycenter while the inclination of the planet has in general no significant influence on the UCO; the most visible exception is for $e = 0.5$ and high inclinations. A more detailed study thereto can be found in [47].

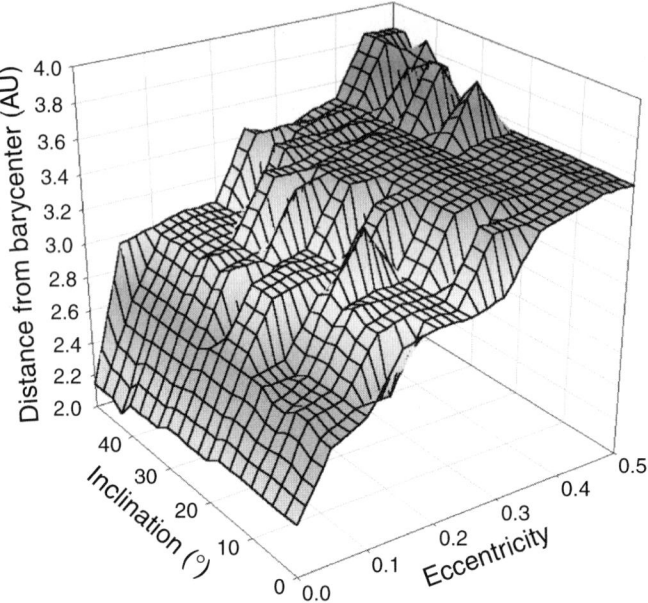

**Fig. 7.5.** The size of the stable zone of P-type motion in a binary with mass-ratio $\mu = 0.5$ depending on the eccentricity of the binary and on the inclination of the planet

## 7.4 Terrestrial Planets in Extra-Solar Planetary Systems

Besides the dynamics of multi-planetary systems and planets in double stars a primary interest is to search for terrestrial-like planets (TP) outside the solar system. In our own system the structure seems to be quite particular: inside the quasi-circular orbits of the gas giants Jupiter, Saturn, Uranus and Neptune four terrestrial planets are orbiting the Sun and other terrestrial-like bodies are satellites of Jupiter (e.g., Europa) and Saturn (e.g., Titan) or are on orbits outside Pluto in the Kuiper belt. Out of the 200 planets known up to now we can observe only large planets from seven earth masses (Gliese 876 d) up to several Jupiter masses. An important question is whether terrestrial planets in extra-solar planetary systems (EPSs) may move there, but our technique is not yet ready to make such observations. Of special interest is whether these planets may move in the so-called habitable zone (HZ) around a host star on stable low eccentric orbits.[8]

---

[8] We will not discuss the complex problem of habitable regions around a host star in detail, because it is still somewhat in contradiction; it depends not only on the dynamical parameters of the orbit of a planet, but also on the astrophysical parameters of the star like the spectral type and the age (Lammer, private communication). A somewhat rough estimate for it is *where water could exist in liquid form on the surface of a planet*; for a more precise definition see [34].

In EPS the "dynamical location" of the HZ depends on where the Jovian planet (gas giant, GG) is moving, and from the configuration we can distinguish four different groups (see Fig. 7.6):

- **G1**: when the GG is very close to the star there could exist stable orbits of TPs for timescales long enough to develop a biosphere.
- **G2**: when this GG moves far away from the central star (like Jupiter) then stable low eccentric orbits for TPs can exist.
- **G3**: when the GG itself moves in the habitable region, a terrestrial-like satellite (like Titan in the system of Saturn) could have a stable orbit.
- **G4**: when the GG itself moves in the habitable region a Trojan-like TP may move on a stable orbit around the Lagrangian equilibrium points $L_4$ or $L_5$.

Besides the extensive study of [42] concerning the stability of orbits of terrestrial planets in extra-solar systems there exist investigations for specific systems for the classes **G1** and **G2**, e.g., [1, 18, 19, 21, 22, 23]. In recent investigations also the dynamical stability of possible terrestrial planets in the 1:1 resonance with the gas giant (**G4**) was studied (e.g., [23]).

We have reasonable theories for the formation of terrestrial planets of all groups in an early stage within the proto-planetary disc; even two planets in a 1:1 mean-motion resonance (**G4**) could be the result of an interaction with the proto-planetary disc [35].

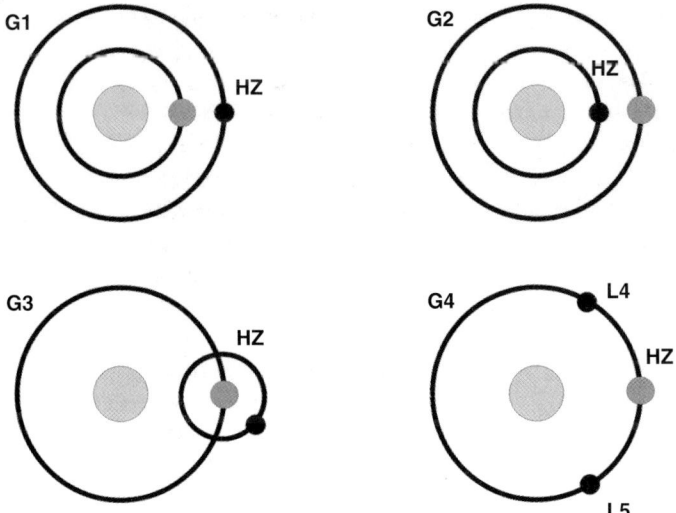

**Fig. 7.6.** Possible habitable regions for terrestrial planets in EPSs (G1 to G4): the *full black circles* show the terrestrial planet, the *light gray full circles* show the gas giant

## 7.5 Theoretical Considerations

For the first three groups there exist many investigations concerning the stability of an additional massless regarded body in the 3D elliptic restricted problem via extensive numerical integrations (e.g., [32, 46, 50]). Quite often for this kind of stability studies the use of supercomputers is appropriate, because of the huge amount of CPU necessary to get good statistical results. Many thousands of orbits are computed simultaneously for a fine grid of initial conditions with such computers.

For **G4** there exist a lot of analytical work concerning the stability of the Lagrangian points in the model of the elliptic restricted problem depending on the mass-ratio of the primaries and the eccentricity of their orbits. A quite important contribution to this topic has been done for cases when the third mass is not regarded as massless [40]. The results of a first-order stability analysis in the framework of the general three-body problem are presented in this book. With $M$ the total mass and $m_1 \leq m_2 \leq m_3$, a mass parameter $R$ was defined as $R = (m_2 + m_3)/M + m_2 m_3/m_1^2 + O(m_2^3 m_3/m_1^4)$. Using these results one can see[9] that in the case of a terrestrial-like planet with a relatively small mass compared to the two primary bodies there is no difference concerning the stability of the equilibrium points. While checking the observed eccentricities of the orbits of the planets in EPSs, the estimated (minimum) mass of the giant planet and the mass of a TP (comparable to the mass of our Earth), it turned out that all planetary systems in the list given by Jean Schneider[10] have stable equilateral Lagrangian points. Surprisingly enough this is also true for planets with large eccentric orbits of the primaries.

To find out the extension of stable regions around the two equilibrium points for the four groups given above purely numerical methods are used. With a suitable integration method (e.g. the Lie integration method, a classical Runge–Kutta, a Bulirsch–Stoer method or a hybrid-symplectic one used by Mercury6 [13]) one solves the equations of motion. For the choice of an appropriate grid of initial conditions one then computes many orbits for a chosen time interval, which is long enough to unveil the stability character of the orbits. Different indicators for the stability of an orbit like the Lyapunov exponents or similar methods ([26, 51]) can be used for that. Another interesting tool is the maximum eccentricity of an orbit achieved during the integration. In all our investigations a stability limit of $e = 0.5$ showed that with such a large eccentricity every orbit interesting for a terrestrial planet in one of the four groups escapes sooner or later because of a close approach to the planet.

In the following we will show – as examples – the results of such investigations for TPs of the groups **G2** and **G4** in the habitable zones of Gl 777A and HD23079.

---

[9] Figure 13, p. 49 in Marchals's book.
[10] http://exoplanet.eu.

## 7.6 Gl 777 A

The first discovery of a planet in Gl 777 A (=HD190360) was reported by Naef et al. [45]. This extra-solar planetary system is a wide binary with a very large separation (3000 AU); for the dynamical investigations of motions close to one star there was no need to take into account the perturbations of the very far companion. The host star is of spectral type G6 IV with $0.9 M_\odot$ and has a planet of mass of $1.33 M_{Jup}$ with a semi-major-axis of 4.8 AU. The large eccentricity ($e = 0.48$) limits the possible region of motion for additional planets to $a < 2.4$ AU (= peri-astron). This is depicted in Fig. 7.7 where one can see that for this system the mean-motion resonances are not important for motions in the habitable zone, because they are all well outside the HZ.

The region of habitability (see Fig. 7.7), where planets could have temperature conditions to allow liquid water on the surface, corresponds roughly to $0.7 < a < 1.3$ AU. The computations were undertaken in a larger region ($0.5 < a < 1.3$) with a grid spacing of $\Delta a = 0.01$ AU and the eccentricity of the known planet was set between $0.4 < e < 0.5$ with a grid size of $\Delta e = 0.01$. The results (Fig. 7.8) of the direct analysis of the largest eccentricity achieved during the orbital evolution for 1 million years show the following: (1) strong vertical lines due to high-order resonances and (2) a growing number of unstable orbits with larger semi-major axes of the terrestrial planet. A similar structure can be observed in the main belt of asteroids due to the perturbations of Jupiter. We checked the eccentricity because it tells us directly the variable distance to the central star and consequently it is a good measure of the differential energy flux (insolation) on the planet. We can therefore determine where the variation of this distance does not exceed 50% which corresponds roughly to an eccentricity of $e = 0.2$.

From these results one can say that for the system Gl 777 A there is quite a good chance that planets will stay within the habitable zone for times which are the necessary condition for developing life in the region with $a < 1$ AU.

## 7.7 The G4 Terrestrial Planet in HD108874

**Fig. 7.7.** Main characteristics of the extra-solar system Gl 777 A. The *light gray region* shows approximately the position of the HZ; the *dark gray bar* indicates how closely the planet approaches the central star in its orbit. The different numbers characterize the mean motion resonances

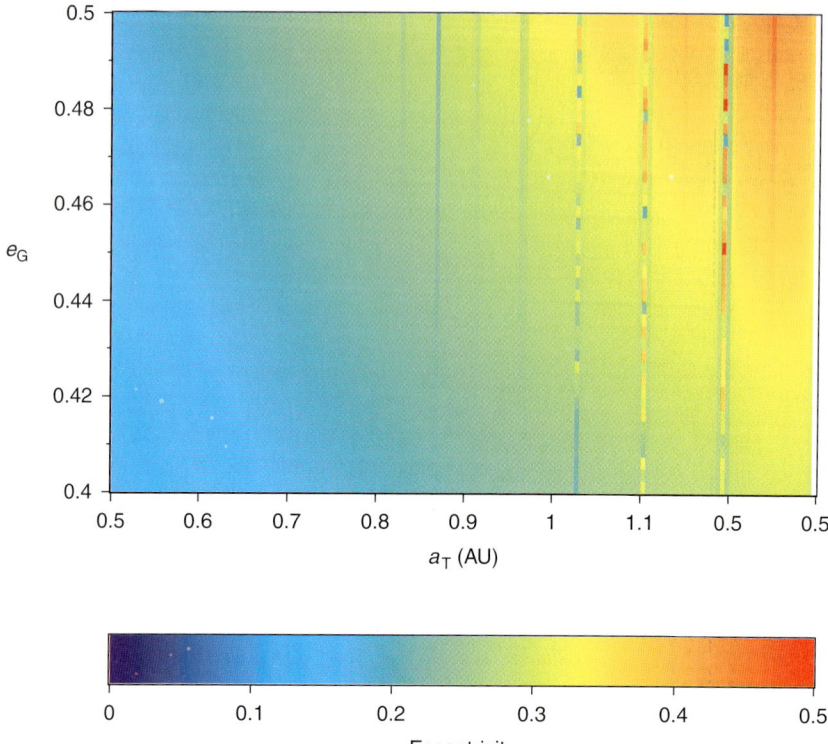

**Fig. 7.8.** Initial condition diagram of semi-major axis of the fictitious terrestrial planet versus the eccentricity of the GG of the extra-solar system Gl 777 A. The *colors* indicate the maximum eccentricity achieved during the integration of 1 million years

Several numerical studies are devoted to find stability regions of type **G4** in EPSs like [22, 23, 24, 53, 54]. There is one good candidate, namely HD108874 ([12]) for hosting a Trojan terrestrial planet. The Sun-like star was known to have only one massive planet with approximately Jupiter mass and an orbit with a small eccentricity in the HZ. Recently a second planet was observed outside the orbit of the already known one ([55]). In Table 7.7 we show the orbital parameters derived from observations with the given error bars.

To take into account the possible observational errors especially of the semi-major axis and the eccentricity of the perturbing outer giant we computed orbits of fictitious Trojan planets in nine different models. We took the nominal values, but also lower and upper bounds according to the given error bars in $e$ and $a$:

- Models **M1×**, namely **M11, M12** and **M13** with $a = 2.43$ AU and three values for the eccentricity $e = 0.18, 0.25$ and $0.42$

**Table 7.7.** Orbital elements of the EPS HD108874

| Name | HD108874 b | HD108874 c |
|---|---|---|
| Mass [$M_{Jup}$] | 1.36($\pm$ 0.13) | 1.018 ($\pm$ 0.03) |
| a (AU) | 1.051 $\pm$ 0.02 | 2.68 $\pm$ 0.25 |
| Orbital period (days) | 395.4($\pm$2.5) | 1605.8($\pm$88) |
| Eccentricity | 0.07($\pm$ 0.04) | 0.25($\pm$ 0.07) |
| $\omega$(deg) | 248.4($\pm$ 36) | 17.3($\pm$ 23) |

- Models **M2**×, namely **M21**, **M22** and **M23** with $a = 2.68$ AU and again $e = 0.18, 0.25$ and $0.42$
- Models **M3**×, namely **M31**, **M32** and **M33** with $a = 2.93$ AU and $e = 0.18, 0.25$ and $0.42$

It turned out that the model **M1x** for all three different eccentricities does not possess any stable orbits of Trojans around $L_4$. For **M12** and **M13** this is due to the chaoticity of the orbits of the two planets themselves; also **M23** is in such a chaotic domain. The orbits of the fictitious Trojans for the largest initial eccentricity ($e = 0.32$) in models **M23** and **M33** also do not have any orbit staying in the vicinity of $L_4$ for timescales larger than 1 Million year. The stable region for the Trojan planets is shown in Fig. 7.9 (upper graph) for **M21**, where we can see that this region around the equilibrium point $L_4$ (located at $a = 1.051$ and $\omega = 308°$) extends from $295° < \omega < 325°$ and $1.035$ AU $< a < 1.075$ AU. For **M22** in Fig. 7.9 (lower graph) we see only some spots of stable orbits which we suspected to disappear for a longer integration. In Fig. 7.10 (upper graph) for **M31** we observe again a slightly larger stable region around the equilibrium point which extends from $290° < \omega < 325°$ and $1.025$ AU $< a < 1.08$ AU. This can be understood that in this model the perturbing planet is farther away from the inner planet and – which is maybe more important – also far away from any low-order MMR. We show the extension in semi-major axes and $\omega$ in Table 7.8 for different models. The stable region in model **M32** almost disappears for a larger eccentricity of the outer planet.

## 7.8 Concluding Remarks for Possible Terrestrial Planets

Up to now many investigations were carried out to examine the stability of additional TPs in different EPSs. In almost all investigations the uncertainties of the orbital elements derived from the observations are not taken into account. Furthermore, new observational runs for the different systems could change the orbital parameters and therefore the stability of a fictitious TP drastically. Then one has to re-explore the phase space of the individual EPS after each modification of the orbital parameters of the discovered giant planet. However, this is not necessary when a series of stability maps are available that help to analyze the dynamical state in an EPS.

(a)

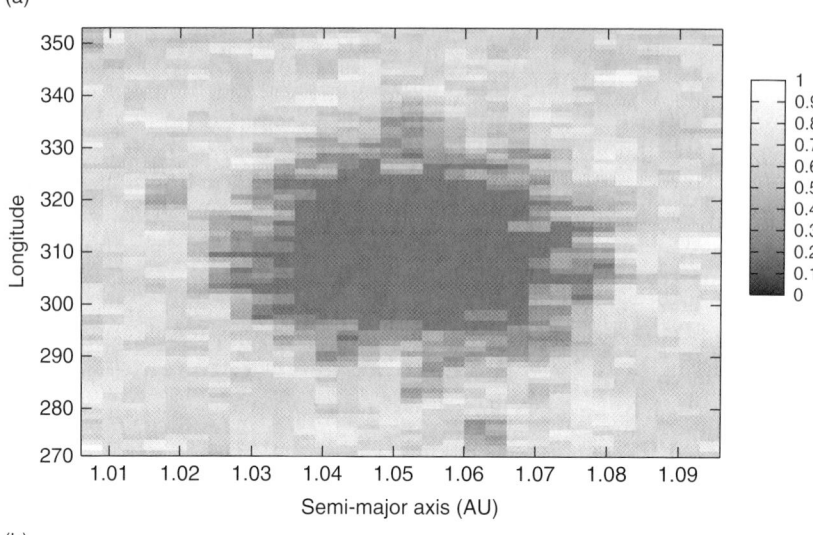

(b)

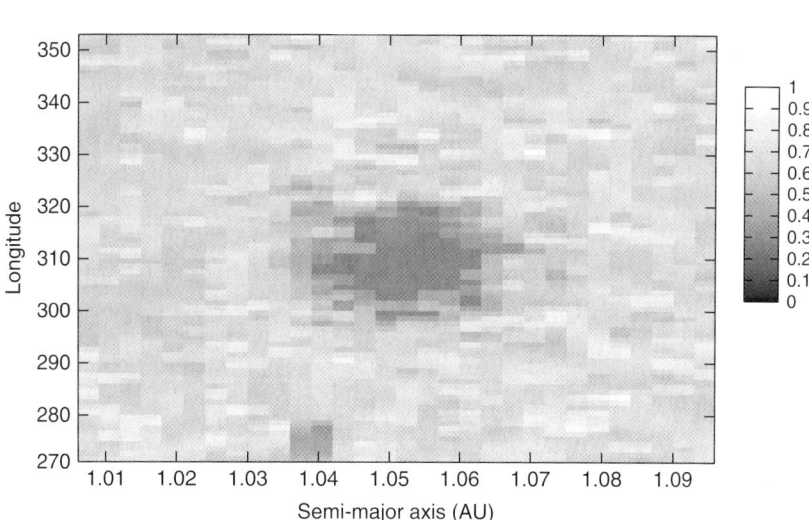

**Fig. 7.9.** Stability regions around $L_4$ in models **M21** and **M22**. The initial semi-major axes ($y$-axis) is plotted versus the initial angular distance from the equilibrium point $L_4$. The *dark regions* around the point $L_4$ ($a = 1.051$ AU, $\omega = 308°$) show that with such initial conditions the maximum eccentricity during the integration time was always $< 0.3$

(a)

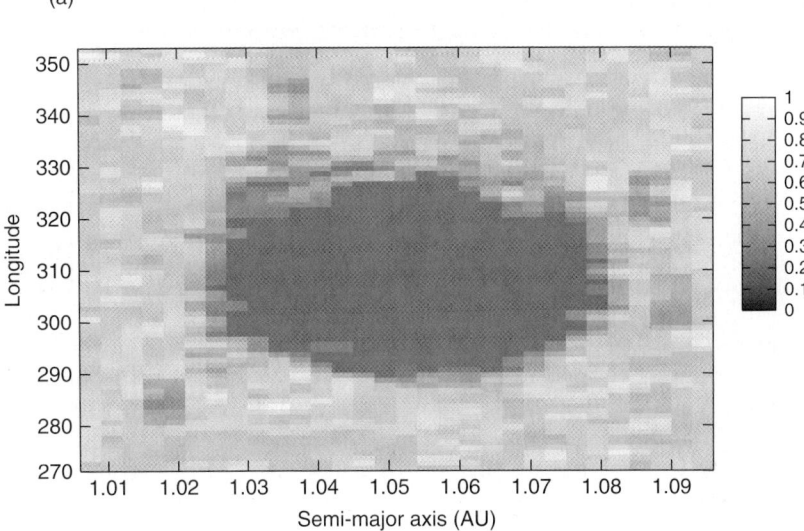

(b)

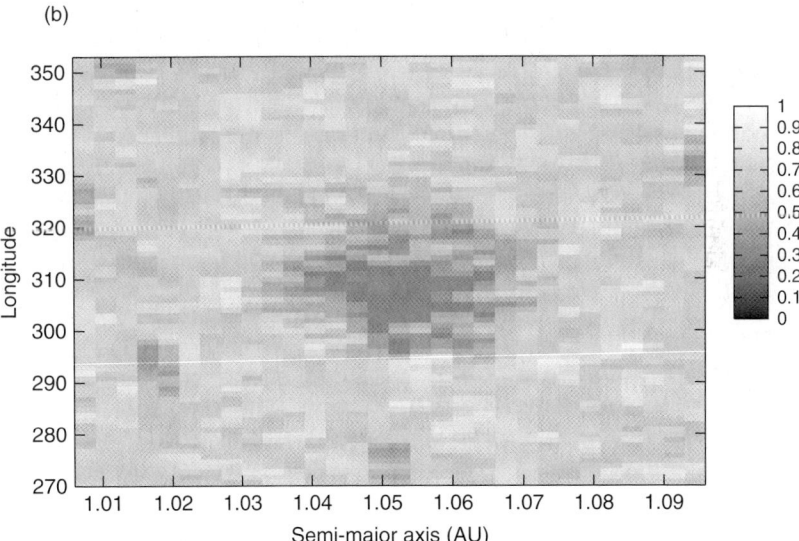

**Fig. 7.10.** Stability regions around $L_4$ in models **M31** and **M32**. The initial semi-major axes ($y$-axis) is plotted versus the initial angular distance from the equilibrium point $L_4$. The *dark regions* around the point $L_4$ ($a = 1.051$ AU, $\omega = 308°$) show that with such initial conditions the maximum eccentricity during the integration time was always $< 0.3$

**Table 7.8.** Extensions of stable regions for different models with respect to the semimajor axes and the angular distance to the Lagrange point $L_4$

| Stability parameter | M21 | M22 | M31 | M32 |
|---|---|---|---|---|
| $a$ (AU) | 2.68 | 2.68 | 2.93 | 2.93 |
| $e$ | 0.18 | 0.25 | 0.18 | 0.25 |
| $\delta\omega(°)$ | 30 | 14 | 35 | 10 |
| $\delta a$ (AU) | 0.04 | 0.015 | 0.045 | 0.01 |

Therefore, we have – in collaboration with our colleagues from the University of Budapest – compiled a catalogue of stability maps for the single-star single-planet systems ([52]). With the aid of this catalogue[11] the stability properties for the **G1** and **G2** habitable zones, for different mass-ratios (between the host star and the giant planet – from 0.0001 to 0.01) *and* for different eccentricities (0.0–0.5) of the giant planet can be determined easily. The dynamical framework is the elliptic restricted three-body problem so that the terrestrial planets' mass was ignored, but this does not restrict the usage of the catalog – as it was explained before. Each plot shows for a fine grid in the initial conditions – semi-major axes of the fictitious TPs versus eccentricity of the giant planet – the orbital state of the respective region. The orbital behavior was classified with different methods, namely the RLI, the FLI and the maximum eccentricity. The catalog consists of more than 90 stability diagrams and helps in the case of any new orbital determination – leading to new elements – that the stability of the systems can be checked immediately without any time-consuming computations. Even if the application is restricted to single-star single-planet systems we have to point out that most of the discovered EPSs join this dynamically simple system.

## Acknowledgments

EP-L wishes to acknowledge the support by the Austrian FWF (Hertha Firnberg Project T122). RD wishes to acknowledge the support by the FWF Project P16024. This study was also supported by the International Space Science Institute (ISSI) and benefits from the team "Evolution of Habitable Planets".

## References

1. Asghari, N., Broeg, C., Carone, L., Casas-Miranda, R., Castro Palacio, J.C., Csillik, I., Dvorak, R., Freistetter, F., Hadjivantsides, G., Hussmann, H., Khramova, A., Khristoforova, M., Khromova, I., Kitiashivilli, I., Kozlowski, S.,

---

[11] http://astro.elte.hu/exocatalogue

Laakso, T., Laczkowski, T., Lytvinenko, D., Miloni, O., Morishima, R., Moro-Martin, A., Paksyutov, V., Pal, A., Patidar, V., Pečnik, B., Peles, O., Pyo, J., Quinn, T., Rodriguez, A., Romano, C., Saikia, E., Stadel, J., Thiel, M., Todorovic, N., Veras, D., Vieira Neto, E., Vilagi, J., von Bloh, W., Zechner, R., and Zhuchkova, E., 2004. *Astron. Astrophys.* **426**, 31.
2. Beaugé, C., Ferraz-Mello, S., and Michtchenko, T., 2003. *ApJ* **598**, 1124.
3. Beaugé, C., Ferraz-Mello, S., and Michtchenko, T., 2003. *MNRAS* submitted.
4. Benest, D., 1988. *A&A* **206**, 143.
5. Benest, D., 1988. *CMDA* **43**, 47.
6. Benest, D., 1989. *A&A* **223**, 361.
7. Benest, D., 1993. *CMDA* **56**, 45.
8. Benest, D., 1996. *A&A* **314**, 983.
9. Benest, D., 1998. *A&A* **332**, 1147.
10. Black, D.C., 1982. *AJ* **87**, 1333.
11. Bois, E., Kiseleva-Eggleton, L., Rambaux, N., and Pilat-Lohinger, E., 2003. *ApJ* **598**, 1312.
12. Butler, R., Marcy, G., Vogt, S. et al., 2002. *AJ* **582**, 455.
13. Chamber, J.E., 1999. *MNRAS* **304**, 793.
14. Dvorak, R., 1984. *CMDA* **34**, 369.
15. Dvorak, R., 1986. *A&A* **167**, 379.
16. Dvorak, R., Froeschlé, Ch., and Froeschlé, C., 1989. *A&A* **226**, 335.
17. Dvorak, R. and Lohinger, E., 1991. In Roy, A.E. (ed.), *Proceedings of the NATO ASI Series*, Plenum Press, New York and London.
18. Dvorak, R., Pilat-Lohinger, E., Funk, B., and Freistetter, F., 2003. *Astron. Astrophys.* **398**, L1–L4.
19. Dvorak, R., Pilat-Lohinger, E., Funk, B. and Freistetter, F., 2003. *Astron. Astrophys.* **410**, L13–L16.
20. Dvorak, R., Pilat-Lohinger, E., Schwarz, R. and Freistetter, F., 2004. *Astron. Astrophys.* **426**, L37–L40.
21. Dvorak, R., Pilat-Lohinger, E., Funk, B., and Freistetter, F., 2003. *A&A* **410**, L13.
22. Érdi, B. and Pál, A., 2003, Dynamics of resonant exoplanetary systems. In Freistetter, F., Dvorak, R., and Érdi, B. (eds), *Proceedings of the 3rd Austrian–Hungarian Workshop on Trojans and Related Topic*, Eötvös University Press, pp. 3–10.
23. Érdi, B., Dvorak, R., Sándor, Zs., and Pilat-Lohinger, E., 2004. *MNRAS* **351**, 1943.
24. Érdi, B. and Sándor, Zs. 2005. Cel. Mech. Dyn. Astron., in press.
25. Ferraz Mello, S. et al.: *Extra-Solar Planetary Systems*, Lect. Notes Phys. **683**, 219–271. Springer, Heidelberg (2005).
26. Froeschlé, C., Lega, E., and Gonczi, R., 1997. *CMDA* **67**, 41.
27. Graziani, F. and Black, D.C., 1981. *ApJ* **251**, 337.
28. Hadjidemetriou, J., 2002. *CMDA* **83**, 141.
29. Hanslmeier, A. and Dvorak, R., 1984. *A&A* **132**, 203.
30. Harrington, R.S., 1977. *AJ* **82**, 753.
31. Holman, M.J., Touma, J., and Tremaine, S., 1997. *Nature* **386**, 254.
32. Holman, M.J. and Wiegert P.A., 1999. *AJ* **117**, 621.
33. Innanen, K.A., Zheng, J.Q., Mikkola, S., and Valtonen, M.J., 1997. *AJ* **113**, 1915.

34. Kasting, J.F, Whitmire D.P., and Reynolds, R.T., 1993. *Icarus* **101**, 108.
35. Laughlin, G. and Chambers, J.E., 2002. *AJ* **124**, 592.
36. Lee, M.H. and Peale, S.J., 2002. *ApJ* **567**, 596.
37. Lee, M.H., 2004. *ApJ* **611**, 517.
38. Lohinger, E. and Dvorak, R., 1993. *A&A* **280**, 683.
39. Lichtenegger, H., 1984. *CMDA* **34**, 357.
40. Marchal, C., 1990. *The Three-Body Problem*, Elsevier, 49.
41. Mayor, M. and Queloz, D., 1995. *Nature* **378**, 355.
42. Menou, K. and Tabachnik, S., 2003. *AJ* **583**, 473.
43. Michtchenko, T. and Malhotra, R., 2004. *Icarus* **168**, 237.
44. Musielak, Z.E., Cuntz, M., Marshall, E.A., and Stuit, T.D., 2005. *A&A* **434**, 355.
45. Naef, D., Mayor, M., Korzennik, S.G., Queloz, D., Udry, S., Nisenson, P., Noyes, R.W., Brown, T.M., Beuzit, J.L., Perrier, C., and Sivan, J.P., 2003. *Astron. Astrophys.* **410**, 1051.
46. Pilat-Lohinger, E. and Dvorak, R., 2002. *CMDA* **82**, 143.
47. Pilat-Lohinger, E., Funk, B., and Dvorak, R., 2003. *A&A* **400**, 1085.
48. Pilat-Lohinger, E., 2005. Dynamics of populations of planetary systems. In Knezevic, Z. and Milani, A. (eds), *Proceedings of IAU Coll. 197*, Cambridge University Press, 71
49. Pilat-Lohinger, E. and Funk, B., 2007, The binary system HD41004 AB. *A&A*, submitted.
50. Rabl, G. and Dvorak, R., 1988. *A&A* **191**, 385.
51. Sandor, Z., Érdi, B., Széll, A., and Funk, B., 2004. *CMDA* **90**, 127.
52. Sandor, Z., Suli., A., Erdi, E., Pilat-Lohinger, E., and Dvorak, R., 2007. *MNRAS* **374**, 1495.
53. Schwarz, R., Pilat-Lohinger, E., Dvorak, R., Érdi, B., and Sándor, Zs., 2005. *Astrobiol. J.*, **5**, 579.
54. Schwarz, R., 2005. Phd thesis, University of Vienna, online database: http://media.obvsg.at/dissd.
55. Vogt, S.S., Butler, P.R., Marcy, W.G., Fischer A.D., Henry, W.G., Laughlin, G., Wright, T.J., and Johnson, A.J., 2005. *AJ* **632**, 638.

# 8

# Origin Theories for the Eccentricities of Extrasolar Planets

Fathi Namouni

CNRS, Observatoire de la Côte d'Azur, BP 4229, 06304 Nice, France
namouni@obs-nice.fr

**Abstract.** Half the known extrasolar planets have orbital eccentricities in excess of 0.3. Such large eccentricities are surprising as it is thought that planets form in a protoplanetary disk on nearly circular orbits much like the current states of the solar system planets.

Possible explanations for the large planetary eccentricities include the perturbations that accompany planet-planet scattering, the tidal interaction between the gas disk and the planets, Kozai's secular eccentricity cycles, the eccentricity excitation during planetary pair migration in mean motion resonance, the perturbations by stellar encounters, stellar-like relaxation that occurs if planets formed through gravitational instability, and the relative acceleration by the stellar jet system of the host star with respect to the companion.

In this chapter, we comment on the relevance and characteristics of the various eccentricity origin theories.

## 8.1 Introduction

Planetary science has known a revolution in the year 1992 when Wolszczan and Frail discovered the first extrasolar planets around a pulsar [68]. This discovery was soon followed in 1995 by the first Jupiter-like planet around the Sun-like star 51 Pegasi [41]. As of this writing we have entered the stage of statistics as more than 200 extrasolar planets are known to orbit Sun-like stars. The revolution that was triggered by these discoveries changed the paradigm of what we think of as a planetary system and what we believe to be the scenarios that led up to the formation of planets. The planet around 51 Pegasi is the prototype of what has become to be known as hot Jupiters: planets with masses comparable to Jupiter's but with periods of a few days and orbits smaller than the tenth of Mercury's distance to the Sun. Planetary orbits also seem to be eccentric. This observation is not surprising since the solar system planets have eccentric orbits except that half the extrasolar planets have eccentricities larger than 0.3 which is much more significant than Jupiter's 0.05. Such large eccentricities are reminiscent of the small body populations

of the solar system that got stirred up by their gravitational interactions with
the larger planets. In this respect, extrasolar planetary eccentricities are more
unusual, witness the median eccentricity of 0.13 of main belt asteroids larger
than 50 km. The planetary revolution did not stop at orbital radii and eccentricities: 10% of known planetary systems belong to binary star systems and
only one planet so far is in a triple stellar system. Multiple planets around a
single star make up about 10% of known planets.

In essence, the planetary revolution heralded the coming of a new planetary
principle: orbital diversity is a rule of planetary formation. Diversity here is
not meant to imply subtle changes but drastic ones with respect to the aspect
of the solar system. This state of affairs has prompted a serious revision of
the theories of planetary formation: hot Jupiters with few-day orbits could
not have formed in situ. Instead they have formed outside a few astronomical
units for a Sun-like star where it is possible for ices to condense and for the
planets to capture large amount of gas from the protoplanetary disk. Only
after they formed, did they travel all the way to meet their current orbits.
It is interesting to note that the concept of radial migration was already
known in the contexts of accretion disks [32] in binary star systems and of
planetary rings [18]. Only before 1995, one could not plausibly contemplate
the prospect of suggesting the existence of massive planets that traveled all
the way from Jupiter's current location just to stop on a close orbit with
a few-day period. The basic aspects of the process of planetary migration
through the tidal interaction of a planet with the gaseous protoplanetary disk
are now well understood [64] yet an important challenge remains: what stops
planetary migration towards the star? The leading contender for stopping
planetary migration is the planet's interaction with the stellar magnetosphere
but a definitive quantitative description is still lacking.

Extrasolar planetary eccentricities have equally resulted in a drastic change
of perception: it is often heard that it is not the extrasolar planets that are
eccentric, rather it is the solar system that lacks eccentricity. This perception
is encouraged by the availability of some simple instabilities that one can set
up in a many-body gravitational system to simulate the generation of the wild
orbits of extrasolar planets. Upon close examination such instabilities as well
as other eccentricity scenarios do not tell the whole story of how extrasolar
planets become eccentric. In fact, just as the features of the planetary migration process yield constraints on the planetary formation scenarios, so do the
various theories of the eccentricity excitation.

It is the aim of this chapter to review the various processes of the origin
of extrasolar planets' eccentricities in the context of planetary formation. We
start by reviewing the properties of extrasolar planetary orbits in Sect. 8.2.
Section 8.3 contains a commentary on the various known theories of eccentricity excitation. Section 8.4 specializes in a recent addition to the eccentricity
theories based on a relationship between the planets and the stellar jet that

is powered by the protoplanetary accretion disk. The final Sect. 8.5 discusses how the eccentricity origin problem may contribute further to the theory of planet formation.

## 8.2 Eccentricity Observations

Extrasolar planets are detected with various observational techniques [54]. The Doppler analysis of the reflex velocity of the host star is by far the most successful technique to date. It is also the technique that has uncovered the large eccentricities of extrasolar planets. If a planet has a circular orbit, the analyzed stellar spectrum yields a sinusoidal oscillation of the stellar reflex motion. If the planet is on an eccentric orbit, the reflex motion as a function of time becomes distorted with respect to a pure sine reflecting the unequal time the star spends in different locations along its orbit around the center of mass of the star–planet system (Fig. 8.1). The discovery of large eccentricity orbits by the Doppler reflex velocity method is due to its ability to detect planets on wider orbits in contrast to that to the transit method. Planets on many-day periods have usually undergone tidal circularization by the host star.

The statistical analysis of extrasolar planet eccentricities reveals very few clues as to the origin of the elongated orbits. For the known sample of 196 planets discovered by the reflex velocity and transit techniques, the median eccentricity is at 0.21 if all planets are counted and at 0.28 if planets with

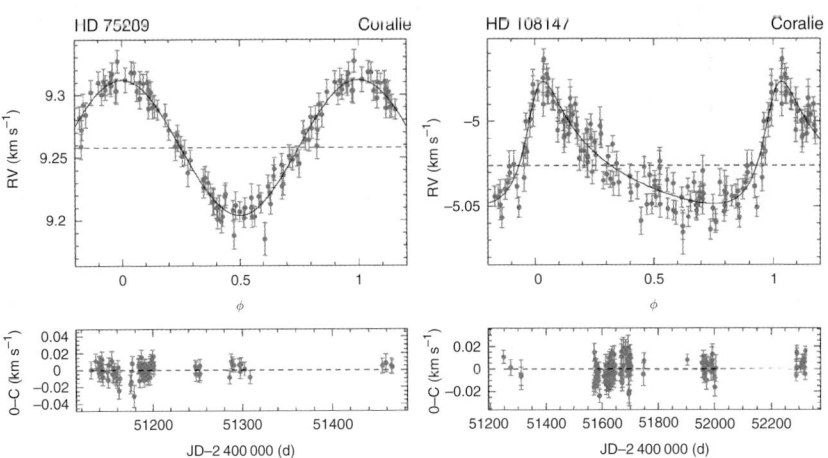

**Fig. 8.1.** Reflex velocity of the stars HD 75289 (**left**) [62] and HD 108147 (**right**) [50]. The planet around HD 75289 has a circular orbit while that around HD 108147 has an orbital eccentricity of 0.5. Pictures taken from the Geneva Extrasolar Planet Search http://obswww.unige.ch/~udry/planet/planet.html

periods smaller than 5 days are excluded because their circular orbits simply reflect tidal circularization. The prevalence of such large eccentricities and the large typical mass of the detected planets (comparable to Jupiter's) has encouraged the comparison of the extrasolar planetary systems to binary star systems. Depending on the methods used, similarities in the eccentricity distribution of both populations can be found [56] or not [21]. What is agreed upon is that there is no correlation between the size of the orbits and their eccentricities in each population, and no striking resemblance of the scatter of both populations orbital size versus eccentricity plane. The size of the orbit usually refers to either the semi-major axis or the pericenter radius. The latter is used to account for those orbits that have not yet had enough time to be circularized—as the pericenter distance is conserved under stellar tides. Finally, eccentricities show a vague correlation with the planetary masses with heavier planets enjoying larger eccentricities.

## 8.3 Eccentricity Origin Theories

Seven known explanations have been put forward to account for the large eccentricities of extrasolar planets. They are: (1) planet–planet scattering, (2) planet–protoplanetary disk interaction, (3) Kozai's secular cycles, (4) excitation through radial migration into a mean motion resonance, (5) stellar encounters, (6) stellar-like $N$-body relaxation and (7) excitation through stellar jet acceleration. In the following, we comment on these possibilities by discussing their instability types, characteristic timescales, their epoch of applicability as well as their advantages and drawbacks.

### 8.3.1 Planet–Planet Scattering

Planet–planet scattering is a simple process to generate eccentric orbits in an $N$-body gravitational system. If a system of two or more planets on planar circular orbits find themselves "initially" closer than is permitted by Chirikov's criterion for the overlapping of mean motion resonances [67], the planets scatter off one another leading to a system with more stable albeit eccentric configurations. Depending on the number, masses and "initial" spacings of the planets, the instability timescale varies between $10^3$ and $10^7$ years [15, 16, 40, 51, 66]. The epoch that is referred to by the adjective "initial" is that of the disappearance of the agent or the conditions that kept the planets from scattering off one another in the first place. This epoch is customarily associated with a significant dispersal of the parent gaseous protoplanetary disk. As well shall point out in the next section, planet–disk interaction is known to primarily erase orbital eccentricities. An additional condition for planet–planet scattering to be operational is the absence of a significant population of smaller bodies such as the primordial asteroid belt. Depending on the mass spectrum in the planetary system, the smaller populations are able to limit the growth of the planetary eccentricities through dynamical friction [4]. This at least how it is believed that the terrestrial planets in the solar system did not acquire large eccentricities [2, 44]. Numerical works that tackle the extrasolar eccentricity problem using planet–planet scattering do not consider the effect of leftover small-body populations after the gaseous disk has dispersed. Planets are set up at a few Hill radii from one another and initial conditions are sampled to reproduce the eccentricity of certain observed systems. The general excitation trend of planet–planet scattering leads to larger eccentricities than the ones observed. What may prove to be a serious problem for planet–planet scattering is the eccentricity distribution obtained in multiple systems that contain Jupiter-mass planets as well as Earth-mass planets. The conservation of angular momentum in this case will force the much smaller planets to have much larger eccentricities than the Jovian planets.

### 8.3.2 Planet–Disk Interaction

A planet embedded in a gaseous disk excites sound waves at the locations of its mean motion resonances within the disk akin to the gravity waves excited by Saturn's satellites in its ring system. The density enhancements at the mean motion resonances act back on the planet resulting in gravitational torque. Two types of resonances contribute to this torque: (1) corotation resonances that primarily affect the semi-major axis and tend to damp any acquired eccentricity and (2) Lindblad resonances that primarily affect the eccentricity and tend to increase it [17, 18]. The torque contribution of the former is larger than the latter's by about 5%. At first sight, planet–disk interaction damps the eccentricity on timescale that depends strongly on the disk's thickness and less strongly on the disk's mass density and the planet's mass [1, 37, 63]. The torques originating from higher order resonances as well as those pertaining to the relative inclination of the planet and the disk do not change the outcome significantly [19, 45]. Only if the corotation torque saturates, can the Lindblad resonances increase the eccentricity [19, 20]. The conditions under which saturation arises are difficult to quantify explaining why an eccentricity increase due to a disk–planet interaction has never been observed in numerical simulations although this might be due to numerical artifacts [38].

### 8.3.3 Secular Kozai Cycles

In his study of asteroids perturbed by Jupiter on high eccentricity and inclination orbits, Kozai [28] showed that the averaging of the interaction potential over the mean motion without expanding the force amplitude in terms of eccentricity and inclination leads to new types of secular resonances. The conservation of the vertical component of angular momentum (vertical refers to the direction of Jupiter's orbital normal) shows that when the orbital eccentricity increases, the inclination decreases. In particular, if objects are set up on inclined but circular orbits, large eccentricities can be achieved as the inclination decreases in its motion around the secular resonance cycle. The application of the Kozai cycle to the eccentricities of extrasolar planets assumes that there is a binary star on a not-too-far inclined orbit that perturbs the planet that formed in a circular orbit in a timescale shorter than the Kozai libration cycle. In essence, the secular Kozai cycle idea transforms the eccentricity problem into an inclination problem. In this sense, the observed planets do not possess a proper eccentricity but one that is forced by the stellar binary and that will always oscillate between its original value, zero, and a maximum value depending on the planetary–binary semi-major axis ratio, the binary's mass and its orbital inclination with respect to the plane on which the planet initially formed. When applied to specific binary star systems with one planet, the Kozai mechanism works fine and helps characterize the orbit and mass of the secondary star required to excite eccentricity [14, 27, 42]. Statistically, Kozai-based excitation of one-planet binary systems tend to yield larger eccentricities than observed [57]. As T Tauri stars form in multiple systems, it is not unreasonable to try and apply the Kozai mechanism to the whole sample of observed extrasolar planets. The problem is that the Kozai cycle is usually destroyed by mutual gravitational interactions. The addition of more planets to the one-planet binary system, forces the precession of the planets' pericenters. If the planets are of comparable mass as it is observed in multiplanet systems, the Kozai cycle is lost.

### 8.3.4 Mean-Motion Resonances

The role of mean motion resonances in exciting orbital eccentricity has its roots in the study of the orbital evolution of Jupiter's and Saturn's regular satellites under planetary and satellite tides [49]. These satellite systems are known to be in or to have crossed mean-motion resonances thereby acquiring forced eccentricities. The combined modeling of the orbital evolution, capture into resonance and the tidal interaction lends valuable bounds on the dissipation factors of Jupiter, Saturn and their satellites. Extrasolar planets form in a gaseous disk that does not dissipate after they acquired most of their masses or else hot Jupiters would not exist. Planet–disk interaction naturally gives rise to orbital migration with different planets in the same system migrating at different rates. This differential migration makes planets in the same system encounter mean-motion resonances. Capture into resonance may occur depending on whether the migration is convergent or divergent (for instance if the outer planet is moving faster or slower than the inner planet). Convergent migration leads to capture into resonance. The subsequent common migration of a planetary pair in resonance pumps up the eccentricities on the migration timescale [31, 69]. Divergent migration does not lead to resonance capture, instead eccentricity jumps are acquired at resonance passage [7, 61].

While convergent migration is certainly the way the known resonant multiple systems have acquired their eccentricities, this excitation method involves a mystery that may shed light on how to halt planet migration in a gas disk. The mystery consists of the observation that convergent migration is far too efficient in exciting eccentricities to the point where in many systems, when capture occurs, migration must stop quickly thereafter or else eccentricities are pumped up to much larger values than those observed. As it is implausible to invoke the dispersal of the gas disk, planetary migration may become ineffective because of the non-linear response of the gas disk to the planet pair. It is interesting to note that when capture occurs, the planetary relative inclinations may be excited as unlike planetary satellites that orbit Jupiter and Saturn, the central potential is keplerian. Consequently, for the same order, eccentricity and inclination resonances are close (but not coincident as the gas disk modifies the pericenter and node precession rates). Planet–disk interaction is not well understood for large eccentricity planets and off-plane (inclined with respect to the disk) orbits. The often used formulas for eccentricity damping from the disk torques have not been verified for eccentric and inclined planets. Divergent migration has the advantage of being applicable to the wider non-resonant multiplanet systems. For a planetary pair, divergent migration requires that the inner planet migrates faster than the outer one. Gap-driven migration (also known as type II) is favorable to such a condition as the migration rate is the viscous timescale of the disk. Divergent migration may take place because viscosity is likely to be a decreasing function of the distance to the star. If however the part of the disk that is located between the two planets is dispersed as when the two planetary gaps merge, the direction and rate of migration may be altered significantly [39].

### 8.3.5 Stellar Encounters

Stellar encounters are common events in star clusters. A planetary system that encounters a star will have its planets feel a tidal force that elongates their orbits. For inner planets that orbit close to the host star, the excitation which lasts for about 1000 years will occur on a secular timescale. Outer planets if they exist will feel a localized impulse somewhere in their orbits. Typical encounter frequencies are of one in $5 \times 10^9$ years while typical encounter parameters are a few hundred astronomical units. Unless planets are way outside the classical planetary region (inside 30 AU), excitation is not efficient [70]. To reverse this conclusion and account for the eccentricities of inner planets, the system must contain several planets with increasing distance and mass from the star in order propagate the stellar tug felt by the outermost planet down to the innermost ones [70].

### 8.3.6 Stellar-Like Relaxation

The qualitative similarity of the eccentricities of extrasolar planets and stellar binaries suggests that planets may form through similar processes as those of multiple stellar systems. If planets formed by gravitational instability, the formation time is so short that the planets find themselves confined to a smaller space than their orbital stability permits. The relaxation of such systems leads to some planetary ejections and many large eccentricity orbits [48]. The applicability of this scenario is limited by two facts: first, the minimum planet mass the gravitational instability allows is a few Jupiter masses. This means that stellar-like relaxation does not work for planets with masses comparable to or smaller than Jupiter's. Second, if a two-phase formation where small planets form through rocky core accretion and the larger ones through gravitational instability [59], then it is likely that the relaxation of the larger planets destroys the smaller planets. This is because the gravitational instability timescales are usually smaller than the planetesimal accumulation timescales. In fact, if large mass planets form through gravitational instability, they are likely to inhibit planetesimal accumulation by clearing the inner disk before planetary embryos are born.

## 8.4 Jet-Induced Excitation

Stellar jets enter the eccentricity excitation problem because of their ubiquity and simplicity [46]. The story of how this works is as follows: although there is disagreement on whether there is a statistically significant resemblance between the eccentricity distributions of extrasolar planets and stellar binary systems, the qualitative similarity is beyond doubt. Those who wish for the similarity to be quantitative, would like to affirm the view that planets are the lower end of the outcome of star formation. This question has been settled observationally in 2005 with two observations: the first is a hot Saturn with a giant rocky core discovered by combining the Doppler reflex velocity method with transit photometry [53]. The second is the imaging of the first planetary candidate which because of bias due to contrast and resolution happens to be a warm distant companion orbiting a young brown dwarf [5]. This proves that planets do not need large rocky cores and may form by gravitational instability. Exit the link between how planets form and their eccentricities.

If planets do not form like binary stars, perhaps they undergo similar excitation processes that lend them similarly elongated orbits. In view of the different physical environments where planets and stars form, the simplest possible excitation process may depend weakly or not at all on the local dynamics of the stellar or planetary companion. Mathematically, this amounts to saying that the acceleration imparted by the process is independent of position and velocity.

Simplicity therefore dictates that the process imparts a constant acceleration that operates during a finite time window. Simplicity also comes with two added advantages: we can already know the excitation timescale and the minimal acceleration amplitude. Dimensional analysis shows that the excitation timescale has to be proportional to $v/A$ where $v$ is the keplerian velocity of the companion around the main star. Further, if the acceleration is to achieve its purpose within the lifetime of the system, $v/A$ must be smaller than about $10^9$ years. This tells us that the acceleration $A > 3 \times 10^{-16}(v/10\,\mathrm{km\,s^{-1}})\,\mathrm{km\,s^{-2}}$.

The process lacks one more attribute: direction. If the acceleration is independent of the formation processes, its direction cannot depend on anything related to the planetary companion such as its orbital plane or the direction from the star to the companion. In an inertial frame related to the planetary or stellar system, we are not left with much choice but the star's rotation axis.

To sum up, what we are looking for is a process that appears everywhere where planet and star formation takes place, acts like a rocket (i.e. with an acceleration that does not depend on the position and velocity of the system) and whose direction is related to the star's rotation axis. The answer is stellar jets [12, 23].

Do planets exist when jets are active? The answer is quite likely: known hot Jupiters have moved close to their host stars because of their interaction with the gas disk. So we know the gas disk was present and had viscosity well after planets finished forming. The gas accreting on the star because of viscosity is the main ingredient along with the magnetic field that threads it needed to launch stellar jets and disk winds. It would therefore be an interesting coincidence that jets shut off when planets appear in the gas disk a few astronomical units away from the star outside the jet launching region.

Are there any observational hints that jet-sustaining disks contain planets? The only possible hint so far is the observation of variable brightness asymmetries in some jet-sustaining disks [8, 55, 65]. The variability timescales of a few days to a few years are so small that they imply either a peculiar stellar activity in the form of single hot spots or the presence of distortions in the disk at the location where the orbital period matches the variability timescale. The first option requires a complex stellar magnetic field that differentiates strongly between the two stellar poles. The second option may be caused distortions in the disk whose origin could be the presence of embedded compact objects.

Do jets have enough strength to build eccentricity? Inferred mass loss rates for known young T Tauri stars lie in the range $\sim 10^{-8} M_\odot\,\mathrm{year}^{-1}$ to $10^{-10} M_\odot\,\mathrm{year}^{-1}$ and may be two orders of magnitude larger depending on the way the rate is measured from the luminosity of forbidden lines [22, 25, 29]. The jet also needs to be asymmetric with respect to the star's equator plane or else there would be no acceleration. Interestingly, a growing number of bipolar jets from young stars [13, 24, 30, 35] are known to be asymmetric as the velocities of the jet and counterjet differ by about a factor of 2. Mass loss processes in young stars therefore yield accelerations:

$$A \sim 10^{-13} \left(\frac{\dot{M}}{10^{-8} M_\odot\,\mathrm{year}^{-1}}\right)\left(\frac{v_e}{300\,\mathrm{km\,s^{-1}}}\right)\left(\frac{M_\odot}{M}\right)\,\mathrm{km\,s^{-2}}, \quad (8.1)$$

where $M$ is the stellar mass and $v_e$ is the outflow's high velocity component. As jets are time-variable processes, the above estimate is only indicative of the epochs at which the rates and velocities are measured. In this sense, it is closer to being a lower bound on what accelerations really are over the jet's lifetime. What is clear is that asymmetric jet acceleration is larger than the minimum amplitude of $10^{-16}(v/10\,\text{km}\,\text{s}^{-1})\,\text{km}\,\text{s}^{-2}$.

For how long can a jet operate? Jet-induced acceleration is technically no different from attaching a rocket to the star and accelerating it very slowly with respect to the outer part of the disk and the planets. As a result, the star acquires a residual velocity that must be smaller than its orbital velocity in the galaxy or else the star is ejected. In fact, there is an even a stronger constraint on the residual velocity from the velocity dispersion in the galaxy. Stars do not have exactly circular orbits in the galaxy. Their motion is slightly distorted or eccentric and such eccentricity is measured as a departure of the galactic orbital velocity from that of circular motion. This velocity dispersion is known for various stellar populations and is of order a few tens of kilometers per second. Since the residual velocity imparted by the jet, $V$ is given by the product $A\tau$ where $\tau$ is the duration of acceleration, imposing that $V < \langle v_g \rangle$ where $\langle v_g \rangle$ is the velocity dispersion in the galaxy yields

$$\tau \leq 10^5 \, \frac{3 \times 10^{-12}\,\text{km}\,\text{s}^{-2}}{A} \, \frac{\langle v_g \rangle}{10\,\text{km}\,\text{s}^{-1}} \, \text{years}. \tag{8.2}$$

This timescale is shorter than the disk's lifetime. In practice, we shall see that shorter times are needed.

Further excitation properties can be deduced by analyzing the effect of the combined jet-induced acceleration and the star's gravitational attraction. As the star's pull decreases with distance, there is a specific location where the latter matches the jet-induced acceleration (that is independent of position and velocity). Outside this radius, the star's pull is weak and orbits escape its gravity. This reveals an interesting feature of jet-induced acceleration: stellar jets are responsible for the outer truncation of circumstellar disks. It is clear that in the interior vicinity of the truncation radius, the orbital perturbations are large as the excitation time becomes comparable to the orbital period. In this region, the keplerian orbits are subject to a sudden excitation; not only the eccentricities are excited but the semi-major axes are also affected leading to inward or outward migration. Well inside the truncation radius, the excitation time is much larger than the orbital period. In this region, eccentricity builds up slowly over a large number of revolutions of the planet around the star and the mean orbital radius remains constant on average. Excitation in this region occurs on secular timescales. Planetary companions mostly fall inside the secular region as they are far inside the truncation radius which is more or less the size of the protoplanetary disk.

## 8.4.1 Secular Jet-Induced Excitation

In the secular region where the excitation time is larger than the orbital period, the dynamics of excitation can be simplified by averaging the acceleration over the orbital period of the companion. For a constant acceleration, the interaction potential is simply $R = \mathbf{A} \cdot \mathbf{x}$ where $\mathbf{x}$ is the position vector. Averaging the interaction potential amounts to averaging the position vector of a pure keplerian motion. A simple calculation shows that $\langle \mathbf{x} \rangle = -3ae\,\mathbf{x}(f = 0)/2r$ where $f$, $a$, $e$ and $r$ are the true anomaly, the semi-major axis, the eccentricity and radius of the keplerian orbit. The direction of $\mathbf{x}$ at pericenter is that of the eccentricity vector $\mathbf{e} = \mathbf{v} \times \mathbf{h}/G(m+M) - \mathbf{x}/|\mathbf{x}|$ where $\mathbf{v}$ is the velocity vector of the companion, $G$ is the gravitational constant, $m$ and $M$ are the masses of the companion and the host star and $\mathbf{h} = \mathbf{x} \times \mathbf{v}$ is the specific angular momentum. This enables us to write the secular potential as

$$\langle R \rangle = -\frac{3}{2}a\,\mathbf{A} \cdot \mathbf{e} = -\frac{3}{2}A\,ae\sin(\varpi - \Omega)\sin I\,, \tag{8.3}$$

where in the last equality, the $z$-direction of the reference frame is chosen along $\mathbf{A}$ and $\varpi$, $\Omega$, $I$, are the longitude of pericenter, longitude of ascending node and the inclination of the orbit. To simplify the excitation problem further, we use the conservation of the component of angular momentum $\mathbf{h}$ along the direction of acceleration as $\mathbf{A} \cdot \dot{\mathbf{h}} = \mathbf{A} \cdot (\mathbf{x} \times \mathbf{A}) = 0$. In the reference frame where $\mathbf{A}$ is along the $z$-direction, the conservation of angular momentum yields $(1-e^2)^{1/2}\cos I = \cos I_0$ where $I_0$ is the initial inclination of the keplerian orbit with respect to the jet-induced acceleration. This relation enables us to eliminate the inclination variable in $\langle R \rangle$ and reduce the problem to an integrable, one dimensional system with

$$\langle R \rangle = -\frac{3}{2}A\,a\,\sqrt{\frac{\sin^2 I_0 - e^2}{1 - e^2}}\,e\sin\omega\,. \tag{8.4}$$

where $\omega = \varpi - \Omega$ is the argument of pericenter. The time evolution of eccentricity and argument of pericenter is obtained from

$$\dot{e} = -\frac{\sqrt{1-e^2}}{na^2\,e}\frac{\partial \langle R \rangle}{\partial \omega}\,, \quad \dot{\omega} = \frac{\sqrt{1-e^2}}{na^2\,e}\frac{\partial \langle R \rangle}{\partial e}\,, \tag{8.5}$$

where $n = \sqrt{G(M+m)/a^3}$ is the companion's mean motion. In this one-dimensional system, $e$ and $\omega$ follow curves of constant $\langle R \rangle$ shown in Fig. (8.2). There are equilibria at $\omega = \pm 90°$ and $e = \sqrt{2}\sin(I_0/2)$ corresponding to $I = \cos^{-1}(\sqrt{\cos I_0})$. The maximum value of $e$ is $\sin I_0$ and corresponds to the cycle of initially circular orbits. For these orbits, $\langle R \rangle = 0$ throughout their cycle implying that the orbits orientation can take only one value $\omega = 0$ modulo 180°.

For time-dependent accelerations and provided that the variation timescale is longer than the orbital period, the eccentricity evolution is given by

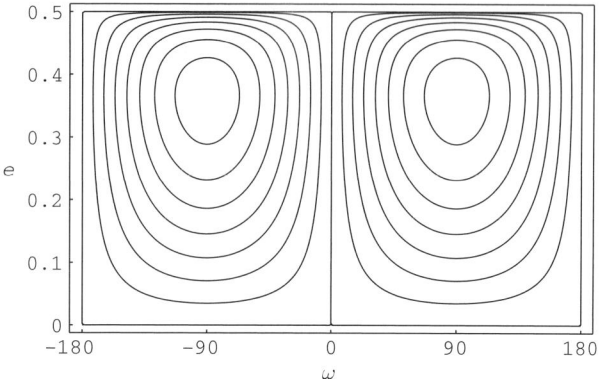

**Fig. 8.2.** Contour plots of the acceleration potential (8.4) in the eccentricity $e$ and argument of pericenter $\omega(°)$ plane. The direction of acceleration makes an angle $I_0 = 30°$ with respect to the companion's angular momentum vector. The time evolution of the two orbits ($e = 0$, $\omega = 0$) and ($e = 0.3$, $\omega = 90°$) is shown in Fig. (8.3).

$$\dot{e} = \frac{3A(t)\varepsilon}{2na} \sqrt{\sin^2 I_0 - e^2}, \qquad (8.6)$$

where $\varepsilon$ is the sign of $\cos\omega$ which is set by the requirement that $e \geq 0$. The solution of (8.6) can be found exactly as

$$e(T) = \left| \sin\left[ \frac{3}{2na} \int_{-\infty}^{T} A(t)\, dt \right] \sin I_0 \right|. \qquad (8.7)$$

The inclination is obtained from $\cos I = \cos I_0 / \sqrt{1 - e(T)^2}$. For strictly constant accelerations (infinite time window), $A(t) = A_0$ and $e$ oscillates between 0 and $\sin I_0$ at the excitation frequency:

$$n_A = \frac{3|A_0|}{na}. \qquad (8.8)$$

Examples of such oscillations that were obtained from the direct integration of the full equations of motion are shown in Fig. (8.3). The good agreement between the secular solution and the results of the numerical integration comes from the fact that $A$ is independent of position and velocity.

To optimize the excitation of a finite eccentricity from an initially circular state, the duration of acceleration needs to be smaller than half the oscillation period: $\tau < \pi na/3|A_0|$. Examples of eccentricity excitation at three different semi-major axes (i.e. three different excitation frequencies) are shown in Fig. 8.4. The dependence of the excitation amplitude on the ratio of the duration to the excitation time is also seen in the same figure. Note that because

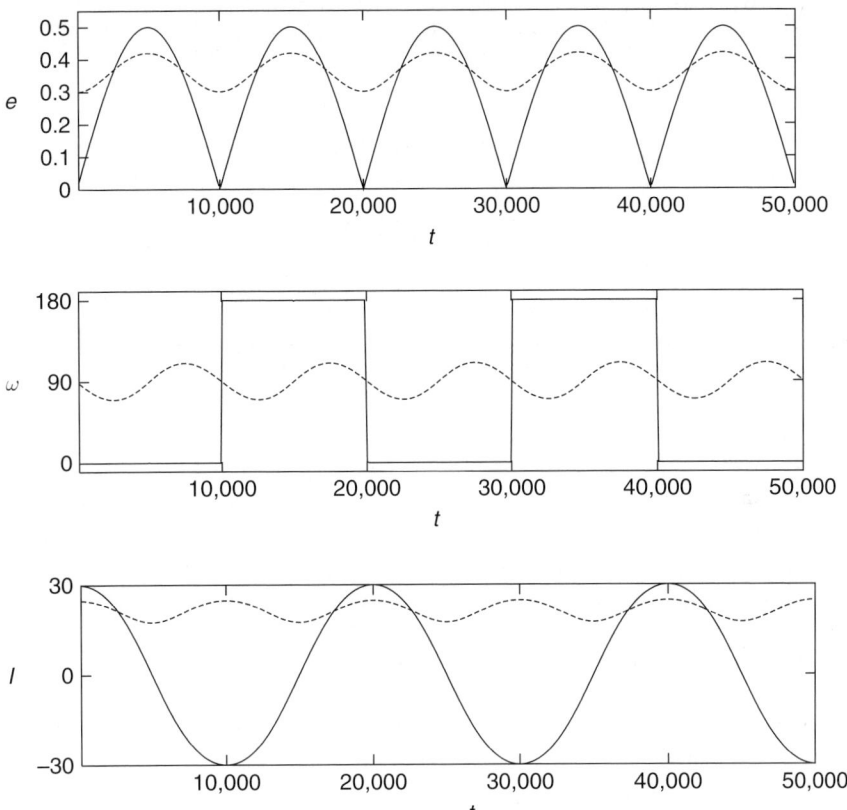

**Fig. 8.3.** Time evolution under a conservative acceleration. The eccentricity $e$, argument of pericenter $\omega(°)$ and inclination $I(°)$ are shown for an initially circular orbit $e = 0$ (*solid*) and an orbit librating about the secular resonance $\omega = 90°$ with an initial eccentricity $e = 0.3$ (*dashed*). The semi-major axis is identical for both orbits and is set to unity. The acceleration corresponds to a period of $10^4$ years at 1 AU. The plots were obtained by the numerical integration of the full equations of motion

eccentricity excitation in the secular region is a slow process compared to the orbital time, the convolution of the dynamics under the constant acceleration $A$ with a finite time window has the effect of shutting off the excitation at some eccentricity value depending on the duration.

The secular excitation through jet-induced acceleration is therefore able to make $e$ reach $\sin I_0$ and is largest if the initial orbital plane contains the direction of acceleration ($I_0 = 90°$). As $\langle R \rangle = 0$ for initially circular orbits, $\omega$ and $\Omega$, remain at zero. This forcing of the pericenter to be perpendicular to the direction of acceleration favors apsidal alignment in multiplanet system.

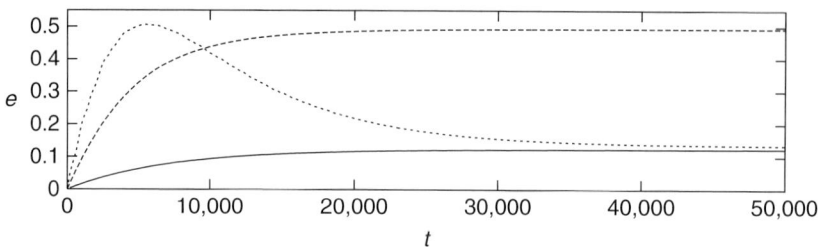

**Fig. 8.4.** Eccentricity excitation by time-dependent constant-direction accelerations. The equations of motion are integrated numerically with an acceleration $A(t) = A_0 H(t) \exp{-t/\tau}$ where $H(t)$ is the Heaviside unit step function and $A_0 = 2.21 \times 10^{-11}\,\mathrm{km\,s^{-2}}$. The oscillation period at 1 AU is $1.11 \times 10^5$ years. The timescale $\tau = 7200$ is chosen so that $V = 5\,\mathrm{km\,s^{-1}}$. The curves correspond the semi-major axes: 1 AU (*solid*), 32 AU (*dashed*) and 128 AU (*dotted*)

If the jet's inclination with respect to the companion's orbital plane is small $I_0 \ll 1$, the maximum eccentricity will be negligible. The $\sin I_0$ limitation is problematic because it is not reasonable to expect stellar jets to be highly inclined with respect to the gas disk where the companions form. Fortunately for the jet-induced excitation theory, there is a natural way out. Jets are known to precess over timescales from $10^2$ to $10^4$ years [10, 11, 58]. The origin of such precession is not known as we lack resolution to probe inside the jet launching region. It is possible that precession is caused by a warp in the disk's plane resulting from interactions with stellar companions as T Tauri stars are known to form in multiple systems. Precession is attractive because it offers the possibility of resonance if the excitation frequency $n_A$ matches the jet precession frequency $\Omega_A$. This in fact is exactly what happens when the eccentricity evolution is derived in the situation where the constant magnitude acceleration rotates at a constant rate. It turns out that the corresponding secular problem is also integrable. The eccentricity and inclination evolution are given by [46]

$$e^2 = \frac{p^2 \sin^2 \alpha}{4\nu_+^2 \nu_-^2} \left[ 2(3+p^2) - 4(1+p\cos\alpha)\,\cos\nu_+ t - 4(1-p\cos\alpha)\,\cos\nu_- t \right.$$
$$+ (1-p^2+\nu_+\nu_-)\cos(\nu_+-\nu_-)t$$
$$\left. + (1-p^2-\nu_+\nu_-)\cos(\nu_++\nu_-)t \right] \tag{8.9}$$

$$\cos I = \frac{1}{2\nu_+^2 \nu_-^2 \sqrt{1-e^2}} \left[ (p^4 - p^2 + 2 + p^2[p^2-3]\cos 2\alpha) \right.$$
$$+ p^2 \sin\alpha^2 (p^2+1+2p\cos\alpha)\,\cos\nu_+ t$$
$$\left. + p^2 \sin\alpha^2 (p^2+1-2p\cos\alpha)\,\cos\nu_- t \right] \tag{8.10}$$

where $\alpha$ is the jet angle with respect to the $z$-axis of the reference frame, $\nu_\pm^2 = p^2 + 1 \mp 2p\cos\alpha$, $p = n_A/2\Omega_A$, and the time $t$ is normalized by $\Omega_A$. The companion's initial orbit is circular and lies in the $xy$-plane.

Nominal resonance is defined where the frequency match, $p = 1$, occurs. It corresponds to a nominal resonant semi-major axis $a_{\text{res}}$ given as

$$a_{\text{res}} \simeq 4 \left(\frac{M+m}{M_\odot}\right) \left(\frac{A}{2 \times 10^{-10} \text{ km s}^{-2}}\right)^{-2} \left(\frac{T_{\text{prec}}}{10^4 \text{ years}}\right)^{-2} \text{ AU}, \quad (8.11)$$

where $T_{\text{prec}} = 2\pi/\Omega_A$. In terms of the resonant semi-major axis, the frequency ratio can be written as $p = \sqrt{a/a_{\text{res}}}$. Far inside resonance ($p \ll 1$), the jet precesses faster than the eccentricity excitation leading to a reduction of the eccentricity amplitude from $\sin\alpha$ to $2p\sin\alpha$. Far outside resonance ($p \gg 1$), the jet's precession is slow compared to the eccentricity excitation so that the latter is described by a constant acceleration without rotation. In the resonance region, the proximity of $p$ to unity increases the denominators of the eccentricity expression (8.9) which leads to eccentricities close to unity. At exact resonance, the eccentricity reaches unity regardless of the jet angle. The width of the region around resonance increases with the jet angle $\alpha$. These features are illustrated in Fig. (8.5) where we plot the expressions (8.9) and (8.10) for a jet angle $\alpha = 1°$, an excitation time $2\pi/n_A = 10^4$ years, and the three values of $p$: 0.05, 0.9 and 1. Finally, we note that as the eccentricity excitation time is $n_A$, no resonant forcing occurs when $\Omega_A = n$ in the secular region ($n_A \ll n$).

### 8.4.2 Sudden Jet-Induced Excitation and Radial Migration

The location where orbits can not longer be retained by the star is where the frequency $n_A$ becomes comparable to the local mean motion $n$ of the companion. Near this limit, the forced periodic oscillations of the semi-major axis $a$ are reinforced by the eccentricity and acquire large amplitudes making the orbits unstable in the long term. Calling $a_{\text{kplr}}$ the semi-major axis of the keplerian boundary where the star where $n_A = n$, we the jet-induced acceleration is expressed as

$$|A_0| \simeq 2 \times 10^{-12} \left(\frac{M+m}{M_\odot}\right) \left(\frac{10^3 \text{ AU}}{a_{\text{kplr}}}\right)^2 \text{ km s}^{-2} \quad (8.12)$$

corresponding to an excitation period $T_A = 2\pi/n_A$:

$$T_A \simeq 10^6 \left(\frac{M+m}{M_\odot}\right)^{1/2} \left(\frac{a_{\text{kplr}}}{10^3 \text{ AU}}\right)^2 \left(\frac{1 \text{ AU}}{a}\right)^{1/2} \text{ years}. \quad (8.13)$$

Figure (8.6) shows an example of an escape orbit of a constant-direction acceleration with $a_{\text{kplr}} = 10^2$ AU and an inclination $I_0 = 30°$. The orbit's initial semi-major axis is 68.5 AU. The characteristics of motion are not strictly keplerian as the companion hovers above the star. Such escape orbits offer an interesting way to expel planets from around their parent stars or equivalently

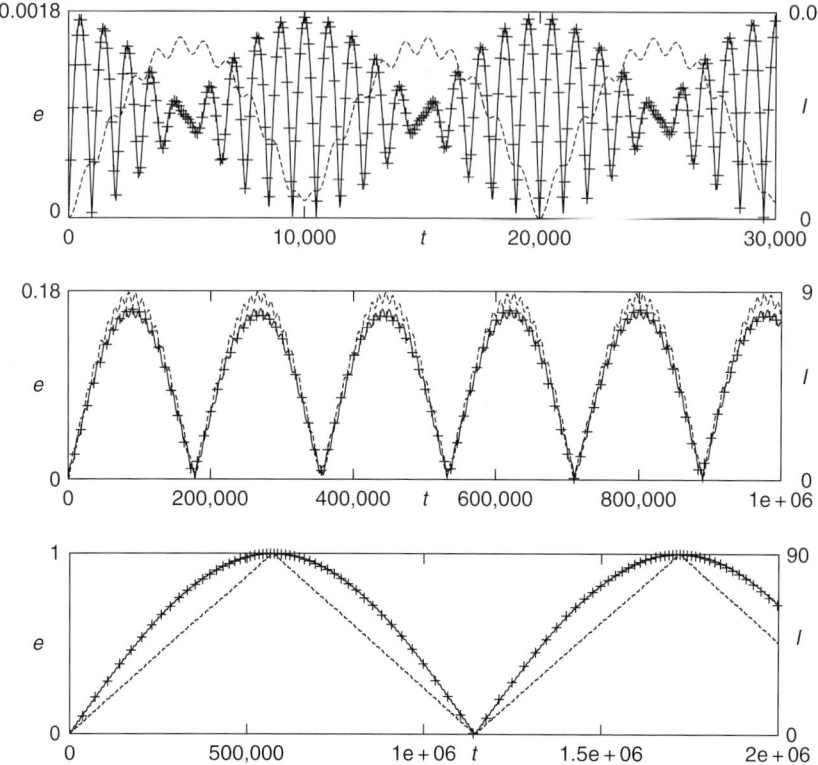

**Fig. 8.5.** Excitation of the eccentricity (*solid*) and inclination (*dashed*) by a nearly perpendicular precessing jet with an angle $\alpha = 1°$. The companion's orbit is located at $a = 1$ AU and evolves from a circular orbit in a plane orthogonal to the jet's precession axis. The acceleration is $A_0 = 2 \times 10^{-10}$ km s$^{-2}$ yielding an excitation time of $2\pi/n_A = 10^4$ years. From *top to bottom*, the panels were obtained from (8.9) to (8.10) with the frequency ratios, $p$: 0.05, 0.9, and 1 – the precession period is $2p \times 10^4$ years. The symbols correspond to the numerical integration of the full equations of motion

to disrupt a binary stellar system. If a companion is formed near the keplerian boundary or is pushed out to it by a possibly remaining inner disk that followed photo-evaporation [26], it could become unbound.

For a realistic jet, the induced acceleration has a finite duration. Accordingly, $a_{\rm kplr}$ varies in time from infinity before the jet's launch to a location determined by the strongest acceleration the jet can provide. Ultimately, the keplerian boundary is pushed out to infinity. The acceleration's finite duration extends the keplerian boundary depending on the ratio of the duration $\tau$ to the excitation time at the keplerian boundary of the equivalent constant-acceleration problem $T_A(a_{\rm kplr})/2$. When $\tau \geq T_A(a_{\rm kplr})/2$, orbits beyond $a_{\rm kplr}$ have enough time to acquire sufficient momentum and escape the

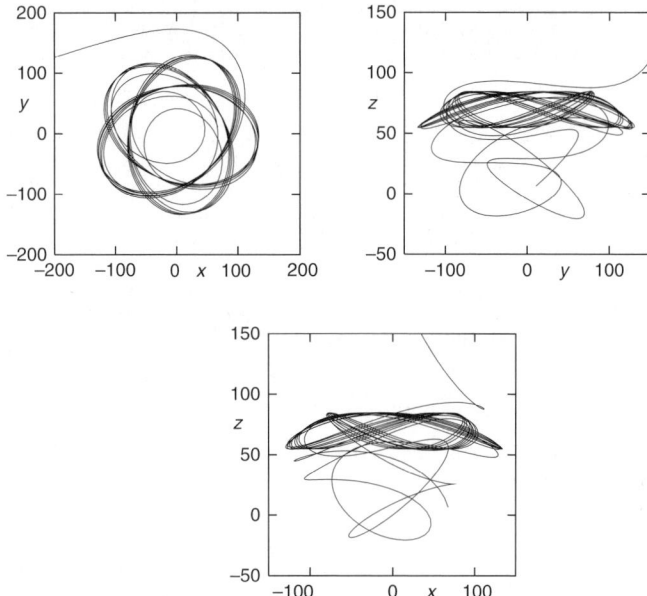

**Fig. 8.6.** Escape of a companion located near the keplerian boundary of a conservative constant-direction acceleration with $I_0 = 30°$. The distances are given in astronomical units. The boundary's semi-major axis is $a_{\text{kplr}} = 100$ AU. Note how the companion hovers above the star before escaping

gravitational pull of the star. When $\tau \leq T_A(a_{\text{kplr}})/2$, the stability region extends beyond $a_{\text{kplr}}$. The new stability boundary is given by the semi-major axis where $\tau \simeq T_A(a_\infty)/2$ which is larger than $a_{\text{kplr}}$ since $T_A$ is a decreasing function of the semi-major axis $a$. The expressions of $T_A$ and the residual velocity $V$, show that $a_\infty \simeq G(M+m)V^{-2}$, the location where the keplerian velocity $v$ matches $V$.

Orbits near the keplerian boundary of a finite duration acceleration that do not escape the pull of the star will end up with elongated orbits whose semi-major axes and eccentricity have changed. This happens because the companion feels an almost instantaneous velocity kick (the orbital period is large compared to $\tau$). The conservation of linear momentum and energy can be combined to find the change in semi-major axis as

$$\frac{1}{a_{\text{f}}} = \frac{1}{a_{\text{i}}} - \frac{2V \sin I_0 \cos \theta}{\sqrt{G(M+m)a_{\text{i}}}} - \frac{V^2}{G(M+m)}, \qquad (8.14)$$

where $a_{\text{i}}$, $a_{\text{f}}$ are the initial and final semi-major axis, $\theta$ is the longitude of the companion along its orbit and $I_0$ is the inclination of the orbital plane with respect to the direction of the residual velocity $\mathbf{V}$. Note that for $I_0 \neq 0$, the final semi-major axis can be larger or smaller than the initial value depending on the longitude of the companion where the velocity pulse if felt.

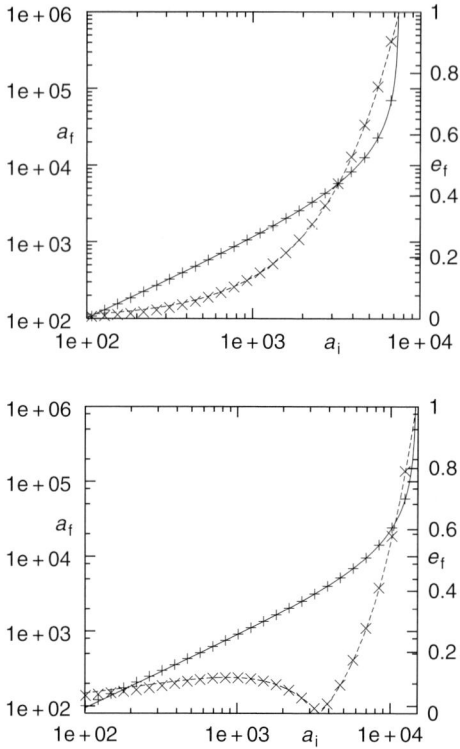

**Fig. 8.7.** Migration and eccentricity excitation near the keplerian boundary of a finite duration acceleration for two inclination values $I_0 = 0$ (**left panel**) and $I_0 = 20°$ (**right panel**). In each panel, the final semi-major axis $a_f$ (AU) (*solid*) and the final eccentricity $e_f$ (*dashed*) are shown as a function of the initial semi-major axis $a_i$ (AU). The parameters are $V = 0.35 \text{ km s}^{-1}$, $a_{\text{kplr}} = 300$ AU and $\tau = 500$ years. The inclined circular orbits were started at the descending node ($\theta = 180°$). For $I_0 = 0°$, $a_\infty = 7341$ AU and for $I_0 = 20°$, $a_\infty = 14{,}542$ AU

Figure (8.7) shows the final semi-major axes and eccentricities at two different inclinations $I_0 = 0$ and $20°$ for an acceleration amplitude corresponding to $a_{\text{kplr}} = 300$ AU and a duration $\tau = 500$ years resulting in a residual velocity $V = 0.35 \text{ km s}^{-1}$. The companions were started at the descending node for $I_0 = 20°$ to illustrate inward and outward migration. Such migration could enhance the delivery of minor bodies to the Oort Cloud and explain the transport of Kuiper belt outliers 2000 CR105 and Sedna (90377).

### 8.4.3 A Test Case: The υ Andromedae Binary System

Multiplanet systems provide good test cases for the excitation by jet-induced acceleration. This is because mutual gravitational interactions cause the eccentricity vectors to precess. If the ensuing precession rates are much faster than the excitation frequencies, the slow build up of eccentricity by jet-induced acceleration will be lost [46, 47]. This situation is similar to that encountered for precessing jets inside the resonance radius $a_{\text{res}}$.

An interesting system for testing the excitation mechanism is that of $\upsilon$ Andromedae [3]. It contains three planets, two of which have their apsidal directions aligned [6, 33, 52] as well as a $0.2 M_\odot$ stellar companion at a projected distance of 750 AU [36]. Numerical simulation can be used to reproduce the planetary orbits from initially circular co-planar orbits to their observed state [43]: $a_{\text{b}} = 0.059$ AU, $e_{\text{b}} = 0.020$, $\omega_{\text{b}} = 241°$, $m_{\text{b}} \sin i = 0.75\, m_{\text{J}}$, $a_{\text{c}} = 0.821$ AU, $e_{\text{c}} = 0.185$, $\omega_{\text{c}} = 214°$, $m_{\text{c}} \sin i = 2.25\, m_{\text{J}}$, $a_{\text{d}} = 2.57$ AU, $e_{\text{d}} = 0.269$, $\omega_{\text{d}} = 247°$ and $m_{\text{d}} \sin i = 2.57\, m_{\text{J}}$. It turns out that mutual planetary perturbations are strong enough to prevent excitation if the acceleration is smaller than $A_0 \sim 10^{-11} \text{km s}^{-2}$. The equivalent smallest keplerian boundary is at $a_{\text{kplr}} \sim 500$ AU. Below this value, the current configuration can be recovered along with the apsidal alignment of the outer two planets. The presence of the stellar companion outside the keplerian boundary leaves us two options: either the excitation by acceleration is ruled out or that the companion was initially inside the boundary and migrated during a sudden excitation (the projected distance of 750 AU does not translate necessarily into a semi-major axis as the companion's orbit is likely to be eccentric).

Figure 8.8 shows a simulation of the jet-induced acceleration of the form $A(t) = A_0 / \cosh(t - t_0)/\tau$ where $A_0 \sim 3 \times 10^{-11}$ km s$^{-2}$ and $\tau = 2000$ years, applied the current planetary system plus a stellar companion on an orbit of semi-major axis $a = 298$ AU, just inside the keplerian boundary of A, $a_{\text{kplr}} \sim 300$ AU. The stellar companion's initial orbit was given an eccentricity $e = 0.3$ in order to decouple its perturbations from the planetary system. In particular, the eccentricity excitation by the Kozai mechanism [28] is not efficient because the corresponding excitation time ($\sim 10^7$ years) is much larger the duration of acceleration and the eccentricity secular frequency of the isolated two-planet system ($\sim 7000$ years). The jet-induced acceleration produces a configuration similar to the observed one with stellar orbital elements: $e = 0.5$, $a = 600$ AU. Apsidal alignment is achieved as the result of the acceleration's strength that maintain the forcing of companion orbits to be perpendicular to the direction of acceleration. Note that only when the acceleration's strength is near maximum and the keplerian boundary nears 300 AU, does the stellar orbit acquire a larger eccentricity.

### 8.4.4 The Solar System

Was the solar system subject to jet-induced acceleration? There are two observations that hint at the dynamical action of the solar system's jet. The first is the inclination of Jupiter's orbital normal by 6° with respect to the Sun's rotation axis. As Jupiter is the more massive planet in the solar system, this implies that either the early protoplanetary disk of the solar system was

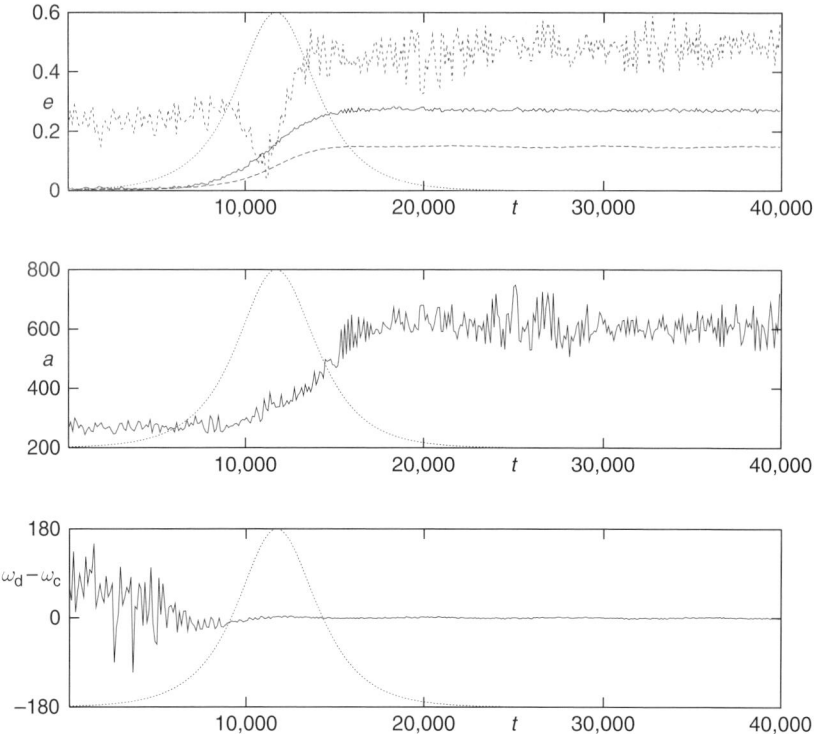

**Fig. 8.8.** Eccentricity excitation, apsidal alignment and binary migration in the ε And system. The acceleration pulse is shown not to scale in all panels. The *first panel* shows the eccentricity excitation of planets d (*solid*) and c (*dashed*) and the eccentricity evolution of ε And B (*short-dashed*) as well as the acceleration pulse normalized to its maximum value (*dotted*). The *second panel* shows the radial migration of ε And B. The *last panel* shows the relative apsidal libration of planets d and c

warped with respect to the Sun's equator plane or that the Jupiter's orbit gained inclination with respect to an early equatorial disk. Both possibilities are consistent with the jet-induced acceleration theory. More recently, the discovery of calcium–aluminum inclusions in the grains of comet Wild 2 suggest that the only possibility for the Jupiter-family comet originating from the early Kuiper belt to contain such high temperature minerals is that they were transported by the solar system's jet. An additional piece of the puzzle is given by outer solar system bodies such as Sedna (90377) that are decoupled from the solar system's planets as their perihelia are larger than Neptune's orbital radius. Such objects could have been transported by the radial migration in the sudden excitation region associated with the jet-induced acceleration.

Why are the solar system's planets eccentricities small? There are three possible reasons for this: first, the resonance radius (8.11) where the jet's precession matches the excitation time could have been outside the planets' orbits. Inside the resonance radius, very little excitation can take place (Fig. 8.5). Second, mutual planetary perturbations could have destroyed the secular eccentricity growth. Third, if the jet angle were to be small and the acceleration duration equally brief, the planets' location near the resonant radius would be of little help to raise their eccentricities.

### 8.4.5 The Unknowns of Jet-Induced Excitation

The unknowns of jet-induced excitation belong to two categories: (1) the unknowns of disk–planet interaction and (2) the unknowns of the time evolution of jets. In the first category comes the issue of the eccentricity damping by the accretion disk. As explained in the section about mean-motion resonances, the damping of eccentricity for large eccentricity and inclination orbits is not currently understood. The advantage of jet-induced excitation with respect to the excitation during migration while in mean-motion resonance is that excitation times can be much smaller than the migration timescale and the viscous timescale of the disk. In this phase, the formed planets gain a substantial inclination that makes them exit the gas disk which should in principle reduce the eccentricity damping significantly. In this context, it is useful to bear in mind that jet-induced acceleration becomes effective in a planetesimal disk only when a few planets are left. The planetesimal mutual gravitational interactions destroy excitation through the random precession of their orbits. When the bodies left in the disk are such that the precession periods due to their mutual perturbations are larger than the excitation time, jet-induced acceleration becomes effective and planets may exit the disk on inclined orbits. For the second category of unknowns, it is safe to say that except for precession, little information is available about the time variations of jets over their entire life span. Perhaps the main advantage of jet-induced excitation is its small set of parameters: amplitude, duration and jet precession frequency. These determine all the features of eccentricity growth or lack thereof. In multiplanet systems, the acceleration subjects all companions, planetary and stellar alike, to the same instability as it is independent of position and velocity. A better knowledge of the time dependence of acceleration can therefore easily confirm or rule out the effect of jet-induced excitation in multiplanet systems.

## 8.5 Concluding Remarks

The various theories of eccentricity excitation are valuable tools to gain insight into one of the most pressing if not the most pressing issue in planet formation theory: the mismatch of protoplanetary disk lifetimes and the timescales of planet formation and planet migration. Protoplanetary disk lifetimes range from $10^5$ to $10^7$ years. To have a viable theory, the planet formation timescale has to be smaller by at least an order of magnitude than these estimates. Planets are believed to form in two ways: (1) rocky core accretion and (2) gravitational instability. The formation of rocky cores precedes gas accumulation on the way to forming gaseous planets. This specific phase is slow. Typical formation timescales range from $10^6$ to $10^7$ years at Jupiter's location [34]. Gravitational instability however has a much smaller timescale of order $10^3$ years at the same location. With all its caveats about the required disk opacity for gravitational fragmentation and the role of shearing instabilities in disrupting a forming protoplanet, formation by gravitational instability looked far more promising than rocky core accretion to explain the existence of the hot Jupiters. This was true until the discovery of a hot Saturn with a 70 Earth mass rocky core orbiting the star HD 149026 [53]—for reference, the mass of Jupiter's rocky core is believed to be 15 Earth masses. The formation of a 70 Earth mass rocky core either as a single body or as a merger of smaller cores is likely to last at least $10^7$ years according to current planetesimal accumulation models. This temporal crisis does not stop at rocky core formation, it worsens because of planet migration. Migration arises from the planet's interaction with the gas disk and has typical timescales of $10^6$ years for an Earth mass core (linear regime, type I) to $10^5$ years for a Jupiter mass planet (non-linear regime, type II) [9, 64]. The duration of the eccentricity excitation phase and its dependence on the mechanism's parameters may help elucidate the time sequence of the events that produced the observed extrasolar planets. In particular, three-dimensional hydrodynamic simulations of disk–planet interactions for eccentric and inclined orbits (off the disk's midplane) may elucidate the problem of the end of resonant excitation in multiple systems. Planet–planet scattering models may benefit from including a wider mass spectrum in the populations of simulated planets. The excitation by Kozai's secular mechanism requires further assessments of the ability to form multiple planets under the stellar companion's perturbations [60]. Similarly, jet-induced excitation was demonstrated for a single planet under the action of a precessing acceleration. The effect of mutual planetary entrainment for planets on both sides of the precession–excitation resonance needs to be investigated and applied to observed extrasolar systems.

# References

1. Artymowicz, P., 1993. *ApJ* **419**, 166.
2. Barge, P. and Pellat, R., 1990. *Icarus* **85**, 481.
3. Butler, R. P., Marcy, G. W., Williams, E., Hauser, H., and Shirts, P., 1997. *ApJ* **474**, L115.
4. Chandrasekhar, S., 1942. *Principles of Stellar Dynamics*, The University of Chicago Press, Chicago.
5. Chauvin, G., Lagrange, A. M., Dumas, C., Zuckerman, B., Mouillet, D., Song, I., Beuzit, J. L., and Lowrance, P., 2005. *AA* **438**, L25.
6. Chiang, E. I., Tabachnik, S., and Tremaine, S., 2001. *AJ* **122**, 1607.

7. Chiang, E. I., Fischer, D., and Thommes, E., 2002. *ApJ* **564**, L105.
8. Cotera, A., Whitney, B. A., Young, E., Wolff, M. J., Wood, K., Povich, M., Schneider, G., Rieke, M., and Thompson, R., 2001. *ApJ* **556**, 958.
9. D'Angelo, G., Kley, W., and Henning, T., 2003. *ApJ* **586**, 540.
10. Davis, C. J., Eislöffel, J., Ray, T. P., and Jenness, T., 1997. *AA* **324**, 1013.
11. Eislöffel, J., Smith, M. D., Davis, C. J., and Ray, T. P., 1996. *AJ* **112**, 2086.
12. Eislöffel, J., Mundt, R., Ray, T. P., and Rodríguez, L. F., 2000. In Mannings, V., Boss, A. P., and Russel, S. S. (eds), *Protostars and Planets IV*, University of Arizona Press, Tucson, p. 815.
13. Fernández, M. and Comerón, F., 2005. *AA* **440**, 1119.
14. Ford, E. B., Kozinsky, B., and Rasio, F. A., 2000. *ApJ* **535**, 385.
15. Ford, E. B., Havlickova, M., and Rasio, F. A., 2001. *Icarus* **150**, 303.
16. Ford, E. B., Rasio, F. A., and Yu, K., 2003. In Deming, D. and Seager, S. (eds), *ASP Conf Series*, **294**, ASP, San Francisco, pp. 181–188.
17. Goldreich, P. and Tremaine, S., 1979. *ApJ* **233**, 857.
18. Goldreich, P. and Tremaine, S., 1980. *ApJ* **241**, 425.
19. Goldreich, P. and Tremaine, S., 1981. *ApJ* **243**, 1062.
20. Goldreich, P. and Sari, R., 2003. *ApJ* **585**, 1024.
21. Halbwachs, J. L., Mayor, M., and Udry, S., 2005. *AA* **431**, 1129.
22. Hartigan, P., Edwards, S., and Pierson, R., 1995. *ApJ* **452**, 736.
23. Hartigan, P., Mundt, J., Reipurth, B., and Morse, J. A., 2000. In Mannings, V., Boss, A., and Russel, S. S. (eds), *Protostars and Planets IV*, University of Arizona Press, Tucson, p. 841.
24. Hirth, G. A., Mundt, R., and Solf, J., 1994. *ApJ* **427**, L99.
25. Hollenbach, D., 1985. *Icarus* **61**, 36.
26. Hollenbach, D. J., Yorke, H. W., and Johnston, D., 2000. In Mannings, V., Boss, A., and Russel, S. S. (eds), *Protostars and Planets IV*, University of Arizona Press, Tucson, p. 401.
27. Holman, M., Touma, J., and Tremaine, S., 1997. *Nature* **386**, 254.
28. Kozai, Y., 1962. *AJ* **67**, 591.
29. Kwan, J. and Tademaru, E., 1995. *ApJ* **454**, 382.
30. Lavalley, C., Cabarit, S., Dougados, C., Ferruit, P., and Bacon, R., 1997. *AA* **327**, 671.
31. Lee, M. H. and Peale, S. J., 2002. *ApJ* **567**, 596.
32. Lin, D. N. C. and Papaloizou, J. C. B., 1979. *MNRAS* **186**, 799.
33. Lissauer, J. J. and Rivera, E. J., 2001. *ApJ* **554**, 1141.
34. Lissauer, J. J. and Stevenson, D. J., 2006. In Reipurth, B., Jewitt, D., and Keil, K. (eds), *Protostars and Planets V*, University of Arizona Press, Tucson, in press.
35. López-Martín, L., Cabrit, S., and Dougados, C., 2003. *AA* **405**, L1.
36. Lowrance, P. J., Kirkpatrick, J. D., and Beichman, C. A., 2002. *ApJ* **572**, L79.
37. Lubow, S. and Ogilvie, G., 2003. *ApJ* **587**, 398.
38. Masset, F. S. and Ogilvie, G. I., 2004. *ApJ* **615**, 1000.
39. Masset, F. S. and Snellgrove, M., 2001. *MNRAS* **320**, L55.
40. Marzari, F. and Weidenschilling, S., 2002. *Icarus* **156**, 570.
41. Mayor, M. and Queloz, D., 1995. *Nature* **378**, 355.
42. Mazeh, T., Krymolowski, Y., and Rosenfeld, G., 1997. *ApJ* **477**, L103.
43. Naef, D., Mayor, M., Beuzit, J. L., Perrier, C., Queloz, D., Sivan, J. P., and Udry, S., 2002. *AA* **414**, 351.

44. Namouni, F., Luciani, J. F., and Pellat, R., 1996. *AA* **307**, 972.
45. Namouni, F., 1998. *MNRAS* **300**, 915.
46. Namouni, F., 2005. *AJ* **130**, 280.
47. Namouni, F. and Zhou, J. L., 2006. *CMDA* **95**, 245.
48. Papaloizou, J. C. B. and Terquem, C., 2001. *MNRAS* **325**, 221.
49. Peale, S. J., 1999. *ARAA* **37**, 533.
50. Pepe, F., Mayor, M., Galland, F., Naef, D., Queloz, D., Santos, N. C., Udry, S., and Burnet, M., 2002. *AA* **388**, 632.
51. Rasio, F., and Ford, E., 1996. *Science* **274**, 954.
52. Rivera, E. J. and Lissauer, J. J., 2001. *ApJ* **530**, 454.
53. Sato, B. and 20 co-authors, 2005. *ApJ* **633**, 465.
54. Schneider, J., 1996. http://www.obspm.fr/planets.
55. Stapelfeldt, K. R. and 15 co-authors, 1999. *ApJ* **516**, L95.
56. Stepinski, T. F. and Black, D. C., 2001. *AA* **371**, 250.
57. Takeda, G. and Rasio, F. A., 2005. *ApJ* **512**, 1001.
58. Terquem, C., Eislöffel, J., Papaloizou, J. C. B., and Nelson, R. P., 1999. *ApJ* **512**, L131.
59. Terquem, C. and Papaloizou, J. C. B., 2002. *MNRAS* **332**, L39.
60. Thébault, P., Marzari, F., Scholl, H., Turrini, D., and Barbieri, M., 2004. *AA* **427**, 1097.
61. Tsiganis, K., Gomes, R., Morbidelli, A., and Levison, H. F., 2005. *Nature* **435**, 459.
62. Udry, S., Mayor, M., Naef, D., Pepe, F., Queloz, D., Santos, N. C., Burnet, M., Confino, B., and Melo, C., 2000. *AA* **356**, 590.
63. Ward, W. R., 1988. *Icarus* **73**, 330.
64. Ward, W. R. and Hahn, J. M., 2000. In Mannings, V., Boss, A., and Russel, S. S. (eds), *Protostars and Planets IV*, University of Arizona Press, Tucson, p. 1135.
65. Watson, A. M., Stapelfeldt, K. R., Krist, J. E., and Burrows, C. J., 2000. *BAAS* **32**, 1481.
66. Weidenschilling, S. and Marzari, F., 1996. *Nature* **384**, 619.
67. Wisdom, J., 1980. *AJ* **85**, 1122.
68. Wolszczan, A. and Frail, D. A., 1992. *Nature* **355**, 145.
69. Yu, Q. and Tremaine, S., 2001. *AJ* **121**, 1736.
70. Zakamska, N. and Tremaine, S., 2004. *AJ* **128**, 869.

# Methods for the Study of the Dynamics of the Oort Cloud Comets I: Modelling the Stellar Perturbations

Marc Fouchard[1,2,5], Christiane Froeschlé[3], Hans Rickman[4] and Giovanni B. Valsecchi[5]

[1] Laboratoire d'Astronomie de Lille/Université de Lille,
    1 impasse de l'Observatoire 59000 Lille, France
[2] Institut de Mécanique Céleste et de Calcul d'Ephémérides, UMR 8028 du CNRS, 75014 Paris, France
    fouchard@imcce.fr
[3] Laboratoire Cassiopée, Université de Nice, CNRS, Observatoire de la Côte d'Azur, BP 4229, FR-06304 Nice, France
    froesch@obs-nice.fr
[4] Uppsala Astronomical Observatory, Box 515, SE-75120 Uppsala, Sweden
    hans@astro.uu.se
[5] INAF-IASF, Via Fosso dell Cavaliere 100, 00133 Roma, Italy
    giovanni@iasf-roma.inaf.it

**Abstract.** In this chapter we present different approximate models of computing the perturbations on the Oort cloud comets caused by passing stars. All these methods are checked against an accurate numerical integration using Everhart's RA15 code. A random sample of stellar passages, simulating those suffered by the solar system, but extrapolated over a time of $10^{(10)}$ years, is created. For each model we measured the errors and their dependence on the encounter parameters.

## 9.1 Introduction

A good knowledge of the present structure of the Oort cloud is one of the clues to understand the formation of the solar system. There is a wide consensus on the fact that the Oort cloud was formed by the residual planetesimals scattered by the giant planets during the period of their formation. Moreover, it is important to follow carefully the history of comets which originated in the Oort cloud, to be able to estimate the total population and the mass of the entire cloud.

There are three major external perturbers which influence the Oort cloud comets and inject them into the planetary region. These are the stellar perturbations, the galactic tidal forces and giant molecular clouds (GMCs). Although a penetrating encounter of the solar system with a GMC is a rare event, it may have considerable effects [18]. However, due to the rarity and the poor knowledge of the circumstances of such encounters [1], they are generally omitted from studies of Oort cloud dynamics.

We concentrate our study on the effects of the two other external perturbers, namely passing stars and the galactic tide. While stellar perturbations occur at random and may be treated as a stochastic process, the galactic tidal force is a quasi–integrable perturbation which acts continuously, changing the cometary orbital elements and in particular the perihelion distance. As shown by many authors [3, 9, 20] the fraction of the population of Oort cloud comets that can become observable would be very small if only the effect of the galactic tide was at work.

Passing stars can randomize the whole population and may cause strong temporary enhancements of the flux of observable comets called comet showers [15], while the galactic tide produces the main part of the steady flux of Oort cloud comets towards the observable region (e.g. [16, 21]). Thus for a general understanding of the origin of LP comets, both pertubers, i.e. passing stars and galactic tide must be modelled.

To get a realistic model of the Oort cloud it is necessary to follow the dynamics of a very large number of test particles ($\simeq 10^6$, to sample as well as possible the available parameter space), over time scales equal to the age of the solar system. Therefore, one has to develop reliable methods that model both the galactic tide and the stellar close encounters with the lowest possible computational cost.

In the present chapter we investigate different algorithms to compute the orbital perturbations of Oort cloud comets caused by passing stars. The reliability and validity of these methods are checked against pure numerical integrations. In an accompanying paper (Chap. 10) we present several fast integrators and mappings that model the effects of the galactic tide. Again, the results computed with those methods will be compared to those obtained using a numerical integration of the equations of motion.

## 9.2 Modelling the Stellar Pertubations

As underlined previously, passing stars may perturb the orbits of Oort cloud comets. Oort [24] considered stellar perturbations as the only source of perturbations that change cometary orbits and inject them into the observable region. However, it has later been shown that the galaxy is another major perturber of cometary orbits. According to Hills [15] stellar perturbations ensure a constant flux of comets with $a \geq 2 \times 10^4$ AU towards the planetary region. As found by Dybczyński [7], star perturbations through the Oort cloud may induce the asymmetry of the distribution of observable comets (see also [8] and citations therein). Moreover Hills [15] showed that a close or penetrating stellar passage through the Oort cloud may deflect large numbers of comets to orbits that enter the planetary region forming what are known as comet showers.

To estimate stellar perturbations Oort [24] used the so-called impulse approximation, introduced in [25] to investigate the influence of stellar encounters on a cloud of meteoroids or comets. This approximation allows to obtain analytical solutions using simplifying assumptions, namely:

- The star velocity is constant and the motion follows a straight line;
- The star velocity is large enough that the comet and the Sun can be considered to be at rest during the stellar passage.

This approximation was used in a large number of papers (e.g., [2, 15, 21, 27, 32]) and has been found to be useful as a quick estimator in numerical Monte Carlo simulations of cometary orbital evolutions (e.g., [5, 12, 13, 16, 17, 22, 26, 28, 31]).

However, some concern over the accuracy of the method arises from simple considerations. For instance, the particularly important, very large impulses tend to be associated with the slowest stellar encounters, for which the assumption of a high stellar speed is not verified, or with very close passages to either the Sun, in which case the neglect of the hyperbolic deflection may lead to large errors, or the comet, where the neglect of even a slight cometary motion may change the results considerably.

The impulse approximation was briefly discussed in [23] in a work that concerned planets rather than comets, and in which an improvement for the case of a very close encounter star–Sun, which took into account the hyperbolic deflection of the stellar trajectory, was introduced.

Later on, Dybczyński [6] introduced a new, improved variant of the impulse approximation that is applicable to any time interval and allows a higher accuracy by treating the stellar orbit exactly. The advantage over the classical impulse approximation was convincingly demonstrated, but there was no discussion of the errors arising from the neglect of the motion of the comet during the stellar passage.

Eggers and Woolfson [10] have shown that for comets moving relatively fast, the impulse approximation breaks down. They developed a sequential method, in which the star path is split up and several partial impulses are computed along the stellar orbit assumed to be rectilinear. The authors found that this method is in good agreement with numerical integrations and also 10–100 times faster; they also suggested that their method may be improved by applying a sequential treatment to Dybczyński's method.

In this chapter we present such a method, that was introduced for the first time in [29]; the comet moves along its changing orbit receiving several impulses from the star which describes a hyperbolic orbit. Our method, the classical impulse approximation, and Dybczyński's algorithm are studied by comparing their results with those obtained in purely numerical integrations of the motion equations for a random sample of stellar encounters, simulating the ones expected for the solar system during $10^{10}$ yr. Each star is allowed to interact with 40 comets placed on Oort cloud orbits with given semi-major axis ($a_0$) and perihelion distance ($q_0$).

In Sect. 9.3 we describe the sample of stellar encounters which may interact with the comets. Section 9.4 is devoted to present the different algorithms, while the results of the computations are presented in Sect. 9.5. Finally the conclusions are given in Sect. 9.6.

## 9.3 Stellar Encounters

Our purpose is to study the errors introduced by the different algorithms as a function of the encounter parameters. We use a distribution of stellar parameters that is as realistic as possible, and to this purpose we have created a random sample of stellar encounters experienced by the solar system during $10^{10}$ yr, according to the frequency distribution given in [14]. The upper limit for the solar impact parameter is $4 \times 10^5$ AU, and we use the 13 stellar categories listed in Table 9.1 of [28] with their corresponding data, for which the total encounter frequency is $10.525\,\pi^{-1}$ pc$^{-2}$Myr$^{-1}$. The total number of encounters in our random sample is thus 395,813.

The stellar speed of a given encounter is computed in the same fashion as in [28]. We combine the apex velocity vectorially with a peculiar velocity vector that is randomly chosen from a 3D gaussian distribution with the prescribed dispersion. The direction of motion of each star is distributed uniformly, since the Oort cloud may be modelled as spherically symmetric; in this way, with respect to a random orientation of the cometary orbit, the stellar encounters are always isotropically oriented. Figure 9.1 shows the frequency distribution of our encounter sample in the parametric plane of approach velocity $V^*$ and solar impact parameter $b^*$.

**Table 9.1.** Computing time spent with different methods for calculating stellar perturbations

| $a_0$ (AU) | $10^4$ | $3 \times 10^4$ | $10^5$ |
|---|---|---|---|
| IEM (s) | $5 \times 10^4$ | $2.3 \times 10^4$ | $1.7 \times 10^4$ |
| CIA (s) | 44 | 45 | 44 |
| DIA* (s) | 37 | 37 | 37 |
| DIA (s) | 130 | 131 | 116 |
| SIA50 (s) | 7 646 | 8 246 | 9 172 |
| SIA250 (s) | 1 470 | 1 557 | 1 764 |
| SIA500 (s) | 779 | 932 | 912 |
| SIA1000 (s) | 476 | 476 | 500 |

The method used is indicated in the left column, and the four SIA calculations are distinguished by the choice of the initial time step. The three columns to the right correspond to different choices of the initial semi-major axis, as indicated in the top line.

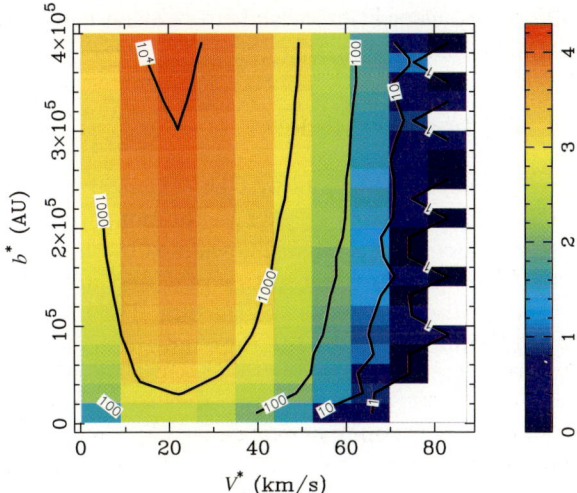

**Fig. 9.1.** The $(V^*, b^*)$ plane is split into equal boxes, and the colour coding shows the number of stellar encounters in each box on a log scale as indicated to the right. Isodensity curves marked by the corresponding numbers are also shown

Our data set for each combination $(a_0, q_0)$ consists of 395,813 stellar encounters, experienced one by one by a sample of 40 comets. For each new star, each comet is put back into its initial orbit with the same starting value of the mean anomaly. The random timing of the encounters means that, each time, the comets find themselves at random orbital positions as the star passes.

## 9.4 The Dynamical Models

In order to estimate the errors of the different methods of computing stellar pertubations we have to compare their results with those obtained with an accurate numerical integration. To this purpose, the Newtonian equations of motion of the comet and the star are integrated in a heliocentric reference frame, using the RA15 [7] integrator. RA15 is a variable step integrator with an external control of the accuracy specified as an input parameter (LL). The value LL = 12 was found sufficient to guarantee a high accuracy of the resulting stellar perturbations.

As explained in [29], since our numerical integration is not extended over an infinite path before and after closest approach, we have performed numerical experiments in order to find the optimal choice of the limiting start and end distance $d_{\lim}$. We found that $d_{\lim} = \sqrt{b^{*2} + r_{\lim}^2}$ with $r_{\lim} = 1.0 \times 10^6$ AU is large enough to define the limiting distance in our study of different dynamical models [29]. In the remainder of this paper, the reference model to compute stellar perturbations, denoted by IEM (integration of the equations of motion),

### 9.4.1 The Classical Impulse Model

In this case, which we hereafter denote by CIA (classical impulse approximation), the comet is held fixed with respect to the Sun, while the star passes with constant velocity along the straight line defined by the impact parameter $b^*$, a unit vector $\hat{\mathbf{b}}^*$ that defines the direction of closest approach and the velocity vector $\mathbf{V}^*$ of the star with respect to the Sun. The impulse of the comet relative to that of the Sun caused by the time-integrated stellar attraction is computed from

$$\Delta \mathbf{v} = \frac{2GM_*}{V^*} \left\{ \frac{\hat{\mathbf{b}}_c}{b_c} - \frac{\hat{\mathbf{b}}^*}{b^*} \right\}; \qquad (9.1)$$

adding this value of $\Delta \mathbf{v}$ to the heliocentric velocity of the comet at the orbital position in question, one obtains a new orbital velocity and thus new values of the orbital elements. Figure 9.2 illustrates the geometry considered.

The CIA computes the perturbation for a time span extending to infinity before and after closest approach; some of the differences between the results of CIA and IEM may result from this. However, this contribution to the total difference should be small, just like the contribution to the perturbation experienced outside $d_{\lim}$. In principle, (9.1) may be modified to account for a finite time interval (see [6, 27]), but the result would be a large increase of the computing time. Since the principal advantage of the CIA is its speed, the use of such a procedure would hence be self-defeating.

### 9.4.2 Dybczyński's Impulse Approximation

In this case, which we denote by DIA, the comet is held fixed with respect to the Sun, while the star describes the part of its heliocentric, hyperbolic orbit

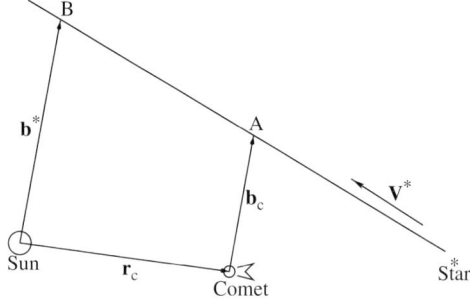

**Fig. 9.2.** Geometry of a stellar passage considered for the classical impulse approximation

that is contained within our limiting distance.[1] Just like the parameters of the straight line of the CIA, this hyperbola is uniquely defined by the parameters of the stellar encounter. Given the position and velocity vectors of the comet with respect to the Sun, the hyperbola characterizing the relative star–comet motion can also be uniquely defined.

As in the case of the CIA, we have to select a position of the comet along its orbit, and once again we pick the one that the unperturbed comet would have at the time of the stellar perihelion passage. Since we choose this time to be the same for the CIA and DIA, the positions of the comet are also the same.

As shown in [6], in a heliocentric frame of reference where the $x$-axis is anti-parallel to the stellar velocity vector, the $y$-axis is parallel to $\hat{\mathbf{b}}^*$ and the $z$-axis completes a right-handed system (this will be referred to as the 'impact frame'), the impulse vector of the comet relative to the Sun is given by

$$\Delta v_x = (a_c \Delta v_{c\xi} + b_c \Delta v_{c\eta})\frac{1}{c_c} - (a_s \Delta v_{s\xi} + b_s \Delta v_{s\eta})\frac{1}{c_s},$$

$$\Delta v_y = (a_c \Delta v_{c\eta} - b_c \Delta v_{c\xi})\frac{b_s - y_c}{c_c b_c} - (a_s \Delta v_{s\eta} - b_s \Delta v_{s\xi})\frac{1}{c_s}, \quad (9.2)$$

$$\Delta v_z = (a_c \Delta v_{c\eta} - b_c \Delta v_{c\xi})\frac{-z_c}{c_c b_c},$$

where $a_s$, $b_s$ and $c_s$ are the semi-axes and focal distance of the heliocentric hyperbola ($c_s^2 = a_s^2 + b_s^2$), and $a_c$, $b_c$ and $c_c$ are the same quantities for the cometocentric hyperbola. All the cometocentric quantities are computed taking into account the heliocentric velocity of the comet. This is an arbitrary choice that deviates from description given in [6], but the results are thereby slightly improved.

The cometary position is described by $y_c$ and $z_c$, and the various $\Delta v$ quantities are given by

$$\Delta v_{s\xi} = \frac{-GM_*}{b_s V_\infty}(\sin \vartheta_{s2} - \sin \vartheta_{s1}),$$

$$\Delta v_{s\eta} = \frac{-GM_*}{b_s V_\infty}(\cos \vartheta_{s1} - \cos \vartheta_{s2}), \quad (9.3)$$

$$\Delta v_{c\xi} = \frac{-GM_*}{b_c V_\infty}(\sin \vartheta_{c2} - \sin \vartheta_{c1}),$$

$$\Delta v_{c\eta} = \frac{-GM_*}{b_c V_\infty}(\cos \vartheta_{c1} - \cos \vartheta_{c2}),$$

where $G$ is the gravitational constant and $M_*$ is the stellar mass, $V_\infty = V^*$ is the stellar speed at infinity and $\vartheta$ is the true anomaly of the star with respect to the Sun or the comet, the indices '1' and '2' denoting the beginning and

---

[1] Dybczyński's method is more general, allowing any time interval to be considered, but this is our choice for comparing the results with the IEM.

end of the computation, respectively. All these quantities are easily computed, and thus the impulse vector is found from the above equations. A great simplification of (9.2) and (9.3) results from considering an infinite time interval like for the CIA, as explained in [6]. We have applied this procedure in some runs for comparison purposes, and we denote these runs by DIA*.

### 9.4.3 The Sequential Impulse Approximation

In this case, which we denote by SIA, the comet is allowed to move along its osculating orbit between the steps of a predefined sequence. At the beginning of the calculation, the star is placed at its proper position at distance $d_{\lim}$ from the Sun. For each step a preselected time interval is used to compute the corresponding change in the true anomaly ($\Delta \vartheta$) of the star's heliocentric or cometocentric orbit, depending on which object is closest to the star. Thus, during the whole calculation, a variable time step is used with the property that it always corresponds to $\Delta \vartheta$ for the orbit of the object that is closest to the star. As a result the time step varies roughly in proportion to the square of the minimum distance from the Sun or comet to the star.

We stress that – in principle – the SIA is capable of any level of precision, but both the precision and the CPU time are determined by the method and choice of parameters to choose the step length. As shown in Sect. 9.5, we have found the above procedure to be appropriate for practical needs.

In practice, the SIA is constructed as follows. After selecting an initial time step $\Delta t_0$, we first compute the initial star–comet distance $d_{c0}$ and the true anomaly step size of the heliocentric orbit of the star:

$$\Delta \vartheta_s = \frac{V^* b^*}{d_{\lim}^2} \Delta t_0 . \qquad (9.4)$$

For each step, we start from the heliocentric position and velocity of the comet and the star in the above-described impact frame. The time is $t$. The following sequence is then stepped through:

- We compute the cometocentric orbital parameters of the star from the positions and velocities, and we find the values of the asterocentric true anomalies of the Sun ($\vartheta_{s1}$) and the comet ($\vartheta_{c1}$).
- We find which body, the Sun or the comet, is closer to the star. If it is the Sun, we compute the time step $\Delta t$ from $\vartheta_{s1}$ and $\Delta \vartheta_s$ given by (9.4). If it is the comet, we compute the cometary true anomaly step from

$$\Delta \vartheta_c = \frac{V^* b_c}{d_{c0}^2} \Delta t_0 \qquad (9.5)$$

and compute the time step $\Delta t$ from $\vartheta_{c1}$ and $\Delta \vartheta_c$ given by (9.5).
- In the former case, if the Sun is closer to the star, we compute the true anomalies at the end of the step from $\vartheta_{s2} = \vartheta_{s1} + \Delta \vartheta_s$, and $\vartheta_{c2} = \vartheta_c(t+\Delta t)$. In the latter case, if the comet is closer to the star, these anomalies are found from $\vartheta_{s2} = \vartheta_s(t+\Delta t)$ and $\vartheta_{c2} = \vartheta_{c1} + \Delta \vartheta_c$.

- We apply Dybczyński's formulae (9.2) and (9.3) to obtain the heliocentric impulse vector of the comet.
- We compute the heliocentric position and velocity vectors of the comet at the middle of the step and add the heliocentric impulse vector to the velocity.
- The heliocentric position and velocity vectors of the comet and the star are computed at the end of the step, thus preparing for the following step.

The calculation ends when the star reaches the heliocentric distance ($d_{\lim}$) after closest approach.

Two important general remarks may be made on the above computation: (1) all the hyperbolic true anomalies are computed via the eccentric anomalies and (2) the inversions of the Kepler equations are made using (4) and (12) of [4] in the elliptic cases, and the hints given in [30] for the hyperbolic cases.

We have used several values for the initial time step $\Delta t_0$ in an attempt to compare them internally as well as with other methods and decide about the best choice.

## 9.5 Results

### 9.5.1 Computing Times

We consider the sample of 395,813 stellar encounters previously defined acting on 40 cometary orbits, having the same initial semi-major axis ($a_0$) and perihelion distance ($q_0$). Computations of the stellar perturbations have been performed using a particular method, for three different values of $a_0$ with always the same $q_0 = 100$ AU. Table 9.1 summarizes the total computing times spent on a 2.8 GHz XEON. The different SIA calculations are distinguished by the following initial time steps: $0.5 \times 10^5$ yr for SIA50, $2.5 \times 10^5$ yr for SIA250, $5 \times 10^5$ yr for SIA500 and $1 \times 10^6$ yr for SIA1000. Thus, the numbers used in the acronyms give the time step in thousands of years.

IEM calculations take more time for smaller values of the cometary semi-major axis, because this makes it more likely for the comet to be close to perihelion during the stellar passage. Since most stellar encounters are distant (see Fig. 9.1), it is in fact the cometary perihelion passages that dominate the use of CPU time. This effect is not seen for SIA, which works directly with the cometary orbital elements, and whose time step is independent of the Sun–comet distance—see (9.5) and (9.7).

Both CIA, DIA* and DIA are extremely quick estimators compared with IEM, and DIA* is just as fast as CIA. Both of these are ∼500 times faster than IEM, while DIA is only ∼200 times faster owing to the calculation of sines and cosines of true anomalies in (9.3). SIA is significantly slower, though this depends on the choice of the initial time step. SIA50 is practically useless, being only a little faster than IEM, but SIA500 and SIA1000 work at least 20–30 times faster than IEM for typical Oort cloud orbits. In fact, for a stellar encounter with a typical speed of 30 km/s, the time required to describe a rectilinear path of $2 \times 10^6$ AU is about 300,000 yr. This generally means only one step for SIA1000 and just a few for SIA500. Of course, the number of steps grows much larger for very slow and/or close stellar encounters.

### 9.5.2 Error Statistics of the Models

For each set of cometary elements $a_0$ and $q_0$, the set of $\sim 16 \times 10^6$ perturbations computed by IEM is divided into four quartiles ($Q_1$, $Q_2$, $Q_3$ and $Q_4$), each containing the same number of perturbations ($\sim 4 \times 10^6$) and such that the 25% smallest perturbations are in $Q_1$, etc. For each model under study and each perturbation in $Q_i$ ($i = 1, 4$), an error $EQ_i$ is computed as

$$EQ_i = \frac{\Delta q_{\text{IEM}} - \Delta q_{\text{mod}}}{\langle |\Delta q_{\text{IEM}}| \rangle_i} \tag{9.6}$$

for $i = 1, 2, 3$, and

$$EQ_4 = \frac{\Delta q_{\text{IEM}} - \Delta q_{\text{mod}}}{\max(|\Delta q_{\text{IEM}}|, \langle |\Delta q_{\text{IEM}}| \rangle_i)}, \tag{9.7}$$

for $i = 4$. Here $\Delta q_{\text{IEM}}$ and $\Delta q_{\text{mod}}$ are the perturbations computed by IEM and the model under study, respectively, and $\langle |\Delta q_{\text{IEM}}| \rangle_i$ is the arithmetic mean of the $|\Delta q_{\text{IEM}}|$ included in $Q_i$.

In [29], it has been shown (see Table II of [29]) that the majority of the perturbations of $\Delta q$ are very small, but the tail of the perturbation distribution extends to very large values. Moreover the models have also been compared against each other (see Table III of [29]). The main results are the following: (1) whatever the model, the average of $EQ$ is always very small, which indicates that none of the models tend to introduce any systematic bias into $\Delta q$; (2) SIA1000 is not much better than DIA except for $a_0 = 10^5$ AU; (3) while the results obtained with DIA* are practically the same as the CIA results, the model DIA is distinctly better than CIA, implying that a large part of the CIA errors arise from the consideration of an infinite time interval.

### 9.5.3 Dependence of the Errors on the Encounter Parameters

To analyse how errors are distributed over the encounter parametric plane, we have divided this plane into cells and studied the error statistics in the cells individually. In Fig. 9.3 we present the results for the case of $a_0 = 3 \times 10^4$ AU with the CIA, DIA and SIA500 models. We divided the encounter sample into four equal parts corresponding to the quartiles of the $|\Delta q|$ distribution as computed by IEM. The left and right panels show the $Q_1$ and $Q_4$ quartiles, respectively, and from the population density level curves we see how the $Q_4$ sample preferentially populates the cells with smaller impact parameters and smaller approach velocities—as expected from (9.1).

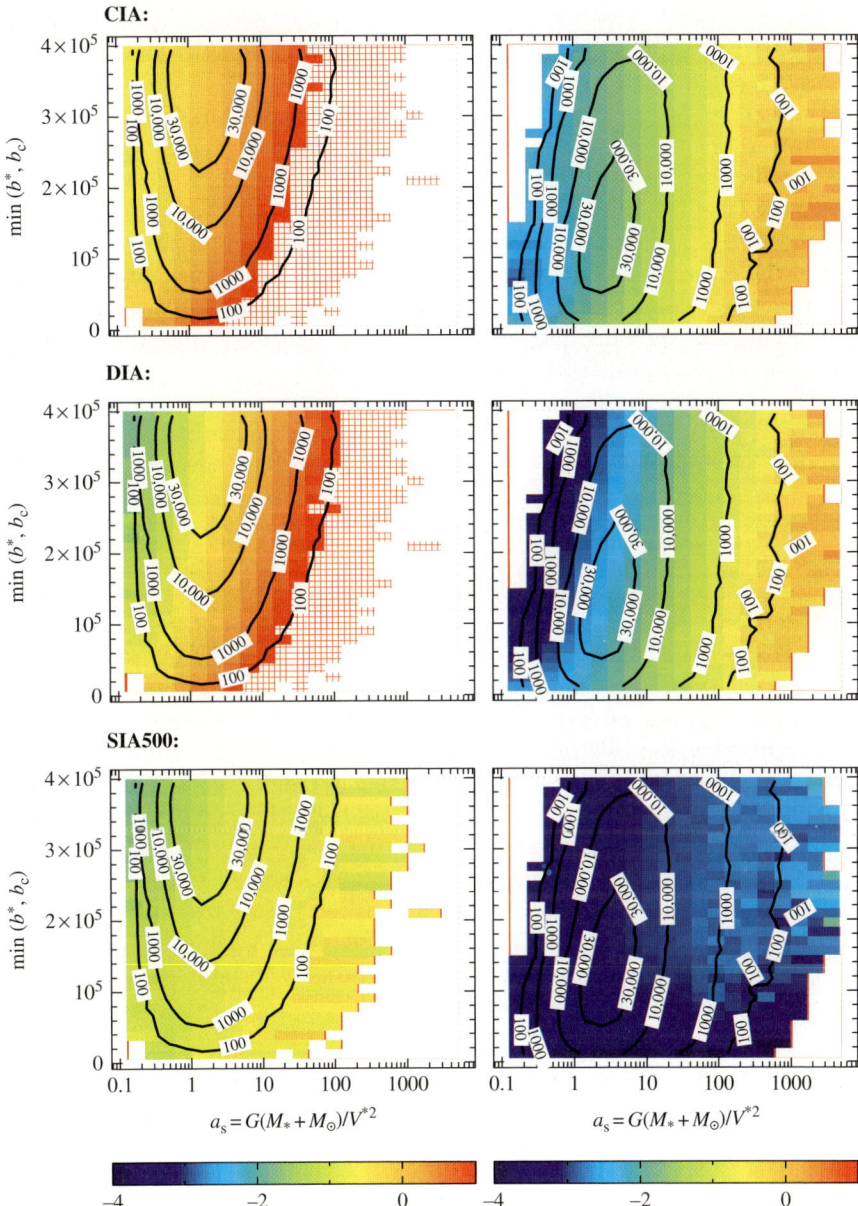

**Fig. 9.3.** Colour diagram of the model errors in the $a_s$, $\min(b^*, b_c)$ plane. The plane is divided into $20 \times 30$ cells. Cells on the same line are gathered until a population of at least 20 stellar passages is reached. The *colour* indicates the value of $^{10}\log|EQ_i|$ that is surpassed by 10% of all the errors $|EQ_i|$ computed in each cell. For the *cross-hatched cells* this value is larger than 1. **Diagrams to the left** correspond to $Q_1$ and **diagrams to the right** correspond to $Q_4$. For each quartile, the isopopulation curves of the corresponding stellar encounters are also shown, indicating the numbers falling in each individual cell. The unit for both axes is AU

The colour coding is chosen to represent the 90$^{th}$ percentile, i.e. the value of |EQ| that is surpassed by 10% of the population in the cell. The orange–red colour range indicates errors larger than 100%.

Our first observation is that such conditions are commonplace for CIA and DIA in the first quartile, whenever the stellar approach velocity is relatively small, while they do not occur at all for the SIA500 model. Next we note that these very large relative errors are confined to $Q_1$ rather than $Q_4$, so that the actual errors in absolute terms are quite small. This is indicative of a component of the errors that is not proportional to the size of the perturbation, and thus becomes relatively important as the perturbations approach zero. Since mainly CIA and DIA are affected, we conclude that neglecting the cometary motion must be an important part of the problem.

For practical purposes it may be more interesting to study the errors of the largest perturbations as shown in the right-hand panels of Fig. 9.3. In this case the CIA and DIA errors occasionally reach the 100% level for the slowest encounters. By contrast, the SIA500 errors are always vanishingly small. It is noteworthy that the largest CIA and DIA errors in $Q_4$ do not occur for the closest encounters but for relatively distant ones with very low velocities—a detailed example will be discussed in Sect. 9.5.4.

A detailed study of the smallest impact parameters of our sample (see Fig. 9.4) shows that DIA yields much better results than CIA—especially for impact parameters less than a few thousand AU.

Neither CIA nor DIA competes in accuracy with SIA500 for these close encounters. However, for impact parameters <1000 AU the computing time is larger with SIA500 than using IEM. Hence, if a high accuracy of the stellar perturbations is desired, one should use IEM for such very rare and close encounters, while a lot of computing time may be saved by using SIA500 or SIA1000 in the rest of the encounter parameter plane.

Changing the value of $a_0$ to either $1 \times 10^4$ or $1 \times 10^5$ AU does not affect the above results very much. The use of CIA or DIA for the inner core of the Oort cloud, where $a_0 < 1 \times 10^4$ AU, especially for stars passing at low velocities, does not appear advisable.

### 9.5.4 Example of a Stellar Passage

In order to illustrate the differences between the models, one stellar passage will now be considered in detail. This example was chosen among those passages that CIA and DIA failed to reproduce. Figure 9.5a shows the geometry of the encounter in the plane of the stellar trajectory. The upper part of the figure lists the ecliptic elements of the stellar and cometary orbits. $M_0$ is the mean anomaly of the comet when the star is at its perihelion; $b_c$ is the value of the cometary impact parameter at the time when the star passes perihelion and taking the heliocentric velocity of the comet into account, as computed for the DIA model; and $\min_{M_0}(b_c)$ is the minimum value of $b_c$ on a sequence of encounters where $M_0$ varies between 0° and 360° (see below).

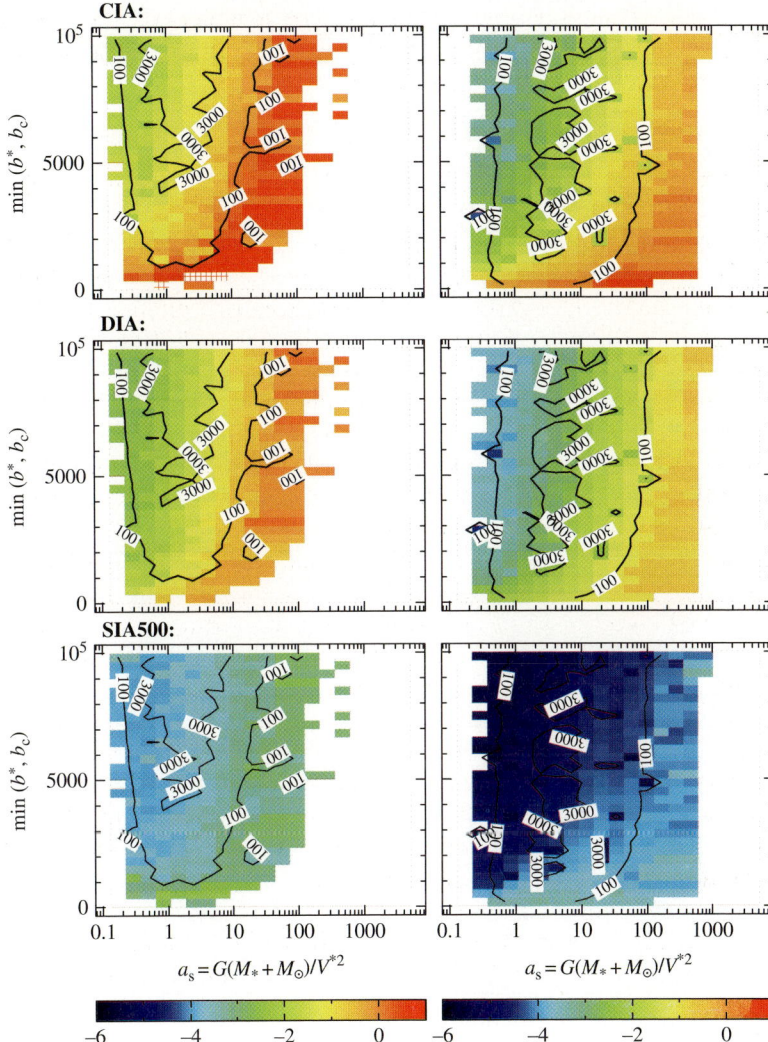

**Fig. 9.4.** Same as Fig. 9.3, but limited to minimum impact parameters less than $10^4$ AU. All these data are contained within the lowest cells of Fig. 9.5

Also shown are the perihelion distance perturbations using the different models. Clearly, SIA500 and IEM are almost equivalent. In order to check if the discrepancy between DIA or CIA and IEM is due to the place where the comet is fixed during the stellar passage (which may yield a large difference when strong perturbations are involved), Fig. 9.5b gives the final perihelion distance of the comet, as $M_0$ varies between $0°$ and $360°$.

Clearly, this encounter cannot be well reproduced by DIA or CIA. This is not due to a close encounter, since both $b^*$ and $b_c$ are always larger than

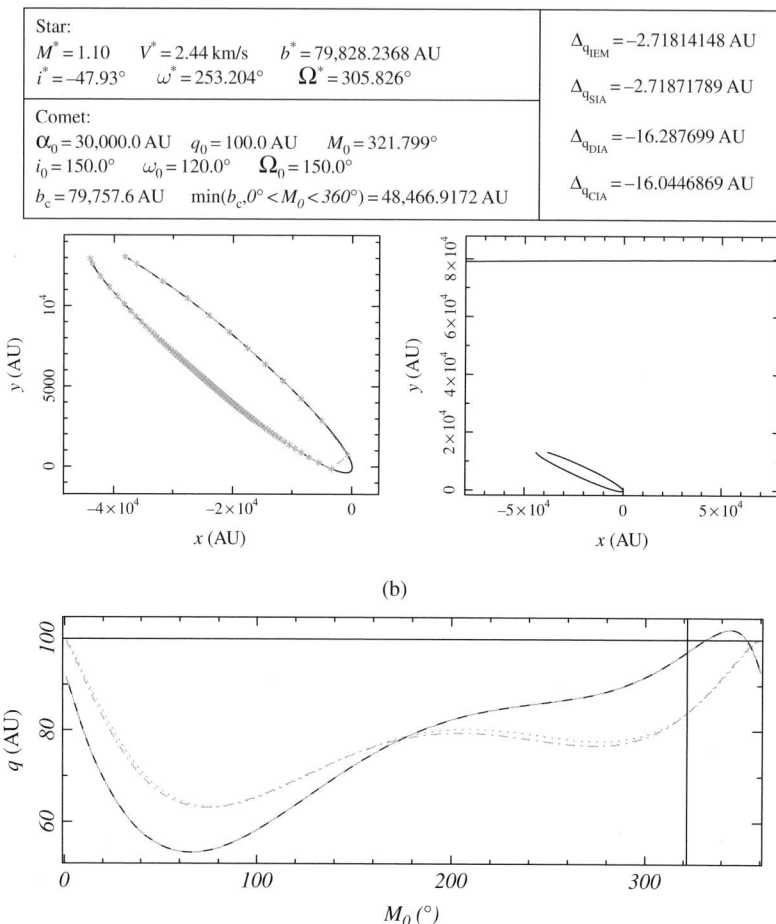

**Fig. 9.5.** (a) **Above**: Ecliptic orbital elements of the star and the comet; $b_c$ is the comet–star impact parameter of the encounter shown, and $\min_{M_0}(b_c)$ is the minimum cometary impact parameter for the stellar passages used to plot part (**b**). To *the right*, perturbations of the perihelion distance computed with different models. **Below**, **left panel**: Cometary orbit during the stellar passage using IEM (*full black line*), and cometary positions computed by SIA500 (*grey asterisks*) joined by a *grey dashed line*, in the plane of the stellar trajectory. **Right panel**: Stellar trajectory and cometary orbit computed by IEM in the same plane. (**b**) Final perihelion distance versus cometary mean anomaly $M_0$ at the time of the perihelion passage of the star using IEM (*full grey curve*), SIA500 (*dashed black curve*), DIA (*dot-dashed grey curve*) and CIA (*dotted grey curve*). *Horizontal line*: initial value of the cometary perihelion distance; *vertical line*: value of $M_0$ used in part (**a**).

48,000 AU, but rather to the small stellar velocity (2.44 km/s). Note from Fig. 9.5a that this causes the comet to perform almost a whole revolution around the Sun during the encounter. It is obvious that keeping the comet fixed at any point in its orbit cannot yield a good approximation to the integrated perturbation. Moreover, due to the large encounter distance, the hyperbolic deflection of the stellar trajectory is negligible, as is the variation of the stellar velocity—hence the CIA and DIA yield practically the same results.

There is an interesting difference between close and distant encounters in the time interval during which the main part of the perturbation builds up. For a close encounter, this is very short, and the motion of the comet is not of prime importance. But for a distant one, if it is very slow like in Fig. 9.5, the motion of the comet is indeed important and may destroy the accuracy of a one-step method. Thus we may explain why the CIA and especially DIA errors are the largest for slow, distant encounters, while for the close encounters in Fig. 9.4 they are not extremely large. Even if the point where one assumes the comet to reside in such a case may not be the best one, keeping the comet fixed is not a bad assumption.

## 9.6 Conclusions

In terms of computer time the CIA and the DIA are extremely quick estimators, since they are respectively $\simeq 500$ and $\simeq 200$ faster than IEM. The computing time of SIA depends on the initial time step as well as the accuracy. For a step choice of $5 \times 10^5$ or $1 \times 10^6$ yr the computations are 20–30 times faster than IEM at the expense of some little loss in accuracy, but the accuracy is significantly better than CIA and DIA. However, for small impact parameters (see Fig. 9.4) the CIA and DIA are not good estimators.

Moreover, for even a distant encounter, when the star velocity is very low the CIA and DIA give wrong results, since the comet is assumed to be fixed during the encounter. CIA (as found previously by Eggers and Woolfson) and DIA are not suitable for the inner core of the Oort cloud.

For simulations of the Oort cloud, CIA gives a reasonably good approximation for a very small computation time, but DIA works significantly better at almost the same amount of time.

The SIA should be used when modelling the inner core of the Oort cloud and the scattered disk. Our choice, according to the best compromise between accuracy and speed is the SIA500, taking into account that when the impact parameter of the star with the Sun or the comet is smaller than 1000 or 2000 AU the IEM is faster than the SIA500.

## References

1. Bailey, M. E., 1983. *MNRAS* **204**, 603.
2. Bailey, M. E., 1986. *MNRAS* **218**, 1.

3. Breiter, S., Dybczyński, P. A., and Elipe, A., 1996. *A&A* **315**, 618.
4. Danby, J. M. A., 1987. *Celest. Mech.* **40**, 303.
5. Duncan, M., Quinn, T., and Tremaine, S., 1987. *AJ* **94**, 1330.
6. Dybczyński, P. A., 1994. *CeMDA* **58**, 139.
7. Dybczyński, P. A., 2002. *A&A* **383**, 1049.
8. Dybczyński, P. A., 2002. *A&A* **396**, 283.
9. Dybczyński, P. A. and Pretka, H., 1996. *EMP* **72**, 13.
10. Eggers, S. and Woolfson, M. M., 1996. *MNRAS* **282**, 13.
11. Everhart, E., 1985. In Carusi, A. and Valsecchi, G. B. (eds), *Dynamics of Comets: Their Origin and Evolution*, Reidel, Dordrecht, p. 185.
12. Fernández, J. A., 1980. *Icarus* **42**, 406.
13. Fernández, J. A., 1982. *AJ* **87**, 1318.
14. García-Sánchez, J., Weissman, P. R., Preston, R. A. et al., 2001. *AJ* **379**, 634.
15. Hills, J. G., 1981. *AJ* **86**, 1730.
16. Heisler, J., 1990. *Icarus* **75**, 104.
17. Heisler, J., Tremaine, S., and Alcock, C., 1986. *Icarus* **70**, 269.
18. Hut, P. and Tremaine, S., 1985. *AJ* **90**, 1548.
19. Levison, H., Dones, L., and Duncan, M. J., 2001. *AJ* **121**, 2253.
20. Maciejewski, A. J. and Pretka, H., 1998. *A&A* **336**, 1065.
21. Matese, J. J. and Lissauer, J. J., 2002. *Icarus* **157**, 228.
22. Mazeeva, O. A., and Emel'yanenko, V. V., 2002. In *Asteroids, Comets, and Meteors*, ACM , pp. 445–448.
23. Morris, D. E. and O'Neill, T. E., 1988. *AJ* **96**, 1127.
24. Oort, J. H., 1950. *Bull. Astron. Inst. Neth.* **11**, 91.
25. Öpik, E. J., 1932. *Proc. Am. Acad. Arts Sci.* **67**, 169.
26. Rémy, F. and Mignard, F., 1985. *Icarus* **63**, 1.
27. Rickman, H., 1976. *BAICz* **27**, 92.
28. Rickman, H., Froeschlé, Cl., Froeschlé, Ch., and Valsecchi, G. B., 2004. *A&A* **428**, 673.
29. Rickman, H., Fouchard, M., Valsecchi, G. B., and Froeschlé, Ch., 2005. *EMP* **97**, 411.
30. Serafin, R. A., 2002. *CeMDA* **82**, 363.
31. Weissman, P. R., 1979. In Duncombe, R. L. (ed.), *Dynamics of the Solar System*, IAU Symp. 81, Reidel, Dordrecht, p. 277.
32. Weissman, P. R., 1980. *Nature* **288**, 242.

# 10

# Methods for the Study of the Dynamics of the Oort Cloud Comets II: Modelling the Galactic Tide

Marc Fouchard[1,2,5], Christiane Froeschlé[3], Sławomir Breiter[4], Roman Ratajczak[4], and Giovanni B. Valsecchi[5] and Hans Rickman[6]

[1] Laboratoire d'Astronomie de Lille/Université de Lille,
   1 impasse de l'Observatoire 59000 Lille, France
[2] Institut de Mécanique Céleste et de Calcul d'Ephémérides, UMR 8028 du CNRS, 75014 Paris, France
   fouchard@imcce.fr
[3] Laboratoire Cassiopée, Université de Nice, CNRS, Observatoire de la Côte d'Azur, BP 4229, FR–06304 Nice, France
   froesch@obs-nice.fr
[4] Astronomical Observatory of A. Mickiewicz University,
   Słoneczna 36, Poznań 60-286, Poland
   breiter@amu.edu.pl,astromek@amu.edu.pl
[5] INAF-IASF, Via Fosso dell Cavaliere 100, 00133 Roma, Italy
   giovanni@iasf-roma.inaf.it
[6] Uppsala Astronomical Observatory, Box 515, SE-75120 Uppsala, Sweden
   hans@astro.uu.se

**Abstract.** In this paper we present several fast integrators and mappings which model the orbital perturbations of Oort cloud comets, caused by the Galactic tide. The perturbations computed with these models are compared with those obtained using an accurate numerical integration using Everhart's RA15 code. In order to have the best compromise between the computing velocity and precision, it is shown that it is necessary to use a **hybrid model** i.e. a combination of two algorithms, according to the values of the semimajor axe $a$ and the eccentricity-$e$ of the considered comet.

## 10.1 Introduction

The present paper follows the accompanying Chap. 9, in which we investigated different algorithms to compute the orbital perturbations of Oort cloud comets caused by passing stars. In this chapter, we present several fast integrators and mappings which modelise the effects of the galactic tide. The results computed with these methods are compared with those obtained using a numerical integration of the equations of motion.

To model the normal and radial components, with respect to the galactic plane, of the galactic tide different methods and mappings are developed in Sect. 10.2. First of all in Sect. 10.2.1 we give the equations of motion of a comet perturbed by the galactic tide. Then the following models are described, namely:

- A symplectic and regularised integrator (Sect. 10.2.2)
- Some averaged Hamiltonian models using, according to the value of the orbital eccentricity $e$, either the Delaunay elements or the "Matese elements" and, for each model, mappings based on the respective Taylor development (Sect. 10.2.3);
- A Lie–Poisson averaged model (Sect. 10.2.4).

In Sect. 10.3 the results of calculations performed using the different models are compared to those obtained by numerical integrations. This comparison allows us to build models which use a composition of two different kinds of integrators in order to increase the velocity of the integrations minimising the loss of accuracy (Sect. 10.4). Section 10.5 is devoted to the conclusions.

## 10.2 Models of Galactic Tide Effects on Cometary Orbits

### 10.2.1 The Cartesian Model

To establish the equations of motion, we consider two different heliocentric frames. The first one is a *rotating frame* with the $\hat{x}'$–axis in the radial direction pointing towards the galactic centre, the $\hat{y}'$-axis pointing transversely along the local circular velocity and $\hat{z}$ completing a right-handed system. The *fixed frame* $(\hat{x}, \hat{y}, \hat{z})$ is such that it coincides with the rotating frame $(\hat{x}', \hat{y}', \hat{z})$ at time $t = 0$ while keeping its axial directions fixed (see Fig. 1 of [8]).

Let us define $\Omega_0$ as the angular velocity about the galactic centre, assuming the Sun to follow a circular orbit (since the motion of the Sun around the galaxy is clockwise in both our frames, $\Omega_0$ is negative, i.e. the vector is directed along $-\hat{z}$). If $\phi_r$ is an angle in the galactic plane measured in the rotating frame from $\hat{x}'$, and $\phi$ the corresponding angle measured in the fixed frame from $\hat{x}$ at time $t$, we have the relation: $\phi = \phi_r + \Omega_0 t$ (see Fig. 1 of [8]). All the final results will be presented in the fixed frame.

The force $\mathbf{F}$ per unit of mass acting on a test particle orbiting the Sun under the influence of the galactic tide is given by (see [10]):

$$\mathbf{F} = -\frac{\mu M_\odot}{r^3}\mathbf{r} + (A-B)(3A+B)x'\hat{x}' - (A-B)^2 y'\hat{y}'$$
$$-[4\pi\mu\rho_\odot - 2(B^2 - A^2)]z\hat{z} , \qquad (10.1)$$

where $x'$, $y'$, $z$ are the coordinates of the comet in the rotating frame, $\mathbf{r}$ is the Sun–comet vector of length $r$, $A$ and $B$ are the Oort constants and $\rho_\odot$ is the local density of the galactic disk in the solar neighbourhood. In the remaining part of the paper we will assume $\rho_0 = 0.1 M_\odot \, \text{pc}^{-3}$ [12] and an angular velocity of the Sun around the galactic centre $\Omega_0 = B - A = -26 \, \text{km s}^{-1} \text{kpc}^{-1}$, with the approximation $A = -B$.

The unit of mass is the solar mass ($M_\odot = 1$), the unit of time is the year, and the unit of length is the astronomical unit (AU). As a consequence, the gravitational constant $\mu$ is equal to $4\pi^2$.

Let $\mathcal{G}_1$, $\mathcal{G}_2$ and $\mathcal{G}_3$ be defined by

$$\begin{aligned}\mathcal{G}_1 &= -(A-B)(3A+B) \\ \mathcal{G}_2 &= (A-B)^2 \\ \mathcal{G}_3 &= 4\pi\mu\rho_\odot - 2(B^2 - A^2)\,.\end{aligned} \qquad (10.2)$$

Then, with the chosen values of $A$, $B$ and $\rho_0$, one has

$$\begin{aligned}\mathcal{G}_2 &= -\mathcal{G}_1 = 7.0706 \times 10^{-16} \, \text{yr}^{-2}, \\ \mathcal{G}_3 &= 5.6530 \times 10^{-15} \, \text{yr}^{-2}, \\ \Omega_0 &= -\sqrt{\mathcal{G}_2}\,.\end{aligned} \qquad (10.3)$$

One may note that the relation $\mathcal{G}_2 = -\mathcal{G}_1$ is a particular case which corresponds to a flat rotation curve of the galaxy, i.e. to a constant tangential velocity of the stars around the galactic centre whatever the star distance to the galactic centre. Numerical experiments have shown that another choice of Oort constants $A$ and $B$ consistent with the observations does not affect the long-term dynamics of the Oort cloud comets [9]. Consequently, some of the models presented here will use the relation $\mathcal{G}_2 = -\mathcal{G}_1$ in order to simplify the equations.

The general equations of motion in Cartesian coordinates are

$$\begin{aligned}\frac{d^2 x}{dt^2} &= -\frac{\mu M_\odot}{r^3} x - \mathcal{G}_1 x' \cos(\Omega_0 t) + \mathcal{G}_2 y' \sin(\Omega_0 t) \\ \frac{d^2 y}{dt^2} &= -\frac{\mu M_\odot}{r^3} y - \mathcal{G}_1 x' \sin(\Omega_0 t) - \mathcal{G}_2 y' \cos(\Omega_0 t) \\ \frac{d^2 z}{dt^2} &= -\frac{\mu M_\odot}{r^3} z - \mathcal{G}_3 z\,,\end{aligned} \qquad (10.4)$$

where $x, y, z$ are the coordinates of the comet in the fixed frame [thus $x' = x \cos(\Omega_0 t) + y \sin(\Omega_0 t)$ and $y' = -x \sin(\Omega_0 t) + y \cos(\Omega_0 t)$]. Equations (10.4) will be referred to as the Cartesian model, also denoted RADAU. The RADAU integrator described by [7] is used at the 15th order, with LL = 12, to integrate (10.4). This integrator was chosen because it is fast, reliable and accurate compared to other non-symplectic integrators.

## 10.2.2 Regularised Symplectic Integrator

This method was first introduced in [5]. For a detailed description of the method one should read the original paper, only the key points are presented here. In the following we recall the key steps.

## Hamiltonian in Cartesian Variables

From (10.1) the complete Hamiltonian is given by

$$\mathcal{H} = \mathcal{H}_0 + \mathcal{H}_1 \tag{10.5}$$

$$\mathcal{H}_0 = \tfrac{1}{2}\left(X^2 + Y^2 + Z^2\right) - \frac{\mu}{r}, \tag{10.6}$$

$$\mathcal{H}_1 = \tfrac{1}{2}\left(\mathcal{G}_1 x'^2 + \mathcal{G}_2 y'^2 + \mathcal{G}_3 z'^2\right), \tag{10.7}$$

where $(X, Y, Z)^{\mathrm{T}}$ is the velocity vector of the comet in the fixed frame.

Introducing the first equation of system (10.3) explicitly, we can rewrite $\mathcal{H}_1$ as

$$\mathcal{H}_1(x, y, z, t) = \tfrac{1}{2}\mathcal{G}_2\left[(y^2 - x^2)C - 2xyS\right] + \tfrac{1}{2}\mathcal{G}_3 z^2, \tag{10.8}$$

where

$$C = \cos\left(2\Omega_0 t\right), \qquad S = \sin\left(2\Omega_0 t\right). \tag{10.9}$$

It is well known that in cometary problems one cannot expect to meet moderate eccentricities of orbits; some kind of regularisation will become unavoidable if a fixed step integrator is to be applied. One of the standard regularising tools is the application of the so-called Kuustanheimo–Stiefel (KS) transformation that turns a Kepler problem into a harmonic oscillator at the expense of increasing the number of degrees of freedom [16]. The approach of [6] is used to set the KS variables in the canonical formalism.

## KS Variables

Leaving aside the in-depth quaternion interpretation of the KS transformation given in [6], we restrict ourselves to the basic set of transformation formulae, treating the KS variables as a formal column vector. In the phase space of the KS coordinates $\mathbf{u} = (u_0, u_1, u_2, u_3)^{\mathrm{T}}$ and KS momenta $\mathbf{U} = (U_0, U_1, U_2, U_3)^{\mathrm{T}}$, the former are defined by means of the inverse transformation

$$\begin{aligned} x &= (u_0^2 + u_1^2 - u_2^2 - u_3^2)/\alpha, \\ y &= 2\left(u_1 u_2 + u_0 u_3\right)/\alpha, \\ z &= 2\left(u_1 u_3 - u_0 u_2\right)/\alpha, \end{aligned} \tag{10.10}$$

where $\alpha$ is an arbitrary parameter with the dimension of a length. A dimension raising transformation cannot be bijective, so the inverse of (10.10) is to some extent arbitrary. Following [6] we adopt

$$\mathbf{u} = \sqrt{\frac{\alpha}{2(r+x)}}\,(0, r+x, y, z)^{\mathrm{T}}, \tag{10.11}$$

for $x \geq 0$, and

$$\mathbf{u} = \sqrt{\frac{\alpha}{2(r-x)}}\,(-z, y, r-x, 0)^{\mathrm{T}}, \tag{10.12}$$

otherwise. A remarkable property of this transformation is that the distance $r$ becomes a quadratic function of $u_i$, namely

$$r = \sqrt{x^2 + y^2 + z^2} = \frac{u_0^2 + u_1^2 + u_2^2 + u_3^2}{\alpha} = \frac{\mathbf{u}^2}{\alpha} . \tag{10.13}$$

The momenta conjugate to $\mathbf{u}$ are defined as

$$\mathbf{U} = \frac{2}{\alpha} \begin{pmatrix} u_0 X + u_3 Y - u_2 Z \\ u_1 X + u_2 Y + u_3 Z \\ -u_2 X + u_1 Y - u_0 Z \\ -u_3 X + u_0 Y + u_1 Z \end{pmatrix} . \tag{10.14}$$

The inverse transformation, allowing the computation of $\mathbf{R} = (X, Y, Z)^{\mathrm{T}}$,

$$\mathbf{R} = \frac{1}{2r} \begin{pmatrix} u_0 U_0 + u_1 U_1 - u_2 U_2 - u_3 U_3 \\ u_3 U_0 + u_2 U_1 + u_1 U_2 + u_0 U_3 \\ -u_2 U_0 + u_3 U_1 - u_0 U_2 + u_1 U_3 \end{pmatrix} , \tag{10.15}$$

can be supplemented with the identity

$$u_1 U_0 - u_0 U_1 - u_3 U_2 + u_2 U_3 = 0 . \tag{10.16}$$

In order to achieve the regularisation without leaving the canonical formalism, we have to change the independent variable from $t$ to a fictitious time $s$ and consider the extended phase space of dimension 10, with a new pair of conjugate variables $(u^*, U^*)$. Thus, in the extended set of canonical KS variables, the motion of a comet is governed by the Hamiltonian function

$$\mathcal{M} = \frac{4\,\mathbf{u}^2}{\alpha^2} \left( \mathcal{K}_0 + U^* + \mathcal{K}_1 \right) = 0 , \tag{10.17}$$

where $\mathcal{K}_0$ and $\mathcal{K}_1$ stand for $\mathcal{H}_0$ and $\mathcal{H}_1$ expressed in terms of the extended KS variables set. The transformation just presented is univalent, hence the respective Hamiltonians will have different functional forms, but equal values: $\mathcal{H}_0 = \mathcal{K}_0$, $\mathcal{H}_1 = \mathcal{K}_1$. Restricting the motion to the manifold of $\mathcal{M} = 0$ is of fundamental importance to the canonical change of independent variable; in practical terms we achieve it by setting

$$U^* = -\mathcal{K}_0 - \mathcal{K}_1 , \tag{10.18}$$

at the beginning of the numerical integration.

Splitting the Hamiltonian function $\mathcal{M}$ into the sum of the principal term $\mathcal{M}_0$ and of a perturbation $\mathcal{M}_1$, we have

$$\mathcal{M}_0 = \tfrac{1}{2}\,\mathbf{U}^2 + (4 U^*/\alpha^2)\,\mathbf{u}^2 , \tag{10.19}$$

$$\mathcal{M}_1 = \frac{4\,\mathbf{u}^2}{\alpha^2} \mathcal{H}_1(x, y, z, t) . \tag{10.20}$$

Although nothing prohibits $u^*$ and $t$ from differing by an additive constant, we do not take advantage from this freedom and will therefore use the symbol $t$ in most instances instead of the formal $u^*$. In the next section we provide equations of motion generated by $\mathcal{M}_0$ and $\mathcal{M}_1$ alone; the complete equations of motion can be quickly obtained by adding the respective right-hand sides.

## Keplerian Motion

The principal virtue of the KS variables consists in their ability to transform the Kepler problem into a four-dimensional oscillator with a constant frequency

$$\omega = 2\frac{\sqrt{2U^*}}{\alpha}. \tag{10.21}$$

In addition one gets the equation for the fictitious time $s$, that may be written as:

$$\frac{ds}{dt} = \frac{\alpha}{4r}. \tag{10.22}$$

Thanks to the introduction of $\alpha$, the fictitious time $s$ has the dimension of time and if we assume

$$\alpha = \frac{2\mu}{|U^*|}, \tag{10.23}$$

orbital periods in $s$ and $t$ will be equal.

For $U^* > 0$, the map $\Phi_0$ representing the solution of the Keplerian motion can be directly quoted from [2]. If $\Delta$ is the fictitious time step, then

$$\Phi_{0,\Delta} : \begin{pmatrix} \mathbf{u} \\ \mathbf{U} \\ U^* \end{pmatrix} \rightarrow \begin{pmatrix} \mathbf{u}\cos\omega\Delta + \mathbf{U}\omega^{-1}\sin\omega\Delta \\ -\mathbf{u}\omega\sin\omega\Delta + \mathbf{U}\cos\omega\Delta \\ U^* \end{pmatrix}. \tag{10.24}$$

Moreover, if $\mathbf{v} = \Phi_{0,\Delta}\mathbf{u}$ and $\mathbf{V} = \Phi_{0,\Delta}\mathbf{U}$ are the final values of variables,

$$\Phi_{0,\Delta} : t \rightarrow t + \frac{2\Delta}{\alpha^2}\left(\mathbf{u}^2 + \frac{\mathbf{U}^2}{\omega^2}\right) + 2\frac{\mathbf{u}^T\mathbf{U} - \mathbf{v}^T\mathbf{V}}{\alpha^2\omega^2}. \tag{10.25}$$

One may easily check that the sum $\mathbf{u}^2 + \mathbf{U}^2\omega^{-2}$ is invariant under $\Phi_0$ and it can be replaced by $\mathbf{v}^2 + \mathbf{V}^2\omega^{-2}$ in practical computations of the Kepler equation (10.25).

It may happen, however, that $U^* < 0$ (when the motion is hyperbolic for instance). A simple modification of $\Phi_0$ in that case amounts to take

$$\omega = 2\frac{\sqrt{-2U^*}}{\alpha}, \tag{10.26}$$

and replacing (10.24) and (10.25) by

$$\Phi_{0,\Delta}: \begin{pmatrix} \mathbf{u} \\ \mathbf{U} \\ U^* \end{pmatrix} \to \begin{pmatrix} \mathbf{u}\cosh\omega\Delta + \mathbf{U}\omega^{-1}\sinh\omega\Delta \\ \mathbf{u}\omega\sinh\omega\Delta + \mathbf{U}\cosh\omega\Delta \\ U^* \end{pmatrix}, \qquad (10.27)$$

and

$$\Phi_{0,\Delta}: t \to t + \frac{2\Delta}{\alpha^2}\left(\mathbf{u}^2 - \frac{\mathbf{U}^2}{\omega^2}\right) - 2\frac{\mathbf{u}^{\mathrm{T}}\mathbf{U} - \mathbf{v}^{\mathrm{T}}\mathbf{V}}{\alpha^2\omega^2}. \qquad (10.28)$$

Similarly to the elliptic case, $\mathbf{u}^2 - \mathbf{U}^2/\omega^2$ is invariant under $\Phi_0$.

## Galactic Tide

The Hamiltonian $\mathcal{M}_1$ has the nice property of being independent of the momenta. Thus a half of the equations of motion have right-hand sides equal to zero, and the remaining right-hand sides are constant.

Accordingly, all KS coordinates are constant, the physical time $t$ does not flow, and the momenta are subjected to a linear "kick":

$$\Phi_{1,\Delta}: \begin{pmatrix} \mathbf{u} \\ t \\ \mathbf{U} \\ U^* \end{pmatrix} \to \begin{pmatrix} \mathbf{u} \\ t \\ \mathbf{U} - \Delta\,\mathbf{F}(\mathbf{u},t) \\ U^* - \Delta\,F^*(\mathbf{u},t) \end{pmatrix}. \qquad (10.29)$$

Mixing Cartesian and KS variables for the sake of brevity, we can represent $\mathbf{F}$ and $F^*$ as

$$\mathbf{F} = \frac{8\,\mathcal{H}_1}{\alpha^2}\,\mathbf{u} + \frac{4\mathbf{u}^2}{\alpha^2}\frac{\partial\mathcal{H}_1}{\partial\mathbf{u}} \qquad (10.30)$$

$$F^* = \frac{4\,\mathbf{u}^2}{\alpha^2}\,\Omega_0\,\mathcal{G}_2\,\xi_3, \qquad (10.31)$$

where

$$\frac{\partial\mathcal{H}_1}{\partial\mathbf{u}} = -\mathcal{G}_2\,\xi_2\,\frac{\partial x}{\partial\mathbf{u}} + \mathcal{G}_2\,\xi_1\,\frac{\partial y}{\partial\mathbf{u}} + \mathcal{G}_3\,z\,\frac{\partial z}{\partial\mathbf{u}} \qquad (10.32)$$

$$\begin{aligned}\xi_1 &= yC - xS, \\ \xi_2 &= xC + yS, \\ \xi_3 &= (x^2 - y^2)\,S - 2\,x\,y\,C\,.\end{aligned} \qquad (10.33)$$

## Symplectic Corrector

One of the advantages offered by the integrators introduced in [11] is a simple definition of a symplectic corrector—an extra stage that improves the accuracy in perturbed motion problems. The symplectic corrector is defined as a solution of the equations of motion generated by

$$\mathcal{M}_c = \{\{\mathcal{M}_0, \mathcal{M}_1\}, \mathcal{M}_1\}, \qquad (10.34)$$

where $\{\ ,\ \}$ is the canonical (or "symplectic") Poisson bracket in the phase space spanned by $\mathbf{u}, t, \mathbf{U}, U^*$. Observing that $\mathcal{M}_0$ is quadratic in $\mathbf{U}$ and linear in $U^*$, we easily obtain

$$\mathcal{M}_c(\boldsymbol{u}, t) = \sum_{i=0}^{3} \left(\frac{\partial \mathcal{M}_1}{\partial u_i}\right)^2 = \mathbf{F}^2 . \tag{10.35}$$

The solution of the equations of motion derived from $\mathcal{M}_c$ results in

$$\Phi_{c,\Delta} : \begin{pmatrix} \mathbf{u} \\ t \\ U_j \\ U^* \end{pmatrix} \rightarrow \begin{pmatrix} \mathbf{u} \\ t \\ U_j - 2\Delta \left(\frac{\partial \mathbf{F}}{\partial u_j}\right) \cdot \mathbf{F} \\ U^* - 2\Delta \left(\frac{\partial \mathbf{F}}{\partial t}\right) \cdot \mathbf{F} \end{pmatrix} . \tag{10.36}$$

In spite of a formally simple form, (10.36) involve rather complicated expressions for the second derivatives of $\mathcal{M}_1$, because

$$\frac{\partial \mathbf{F}}{\partial u_j} \cdot \mathbf{F}(\mathbf{u},t) = \sum_{i=0}^{3} \frac{\partial^2 \mathcal{M}_1}{\partial u_i \partial u_j} \frac{\partial \mathcal{M}_1}{\partial u_i}, \tag{10.37}$$

$$\frac{\partial \mathbf{F}}{\partial t} \cdot \mathbf{F}(\mathbf{u},t) = \sum_{i=0}^{3} \frac{\partial^2 \mathcal{M}_1}{\partial u_i \partial t} \frac{\partial \mathcal{M}_1}{\partial u_i} . \tag{10.38}$$

One may find a detailed method to compute the Hessian matrix of $\mathcal{M}_1$ in [5].

**Laskar–Robutel Integrators**

The composition methods of [11] differ from usual recipes because, regardless of the number of "stages" involved in one step, they all remain second-order integrators according to the formal estimates. However, if the Hamiltonian has been split into a leading term and a perturbation having a small parameter $\varepsilon$ as a factor, the truncation error of the integrator is $\max(\varepsilon^2 h^3, \varepsilon h^m)$ where $m$ is the number of stages involved in one step. The second term of this sum is similar to classical composition methods errors, and the first can be quite small for weakly perturbed problems. At the expense of the $\varepsilon^2 h^3$ term in the error estimate, the authors were able to avoid backward stages that degrade numerical properties of usual composition methods. The use of a corrector improves the integrator by reducing the truncation error: its first term drops to $\varepsilon^2 h^5$.

Following the recommendation of [11], and after having performed numerical tests (see [5]), the best integrator is obtained using the following composition for each single step of size $h$:

$$\Phi_h = \Phi_{c,q} \circ \Phi_{1,d_1} \circ \Phi_{0,c_2} \circ \Phi_{1,d_2} \circ \Phi_{0,c_3} \circ$$
$$\circ \Phi_{1,d_2} \circ \Phi_{0,c_2} \circ \Phi_{1,d_1} \circ \Phi_{c,q} , \qquad (10.39)$$

where
$$\begin{array}{ll} d_1 = h/12, & d_2 = (5/12)\,h , \\ c_2 = (1/2 - \sqrt{5}/10)\,h, & c_3 = h/\sqrt{5} , \\ q = -h^3(13 - 5\sqrt{5})/288 . & \end{array} \qquad (10.40)$$

This integrator will be referred as LARKS.

## LARKS Step Size Choice

The Hamiltonian error of LARKS, based on the above composition, is proportional to $\varepsilon^2 h^4$ (see [5]). Observing that $\varepsilon \propto ar^2$, where $a$ is the semi-major axis of a comet, and $r$ is the Sun–comet distance, we look for the step size selection rule that gives a similar precision for a wide range of initial conditions. This can be achieved if the product

$$K = \varepsilon^2 h^4 \qquad (10.41)$$

has similar values for all comets to be studied. Thus, finding some optimum step size $h_o$ for a given semi-axis $a_o$, and then launching the integration for a different semi-axis $a_1$, we adjust the step size and use

$$h_1 = h_o \left(\frac{a_o}{a_1}\right)^{3/2} , \qquad (10.42)$$

when the orbit is elliptic ($a_1 > 0$) and using $a_1$ instead of $r$. For a hyperbolic orbit the step size is adjusted through

$$h_1 = h_o \left(\frac{a_o^{3/2}}{|a_1|^{1/2} r}\right) . \qquad (10.43)$$

In the test described in this section, we set $h_o$ as $1/20$ of the Keplerian period implied by $a_o = 50,000$ AU and adjusted the step according to (10.42) or (10.43) for other orbits. For elliptical orbits, in order to avoid numerical resonance between the step size and the orbital period [19], we do not use a step size larger than $1/20$ of the Keplerian period, even if it might be allowed by (10.42).

## Stop Time for LARKS

The fact that the fictitious time $s$ is the independent variable is an inevitable issue associated with the use of the KS variables regularisation. What happens if one wants to obtain the state of a comet at some particular final epoch of the physical time $t$? This problem appears if one wants to stop the integration as close as possible to a precise value $T_f$ of the real time.

A method which turns out to be effective, whatever the dynamics, is the following. Let $f$ be the function defined by $f(s) = t - T_f$. Thus the problem is to solve $f(s) = 0$. Let $f_p$ and $s_p$ be the values of $f$ and $s$ before some step, and $f_a$ and $s_a$ be the values after this step. The integration stops as soon as $f_a > 0$. From this point, the method is built according to an iterative process which evaluates $f_p, s_p$ and $f_a, s_a$, such that the solution $s_s$ is always between $s_p$ and $s_a$.

For each step, one computes the derivatives $df_p$ and $df_a$ of $f$ in $s_p$ and $s_a$, respectively, using (10.22) and (10.13). Consequently, one can easily compute the equation of the tangent to $f$ in $s_p$ and $s_a$. Let $s_m$ be the value of $s$ for which the two tangents intersect. If $s_m$ is not between $s_p$ and $s_a$, the next guess $s_g$ of the solution $s_s$ is computed using a linear approximation of $f$ or a bisection method between $s_p$ and $s_a$. The choice is made according to the method which makes the most important reduction of the interval $[s_p, s_a]$. Otherwise, if $s_m$ is between $s_p$ and $s_a$, then the next guess $s_g$ is given by the intersection of the tangent at $f$ in $s_p$ (resp. in $s_a$) with the abscissa axis if it lies between $s_p$ and $s_m$ (resp. between $s_m$ and $s_a$).

One stops the iterative process as soon as $f(s_g)$ is close enough to 0, that is $t_g - T_f \approx 0$, where $t_g$ is the value of the real time obtained for $s_g$.

**Stop at Perihelion for LARKS**

In the framework of Oort cloud comets dynamics, it may be necessary to suspend the integration of a comet at its perihelion. When an integrator like RADAU is used, then the step size is very small when the comet passes through its perihelion, thus it is quite easy to stop at the cometary perihelion only by checking the evolution of the Sun–comet distance. This is not the case for LARKS, which may have a large step size even when the comet is at its perihelion, hence the evolution of the Sun–comet distance is not sufficient.

However, when a comet is near its perihelion, one may neglect the perturbative part due to the galactic tide. Consequently, as it has been already noted, the motion in the KS variables is simply a harmonic oscillator. Using this property it is very easy to stop the integration exactly at the cometary perihelion. Indeed, if $(\mathbf{u}, \mathbf{U})$ are the KS variables of a comet, then $\mathbf{u} \cdot \mathbf{U} = 0$ when the comet is at its perihelion. Thus, let $(\mathbf{u}_p, \mathbf{U}_p)$ be the KS variables before some step and $(\mathbf{u}_a, \mathbf{U}_a)$ after this step. When $\mathbf{u}_p \cdot \mathbf{U}_p < 0$ and $\mathbf{u}_a \cdot \mathbf{U}_a > 0$ it means that the comet went through its perihelion during the step. When $U^* > 0$, using (10.24), the step length $h$ which should be performed from $(\mathbf{u}_p, \mathbf{U}_p)$ to the exact perihelion is given by

$$h = \frac{1}{2\omega} \tan^{-1}\left(\frac{2\omega \mathbf{u}_p \cdot \mathbf{U}_p}{\omega^2 u_p^2 - U_p^2}\right). \qquad (10.44)$$

When $U^* < 0$, using (10.25), the step length is given by

$$h = \frac{1}{4\omega} \log\left(\frac{1+X}{1-X}\right), \tag{10.45}$$

where

$$X = \frac{2\omega \mathbf{u}_\mathrm{p} \cdot \mathbf{U}_\mathrm{p}}{\omega^2 u_\mathrm{p}^2 + U_\mathrm{p}^2}. \tag{10.46}$$

### 10.2.3 The Hamiltonian Models

In this section we will give the Hamiltonian equations of motion derived from the average Hamiltonian, using two sets of variables according to the value of the orbital eccentricity $e$ (see [9]).

**The Hamiltonian Model with Delaunay Variables**

The complete Hamiltonian given by

$$\mathcal{H} = -\frac{\mu}{2a} + \mathcal{G}_1 \frac{x'^2}{2} + \mathcal{G}_2 \frac{y'^2}{2} + \mathcal{G}_3 \frac{z^2}{2} \tag{10.47}$$

may be written using the Delaunay's variables: $L = \sqrt{\mu a}$, $G = \sqrt{\mu a(1-e^2)}$, $H = G \cos i$, $\ell = M$, $g = \omega$ and $h = \Omega$, where $a, e, i, M, \omega, \Omega$ are the cometary semi-major axis, eccentricity, inclination, mean anomaly, argument of perihelion and longitude of node (all the angles being measured in the fixed galactic frame). The mean anomaly being a fast variable with respect to the other ones, the Hamiltonian is averaged over $\ell$.

Then, one writes the averaged Hamiltonian equations of motion, which gives:

$$\left\langle \frac{dL}{dt} \right\rangle = 0 \tag{10.48}$$

$$\left\langle \frac{dG}{dt} \right\rangle = -\frac{5L^2}{2\mu^2}(L^2 - G^2)\left\{ \cos g \sin g \left[ \mathcal{G}_3\left(1 - \frac{H^2}{G^2}\right) \right.\right.$$
$$+ (\mathcal{G}_1 \sin^2 h_\mathrm{r} + \mathcal{G}_2 \cos^2 h_\mathrm{r}) \frac{H^2}{G^2} - \mathcal{G}_1 \cos^2 h_\mathrm{r} \tag{10.49}$$
$$\left. - \mathcal{G}_2 \sin^2 h_\mathrm{r} \right] - (\mathcal{G}_1 - \mathcal{G}_2)(\cos^2 g - \sin^2 g) \cos h_\mathrm{r} \sin h_\mathrm{r} \frac{H}{G} \right\}$$

$$\left\langle \frac{dH}{dt} \right\rangle = \frac{L^2}{2\mu^2}(\mathcal{G}_1 - \mathcal{G}_2)\left\{ 5(L^2 - G^2)\frac{H}{G} \cos g \sin g (\cos^2 h_\mathrm{r} - \sin^2 h_\mathrm{r}) \right.$$
$$+ \sin h_\mathrm{r} \cos h_\mathrm{r} \tag{10.50}$$
$$\left. \cdot \left[ G^2 - H^2 + 5(L^2 - G^2)\left( \cos^2 g - \sin^2 g \frac{H^2}{G^2} \right) \right] \right\}$$

$$\left\langle \frac{dg}{dt} \right\rangle = \frac{L^2 G}{2\mu^2} \Big\{ \mathcal{G}_3 \Big[ 1 - 5\sin^2 g \Big( 1 - \frac{L^2 H^2}{G^4} \Big) \Big]$$
$$+ (\mathcal{G}_1 \cos^2 h_r + \mathcal{G}_2 \sin^2 h_r)(1 - 5\cos^2 g) \qquad (10.51)$$
$$- 5(\mathcal{G}_1 \sin^2 h_r + \mathcal{G}_2 \cos^2 h_r)\frac{L^2 H^2}{G^4} \sin^2 g$$
$$+ 5(\mathcal{G}_1 - \mathcal{G}_2) \cos g \sin g \cos h_r \sin h_r (G^2 + L^2)\frac{H}{G^3} \Big\}$$

$$\left\langle \frac{dh}{dt} \right\rangle = \frac{L^2}{2G\mu^2} \Big\{ (\mathcal{G}_1 \sin^2 h_r + \mathcal{G}_2 \cos^2 h_r - \mathcal{G}_3)$$
$$[G^2 + 5(L^2 - G^2)\sin^2 g]\frac{H}{G} - 5(\mathcal{G}_1 - \mathcal{G}_2) \qquad (10.52)$$
$$(L^2 - G^2) \cos g \sin g \cos h_r \sin h_r \Big\},$$

where $h_r = h - \Omega_0 t$ at time $t$.

The quantity $L$, i.e. the semi-major axis, is obviously conserved since the mean anomaly does not appear in the averaged Hamiltonian. Furthermore, one may note that, when the radial component of the tide is neglected, i.e. when $\mathcal{G}_1 = \mathcal{G}_2 = 0$, then $H$ is conserved. In this case the dynamics is completely integrable. Many papers were devoted to this peculiar case: for instance [1, 3, 10, 14, 15].

## The Hamiltonian Model with Matese Elements

When $e \simeq 1$, (10.49)–(10.52) become singular. In order to remove this singularity, we adopt the variables: $L$, $\Theta = H_b$, $H$, $M$, $\theta = b$ and $\lambda = l$, with $H_b = -\sqrt{\mu a(1-e^2)} \cos\alpha$ and $H = \sqrt{\mu a(1-e^2)} \cos i$. Here $b$ and $l$ are the latitude and longitude of perihelion of the comet, and $\alpha$ is the angle between the orbital plane and the plane orthogonal to the galactic plane and passing through the perihelion and the galactic poles, measured from the south galactic pole to the cometary velocity (see Fig. 1 of [9]).

This set of elements will be referred to as Matese elements since it first appeared in [15]. Similar elements have been used elsewhere in the literature in order to remove the singularity at $e = 1$: see [18] for elliptic collision orbits and [17] for hyperbolic collision orbits. The Matese elements are slightly different from those used in the quoted papers, but the procedure to define them is similar. These can be shown to be canonical.

One substitutes $x'$, $y'$ and $z$ by the Matese elements in (10.47), and averages with respect to the mean anomaly. Then, the Hamiltonian equations of motion are:

$$\left\langle \frac{dL}{dt} \right\rangle = 0$$

$$\left\langle \frac{d\Theta}{dt} \right\rangle = \frac{L^2}{2\mu^2} \Big[\cos\theta \sin\theta(-4\Theta^2 + 5L^2)(\mathcal{G}_1 \cos^2\lambda_r + \mathcal{G}_2 \sin^2\lambda_r)$$

$$-\mathcal{G}_1 \left(\Theta \cos\theta \cos\lambda_r + H \sin\lambda_r \frac{\sin\theta}{\cos^2\theta}\right)$$

$$\cdot \left(\Theta \sin\theta \cos\lambda_r + H \sin\lambda_r \frac{1}{\cos\theta}\right)$$

$$-\mathcal{G}_2 \left(\Theta \cos\theta \sin\lambda_r - H \cos\lambda_r \frac{\sin\theta}{\cos^2\theta}\right)$$

$$\cdot \left(\Theta \sin\theta \sin\lambda_r - H \cos\lambda_r \frac{1}{\cos\theta}\right)$$

$$+\mathcal{G}_3 \frac{\sin\theta}{\cos^3 b}(5(\Theta^2 - L^2)\cos^4\theta + 4H^2)\Big]$$

$$\left\langle \frac{dH}{dt} \right\rangle = \frac{L^2(\mathcal{G}_1 - \mathcal{G}_2)}{2\mu^2 \cos^2\theta} \Big[(-4\Theta^2 \cos^2\theta - 4H^2 + 5L^2 \cos^2\theta)\cos^2\theta \cos\lambda_r \sin\lambda_r$$

$$+(\Theta \cos\theta \sin\theta \sin\lambda_r - H \cos\lambda_r)(\Theta \cos\theta \sin\theta \cos\lambda_r + H \sin\lambda_r)\Big]$$

$$\left\langle \frac{d\theta}{dt} \right\rangle = \frac{L^2}{2\mu^2 \cos^2\theta} \Big\{-4\Theta \cos^4\theta \left(\mathcal{G}_1 \cos^2\lambda_r + \mathcal{G}_2 \sin^2\lambda_r\right)$$

$$+\mathcal{G}_1(\Theta \cos\theta \sin\theta \cos\lambda_r + H \sin\lambda_r)\cos\theta \sin\theta \cos\lambda_r$$

$$+\mathcal{G}_2(\Theta \cos\theta \sin\theta \sin\lambda_r - H \cos\lambda_r)\cos\theta \sin\theta \sin\lambda_r$$

$$+\mathcal{G}_3 \Theta \cos^2\theta(1 - 5\sin^2\theta)\Big\}$$

$$\left\langle \frac{d\lambda}{dt} \right\rangle = \frac{L^2}{2\mu^2 \cos^2\theta} \Big\{-4H \cos^2\theta \left(\mathcal{G}_1 \cos^2\lambda_r + \mathcal{G}_2 \sin^2\lambda_r\right)$$

$$+\mathcal{G}_1 \sin\lambda_r(\Theta \cos\theta \sin\theta \cos\lambda_r + H \sin\lambda_r)$$

$$-\mathcal{G}_2 \cos\lambda_r(\Theta \cos\theta \sin\theta \sin\lambda_r - H \cos\lambda_r) - 4\mathcal{G}_3 H \sin^2\theta\Big\}$$

The singularity at $\cos\theta = 0$ is evident, but these equations show that the singularity at $e = 1$ has indeed disappeared.

## The Mappings

The two above averaged models are already faster than the Cartesian model (see [8]). However, in order to enhance their efficiency, one may consider the truncated Taylor development of their solution. More precisely, one writes any of the averaged models in the form:

$$\frac{d\mathbf{x}}{dt} = \mathbf{f}(\mathbf{x}_r) \,, \tag{10.53}$$

where $\mathbf{x}$ is the vector defined by the Delaunay or the Matese elements, and the subscript r means that the longitude is measured in the rotating frame.

Then, the truncated Taylor development at order $N$ of the solution of this equation is

$$\mathbf{x}(T+\Delta T) = \mathbf{x}(T) + \sum_{n=1}^{N} \frac{d^n \mathbf{x}}{dt^n}\bigg|_T \frac{\Delta T^n}{n!} + \mathcal{O}(\Delta T^{N+1}), \qquad (10.54)$$

where the subscript $T$ indicates that the quantities are computed at time $T$.

Taking $\Delta T$ equal to one orbital period of the comet (which is consistent with having averaged the Hamiltonian) this development provides us an easy way to derive mappings of different orders.

Experiments have shown (see [18]) that the mappings of order 3 give the best compromise between precision and velocity. However, one should be very careful in using these mappings, due to the singularities of the two averaged models. Indeed, the effects of the singularity when $e = 1$ for the model using the Delaunay variables and the singularity when $\cos b = 0$ for the model using the Matese variables are enhanced when one uses the truncated Taylor development of their solution.

It turns out that the mapping using the Delaunay elements may be safely used for eccentricity smaller than 0.999, otherwise the mapping using the Matese elements is more precise. The composition of this two mappings will be referred as the MAPP model.

### 10.2.4 The Lie–Poisson Model

This model is described in detail in [5]. One should refer to this paper for a full description of the method.

**Equations of Motion**

The integrators presented in the previous sections solve the equations of motion in the fixed reference frame, where the radial component of the galactic tide is explicitly time-dependent. Our second method can be more conveniently discussed in the rotating heliocentric reference frame $(\hat{x}', \hat{y}', \hat{z})$. The present model uses also the first equation of (10.3) explicitly, which simplifies the results drastically.

The Hamiltonian function for a comet subjected to the galactic tide in the rotating frame is given by

$$\mathcal{H} = \mathcal{H}_0 + \mathcal{H}_1, \qquad (10.55)$$

$$\mathcal{H}_0 = \frac{1}{2}\left(X'^2 + Y'^2 + Z^2\right) - \frac{\mu}{(x'^2 + y'^2 + z^2)^{1/2}}, \qquad (10.56)$$

$$\mathcal{H}_1 = \Omega_0\left(y' X' - x' Y'\right) + \frac{1}{2}\left(\mathcal{G}_2\left(y'^2 - x'^2\right) + \mathcal{G}_3 z^2\right), \qquad (10.57)$$

where $(x', y', z)$ and $(X', Y', Z)$ are the position and velocity of the comet in the rotating frame.

As in Sect. 10.2.3, one may average the Hamiltonian $\mathcal{H}$ with respect to the mean anomaly $\ell$. The averaged Hamiltonian $\langle\mathcal{H}_1\rangle$ is now expressed in terms of the Laplace vector $\mathbf{e}$ and a scaled angular momentum vector $\mathbf{h}$ instead of canonical elements as the Delaunay or the Matese ones. Their components are related to the Keplerian orbit elements

$$\mathbf{e} \equiv \begin{pmatrix} e_1 \\ e_2 \\ e_3 \end{pmatrix} = e \begin{pmatrix} \cos\omega\cos\Omega_r - c\sin\omega\sin\Omega_r \\ \cos\omega\sin\Omega_r + c\sin\omega\cos\Omega_r \\ s\sin\omega \end{pmatrix}, \quad (10.58)$$

$$\mathbf{h} \equiv \begin{pmatrix} h_1 \\ h_2 \\ h_3 \end{pmatrix} = \sqrt{1-e^2} \begin{pmatrix} s\sin\Omega_r \\ -s\cos\Omega_r \\ c \end{pmatrix}, \quad (10.59)$$

where $e$ is the eccentricity, $s = \sin i$, $c = \cos i$. Recalling that in the rotating frame the momenta $X'$ and $Y'$ are not equal to velocities $\frac{dx'}{dt}$ and $\frac{dy'}{dt}$ (this fact can be immediately deduced from the canonical equations $\frac{dx'}{dt} = \partial\mathcal{H}/\partial X'$ and $\frac{dy'}{dt} = \partial\mathcal{H}/\partial Y'$), we assume that the usual transformation rules between Keplerian elements and position/velocity are used with the velocities directly substituted by the momenta. With this approach the Keplerian motion in the rotating frame is described by the means of orbital elements that are all constant except for $\Omega_r$ which reflects the frame rotation ($\frac{d\Omega_r}{dt} = -\Omega_0$).

Using the "vectorial elements" $\mathbf{h}$ and $\mathbf{e}$, letting $n$ stand for

$$n = \sqrt{\frac{\mu}{a^3}}, \quad (10.60)$$

and changing the independent variable from time $t$ to $\tau_1$, such that

$$\frac{d\tau_1}{dt} = \frac{\mathcal{G}_3}{n}, \quad (10.61)$$

one obtains the averaged Hamiltonian $\langle\mathcal{H}\rangle$, given by

$$\langle\mathcal{H}\rangle = n\,a^2 \left[\tfrac{5}{4}e_3^2 + \tfrac{1}{4}h_1^2 + \tfrac{1}{4}h_2^2 + \right.$$
$$\left. + \nu\left(-\tfrac{5}{4}e_1^2 + \tfrac{5}{4}e_2^2 + \tfrac{1}{4}h_1^2 - \tfrac{1}{4}h_2^2 - n\,\Omega_0^{-1}h_3\right)\right], \quad (10.62)$$

where, all the constant terms have been dropped and, using the usual approximation $\Omega_0 = -\sqrt{\mathcal{G}_2}$, we introduced a dimensionless parameter

$$\nu = \frac{\Omega_0^2}{\mathcal{G}_3} = \frac{\mathcal{G}_2}{\mathcal{G}_3}. \quad (10.63)$$

The vectorial elements can be used to create a Lie–Poisson bracket

$$(f;g) \equiv \left(\frac{\partial f}{\partial \mathbf{v}}\right)^{\mathrm{T}} \mathbf{J}(\mathbf{v}) \frac{\partial g}{\partial \mathbf{v}}, \quad (10.64)$$

with the structure matrix

$$\mathbf{J}(\mathbf{v}) = \begin{pmatrix} \hat{\mathbf{h}} & \hat{\mathbf{e}} \\ \hat{\mathbf{e}} & \hat{\mathbf{h}} \end{pmatrix} . \tag{10.65}$$

The "hat map" of any vector $\mathbf{x} = (x_1, x_2, x_3)^{\mathrm{T}}$ is defined as

$$\hat{\mathbf{x}} = \begin{pmatrix} 0 & -x_3 & x_2 \\ x_3 & 0 & -x_1 \\ -x_2 & x_1 & 0 \end{pmatrix} . \tag{10.66}$$

This matrix is known as the vector product matrix, because

$$\hat{\mathbf{x}} \, \mathbf{y} = \mathbf{x} \times \mathbf{y} . \tag{10.67}$$

Using the Lie–Poisson bracket (10.64) we can write equations of motion for the vectorial elements

$$\mathbf{v} = (h_1, h_2, h_3, e_1, e_2, e_3)^{\mathrm{T}} \tag{10.68}$$

in the non-canonical Hamiltonian form

$$\mathbf{v}' = (\mathbf{v}; \mathcal{K}) , \tag{10.69}$$

where derivatives with respect to $\tau$ are marked by the "prime" symbol and the scaled Hamiltonian

$$\mathcal{K} = -\frac{\langle \mathcal{H} \rangle}{n \, a^2} . \tag{10.70}$$

Writing (10.69) explicitly, we obtain

$$h_1' = -\frac{5}{2}(1-\nu) \, e_2 \, e_3 + \frac{1-\nu}{2} h_2 \, h_3 + \frac{n\nu}{\Omega_0} h_2 , \tag{10.71}$$

$$h_2' = \frac{5}{2}(1+\nu) \, e_1 \, e_3 - \frac{1+\nu}{2} h_1 \, h_3 - \frac{n\nu}{\Omega_0} h_1, \tag{10.72}$$

$$h_3' = \nu \, (h_1 \, h_2 - 5 \, e_1 \, e_2), \tag{10.73}$$

$$e_1' = -\frac{4+\nu}{2} h_2 \, e_3 + \frac{5}{2} \nu \, h_3 \, e_2 + \frac{n\nu}{\Omega_0} e_2, \tag{10.74}$$

$$e_2' = \frac{4-\nu}{2} h_1 \, e_3 + \frac{5}{2} \nu \, h_3 \, e_1 - \frac{n\nu}{\Omega_0} e_1, \tag{10.75}$$

$$e_3' = \frac{1-4\nu}{2} h_1 \, e_2 - \frac{1+4\nu}{2} h_2 \, e_1. \tag{10.76}$$

Substituting $\nu = 0$, the readers may recover the correct form of the galactic disk tide equations published in [3, 4]. Equations (10.71)–(10.76) admit three integrals of motion: apart from the usual conservation of the time-independent Hamiltonian $\mathcal{K} = \mathrm{const}$, two geometrical constraints

$$\mathbf{h}\cdot\mathbf{e}=0, \qquad h^2+e^2=1, \tag{10.77}$$

are respected thanks to the properties of the Lie–Poisson bracket (10.64). Indeed, both quadratic forms are the Casimir functions of our bracket, i.e.

$$(\mathbf{h}\cdot\mathbf{e};\,f)=(h^2+e^2;\,f)=0, \tag{10.78}$$

for any function $f$, hence in particular for $f=\mathcal{K}$.

**Lie–Poisson Splitting Method**

The Hamiltonian $\mathcal{K}$ can be split into a sum of three non-commuting terms

$$\mathcal{K}=\mathcal{K}_1+\mathcal{K}_2+\mathcal{K}_3, \tag{10.79}$$

$$\mathcal{K}_1=\frac{5}{4}\nu e_1^2-\frac{1+\nu}{4}h_1^2, \tag{10.80}$$

$$\mathcal{K}_2=-\frac{5}{4}\nu e_2^2-\frac{1-\nu}{4}h_2^2, \tag{10.81}$$

$$\mathcal{K}_3=-\frac{5}{4}e_3^2+\frac{n\nu}{\Omega_0}h_3. \tag{10.82}$$

Each of the terms $\mathcal{K}_i$ is in turn a sum of two components that commute, because it can be easily verified that $(e_j;\,h_j)=0$ for all $j\in\{1,2,3\}$. In these circumstances, we can approximate the real solution

$$\mathbf{v}(\tau)=\exp{(\tau\,L)}\,\mathbf{v}(0), \tag{10.83}$$

where $L\,f\equiv(f;\,\mathcal{K})$, using a composition of maps

$$\Psi_{i,\tau}:\ \mathbf{v}(0)\to\mathbf{v}(\tau)=\exp{(\tau\,L_i)}\,\mathbf{v}(0), \tag{10.84}$$

where $L_i\,f\equiv(f;\,\mathcal{K}_i)$ for $i=1,2,3$. Each $\Psi_{i,\tau}$ is in turn a composition of two maps

$$\Psi_{i,\tau}=E_{i,\tau}\circ H_{i,\tau}=H_{i,\tau}\circ E_{i,\tau}, \tag{10.85}$$

generated by the $e_i$ and $h_i$ related terms of $\mathcal{K}_i$.

**The Contribution of $\mathcal{K}_1$**

The two terms of $\mathcal{K}_1$ generate equations of motion

$$\mathbf{v}'=\left(\mathbf{v};\,\tfrac{5}{4}\nu\,e_1^2\right)=\tfrac{5}{2}e_1\nu\begin{pmatrix}\mathbf{0}&\mathbf{Y}_1\\\mathbf{Y}_1&\mathbf{0}\end{pmatrix}\mathbf{v}, \tag{10.86}$$

and

$$\mathbf{v}'=\left(\mathbf{v};\,-\tfrac{1}{4}(1+\nu)\,h_1^2\right)=-\tfrac{1}{2}h_1\,(1+\nu)\begin{pmatrix}\mathbf{0}&\mathbf{Y}_1\\\mathbf{Y}_1&\mathbf{0}\end{pmatrix}\mathbf{v}, \tag{10.87}$$

where
$$\mathbf{Y}_1 = \begin{pmatrix} 0 & 0 & 0 \\ 0 & 0 & -1 \\ 0 & 1 & 0 \end{pmatrix} . \tag{10.88}$$

The composition of these two maps results in
$$\Psi_{1,\tau} : \mathbf{v} \to \begin{pmatrix} \mathbf{M}_1 & \mathbf{N}_1 \\ \mathbf{N}_1 & \mathbf{M}_1 \end{pmatrix} \mathbf{v} , \tag{10.89}$$

where
$$\mathbf{M}_1 = \begin{pmatrix} 1 & 0 & 0 \\ 0 & c_{11}c_{12} & -c_{11}s_{12} \\ 0 & c_{11}s_{12} & c_{11}c_{12} \end{pmatrix} , \quad \mathbf{N}_1 = \begin{pmatrix} 0 & 0 & 0 \\ 0 & s_{11}s_{12} & s_{11}c_{12} \\ 0 & -s_{11}c_{12} & s_{11}s_{12} \end{pmatrix} . \tag{10.90}$$

$$\psi_{11} = \tfrac{5}{2} e_1 \nu \tau \quad \text{and} \quad \psi_{12} = \tfrac{1}{2}(1+\nu) h_1 \tau , \tag{10.91}$$

introducing
$$c_{ij} = \cos \psi_{ij}, \quad s_{ij} = \sin \psi_{ij} . \tag{10.92}$$

### The Contribution of $\mathcal{K}_2$

The equations of motion derived from the two terms of $\mathcal{K}_2$ are
$$\mathbf{v}' = \left(\mathbf{v}; -\tfrac{5}{4}\nu e_2^2\right) = \tfrac{5}{2}\nu e_2 \begin{pmatrix} \mathbf{0} & \mathbf{Y}_2 \\ \mathbf{Y}_2 & \mathbf{0} \end{pmatrix} \mathbf{v} , \tag{10.93}$$

and
$$\mathbf{v}' = \left(\mathbf{v}; -\tfrac{1}{4}(1-\nu) h_2^2\right) = \tfrac{1}{2}(1-\nu) h_2 \begin{pmatrix} \mathbf{0} & \mathbf{Y}_2 \\ \mathbf{Y}_2 & \mathbf{0} \end{pmatrix} \mathbf{v} , \tag{10.94}$$

where
$$\mathbf{Y}_2 = \begin{pmatrix} 0 & 0 & 1 \\ 0 & 0 & 0 \\ -1 & 0 & 0 \end{pmatrix} . \tag{10.95}$$

Composing the two maps we obtain
$$\Psi_{2,\tau} : \mathbf{v} \to \begin{pmatrix} \mathbf{M}_2 & \mathbf{N}_2 \\ \mathbf{N}_2 & \mathbf{M}_2 \end{pmatrix} \mathbf{v} , \tag{10.96}$$

where
$$\mathbf{M}_2 = \begin{pmatrix} c_{21}c_{22} & 0 & -c_{21}s_{22} \\ 0 & 1 & 0 \\ c_{21}s_{22} & 0 & c_{21}c_{22} \end{pmatrix} , \quad \mathbf{N}_2 = \begin{pmatrix} s_{21}s_{22} & 0 & c_{22}s_{21} \\ 0 & 0 & 0 \\ -c_{22}s_{21} & 0 & s_{21}s_{22} \end{pmatrix} , \tag{10.97}$$

$$\psi_{21} = \tfrac{5}{2}\nu e_2 \tau \quad \text{and} \quad \psi_{22} = -\frac{h_2(1-\nu)}{2}\tau . \tag{10.98}$$

## The Contribution of $\mathcal{K}_3$

The equations of motion derived from the two terms of $\mathcal{K}_3$ are

$$\mathbf{v}' = \left(\mathbf{v}; -\tfrac{5}{4} e_3^2\right) = \tfrac{5}{2} e_3 \begin{pmatrix} \mathbf{0} & \mathbf{Y}_3 \\ \mathbf{Y}_3 & \mathbf{0} \end{pmatrix} \mathbf{v}, \qquad (10.99)$$

and

$$\mathbf{v}' = \left(\mathbf{v}; h_3\, n\, \nu\, \Omega_0^{-1}\right) = -\frac{n\,\nu}{\Omega_0} \begin{pmatrix} \mathbf{0} & \mathbf{Y}_3 \\ \mathbf{Y}_3 & \mathbf{0} \end{pmatrix} \mathbf{v}, \qquad (10.100)$$

where

$$\mathbf{Y}_3 = \begin{pmatrix} 0 & -1 & 0 \\ 1 & 0 & 0 \\ 0 & 0 & 0 \end{pmatrix}. \qquad (10.101)$$

Composing the two maps we obtain

$$\Psi_{3,\tau} : \mathbf{v} \to \begin{pmatrix} \mathbf{M}_3 & \mathbf{N}_3 \\ \mathbf{N}_3 & \mathbf{M}_3 \end{pmatrix} \mathbf{v}, \qquad (10.102)$$

where

$$\mathbf{M}_3 = \begin{pmatrix} c_{31} c_{32} & c_{31} s_{32} & 0 \\ -c_{31} s_{32} & c_{31} c_{32} & 0 \\ 0 & 0 & 1 \end{pmatrix}, \quad \mathbf{N}_3 = \begin{pmatrix} s_{31} s_{32} & -c_{32} s_{31} & 0 \\ c_{32} s_{31} & s_{31} s_{32} & 0 \\ 0 & 0 & 0 \end{pmatrix}, \qquad (10.103)$$

$$\psi_{31} = \frac{5}{2} e_3 \tau \quad \text{and} \quad \psi_{32} = \frac{n\,\nu}{\Omega_0} \tau. \qquad (10.104)$$

### 10.2.5 The Lie–Poisson Method of Order 2

The composition methods of [11] cannot be used for our Lie–Poisson splitting method, because the Hamiltonian function has been partitioned into three terms. Moreover, none of the terms can be qualified as a small perturbation. In these circumstances, the principal building block can be a "generalised leapfrog"

$$\Psi_\Delta = \Psi_{1,\Delta/2} \circ \Psi_{2,\Delta/2} \circ \Psi_{3,\Delta} \circ \Psi_{2,\Delta/2} \circ \Psi_{1,\Delta/2}. \qquad (10.105)$$

This Lie–Poisson method, called LPV2, is a second-order method with a local truncation error proportional to the cube of the step size $\Delta^3$. Although we use LPV2 as a final product in this paper, it can be used as a building block for higher-order methods. A collection of appropriate composition rules can be found in [13].

In practice, the step size $\Delta$ will be set equal to one orbital period.

## 10.3 Comparisons Between the Different Models

In order to compare the reliability and speed of the integrators we performed the following experiment: 400,000 sets of initial orbital elements were randomly chosen in a specified range, under the condition that their respective distribution is uniform, i.e.

- The initial semi-major axes in the range $3000 \leq a_0 \leq 10^5$ AU, with distribution uniform in $\log_{10} a_0$
- The initial eccentricity in the range $0 \leq e_0 \leq 0.9999$, with a uniform distribution
- The initial inclination $i_0$ such that $-1 \leq \cos i_0 \leq 1$, with a uniform distribution
- The initial argument of the perihelion, the longitude of the ascending node and the initial mean anomaly (where needed) in the range from 0 to $2\pi$, with a uniform distribution

Using this set of elements, we integrated the equations over one cometary period using LARKS, LPV2, MAPP and compared the results with those obtained with the Cartesian model.

For the stopping time in LARKS, the method described at the end of Sect. 10.2.2 until $|T_f - t| < 10^{-3}$ yr was used.

The relative error in the comet position $E_\mathrm{p}$ was defined as

$$E_\mathrm{p} = \left| \frac{q_\mathrm{mod} - q_\mathrm{R}}{q_0} \right|, \qquad (10.106)$$

where $q_\mathrm{mod}$ and $q_\mathrm{R}$ denote the value of the perihelion distance at the end of the integration of one period computed by the tested integrator and by the RADAU, respectively, and $q_0$ is the initial value of the perihelion distance.

Then, the $e_0$–$\log_{10} a_0$ plane is divided into $60 \times 70$ cells. In each cell we record the maximum value $E_\mathrm{max}$ reached by the error $E_\mathrm{p}$ for the initial conditions belonging to the cell.

The results obtained for the three models are shown in Fig. 10.1. The MAPP and the LPV2 models, both used with a step size equal to the unperturbed Keplerian period, are equivalent as far as the accuracy is concerned. Indeed, the best analytical fit of the level curve $E_\mathrm{p} = 0.01$ is given by:

$$a_c = 10^{4.748 \pm 0.004}(1-e)^{0.182 \pm 0.006}, \qquad (10.107)$$

for the MAPP model, and

$$a_c = 10^{4.751 \pm 0.003}(1-e)^{0.185 \pm 0.005}, \qquad (10.108)$$

for the LPV2 model. These two equations may be considered as identical within the error bounds of the exponents.

For both models the error is essentially due to the averaging of the equations of motion with respect to the mean anomaly. Conversely, the LARKS method is highly reliable in the whole phase space domain under study, since the error never exceed 0.01. The effect of the time step selection rule (10.42)

# 10 Methods for the Study of the Dynamics of the Oort Cloud Comets II

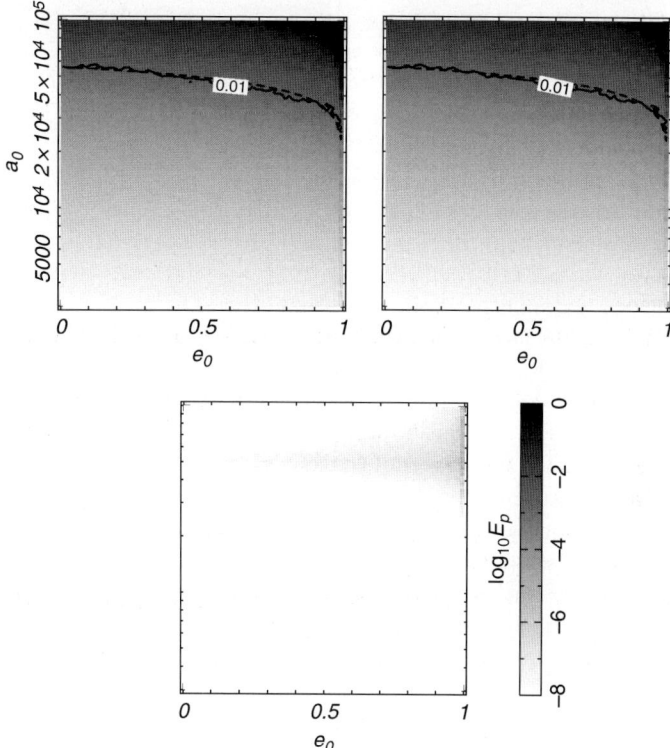

**Fig. 10.1.** Maximum error $E_p$ [(see 10.106)] in each cell of the $e_0$–$a_0$ plane for the models MAPP (**top left**), LPV2 (**top right**) and LARKS (**bottom**). The *solid line curves* correspond to $E_p = 0.01$ and the *dotted curves* are the best fits of the level curves

is clearly visible above $a_0 = 50,000$ AU; the reliability of LARKS is almost conserved when $a_0$ increases.

Speaking about the computation times required to perform all the integrations, the MAPP, LPV2 and LARKS needed 5.5, 1.8 and 99 s, whereas the RA15 integration took 1820 s. That is, LPV2 is three times faster than MAPP, and almost 40 times faster than LARKS, and LARKS is almost 20 times faster than RA15.

## 10.4 Hybrid Integrators

### 10.4.1 Definition

In order to have the best compromise between velocity and precision, one can consider hybrid models which use the fastest accurate model according to the values of the cometary eccentricity and semi-major axis.

In [9] the hybrid model MAPP + RADAU was introduced and applied to reproduce the effects of the galactic tide on the dynamics of $10^6$ comets over 5 Gyr. This hybrid model was such that MAPP was used below the analytical fit of the $E_p = 0.01$ level curve given by (10.107), otherwise RADAU was used.

This hybrid model became obsolete since [5] where LPV2 and LARKS have been introduced. Indeed, the hybrid model LPV2 + LARKS, where LPV2 is used below the analytical fit of the $E_p = 0.01$ level curve given by (10.108), is much faster than MAPP + RADAU and has the same accuracy. In the special case of a galactic potential such that $A \neq -B$, i.e. $\mathcal{G}_2 \neq -\mathcal{G}_1$, LARKS is easily generalisable whereas LPV2 is not. Thus in such a case the hybrid model MAPP + LARKS may be used.

Let us consider the integration of a comet with a hybrid model, say LARKS + LPV2. The oscillation of the eccentricity with time may lead to repeating shifts between the use of LARKS and the use of LPV2. However, when LARKS is used the semi-major axis oscillates with a period equal to the orbital period of the comet (this may be easily understood from the fact that in the Hamiltonian formalism, the mean anomaly and $L = \sqrt{\mu a}$ are conjugate coordinates), whereas it is a constant of motion when LPV2 is used. Indeed, for the LPV2 model the mean anomaly cancels out.

Since LPV2 is applied for an integer number of orbital periods, one may just record the value of the mean anomaly at the beginning of a sequence where LPV2 is used, and restore the mean anomaly value at the end of the sequence. Consequently, when one shifts from LPV2 to LARKS, the memory of the original orbit is conserved, as far as the averaging is neglected.

However, a shift from LARKS to LPV2 occurs for an arbitrary value of the mean anomaly, thus it occurs for an arbitrary value of the semi-major axis in the interval of its oscillations. Consequently, from one such shift to another one, the LPV2 model will be applied to different averaged orbits since the semi-major axis is different. After many shifts, a drift on the semi-major axis value may be observed.

An easy way to remove this drift is to allow the shift between LARKS and LPV2 only when the comet is exactly at its perihelion. Indeed, in this way, the shift occurs always at a precise time of the semi-major axis oscillation, thus the memory of the real orbit may be conserved when many shifts from LARKS to LPV2 are performed.

## 10.5 Conclusion

Different models of the galactic tide have been presented. The first one, called LARKS, is a symplectic integrator which uses the Kuustanheimo–Stiefel (KS) transformation to regularise the equations of motion. This model turns out to be reliable over the whole phase space and almost 20 times faster than a non-symplectic integrator using Cartesian coordinates (RADAU). The two other models are based on the averaging of the equations of motion with respect to the mean anomaly. One, which is called MAPP, which uses the Taylor development at order three of the solutions of the averaged equations of motion. In this case the equations are written using the Hamiltonian formalism with two different sets of canonical variables according to the value of the eccentricity. The second averaged model, called LPV2, considers the equations of motion using the normalised Laplace and angular momentum vectors. Then a Lie–Poisson integrator of order 2 is used by splitting the Hamiltonian into three parts.

As regards the accuracy, these two models are equivalent, but they are reliable only in a limited domain of the phase space. These models are both faster than LARKS, but LPV2 is three times faster than MAPP and 40 times faster than LARKS. The main advantage of MAPP is that it is more didactic. MAPP is also more general since it may consider any kind of radial component of the tide—but with the assumption that the tide is axi-symmetric—whereas LPV2 requires $\mathcal{G}_2 = -\mathcal{G}_1$.

The best hybrid model can be defined as a combination between the use of LPV2 and LARKS according to the value of the cometary eccentricity and semi-major axis. For instance, if one wants a confidence level of 1% on the perihelion distance variation over one cometary period, one may use the analytical equation given by

$$a_c = 10^{4.751}(1-e)^{0.185} \tag{10.109}$$

to define the upper frontier of the domain where LPV2 may be used.

If one wants a confidence level of 0.1% one may consider the value

$$a_c = 10^{4.570}(1-e)^{0.176} \tag{10.110}$$

as the upper limit of the domain where LPV2 may be used. However, the use of (10.110) will slow down the integrations since it reduces the domain of application of LPV2. It may be also softer to use RADAU rather than LARKS for hyperbolic orbits or when the semimajor axis is greater than 100 000 all. Indeed the loss in computation time will be small whereas the accuracy of LARKS is less guaranteed. The above hybrid model may be used for any long-term simulations of the Oort cloud comets dynamics under the effects of the galactic tide.

# References

1. Brasser, R., 2001. *MNRAS* **324**, 1109.
2. Breiter, S., 1998. *CeMDA* **71**, 229.
3. Breiter, S. and Ratajczak, R., 2005. *MNRAS* **364**, 1222.
4. Breiter, S. and Ratajczak, R., 2006. *MNRAS* **367**, 1808.
5. Breiter, S., Fouchard, M., and Ratajczak, R., 2007. *MNRAS* **377**, 1151.
6. Deprit, A., Elipe, A., and Ferrer, S., 1994. *CeMDA* **58**, 151.

7. Everhart, E., 1985. In Carusi, A. and Valsecchi G. B. (eds), *Dynamics of Comets: Their Origin and Evolution*, Reidel, Dordrecht, p. 185.
8. Fouchard, M., Froeschle, Ch., Matese, J. J., and Valsecchi, G. B., 2005. *CeMDA* **93**, 231.
9. Fouchard, M., Froeschle, Ch., Valsecchi, G. B., and Rickman, H., 2006. *CeMDA* **95**, 299.
10. Heisler, J. and Tremaine, S., 1986. *Icarus* **65**, 13.
11. Laskar, J. and Robutel, P., 2001. *CeMDA* **80**, 39.
12. Levison, H., Dones, L., and Duncan, M. J., 2001. *AJ* **121**, 2253.
13. McLachlan, R. I. and Quispel, G. R. W., 2002. *Acta Numer.* **11**, 341.
14. Matese, J. J. and Whitman, P. G., 1989. *Icarus* **82**, 389.
15. Matese, J. J. and Whitman, P. G., 1992. *CeMDA* **54**, 13.
16. Stiefel, E. L. and Scheifele, G., 1971. *Linear and Regular Celestial Mechanics*, Springer-Verlag, Berlin.
17. Tommei, G., 2006. *CeMDA* **94**, 176.
18. Tremaine, S., 2001. *CeMDA* **79**, 231.
19. Wisdom, J. and Holman, M., 1992. *AJ* **104**, 2022.

# 11

# Special Features of Galactic Dynamics

Christos Efthymiopoulos, Nikos Voglis[†] and C. Kalapotharakos

Research Center for Astronomy and Applied Mathematics,
Academy of Athens Soranou Efessiou 4, 115 27 Athens, Greece
cefthim@academyofathens.gr
nvogl@academyofathens.gr
ckalapot@phys.uoa.gr

**Abstract.** This is an introductory article to some basic notions and currently open problems of galactic dynamics. The focus is on topics mostly relevant to the so-called 'new methods' of celestial mechanics or Hamiltonian dynamics, as applied to the ellipsoidal components of galaxies, i.e., to the elliptical galaxies and to the dark halos and bulges of disk galaxies. Traditional topics such as Jeans theorem, the role of a 'third integral' of motion, Nekhoroshev theory, violent relaxation, and the statistical mechanics of collisionless stellar systems are first discussed. The emphasis is on modern extrapolations of these old topics. Recent results from orbital and global dynamical studies of galaxies are then shortly reviewed. The role of various families of orbits in supporting self-consistency, as well as the role of chaos in galaxies, are stressed. A description is then given of the main numerical techniques of integration of the N-body problem in the framework of stellar dynamics and of the results obtained via N-Body experiments. A final topic is the secular evolution and self-organization of galactic systems.

## 11.1 Introduction

The present lecture notes are an introduction to selected topics of *Galactic Dynamics*. The focus is on topics that we consider more relevant to the main theme of this workshop, *Celestial Mechanics*. This is not intended to be a review article. In fact, any of the topics below could be the subject of a separate review. Only the main ideas and notions are introduced, as well as some important currently open problems in each topic. Some relevant results from our own research are also presented. We discuss topics related mostly to the so-called *ellipsoidal components* of galaxies. These are (a) the dark halos of both elliptical and disk galaxies, (b) the luminous matter in elliptical galaxies, and (c) the bulges of disk galaxies. We shall only occasionally refer to the dynamics of disks, bars or spiral structure. These are important chapters of galactic dynamics which, however, go beyond the limits of the present article.

The fact that galactic (or stellar) dynamics and celestial mechanics share many common concepts, tools and methods of study is nowadays widely recognized in the community of dynamical astronomers. The connection of the two disciplines is transparent in recent advanced textbooks such as Contopoulos' *Order and Chaos in Dynamical Astronomy* [61] or Boccaletti and Pucacco *Theory of Orbits* [36] (other standard references for galactic dynamics are Binney and Tremaine [33] or Bertin [25]). However, this connection was not always recognized. Until the 1960s, the two fields emphasized rather different aspects of study, Celestial Mechanics focusing mostly on analytical expansions of perturbation theory in few body-type problems [135, 295], and Galactic Dynamics focusing on the properties of the distribution function of stellar systems composed by a large number of bodies [44, 239]. The shift of paradigm in the two fields can be traced in academic events like a celebrated 1964 Thessaloniki IAU Symposium [56], see the description in [62]).

We would like to point out one more guiding element of the exposition of ideas followed below. In his talk at the beginning of this meeting, A. Morbidelli has presented his view of the division of the problems of Celestial Mechanics into open, i.e. unresolved, and closed, i.e. resolved, problems. In Galactic Dynamics the very nature of problems does not permit such coarse classifications. We could claim, instead, that all practically interesting problems are still largely open. The main obstruction to closing problems is the lack of sufficient observational data, which, in many cases, is due to our fundamental inability to obtain such data. Let us give one trivial example: from the image of a galaxy in the sky it is impossible to deduce the *shape* of the galaxy without additional dynamical arguments. Such arguments are to an extent amenable to a posteriori observations, but the mapping of dynamics to such observations is usually non-unique. Similarly, the determination of the *pattern speed* of a spiral or barred disk galaxy requires a set of dynamical assumptions going well beyond the form of the underlying gravitational potential (the latter can in principle be determined by the observed rotation curve or distribution of matter in the galaxy). Since mankind cannot observe galaxies from different viewpoints, or for times relevant to galactic timescales, these fundamental constraints will remain with us and require a rather large effort in dynamical modelling needed to constrain uncertainties and explain even the simplest available observations of any particular galaxy. We let apart the fact that large amounts of matter in a galaxy, with dominant dynamical role, are either non-detectable by direct observational means (e.g. central black holes or the dark matter) or subject to non-gravitational interactions (e.g. gas, dust or star formation and evolution), that seriously complicate the dynamics.

As we shall see in the next section, from the stellar dynamical point of view the most general information regarding a stellar system is contained in its phase space density or distribution function $f(\mathbf{x}, \mathbf{v}, t)$. This function accounts for all kinds of photometric or kinematical data that can be observationally determined. Furthermore, we can use $f(\mathbf{x}, \mathbf{v}, t)$ to derive dynamical properties of the system that cannot be directly observed. The equilibria of galaxies are described by time-independent forms of $f$, while evolving galaxies, stellar dynamical instabilities or density waves are described by time-dependent forms of $f$. We may thus state that the determination of the distribution function of galaxies constitutes the *central goal* of galactic dynamics. The presentation below emphasizes this point of view, by focusing on *dynamical methods* of study of the distribution function. Other methods, that seek to determine the distribution function from the observational data via 'inversion' algorithms, are not presented here (see [87] for a review).

The presentation is organized as follows: Sect. 11.2 presents some basic notions of galactic dynamics such as the concept of relaxation time, Jeans' theorem, third integral of motion, etc. In Sect. 11.3 we present the statistical mechanical approach to the study of the distribution function, by dealing mostly with the theory of violent relaxation and with its modern modifications. Section 11.4 deals with the orbital approach. We present the main types of orbits encountered in spherical, axisymmetric or triaxial systems, and discuss the methods of 'global dynamics' and of 'self-consistent modelling' of galaxies both of which occupy an important place in current research. Section 11.5 focuses on the $N$-body method. We describe the main techniques to integrate the $N$-body problem when $N$ is large, and discuss recent results from global dynamical studies of galactic systems from $N$-body simulations.

## 11.2 Basic Notions

### 11.2.1 Time of Relaxation

The stellar dynamical study of galaxies is simplified by approximating these systems as *collisionless* $N$-body systems, i.e. by assuming that the stars 'feel' a mean field gravitational potential $\Phi(\mathbf{x}, t)$, and by ignoring the granularity of the field due to the point mass distribution of matter. This approach is justified by the remark that in galaxies the so-called two-body relaxation time $T_R$, i.e. the time needed in order that close encounters significantly affect an otherwise smooth stellar orbit, is much larger than the Hubble time of the Universe [44, 286]. An order of magnitude calculation of the two-body relaxation time can be based on considering deflections of the orbit of a star that moves in a nearly homogeneous sea of other stars (Fig. 11.1). Let $\mathbf{v_0}$ be the velocity of the test star at a particular moment when the impact parameter of its close encounter with a second star is equal to $b$. Neglecting the attraction by other stars, the angle of deflection $\psi$ after the encounter is readily found:

$$\tan\left(\frac{\psi}{2}\right) = \frac{mG}{bv_0^2} , \qquad (11.1)$$

where $m$ is the mass of the attracting star and $G$ Newton's constant of gravity. Practically all the angles $\psi$ of successive scattering events are small, since impact parameters are in general big. For example, the probability that a second star passes in the vicinity of the sun at a distance of the order of 10,000 AU is about one event in the galaxy's lifetime (see presentation by

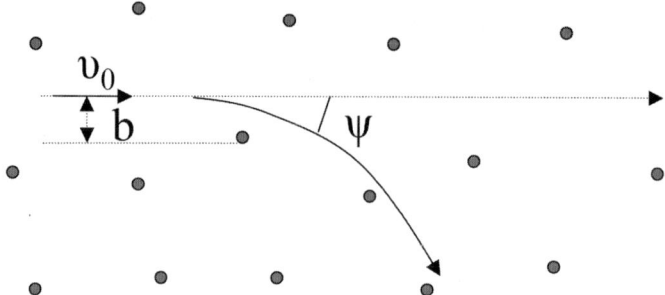

**Fig. 11.1.** Deflection of a test particle (*star*) moving in a homogeneous sea of other particles

M. Fouchard et al. in the same volume). This minimum impact parameter $b_{\min} \approx 10{,}000$ AU is of order $b_{\min} \sim D/N^{1/3}$, where $D$ is the typical length scale (e.g. diameter) of the galaxy and $N$ the number of stars in it. We may also set a maximum impact parameter $b_{\max} \sim D$. We may thus estimate an upper bound for the cumulative deflection angle after a large number of encounters, within a time interval $T$, by squaring (11.1) (with $\tan(\psi/2) \simeq \psi/2$) and summing over the number of stars contained in a differential cylindrical volume of radius $b$, width $db$ and length $v_0 T$:

$$\psi^2_{\text{cum}} = \sum_T \psi^2 \approx \int_{b_{\min}}^{b_{\max}} db\, 2\pi b v_0 T \rho \frac{4m^2 G^2}{b^2 v_0^4} \tag{11.2}$$

where $\rho$ is the mean density. Setting typical values for the density $\rho \sim mN/D^3$ and stellar velocity $v_0^2 \sim GNm/D$, we find from the above formula that the cumulative deflection will become of order unity (usually we request $\psi_{\text{cum}} = \pi/2$) when $T = T_{\text{R}}$ becomes equal to

$$T_{\text{R}} \approx \frac{0.1 N}{\ln N} T_{\text{D}} \tag{11.3}$$

where $T_{\text{D}} \sim D/v_0$ is the typical dynamical time or period of a typical orbit across the galaxy. Setting $T_{\text{D}} \sim 10^8$ yr and $N \sim 10^{10}$–$10^{13}$, we find $T_{\text{R}} \sim 10^{15}$–$10^{18}$ yr, i.e. at least five orders of magnitude larger than the Hubble age of the Universe $T_{\text{H}} \sim 10^{10}$ yr. We conclude that close encounters cannot affect the dynamics in timescales comparable to the present lifetime of a galaxy.

Due to the above calculation of the relaxation time, the basic paradigm for galaxies is a collisionless stellar system in which the collisionless Boltzmann equation applies (Sect. 11.2.3). However, the true nature of relaxation depends also somewhat on what region of the galaxy we consider as well as on the properties of the system's stellar orbits. For example, the above analysis is not precise at the centres of galaxies, especially when the latter are occupied by large central mass concentrations. Furthermore, if a system has a large degree of stochasticity, i.e. many orbits with Lyapunov times smaller or equal to the Hubble time, then the two-body relaxation time for such a system is drastically reduced, perhaps by more than three orders of magnitude [130, 249]. This is because an initially small deflection, caused by a two-body encounter, is amplified by the mechanism of exponential deviations of nearby orbits due to positive Lyapunov exponents. This may have affected systems that are 'granular', for example galaxies containing a high percentage of globular clusters [308]. The extent to which such phenomena appear in real galaxies is not yet fully known.

### 11.2.2 Distribution Function

The most basic quantity in stellar systems is the fine-grained distribution function

$$f(\mathbf{x}, \mathbf{v}, t) = \lim_{\mathrm{d}^6 \mu \to 0} \frac{\mathrm{d}m(\mathbf{x}, \mathbf{v}, t)}{\mathrm{d}^3 \mathbf{x} \, \mathrm{d}^3 \mathbf{v}} \tag{11.4}$$

yielding the mass $\mathrm{d}m(\mathbf{x}, \mathbf{v}, t)$ contained at time $t$ within an infinitesimal phase space volume $\mathrm{d}^6 \mu = \mathrm{d}^3 \mathbf{x} \, \mathrm{d}^3 \mathbf{v}$ centred around any point $(\mathbf{x}, \mathbf{v})$ of the 6D phase space of stellar motions (called the $\mu$-space in statistical mechanics). In the $N$-body approximation the mass $\mathrm{d}m(\mathbf{x}, \mathbf{v}, t)$ can be considered proportional to the number of particles, i.e. stars or fluid elements of the dark matter, within the volume $\mathrm{d}^3 \mathbf{x} \, \mathrm{d}^3 \mathbf{v}$. Furthermore, it is often convenient to introduce a coarse-grained distribution function

$$F(\mathbf{x}, \mathbf{v}, t) = \frac{1}{\Delta^3 \mathbf{x} \, \Delta^3 \mathbf{v}} \int_{\Delta^3 \mathbf{x} \, \Delta^3 \mathbf{v}} f(\mathbf{x}, \mathbf{v}, t) \mathrm{d}^3 \mathbf{x} \, \mathrm{d}^3 \mathbf{v} \tag{11.5}$$

which gives the average of the fine-grained distribution function $f$ in small, but not infinitesimal volume elements $\Delta^3 \mathbf{x} \, \Delta^3 \mathbf{v}$ around the phase space points $(\mathbf{x}, \mathbf{v})$. Contrary to the fine-grained distribution $f$, the value of the coarse-grained distribution $F$ depends on the particular choice of partitioning of the phase space by which the volume elements $\Delta^3 \mathbf{x} \, \Delta^3 \mathbf{v}$ are defined. This fact has some interesting implications in the modelling process of a galaxy, discussed in Sect. 11.3 below.

The distribution function can be used to derive several other useful quantities. For example, the spatial mass density $\rho(\mathbf{x}, t)$ of the system is given by the integral of the d.f. $f$ over velocities, e.g. (in Cartesian coordinates)

$$\rho(\mathbf{x}, t) = \int_{-\infty}^{\infty} \int_{-\infty}^{\infty} \int_{-\infty}^{\infty} f(\mathbf{x}, \mathbf{v}, t) \mathrm{d}v_x \, \mathrm{d}v_y \, \mathrm{d}v_z \ . \tag{11.6}$$

The latter quantity, $\rho(\mathbf{x}, t)$, can be used in turn to calculate the gravitational potential $\Phi(\mathbf{x}, t)$ via Poisson's equation:

$$\nabla^2 \Phi(\mathbf{x}, t) = 4\pi G \rho(\mathbf{x}, t) \ . \tag{11.7}$$

The orbits of stars are given by the Hamiltonian

$$H(\mathbf{x}, \mathbf{p}, t) \equiv \frac{\mathbf{p}^2}{2} + \Phi(\mathbf{x}, t) \tag{11.8}$$

setting, for simplicity, $\mathbf{p} = \mathbf{v}$ in Cartesian coordinates and the average stellar mass equal to unity. We often consider galaxies in *steady-state equilibrium* (Sect. 11.2.3), in which case we drop the explicit dependence of $f$ on the time $t$:

$$H(\mathbf{x}, \mathbf{p}) \equiv \frac{\mathbf{p}^2}{2} + \Phi(\mathbf{x}) . \tag{11.9}$$

Assuming a nearly constant mass-to-light ratio, the observable photometric or kinematic profiles of a galaxy can be deduced from various moments of $f$. For example, if the axis $x$ is identified to the direction of the line of sight, the surface density at any point $\mathbf{R} \equiv (y, z)$ of the plane of projection normal to $x$ is given by

$$\Sigma(\mathbf{R}) = \int_{-\infty}^{\infty} \rho(x, \mathbf{R}) dx \tag{11.10}$$

with $\rho$ given by (11.6). The quantity $\Sigma(\mathbf{R})$ can be compared to observed surface brightness profiles. On the other hand, the line-of-sight velocity distribution at a particular point $\mathbf{R}$ of the same plane of projection is given by

$$\text{LOSVD}(\mathbf{R}, v_x) = \frac{1}{\Sigma(R)} \int_{-\infty}^{\infty} \int_{-\infty}^{\infty} \int_{-\infty}^{\infty} dx \, dv_y \, dv_z \, f(x, \mathbf{R}, v_x, v_y, v_z) \tag{11.11}$$

and the latter quantity can be compared to the profiles of spectral lines determined also observationally. Via the line-of-sight velocity distributions we can determine mean velocity profiles,

$$\mu(\mathbf{R}) = \int_{-\infty}^{\infty} v_x \, \text{LOSVD}(\mathbf{R}, v_x) dv_x \tag{11.12}$$

and velocity dispersion profiles

$$\sigma^2(\mathbf{R}) = \int_{-\infty}^{\infty} (v_x - \mu(\mathbf{R}))^2 \text{LOSVD}(\mathbf{R}, v_x) dv_x . \tag{11.13}$$

Also related to observations is the concept of *velocity ellipsoid*. This is an ellipsoid in velocity space assigned to every point $\mathbf{x}$ of ordinary space. Fixing an orthogonal coordinate system, say, Cartesian axes $x_1 = x, x_2 = y, x_3 = z$, we calculate the second moments

$$\sigma_{ij}^2(\mathbf{x}) = \frac{1}{\rho(\mathbf{x})} \int_{-\infty}^{\infty} (v_i - V_i)(v_j - V_j) f(\mathbf{x}, \mathbf{v}) d^3\mathbf{v} \tag{11.14}$$

where the indices $i, j$ run the values 1, 2 or 3, $V_i$ is the mean velocity in the $i$th direction at the point $\mathbf{x}$ and the integral denotes a triple integral with respect to the velocities. The $3 \times 3$ matrix $\sigma$, with elements $\sigma_{ij}$, is symmetric, thus it has three real eigenvalues, say $\sigma_1$, $\sigma_2$ and $\sigma_3$ and unit eigenvectors $\mathbf{e}_{\sigma,k}$, $k = 1, 2, 3$. The velocity ellipsoid is defined by the equation

$$\sum_{k=1}^{3} \frac{(\mathbf{v} \cdot \mathbf{e}_{\sigma,k})^2}{\sigma_k^2} = 1 \, . \tag{11.15}$$

The shape of the velocity ellipsoid at a point $\mathbf{x}$ gives the dispersion of the distribution of velocities in different local directions of motion. In particular, a system is called *isotropic* at the point $\mathbf{x}$ if the velocity ellipsoid at $\mathbf{x}$ is a sphere, otherwise it is called *anisotropic*. In the case of anisotropic systems, we further distinguish systems with two or three unequal axes of the velocity ellipsoid. This distinction is important, because it allows one to link the kinematic observations available for a particular system to dynamical features of the same system. For example, the observation that the velocity ellipsoid in the solar neighbourhood has three unequal axes led to the discovery that the stellar orbits in the solar neighbourhood are subject to a 'third integral' [54], besides the energy and angular momentum integrals.

### 11.2.3 Stellar Dynamical Equilibria: Old and New Versions of Jeans' Theorem

The basic equation governing the time evolution of the distribution function $f$ in collisionless stellar systems is Liouville's equation implemented in the $\mu$-space of motion of the Hamiltonian (11.8), otherwise called Boltzmann's equation (or Vlasov's equation in plasma physics):

$$\frac{\mathrm{d}f}{\mathrm{d}t} = \frac{\partial f}{\partial t} + \mathbf{p}\frac{\partial f}{\partial \mathbf{x}} - \frac{\partial \Phi}{\partial \mathbf{x}}\frac{\partial f}{\partial \mathbf{p}} = 0 \tag{11.16}$$

where we have adopted the notation $\mathbf{p} \equiv \mathbf{v}$ for canonical momenta, i.e. consider stellar masses equal to unity. Equation (11.16) states that the mass contained within any infinitesimal volume $\mathrm{d}^6\mu$ that travels in phase space along the orbits corresponding to the potential $\Phi$ (determined by (11.7)) is preserved. Furthermore, the measure of the volume $\mathrm{d}^6\mu$ is also preserved (Liouville's theorem). Now, the morphological regularity and the commonly observed characteristics of most galaxies suggest that the majority of these systems are close to a state of statistical equilibrium. Thus, we often look for *steady-state* solutions of (11.16) that do not exhibit an explicit dependence of $f$ on time. Setting $\partial f/\partial t = 0$ in (11.16) yields

$$\mathbf{p}\frac{\partial f}{\partial \mathbf{x}} - \frac{\partial \Phi}{\partial \mathbf{x}}\frac{\partial f}{\partial \mathbf{p}} = \{f, H\} = 0 \tag{11.17}$$

where $\{\cdot, \cdot\}$ denotes the Poisson bracket operator.

Despite its formal simplicity, the physical content of (11.17) is remarkable. Consider a *fixed* phase volume $\mathrm{d}^6\mu$ centred at some phase space point $(\mathbf{x}, \mathbf{p})$ of a galaxy in steady-state equilibrium. The stars follow orbits determined by the Hamiltonian (11.9). The orbits remain smooth in the course of time, because there are no short range stochastic force terms affecting the stars, similar, for example, to collisions in a perfect gas. Nevertheless, a detailed equilibrium is established in the phase space, i.e. if (11.17) is valid the number of stars leaving the volume $\mathrm{d}^6\mu$ at any moment $t$ must be equal to the number of stars entering the same volume. Furthermore, the gravitational potential determining the orbits is given by (11.7), which involves also the positions of the stars. This means that the motions of the stars are combined in such a way so as to reproduce the same macroscopic distribution of matter continually in time. For this reason, galactic equilibria are called *self-consistent*, i.e. supported solely by the orbits of stars within the system. It is a great theoretical challenge to understand the processes by which nature forms such remarkable systems.

Consider a system in steady-state equilibrium and suppose that the mathematical form of the function $f(\mathbf{x},\mathbf{p})$ was given. Then, according to (11.17), the function $f$ constitutes an integral of the motion in involution with the Hamiltonian. If, on the other hand, we know by independent means a complete set of functionally independent integrals of motion $I_1, I_2, \ldots$ under the Hamiltonian flow of $H$, it follows that $f$ is necessarily a composite function of the phase space canonical variables $(\mathbf{x},\mathbf{p})$ through one or more of the integral functions $I_1, I_2, \ldots$. That is

$$f(\mathbf{x},\mathbf{p}) \equiv f(I_1(\mathbf{x},\mathbf{p}), I_2(\mathbf{x},\mathbf{p}),\ldots) . \tag{11.18}$$

The last equation is known as *Jeans' theorem* of stellar dynamics [154].

Although fundamental in theory, Jeans' theorem, in the above general form, is of limited usefulness, because it specifies neither (a) which integrals out of the set $I_1, I_2, \ldots$ should actually appear as arguments in the distribution function of a specific system, nor (b) the explicit form of the dependence of $f$ on these integrals. Regarding point (a), a 'strong' Jeans' theorem proved by Lynden-Bell [194] asserts that only *isolating* integrals can be arguments of the function $f$. An integral $I_i$ is called isolating if the constant value condition $I_i(\mathbf{x}(t),\mathbf{p}(t)) = c_i$ defines a manifold in phase space of dimension lower than the phase space dimension (equal to six for three dimensional systems). If we have a set of $M$ isolating integrals $I_1, I_2, \ldots I_M$, any orbit $(\mathbf{x}(t),\mathbf{p}(t))$ is restricted on a sub-manifold of phase space which is the intersection of all the manifolds defined by the constant value conditions $I_i(\mathbf{x}(t),\mathbf{p}(t)) = c_i, i = 1, 2, \ldots, M$.

A case of particular interest is when the Hamiltonian of motion $H$ is integrable in the Arnold–Liouville sense. In three degrees of freedom systems this means that there are three functionally independent integrals ($H$ itself can be taken as one of them) which are mutually in involution, namely

$$\{I_i, I_j\} = \sum_{k=1}^{3} \frac{\partial I_i}{\partial x_k}\frac{\partial I_j}{\partial p_k} - \frac{\partial I_i}{\partial p_k}\frac{\partial I_j}{\partial x_k} = 0 . \tag{11.19}$$

In that case, the Arnold–Liouville theorem (see e.g. [13] or [122]) asserts that if the manifolds defined by the constant value conditions $I_i(\mathbf{x}(t), \mathbf{p}(t)) = c_i$, $i = 1, 2, 3$ are compact, then they are topologically equivalent to 3-tori. The integrals $I_i$, $i = 1, 2, 3$ are isolating and the strong Jeans' theorem takes the following form: *if the Hamiltonian of a collisionless stellar system in steady-state equilibrium is Arnold–Liouville integrable, the fine-grained distribution function f has constant value at all the points $(\mathbf{x}, \mathbf{p})$ of an invariant torus of the system.*

We examine below some simple examples of application of the strong Jeans' theorem in stellar dynamics.

**Spherical Systems**

A spherical system in equilibrium is the simplest model of a galactic system. This model is not very realistic, but it serves (a) to introduce some basic concepts, and (b) as a starting point for the analysis of more realistic systems. In spherical coordinates, the distribution function depends on $r$ and on the three velocity components $v_r = \dot{r}$, $v_\theta = r\dot{\theta}$, $v_\phi = r\sin\theta\dot{\phi}$, namely $f \equiv f(r, v_r, v_\theta, v_\phi)$. The mass density depends only on $r$, $\rho \equiv \rho(r)$. The orbits are determined by a spherical potential $\Phi(r)$, given by the solution of (11.7):

$$\Phi(r) = -\frac{GM(r)}{r} - \int_r^\infty \frac{G\,\mathrm{d}m(r')}{r'}$$

$$= -\frac{G}{r}\int_0^r 4\pi r'^2 \rho(r')\mathrm{d}r' - G\int_r^\infty 4\pi r' \rho(r')\mathrm{d}r' . \quad (11.20)$$

The orbits obey three isolating integrals of motion in involution, namely the energy $E = H$, and the components of the angular momentum $p_\theta = r^2\dot{\theta}$ and $p_\phi = r^2\sin^2\theta\dot{\phi}$. The angular momentum vector $\mathbf{L} = \mathbf{r} \times \mathbf{v}$ is constant and an orbit is restricted on the plane normal to $\mathbf{L}$. The modulus $L = |\mathbf{L}|$ is an integral in involution with $p_\phi$, and the triplet $(E, L, p_\phi)$ is the usual choice of integrals in the study of spherical systems.

According to the strong Jeans' theorem, the general form of the distribution function $f(r, v_r, v_\theta, v_\phi)$ in equilibrium can only be a composite function:

$$f \equiv f(E(r, v_r, v_\theta, v_\phi), L(r, v_r, v_\theta, v_\phi), p_\phi(r, v_r, v_\theta, v_\phi)) . \quad (11.21)$$

Further restrictions in the form of $f$ can be imposed on the basis of the kinematical properties of the system under study. For example, if the system has no preferential kinematical axis (e.g. an axis of rotation), the integral $p_\phi$ cannot appear as an argument in $f$. This implies that there is equal probability to find a star moving in a plane of any possible orientation with respect to the galactic frame of reference. This is applicable, e.g. to the spherical limit of giant elliptical galaxies, since there is evidence that the these galaxies are *not* rotationally supported against gravity [27, 77, 151] but they are 'hot systems' with small or no rotation, in which gravity is balanced by the distribution of velocities in random directions [28, 29]. In the spherical limit, we use distribution functions of the form $f(E)$ or $f(E, L)$. If $f \equiv f(E)$ the galaxy is called isotropic. The expression for the orbital energy $E = v_r^2/2 + v_\theta^2/2 + v_\phi^2/2 + \Phi(r)$ yields a symmetric dependence of $f$ on either of the three velocity components. This implies equal axes of the velocity ellipsoid $\sigma_r^2 = \sigma_\theta^2 = \sigma_\phi^2$. On the other hand, if $f \equiv f(E, L)$ the system is called anisotropic. The appearance of $L = r\sqrt{v_\theta^2 + v_\phi^2}$ in $f$ breaks the symmetry of the functional dependence of $f$ on $v_r$ and $v_\theta$ (or $v_\phi$). The velocity ellipsoid has two equal axes $\sigma_r^2 \neq \sigma_\theta^2 = \sigma_\phi^2$. Since every orbit is confined to a plane, we consider the total velocity in the transverse direction of motion $v_t^2 = v_\theta^2 + v_\phi^2$ and define the *anisotropy parameter* $\beta(r)$ [[33], p. 204]:

$$\beta(r) = 1 - \frac{\sigma_r^2(r)}{2\sigma_t^2(r)} \tag{11.22}$$

with

$$\sigma_r^2(r) = \frac{1}{\rho(r)} \int_0^{\sqrt{-2\Phi(r)}} \int_0^{\sqrt{-2\Phi(r)-v_r^2}} f v_r^2 v_t \, dv_r \, dv_t$$

and

$$\sigma_t^2(r) = \frac{1}{\rho(r)} \int_0^{\sqrt{-2\Phi(r)}} \int_0^{\sqrt{-2\Phi(r)-v_r^2}} f v_t^3 \, dv_r \, dv_t.$$

The limits of integration in the above equations are imposed by the consideration of only bound orbits ($E < 0$). The parameter $\beta$ is a function only of $r$. In practice we find that realistic systems are nearly isotropic in their central parts, $\beta(r) \to 0$ as $r \to 0$, and *radially* anisotropic in their outer parts, i.e. $\beta(r) \to 1$ for $r$ large. This means that there is a predominance of radial orbits in the outer part of the galaxy, i.e. orbits with a large difference between the apocentric and pericentric distances. This phenomenon is linked to the relaxation process of galaxies (Sect. 11.3). In particular, this is the expected final behaviour of systems subject to a phase of 'collapse' [103], and this behaviour is confirmed by $N$-body experiments of violent relaxation [3, 320].

**Axisymmetric Systems and the 'Third Integral' of Motion**

The Hamiltonian of motion in an axisymmetric galaxy can be written in cylindrical canonical variables $(R, \phi, z, p_R, p_\phi, p_z)$:

$$H \equiv \frac{p_R^2}{2} + \frac{p_\phi^2}{2R^2} + \frac{p_z^2}{2} + \Phi(R, z) \tag{11.23}$$

where $z$ is the axis of symmetry, and $p_R = \dot{R}$, $p_\phi = R^2\dot{\phi}$, $p_z = \dot{z}$. Since the azimuthal angle $\phi$ is ignorable, the canonical momentum $p_\phi$ is a second integral of motion, besides the energy $E = H$. This can be identified to the $z$-projection of the angular momentum vector $p_\phi = L_z$. The study of orbits can be simplified by considering only the motion on the meridional plane $(R, z)$

$$\ddot{R} = -\frac{\partial \Phi}{\partial R} + \frac{L_z^2}{R^3}, \quad \ddot{z} = -\frac{\partial \Phi}{\partial z}. \tag{11.24}$$

The form of these equations implies that (11.23) can be viewed as a two degrees of freedom Hamiltonian, where $p_\phi$, replaced by $L_z$, is considered as a parameter. The angular motion is readily found via $\dot{\phi} = L_z/R^2$. The orbits on the equatorial plane are defined by a central potential $\Phi(R,0)$ (provided that the system is symmetric with respect to the equatorial plane, i.e. the function $\Phi(R,z)$ is even with respect to $z$).

If we consider circular orbits in the equatorial plane for a particular value of $L_z$, the circular radius is given by the root of the equation:

$$-\frac{\partial \Phi(R_c,0)}{\partial R} + \frac{L_z^2}{R_c^3} = 0. \tag{11.25}$$

The circular orbit appears as an equilibrium point on the meridional plane, at $R = R_c$, $\dot{R} = 0$. If we expand the Hamiltonian with respect to this point we get (ignoring a constant term $L_z^2/2R_c^2$)

$$H = \frac{1}{2}\left(p_Y^2 + p_z^2 + \omega_Y^2 Y^2 + \omega_z^2 z^2\right) + \sum_{k=3}^{\infty} H^{(k)}(Y,z;L_z) \tag{11.26}$$

where

$$Y = R - R_c, \quad \omega_Y^2 = \frac{\partial^2 \Phi(R_c,0)}{\partial^2 R} + \frac{3L_z^2}{R_c^4}, \quad \omega_z^2 = \frac{\partial^2 \Phi(R_c,0)}{\partial^2 z},$$

and the functions $H^{(k)}(Y,z;L_z)$ are polynomials of degree $k$ in the variables $Y,z$, depending also on $L_z$ as a parameter.

The Hamiltonian (11.26) has a particular place in the history of both galactic dynamics and dynamical systems theory because (a) it is the first Hamiltonian for which a 'third integral' of motion was calculated [54] and (b) its third-order truncation yields the Hénon–Heiles [139] Hamiltonian that has served as a prototype of many studies in non-linear Hamiltonian dynamical systems.

Special forms of the third integral, e.g. quadratic in the velocities, were considered by various authors (see references in [239]). On the other hand, Contopoulos [54] explored the question of whether a third integral of motion $I$, besides $H$ and $L_z$ can be constructed *algorithmically* for the Hamiltonian (11.26). The existence of $I$ implies that all the orbits are regular (no chaos is present). Furthermore, according to Jeans' theorem the integral can possibly appear as an argument in the distribution function. Recalling arguments similar to the spherical case, we then find that if $f$ depends on $I$ the velocity ellipsoid at any point of ordinary space has unequal axes $\sigma_R \neq \sigma_z$, while if $f$ does not depend on $I$ the dispersions are equal $\sigma_R = \sigma_z$. The observational data in our own galaxy, in the solar neighbourhood, favoured the former case to be true.

Contopoulos [54] combined two earlier methods of Whittaker [337] and Cherry [48, 49] in order to show that, in the so-called non-resonant case, when the frequencies $\omega_Y, \omega_z$ are incommensurable, an integral can be *formally* constructed in the form of a polynomial series in the canonical variables, by an algorithm which is significantly simpler than the use of canonical transformations as in the Birkhoff–von Zeipel method [34], widely used in Celestial Mechanics. Given that such formal series are, in general, *not* convergent [277], the above series does not represent a real third integral of the system. However, we shall see below that the series has an *asymptotic* behaviour. Namely, if we define a remainder $R^{(n)}$ for the series at the $n$th order of truncation, the remainder initially decreases as $n$ increases, giving the impression that the series is convergent. However, after an optimal order $n_{opt}$ the remainder becomes an increasing function of $n$ (Fig. 11.2), implying divergence of the series. If we truncate the integral series at the order $n_{opt}$, we obtain a function $I = I^{(2)} + I^{(3)} + \ldots + I^{(n_{opt})}$ which is an approximate integral of motion, in the sense that the time variations $dI/dt$ are quite small, of order $R^{(n_{opt})}$. The apparent improvement of the accuracy of the integral as $n$ increases (below $n_{opt}$) was checked by a computer program that calculated the series [67]. This was confirmed later by Gustavson [131] with a calculation of the Birkhoff series in the Hénon–Heiles Hamiltonian.

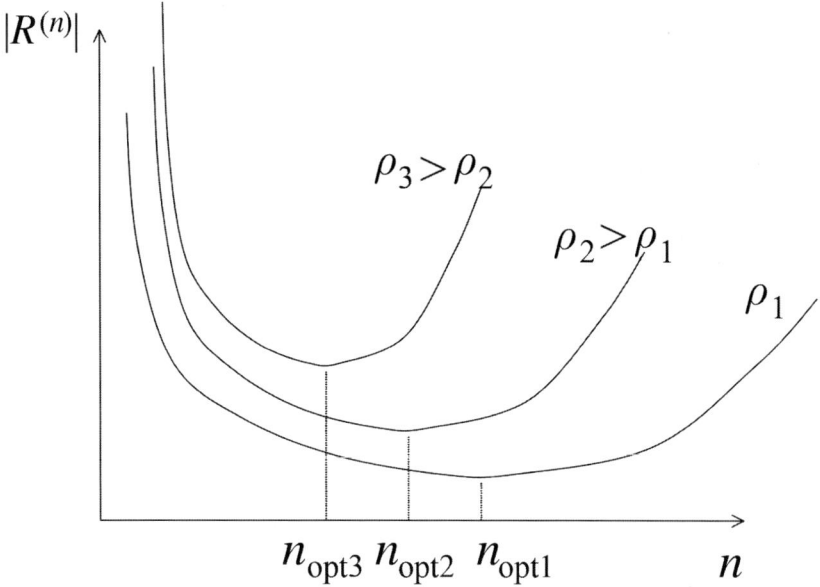

**Fig. 11.2.** Asymptotic behaviour of the 'third integral'. The size of the remainder $|R^{(n)}|$ of the integral series as a function of the order of truncation $n$ for three different effective distances from the equilibrium point

While in the non-resonant case the calculation of the third integral by the direct method of Contopoulos is simpler than by the Birkhoff normal form, the situation is reversed in the case of resonant integrals, i.e. when the frequencies satisfy a commensurability relation $m_1\omega_Y + m_2\omega_z = 0$, with $m_1, m_2$ integers. A direct method to construct a resonant integral without using a normal form was given by Contopoulos [55] and exploited in the case of particular resonances by Contopoulos and Moutsoulas [67]. However, this method involves a 'back and forth' algorithm between successive orders of truncation, which is an essential complication. The discrete analogue of the direct method for symplectic mappings was given by Bazzani and Marmi [20], in the non-resonant case, and by Efthymiopoulos [98] in the resonant case. However, all these direct methods are currently superseded by the use of the Birkhoff method via Lie canonical transformations [90, 123, 150, 315], which is the simplest method to implement in the computer [120].

The Lie method of construction of a third integral was implemented in axisymmetric galaxies by Gerhard and Saha [117]. These authors studied various constructive methods of the canonical perturbation theory. A particular method is to express the Hamiltonian in the action–angle variables of the *spherical* part of the potential, since analytical expressions yielding the action–angle variables in terms of the usual canonical variables are explicitly known in that case. The Lie method can then be used in order to construct a formal third integral, besides the energy and $L_z$. The so-obtained expressions represented a well-preserved integral if the system's axial ratio was greater than 0.5. Further models of this type were given by Dehnen and Gerhard [85], starting from the spherical isochrone model (Sect. 11.3.4) to represent the unperturbed system. On the other hand, Matthias and Gerhard [205] tested whether the boxy elliptical galaxy NGC 1600 is better fitted by a two-integral or three-integral model. They found that three-integral models better reproduce the available kinematic data for the galaxy. This conclusion was confirmed in subsequent studies ( [42, 51, 113, 115, 316] and [ [87], Sect. 4 and references therein]). We will return below to the question of the choice between two-integral or three-integral axisymmetric models, when discussing relevant results from $N$-body simulations.

Besides the harmonic oscillators, or the spherical model, there are other integrable axisymmetric models that can serve as starting models for the construction of formal third integrals. For example, Petrou [247] constructed a third integral starting from an axisymmetric model of the form $\Phi(r,\theta) = \Phi_0(r) + \Phi_1(\theta)/r^2$ (in polar coordinates) which is known to be integrable [[125], p. 457]. Another possible choice is an axisymmetric Stäckel potential [88, 293]. Other models based on local Stäckel fits are reviewed in Dejonghe and Bruyne [87].

When the potential has a central cusp, a convenient method to calculate third integrals is the semi-analytical (or semi-numerical) method. Essentially, this means to start with a plausible model in which action–angle variables are explicitly constructed, and then to introduce canonical transformations to new action–angle variables with generating functions specified through a numerical criterion. This criterion can be based on either the 'theoretical' Hamiltonian flow (found by the normal form) fitting the true Hamiltonian flow of the system, or the theoretical tori, viewed as geometrical objects, fitting the real tori of the system. Such fitting methods were introduced in galactic dynamics by Ratcliff et al. [257], McGill and Binney [207] and Kent and de Zeeuw [171].

**(Non-)convergence Properties of the 'Third Integral': The Theory of Nekhoroshev**

The optimal order of truncation $n_{\text{opt}}$ of the third integral series, as well as the size of the optimal remainder $R^{(\text{opt})}$ are questions that can be examined in the framework of the theory of Nekhoroshev [23, 190, 236, 256], implemented, in particular, in the case of elliptic equilibria [105, 121, 132, 238]. This theory states that as the parameter $\varepsilon$ that quantifies the perturbation of the system from an integrable system decreases, the size of the optimal remainder becomes *exponentially small* in $1/\varepsilon$, that is

$$R_{\text{opt}} = O\left(\exp\left(-\frac{1}{\varepsilon^p}\right)\right) \tag{11.27}$$

where the exponent $p$ depends on the number of degrees of freedom of the system under study. Conversely, approximate integrals of the type of the 'third integral' retain almost constant values for times exponentially long in $1/\varepsilon$, that is, $T_{\text{Nek}} = O(\exp(1/\varepsilon^p))$. In galactic-type Hamiltonians such as (11.26), the effective perturbation $\varepsilon$ is identified to the average distance $\rho$ of an orbit from the elliptic equilibrium. Thus, without being able to prove the existence of an exact third integral for the orbits $R(t), z(t)$ on the meridional plane, we can assert that, even if such orbits are chaotic, an orbit will behave effectively like regular for a time exponentially long in $1/\rho$, where $\rho$ is the distance of the orbit from the equilibrium $R = R_c, z = 0$.

A heuristic derivation of the formula $R_{\text{opt}} = O(\exp(-1/\rho))$, based on the use of integrals calculated by the direct method, can be given following a theorem by Giorgilli [121]. We make the derivation in action–angle variables $(J, \phi)$. We set $Y = \sqrt{2J_1/w_1}\sin\phi_1$, $p_Y = \sqrt{2w_1 J_1}\cos\phi_1$, $z = \sqrt{2J_2 w_2}\sin\phi_2$, $p_z = \sqrt{2w_2 J_2}\cos\phi_2$, and $w_1 \equiv w_Y$, $w_2 \equiv w_z$. The Hamiltonian (11.26) takes the form

$$H = \omega_1 J_1 + \omega_2 J_2 + \sum_{k=3}^{\infty} H^{(k)}(J_1, J_2, \phi_1, \phi_2) \tag{11.28}$$

where the functions $H^{(k)}$ are of degree $k/2$ in the actions and contain trigonometric terms of the form $e^{i(k_1\phi_1 + k_2\phi_2)}$ with $k_1, k_2$ integers, $|k_1| + |k_2| \leq k$ and of the same parity as $k$. We look for a third integral as a series yielding a correction to the action $J_1$ or $J_2$ (each of the actions is an exact integral in the harmonic oscillator limit of (11.26)). We thus set

$$I = J_1 + \sum_{k=3}^{\infty} I^{(k)}(J_1, J_2, \phi_1, \phi_2),$$

the functions $I^{(k)}$ satisfying the same properties as the functions $H^{(k)}$ (we set $I^{(2)} = J_1$). The integral is calculated by splitting the condition $\{I, H\} = 0$ for the integral I to terms of equal order. This yields the relation

$$\{I^{(k)}, H^{(2)}\} = -\sum_{s=2}^{k-1} \{I^{(s)}, H^{(k+2-s)}\} \qquad (11.29)$$

with $H^{(2)} = \omega_1 J_1 + \omega_2 J_2$. Equation (11.29) can be solved recursively to yield $I^{(k)}$ in the $k$th step from the terms $I^{(s)}$, $s = 2, \ldots, k-1$ determined in the previous steps. If we express the terms $I^{(s)}$ in sums of Fourier terms of the form $J_1^{s_1/2} J_2^{s_2/2} e^{i(k_1\phi_1 + k_2\phi_2)}$, we readily see that the algebraic nature of the direct scheme (11.29) is quite similar to that of the Birkhoff–von Zeipel normal form scheme: each Fourier term is an *eigenfunction* of the linear differential operator $\{\cdot, H^{(2)}\}$ with eigenvalue equal to $-i(k_1\omega_1 + k_2\omega_2)$, that is

$$\{J^{s/2} e^{ik\cdot\phi}, H^{(2)}\} = -i(k\cdot\omega) J^{s/2} e^{ik\cdot\phi}$$

where we use the abbreviations $J^{s/2} \equiv J_1^{s_1/2} J_2^{s_2/2}$, $k \equiv (k_1, k_2)$, $\omega \equiv (\omega_1, \omega_2)$. This implies that the solution of (11.29) for $I^{(k)}$ yields precisely a sum of the same Fourier terms as in the r.h.s. of the same equation, each term being divided by the divisor $k\cdot\omega$. The presence of divisors is important because, for generic incommensurable frequency vectors $\omega$ there are integer vectors $k$ that can be found, which render the product $k\cdot\omega$ a *small divisor*. For example, from number theory it is known [24] that most irrationals satisfy diophantine conditions of the form

$$|k \cdot \omega| \geq \frac{\gamma}{|k|^\tau} \qquad (11.30)$$

with $\gamma$ an O(1) constant and $\tau$, the diophantine exponent, depending on the number of degrees of freedom ($\tau \geq n - 1$). This means that, as $|k|$ increases, the minimum size of divisors appearing in the recurrent solution of (11.29) decreases, i.e. the divisors become smaller and smaller. Furthermore, as one repeatedly implements the recurrence relation, such small divisors *accumulate* in the form of products in the denominators of the various integral terms. That is, there are Fourier terms $f^{(k)}$ in $I^{(k)}$ with an accumulation of divisors yielding a size

$$||f^{(k)}|| \sim \frac{F^{(k)}}{a_3 a_4 \ldots a_k} \qquad (11.31)$$

with divisors $a_s$, $s = 3, 4, \ldots, k$ satisfying $a_s \sim 1/s^\tau$ according to (11.30). The numerator $F^{(k)}$ in (11.31) can be estimated by the remark that, for any term $I^{(s)}$, the Poisson bracket in the r.h.s. of (11.29) means to take the derivatives $\partial I^{(s)}/\partial J$, or $\partial I^{(s)}/\partial \phi$, which both cause the appearance of a factor O(s) in front of the corresponding Fourier terms of $I^{(s)}$. Hence, the repeated action of Poisson brackets, up to order $k$ creates a factor $F^{(k)} \sim O(3)O(4)\ldots O(k) \sim k!$ in the numerator of the Fourier terms of $I^{(k)}$ (see [99] for a more detailed analysis). Putting these remarks together, the size of Fourier terms (11.31) can be estimated as $||f^{(k)}|| \sim k!^{\tau+1}$. If we now consider an orbit of effective distance $\rho$ from the equilibrium, we have $J^{s/2} \sim \rho^s$ for this orbit, so that the value of the remainder of the formal series at the $k$th order of truncation can be estimated as:

$$R^{(k)} \sim k!^{\tau+1} \rho^k . \tag{11.32}$$

The estimate (11.32) contains the essential result regarding the asymptotic character of formal series: using Stirling's formula $k! \sim (k/e)^k$, for large k, we have $R^{(k)} \sim (k^{\tau+1}\rho/e^{\tau+1})^k$. We then want to check whether the remainder decreases or increases as the order $k$ of calculation of the formal integral increases. We see immediately that as long as $k << e/\rho^{1/(\tau+1)}$, the remainder decreases with $k$, while if $k >> e/\rho^{1/(\tau+1)}$ the remainder increases with $k$. Thus the optimal order is an order $n_{opt}$ where the remainder is minimum, which can be estimated as $n_{opt} \sim e/\rho^{1/(\tau+1)}$. Inserting this in (11.32) we find the value of the remainder at the optimal order of truncation $R_{opt} \sim \exp(-n_{opt}) \sim \exp(-1/\rho^{(1/(\tau+1))})$, which leads to Nekhoroshev's formula of exponentially small time variations of the truncated integral $I = I^{(2)} + I^{(3)} + \ldots + I^{(n_{opt})}$.

In generic nearly integrable Hamiltonian systems of the form $H(J,\phi) = H_0(J) + \varepsilon H_1(J,\phi)$ the Nekhoroshev theory is much more complicated than in the simple case of elliptic equilibria. The main complication is that the frequencies $\omega(J) = \partial H/\partial J$ depend on the actions, a fact that renders necessary the separate treatment of several non-resonant or resonant domains that coexist in the space of actions. This treatment is the so-called geometric part of Nekhoroshev theorem (see [229] for an instructive introduction and [122] for a rigorous but still pedagogical proof). On the other hand, the analytical part of the theorem is treated more easily if we avoid dealing with repeated Poisson brackets, as above, acting on the series terms of successive orders of normalization. This is done by setting from the start a number of assumptions regarding the analyticity properties of the Hamiltonian under consideration in a complexified space of actions and angles and by using various forms of the Cauchy theorem for analytical functions. This simplifies considerably the proof of the analytical part of the theorem.

Nevertheless, it seems that when one wants to find realistic estimates as regards the optimal order of truncation and the optimal value of the remainder, one has to rely on the classical methods of analysis of series convergence. The first systematic exploitation of these questions, referring to the method of Birkhoff series, was made by Servizi et al. [272], who calculated 'pseudoradii of convergence' for the Birkhoff normal form in symplectic mappings representing the Poincaré surface of section of 2D Hamiltonian systems. Kaluza and Robnik [160] found that there was no indication of divergence of the formal series below the order $n = 15$ in the Hénon–Heiles model. A particular application in the problem of stability of the Trojan asteroids [124] showed that the optimal order of truncation of the integrals in this case is beyond $n = 32$ (in some cases we find $n_{\text{opt}} > 60$, [101]). But a precise treatment of the problem was made only very recently [63, 99]. In these works scaling formulae are given yielding the optimal order of truncation as a function of the distance from the elliptic equilibrium and of the number of degrees of freedom. These formulae are derived theoretically and verified by computer algebraic calculations. A recent application in the case of galactic potentials was given by Belmonte et al. [22].

The estimate $n_{\text{opt}} \sim e/\rho^{1/(\tau+1)}$ implies that the optimal order of truncation is smaller, and the value of the optimal remainder is larger, for larger $\rho$. This behaviour is shown schematically in Fig. 11.2. On the other hand, when $\rho$ surpasses a threshold value $\rho_c$, at which $n_{\text{opt}}$ approaches the lowest possible value $n_{\text{opt}} = 3$, there is no more meaning in calculating a third integral $I$, since the series will be divergent from the start. This situation corresponds physically to the fact that for $\rho > \rho_c$, or energy $E > E_c \sim \rho_c^2$, the majority of orbits in phase space are *chaotic*. In fact, in generic Hamiltonian systems of the form (11.26), some degree of chaos exists in the phase space of motions for arbitrarily small values of the energy. When regular and chaotic orbits coexist, the system is said to have a *divided* phase space [[61], pp. 17–19]. However, for values $E < E_c \sim \rho_c^2$, the largest measure in phase space is occupied by regular orbits, lying on invariant tori, while for $E > E_c$ it is occupied by chaotic orbits. In the Hénon–Heiles system, for example, $E_c = 1/6$.

The occurrence of a divided phase space, which is a generic phenomenon, renders problematic the implementation of Jeans' theorem in realistic stellar systems because there is no uniform answer regarding the number and the form of integrals (or approximate integrals) which are preserved in different regions of the phase space. We shall come back to this question in Sect. 11.2.5.

**Triaxial Systems**

The paradigm of integrable triaxial galactic potential models are ellipsoidal Stäckel potentials [81, 95, 179, 195, 289, 290]:

$$\Phi(\lambda, \mu, \nu) = -\frac{F_1(\lambda)}{(\lambda - \mu)(\lambda - \nu)} - \frac{F_2(\mu)}{(\mu - \nu)(\mu - \lambda)} - \frac{F_3(\nu)}{(\nu - \lambda)(\nu - \mu)} \quad (11.33)$$

where $(\lambda, \mu, \nu)$ are the so-called ellipsoidal coordinates. These can be related to Cartesian coordinates $(x, y, z)$ via the three solutions for $u$ of the equation

$$\frac{x^2}{u - a^2} + \frac{y^2}{u - b^2} + \frac{z^2}{u - c^2} = 1 \quad (11.34)$$

where the constants $a^2 \geq b^2 \geq c^2$ represent the axes of concentric ellipsoids. The form of the two integrals, besides the Hamiltonian, is given, e.g. in

[60] where the main types of orbits are also analysed. A case of particular interest for galactic dynamics is the *perfect ellipsoid*. The density is given by

$$\rho = \frac{\rho_0}{(1+m^2)^2} \qquad (11.35)$$

where $m$ is the ellipsoidal radius defined as:

$$m^2 = \frac{x^2}{a^2} + \frac{y^2}{b^2} + \frac{z^2}{c^2}. \qquad (11.36)$$

The form of the integrals in that case is given, e.g. in de Zeeuw and Lynden-Bell [81]. The density function (11.35) belongs to a more general class of density functions that can serve as models of triaxial galaxies

$$\rho = \frac{\rho_0}{(1+m^2)^q}. \qquad (11.37)$$

However, a numerical study [308] indicated that for $q > 0$, only the value $q = 2$ of the perfect ellipsoid yields an integrable system, whereas other values yield systems containing stochastic orbits with positive Lyapunov exponents.

The use of an ellipsoidal radius $m$ is an easy method to 'produce' triaxial systems from known spherical systems, namely if one has a given potential or density function for the spherical system $\Phi(r), \rho(r)$, one obtains a triaxial system by replacing $r$ with $m$ in *either* the potential $\Phi$ *or* the density $\rho$. One then has to solve again Poisson's equation for the missing function. Examples of this type of models are reviewed in Merritt [[213], Sect. 1].

If the potential near the centre of a triaxial galaxy is close to harmonic, one may try to calculate approximate integrals of motion of the type of the 'third integral', namely, expanding the potential as:

$$\Phi(x,y,z) = \frac{1}{2}(\omega_x^2 x^2 + \omega_y^2 y^2 + \omega_z^2 z^2) + \sum_{k=3}^{\infty} P_k(x,y,z) \qquad (11.38)$$

where the functions $P_k$ are polynomial of degree $k$ in the Cartesian coordinates $x, y, z$, one looks for approximate integrals of the form

$$I_x - \frac{1}{2}\omega_x^2 x^2 + \ldots \qquad (11.39)$$

and similarly for $I_y$, $I_z$. Such integrals can be constructed either by the direct method or by the Birkhoff normal form. If resonances are present among the frequencies, one may look for resonant integrals that are given as linear combinations of the actions $J_x$, $J_y$ and $J_z$ with integer coefficients [22, 82, 315]. As in the two degrees of freedom case, the validity of the approximation of such integrals is determined by the theory of Nekhoroshev. A particular example was studied in Contopoulos et al. [65]. It was found that there are cases where (a) *two* integrals, (b) *no* integrals and (c) *only one* integral besides the Hamiltonian, appear to be well-preserved along the orbits' flow. Case (c) is the most interesting, because it contradicts a claim by Froeschlé and Scheidecker [111] that the number of preserved integrals besides the energy is either two or zero, that is, the orbits either lie on 3D invariant tori of the phase space or they are completely chaotic. There was a recent revival of interest in this issue after the remark [314] that the case of preservation of one more integral besides the energy (or the Jacobi constant in rotating systems) may be associated with the phenomenon of 'stable chaos' [224, 225] that is well known in Celestial Mechanics. Besides the differences in the form of local velocity ellipsoids, that were discussed above, the question of other consequences of the number and form of preserved integrals on the dynamical structure of galactic systems is still open.

**Jeans' Theorem in Systems with Divided Phase Space**

As already mentioned, the occurrence of a divided phase space, which is a generic phenomenon in stellar systems apart from the idealized spherical or Stäckel cases, renders problematic the implementation of Jeans' theorem in realistic stellar systems. This is because (a) it is not clear how to incorporate approximate integrals of the form of the 'third integral' in the arguments of the distribution function, (b) such integrals have different expressions when *resonances* are present, each resonance being characterized by its own form of resonant integral and (c) the integrals are not valid for chaotic orbits, which, however, coexist with the regular orbits within any hypersurface of the phase space defined by a constant energy condition.

As regards the form of the distribution function in the chaotic subdomain of the phase space, theorem of Arnold [12] suggests that in generic Hamiltonian systems of more than two degrees of freedom there is an a priori topological possibility for $O(1)$ excursions of chaotic orbits in phase space, even if the system differs from an integrable system by an arbitrarily small perturbation $O(\varepsilon)$. Such excursions are possible through heteroclinic chains that span the whole interconnected chaotic subset of the phase space, i.e. the *Arnold web*. Furthermore, if $\varepsilon$ is large enough, there are large chaotic domains formed by the 'resonance overlap' mechanism [50, 57, 264]. In that case, the results of numerical integrations [69] indicate that the transport of chaotic orbits is efficient enough so as to create a uniform measure throughout any connected chaotic domain. On the other hand, as $\varepsilon \to 0$, the resonance overlap mechanism almost disappears and the transport of chaotic orbits through the Arnold web occurs in a timescale characteristic of Arnold diffusion. The latter is much slower than any timescale of relevance to galactic dynamics, as exemplified in a number of studies [119, 134, 181]. The slowness of Arnold diffusion has the consequence that there may be considerable deviations of the phase space density from a uniform measure in the chaotic subdomain of the phase space. Such deviations are opposed to the validity of Jeans' theorem, i.e. that the distribution function $f(E)$ is constant within the chaotic subdomain of any hypersurface of constant energy. In that sense, the latter statement should be true only in integrable isotropic systems such as the spherical systems considered in Sect. 11.2.3.

This is precisely the claim made by Binney [30] in a paper that initiated a fruitful discussion on the interconnection between *global phase space dynamics*, on the one hand, and the form of the distribution function, on the other hand. In particular, an important line of research in galactic dynamics in the last three decades has been the detailed exploration of the various types of regular or chaotic orbits that coexist in a galaxy, as well as their relative statistical importance in creating building blocks of the self-consistent distribution function of the system. This research on *self-consistent models* of galaxies, discussed in Sect. 11.4 below, is today a very active area. Furthermore, an even more powerful line of research on the same problem has appeared in recent years: exploring the orbital content of systems resulting from *N-body simulations*. This was made possible after the use of appropriate 'smooth potential' techniques of simulation of the $N$-body problem that yield smooth solutions of the equations of motion and of the variational equations for stellar orbits. In Sect. 5, we refer to the main results of this approach which yields the closest approximations to the study of realistic stellar systems, since the equilibria reached in $N$-body simulations are by definition (a) self-consistent and (b) stable. The above methods are quite powerful and have yielded some important results towards understanding the equilibria of systems with divided phase space.

Our basic understanding today is that there are two types of orbits that play a major role in the equilibria of galaxies. These are (a) the regular orbits and (b) chaotic orbits exhibiting significant chaotic diffusion over times comparable to the Hubble time. In particular, the orbits in the chaotic subdomain are important if they can spread and produce an almost uniform measure in this domain at times comparable to the age of the system. On the other hand, weakly chaotic orbits that exhibit 'stickiness' phenomena [58, 100, 168] play a role similar to the role of regular orbits. Such differences can be quantified by the introduction of appropriate measures of the inverse Lyapunov number, i.e. the Lyapunov time of the orbits (e.g. [327]).

In the context of the above discussion, we can mention a proposal of a new form of Jeans' theorem by Merritt [213, 217], that is applicable to systems with a divided phase space: 'The phase-space density of a stationary stellar system must be constant within every well-connected region'. The definition of 'well-connected' is '... one that cannot be decomposed into two finite regions such that all trajectories lie on either one or the other (what the mathematicians call "metric transitivity")' [213].

In the idealized case of a phase fluid set from the start to satisfy the above condition, the above version of Jeans' theorem corresponds essentially to the preservation of the phase space density under the system's Hamiltonian flow. In practice, however, a definition of 'well-connected' region such as the above one, i.e. based only on topological arguments, may not be so convenient in describing galactic equilibria. We can give the following qualitative argument: suppose the 6D phase space of a galactic system is represented schematically as the $(Q, P)$ space of Fig. 11.3. Suppose also that the system's Hamiltonian $H$ differs from an integrable Hamiltonian, with exact integrals $I_1, I_2, I_3$, by an arbitrarily small perturbation of order $O(\varepsilon)$. According to Nekhoroshev theorem, if $\varepsilon$ is below a threshold, there are approximate integrals $\tilde{I}_1, \tilde{I}_2, \tilde{I}_3$ that have variations of order $O(\exp(-1/\varepsilon))$ over timescales of order $O(\exp(1/\varepsilon))$, i.e. much longer than the age of the galaxy. Thus, for all practical purposes, we may describe the system by a distribution function depending on these approximate integrals $f(\tilde{I}_1, \tilde{I}_2, \tilde{I}_3)$. Consider now two different regions of $(Q, P)$, region A and region B (Fig. 11.3), with an $O(1)$ separation in phase space (in units normalized to the overall extent of the phase space in $P$ and $Q$). As a consequence, the values of $\tilde{I}_i$, which are functions of the variables $(Q, P)$, will in general also have an $O(1)$ difference in the two regions, that is $|\tilde{I}_i(A) - \tilde{I}_i(B)| = O(1)$. Since these integrals are arguments of the distribution function, it follows that the same order of the difference will also appear in $f$, that is

$$|f(A) - f(B)| = O(1) . \tag{11.40}$$

Given, now, that the Nekhoroshev theorem for approximate integrals is valid in open domains of the phase space, it follows that (11.40) is valid for $f(A), f(B)$ standing for the value of the distribution function at *any* pair of points inside the regions A and B, respectively, provided that the approximate integrals $\tilde{I}_i$ are well preserved in both regions. On the other hand, according to the KAM theorem [11, 173, 230], there is a chaotic subset of measure $O(\varepsilon)$

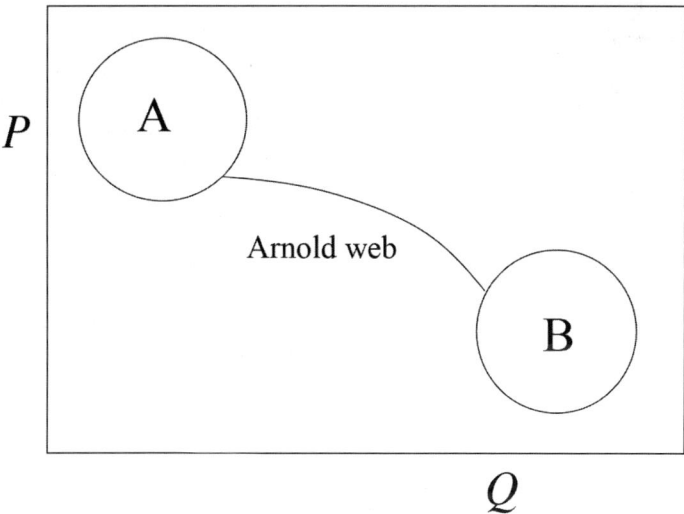

**Fig. 11.3.** A schematic representation of the phase space. Regions A and B communicate via the Arnold web

in region A, which is the complement of the invariant tori of A, and a similar subset in region B. Suppose that the two subsets communicate via the Arnold web. Then, according to the previous definitions, the two subsets belong to one 'well-connected' chaotic region and we should have $f(A) - f(B) = 0$ for any pair of points in A and B belonging to this region. Thus we see that if we use the approximate integrals $\tilde{I}_i$ as arguments in the distribution function we reach a different conclusion (11.40) than if we use the concept of well-connectedness. This is because the integrals $\tilde{I}_i$ are not exact, but they are preserved for times of the order of the Nekhoroshev time $t_{\text{Nek}}$. Thus the equalization of $f(A)$ and $f(B)$ in the chaotic subset can happen only after a time $t > t_{\text{Nek}}$, which is much larger than the age of the system.

The above example shows that a more pragmatic definition of what 'well-connected' means is required in the case of galaxies, in order to take into account the fact that the topological well-connectedness may not have always practical dynamical implications for the equilibria of galaxies. This is because the lifetime of galaxies is much smaller than the typical Nekhoroshev time.

A numerical example of the form of the distribution function in systems with divided phase space was given by analysing the orbits and approximate integrals in the phase space of systems produced by $N$-body simulations [64, 70, 97, 102]. Figure 11.4 [64] shows one example of a nearly prolate system. This system resulted from a collapse simulation with cosmological initial conditions [102]. The self-consistent gravitational potential is calculated by the self-consistent field code of Allen et al. [5]. If we ignore triaxial terms, the potential can be expanded in a polynomial series in the $(R, z)$ variables, namely:

$$\Phi(R, z) = \sum_{k=0}^{8} \sum_{l=0}^{8} g_{kl} R^{2k} z^{2l} \qquad (11.41)$$

where $z$ is the long axis of the system and the coefficients $g_{kl}$ are specified numerically, via the code potential. The form of the potential (11.41) is such that a third integral can be calculated in the form of series. We calculate

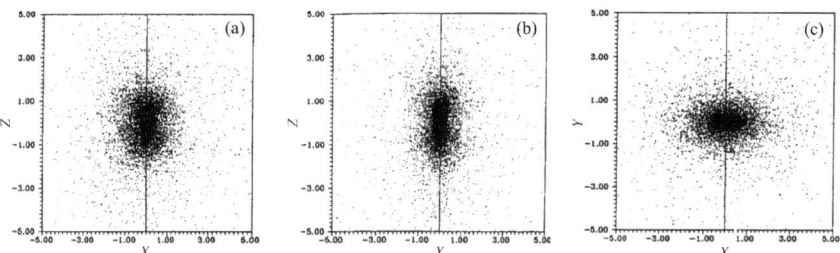

**Fig. 11.4.** Projection of the final state of an $N$-body collapse experiment in the three planes (**a**) $X$–$Z$, (**b**) $Y$–$Z$, (**c**) $X$–$Y$ of ordinary space. The system is nearly prolate, with one long axis ($Z$) and two short axes ($X,Y$) (after [64])

a different integral for box orbits (non-resonant integral) or for higher-order resonant orbits (e.g. 1:1 resonance for tube orbits).

The question, now, is whether such integrals should appear as arguments in the distribution function of the system. The answer is affirmative, as indicated by Fig. 11.5. Figure 11.5a shows a Poincaré surface of section $(R, \dot{R})$ for an energy $E = -1.6 \times 10^6$ (in the $N$-body units) which is close to the central value of the potential well $(-2 \times 10^6)$, and angular momentum $L_z$ close to zero. We then integrate the orbits of the *real* particles of the $N$-body system with energies in a bin centred at the above value of $E$, until each orbit intersects the Poincaré section for the first time. By this numerical process, a particle

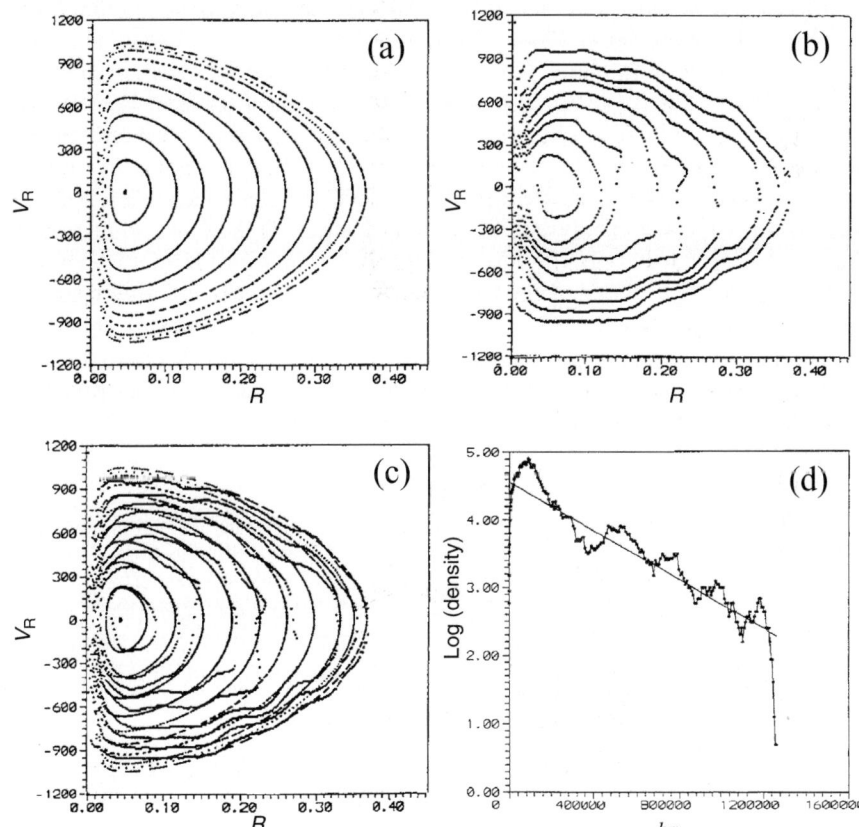

**Fig. 11.5.** (a) The Poincaré surface of section of the system of Fig. 11.4 for energy $E = -1.6 \times 10^6$ (in the $N$-body units) and $L_z$ very close to zero, $L_z = 0.045$. (b) The equidensity contours of the distribution of the real particles in the Poincaré section, for energies within a bin $\Delta E = 2 \times 10^4$ around the value $E$ of (a) and angular momentum $|L_z| \leq 0.09$. (c) The plots (a) and (b) together. (d) Exponential dependence of the distribution function on the value of the third integral along the invariant curves of (a) (after [64])

located on an invariant torus of the system, that corresponds to a particular value of the third integral $I$, is transferred to a point on an *invariant curve* of the section $(R, \dot{R})$ where the section is intersected by the torus. This also means that if the phase space density (distribution function $f$) depends on $I$, the *surface density* of points in the section $(R, \dot{R})$ will also be stratified in such a way that the *equidensities* should coincide with the invariant curves corresponding to different label values of $I$. Precisely, this is what we see in Fig. 11.5b. Namely, the equidensity contours of the distribution of the real particles in the surface of section have a good coincidence with the invariant curves (shown together in Fig. 11.5c). This is a numerical indication that the integral $I$ should, indeed, be included as an argument in $f$ [70]. We have calculated numerically the dependence of the surface density $f_S$ on $I$ and found it to be exponential (Fig. 11.5d).

For larger energies, the divided nature of the phase space is clearly manifested (Fig. 11.6a). In particular, besides the region of invariant curves corresponding to box orbits (A), we distinguish a second island around the 1:1 resonance (B) as well as a connected chaotic domain (C) separating the two regular domains. Some finer details, e.g. secondary resonances (D) are distinguished but they are not dynamically so important. If, now, we compare this figure to the equidensity plot of the distribution of particles (Fig. 11.6b,c), the tendency to have a distribution stratified according to the underlying phase space structure is again visible to a large extent. This indicates that both the non-resonant third integral, yielding the tori of region (A), and the resonant integral yielding the tori of (B), should appear locally as arguments of the distribution function $f$ (the dependence of $f$ on $I$ in region A is again exponential, Fig. 11.6d).

For still larger energies, the chaotic domain occupies a large volume of the phase space (Fig. 11.7a). We find that the phase-space density of the real particles in the connected chaotic domain is nearly constant. Figure 11.7b shows the density on the Poincaré section along a constant line $R = 0.65$, as a function of $\dot{R}$. The variations shown in Fig. 11.7b are *not* completely due to the sampling noise, but, in general, they are small enough so as to allow us to characterize the density as nearly constant in the connected chaotic domain (C). In this case Merritt's version of Jeans' theorem is applicable.

The nature of the above questions prevents one from making clearcut statements as per what phenomena introduced by regular or chaotic orbits should be considered as dynamically important. Let us notice, however, that the galaxies are quite complex systems and such questions have not yet been fully explored even in simple toy models of basic research in Hamiltonian dynamical systems. We mention one example which is of particular importance in the study of global dynamics of galaxies: the distinction between Arnold diffusion [12] and resonance overlap diffusion [50, 57, 264] and the role of these two types of diffusion in galaxies. The difference between the two types of diffusion is topological, but it is also a difference in the diffusion rate. As regards the rate of Arnold diffusion, there is a general belief that this should

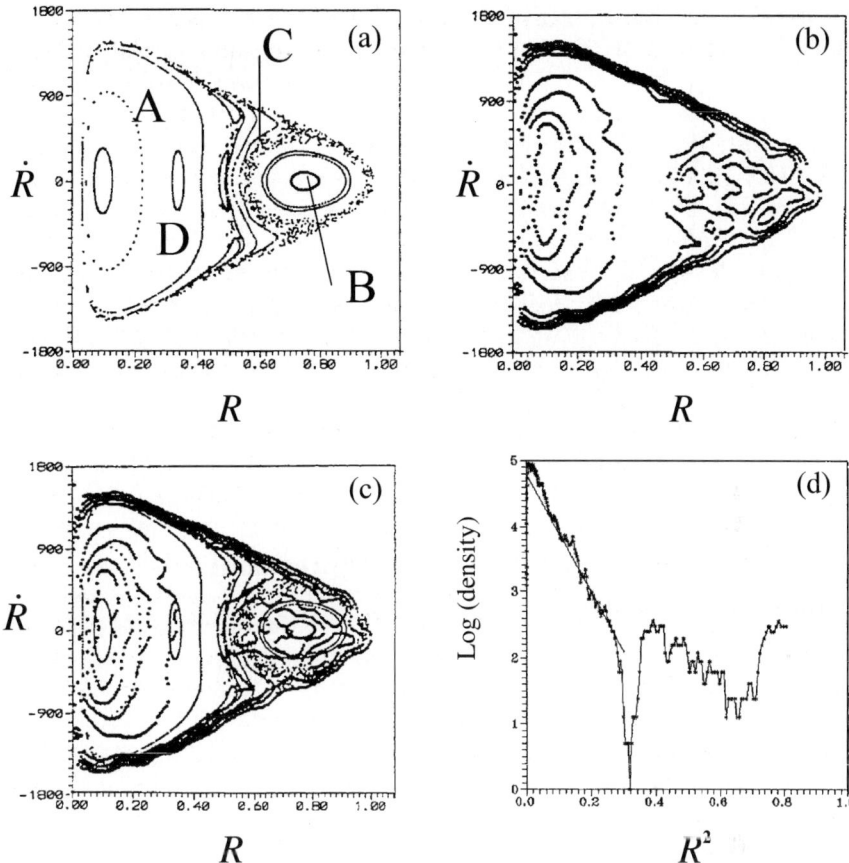

**Fig. 11.6.** As in Fig. 11.5, for a different value of the energy ($E = -9 \times 10^5$). The region ($A$) corresponds to box orbits, ($B$) to loop orbits (1:1 resonance), ($C$) chaotic orbits and ($D$) a secondary resonance inside the domain of box orbits. In ($d$) we use the approximation $I \propto R^2$ (after [64])

be connected to Nekhoroshev theorem and that, hence, it is very slow to be of any importance in galaxies. This was partly verified recently by the interesting numerical experiments of Froeschlé and his collaborators [110, 133, 134, 188]. More work is requested in order that such simulations help us clarify questions such as what is a pragmatic definition of 'well-connected' domains of phase space and how to implement such ideas in galactic dynamics.

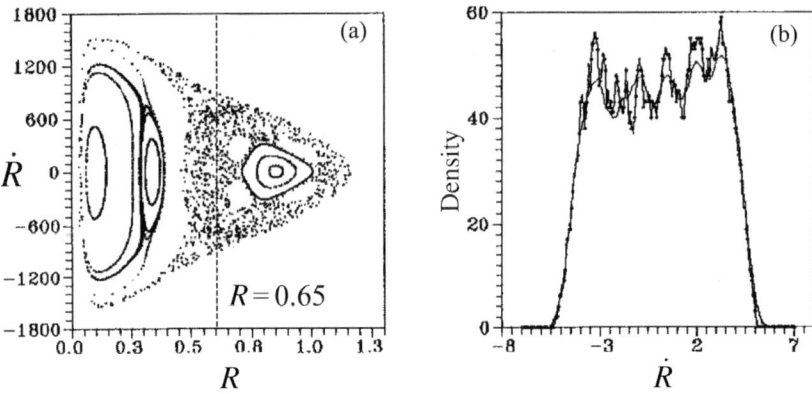

**Fig. 11.7.** (a) As in Fig. 5a for energy $E = -8 \times 10^5$. (b) the density of real particles of the N-Body system in a slice around the line $R = 0.65$ is almost constant (after [64])

## 11.3 The Statistical Mechanical Approach: Violent Relaxation

### 11.3.1 Observational Evidence of the Equilibrium State Assumption

The smoothness of observed photometric profiles suggests that at least the spheroidal components of galaxies are in a form of statistical equilibrium. The surface brightness profiles of many elliptical galaxies are well-fitted by the de Vaucouleurs' [78] $R^{1/4}$ law (Fig. 11.8a):

$$I(R) = I_e \exp\left(-7.67[(R/R_e)^{1/4} - 1]\right) \tag{11.42}$$

where $I_e$ is the value of the surface brightness (in mag/arcsec$^2$) at the radius $R_e$ of a disk in the plane of projection containing half of the total light. In a number of galaxies this relation has been verified in a range up to ten magnitudes [79]. The profiles of bulges and of some ellipticals follow a similar law, namely the Sersic $R^{1/n}$ law [270, 271]. On the other hand, the central profiles of elliptical galaxies were reliably observed by the Hubble space telescope. It was found that the profiles have central cusps, i.e. the surface brightness grows as a power-law in the centre $I(R) \propto R^{-\gamma}$, $\gamma > 0$ [71, 106, 186]. There are two groups of observed central profiles (Fig. 11.8b), namely (a) shallow profiles ($\gamma \leq 0.2$) and (b) abrupt profiles ($\gamma \sim 1$). However, as emphasized by Merritt [212], even shallow profiles in the surface brightness correspond to power-law cusps in the 3D density profile $\rho(r) \propto r^{-a}$ with power exponents $a > 1$. This means that at least the centres of galaxies deviate considerably from simple isothermal models with a Boltzmann-type distribution function such as the King models [172].

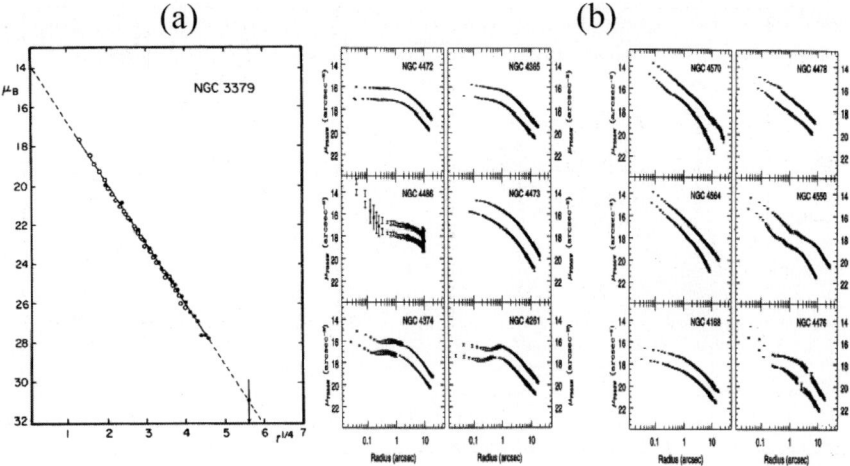

**Fig. 11.8.** (a) Fitting of the elliptical galaxy NGC3379 by de Vaucouleurs' law (after [79]). (b) Shallow (**left**) and cuspy (**right**) profiles of the cores of elliptical galaxies (after [106])

$$f(\mathbf{x}, \mathbf{p}) = A \exp\left[-\beta \left(E(\mathbf{x}, \mathbf{p}) - E_0\right)\right] = A \exp\left[-\beta \left(\frac{\mathbf{p}^2}{2} + \Phi(\mathbf{x}) - E_0\right)\right] \tag{11.43}$$

with $A, \beta, E_0$ constants, or their non-isotropic generalizations [223]. The latter models are characterized by flat density profiles at the centre [[33], p. 234]. Thus, the nature of statistical equilibrium of galaxies should be quite different to the isothermal equilibrium. Furthermore, since the time of two-body relaxation is much larger than the age of the Universe, galaxies had no time to approach such an equilibrium. The very fact that galaxies are statistically relaxed systems seems, at first, to be a paradox (the so-called 'Zwicky's paradox').

### 11.3.2 The Theory of Violent Relaxation

A way out of the paradox developed gradually in the 1960s, after a systematic study of the hypothesis that in the early phase of galaxy formation, galaxies were subject to a sort of 'violent relaxation' [196] caused by the collapse and ultimate merger of clumps of matter produced by the non-linear evolution of initially small density inhomogeneities in the early Universe. We can mention in this context an influential paper by Eggen et al. [103] under the characteristic title 'Evidence from the Motions of Old Stars that the Galaxy Collapsed', as well as one of the first numerical simulations of a spherical gravitational system in the computer by Hénon [137].

The theoretical foundations of the statistical mechanics of violent relaxation were set by Lynden-Bell [196], using a continuum approach for the distribution function, and re-derived by Shu [274] with a particle approach to the same distribution. These analytical studies are now considered classical, despite the fact that the so-derived equilibrium distribution functions are

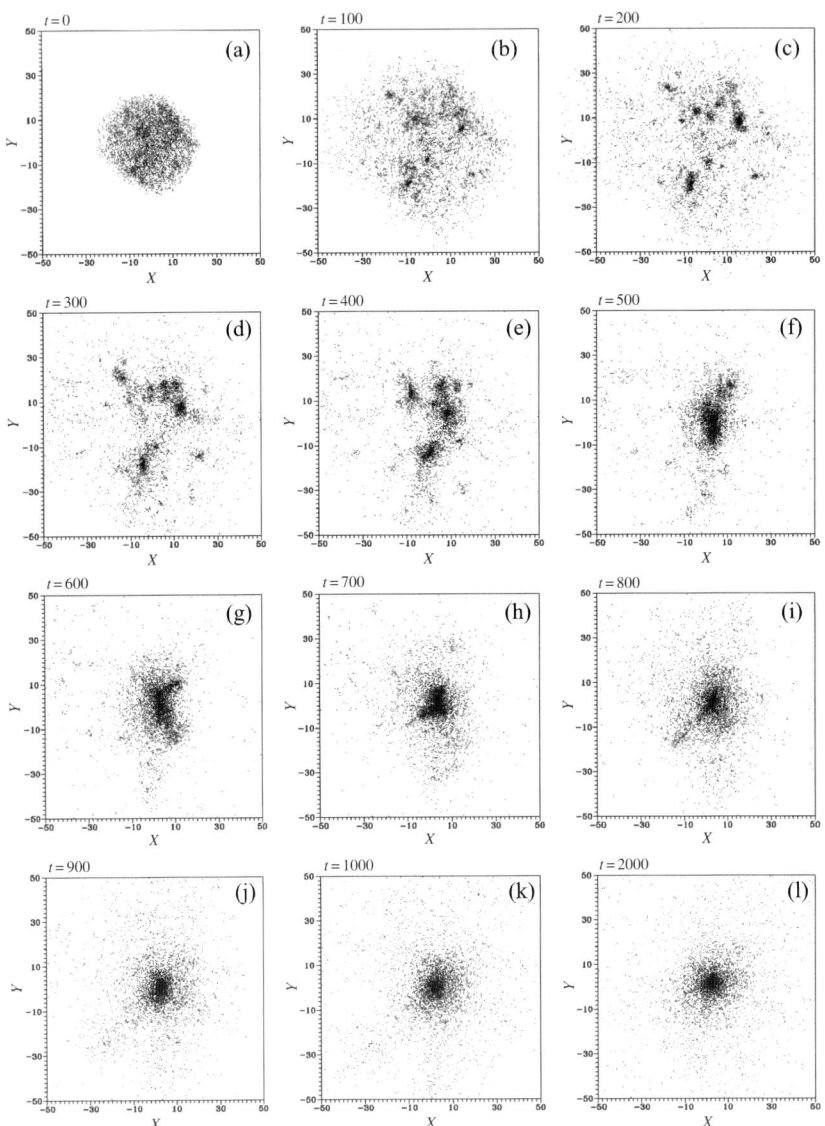

**Fig. 11.9.** $N$-body simulation of the collapse and violent relaxation of a nearly spherical mass (protogalaxy). The initial conditions correspond to a hierarchical clustering scenario (clumpy initial conditions) in a $\Lambda$-CDM expanding Universe (after [102])

far from able to account for the properties of systems produced by realistic
$N$-body simulations or for the data of observed galaxies in the sky.

An example of such a collapse process is shown in Fig. 11.9 [102]. This is an
$N$-body simulation of an isolated system containing one galactic mass represented with 5616 particles. The mass is initially contained in a nearly spherical
subvolume of the Universe. The particles are assigned positions and velocities
in agreement with the general Hubble expansion of the Universe in a $\Lambda$-CDM
scenario, but the position and velocity vectors of each particle are perturbed
according to a prescription for the spectrum of density perturbations in the
Universe at the moment of decoupling, and following a well-known technique
of translating these perturbations into perturbations of positions and velocities introduced by Zel'dovich [341]. As seen in Fig. 11.9, the system initially
expands following the general expansion of the Universe (Fig. 11.9a,b), but the
extra gravity due to local overdensities results in a gradual detachment of the
system from the average Hubble flow, so that the system reaches a maximum
expansion radius (Fig. 11.9c) and then begins to collapse. At the initial phase
of collapse, small subclumps are formed within the spherical volume which collapse to local centres forming larger bound objects (Fig. 11.9d,e). However,
these clumps also collapse towards a common centre of gravity (Fig. 11.9f),
until the overall system relaxes, after a phase of rebound, to a final equilibrium
(Fig. 11.9g–l).

There is a wide variety of initial conditions that lead to the above type of
relaxation process. For example, a currently popular scenario of formation of
the elliptical galaxies via the merger of spiral galaxies [16, 17, 40, 116, 142, 232,
235, 301] corresponds to a case where, instead of many clumps, as in Fig. 11.9,
we have only two major clumps corresponding to the dark halos of the spiral
galaxies. In that case the presence of gas dynamical processes must be taken
into account. Nevertheless, the main process driving the system towards a final
equilibrium state is again a violent relaxation process, although the initial
conditions and the detailed time evolution of the system are different than in
the case of a simple collapse or a multiple merger event.

The statistical mechanical theory of violent relaxation aims, precisely, at
justifying theoretically the tendency of such systems to settle down to an equilibrium, and to find the form of the distribution function $f$ at this equilibrium.

A simplified version of the main steps in the derivation of Lynden-Bell's
statistics is the following:

1. We consider a compact $\mu$-space (i.e. we consider that escapes are negligible), and implement a coarse-graining process by dividing the $\mu$-space in
   a number of, say, $N$ *macrocells* of equal volume (Fig. 11.10) labelled by
   an index $i = 1, 2, \ldots, N$. We further divide each macrocell into a number of *microcells* that may or may not be occupied by *elements of the
   Liouville phase flow* of the stars moving in $\mu$-space. In Fig. 11.10 these
   phase elements are shown by dark squares that occupy some microcells
   within each macrocell.

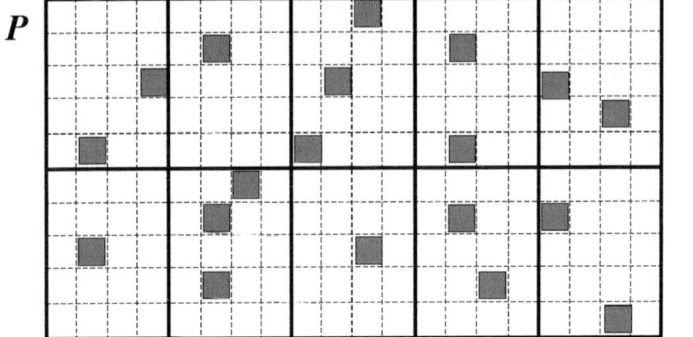

**Fig. 11.10.** A simple schematic representation of the phase space $(Q,P)$ of a violently relaxing system. The *grey squares* represent elements of the phase fluid. The phase space is partitioned into microcells (*fine grid of dashed lines*) and macrocells (*coarse grid of bold lines*)

2. We adopt the *equal a priori probability* assumption, namely we assume that each element of phase flow has equal a priori probability to be found in any of the macrocells of Fig. 11.10. As the system evolves in time, each phase element travels in phase space by respecting this assumption. We should note that, because of the phase mixing, the form of the phase elements also changes in time. However, this deformation does not change the volume of an element. We can thus proceed in counting the number of phase elements in each macrocell by keeping the simple schematic picture of Fig. 11.10.

3. We denote by $n_i$ the occupation number of the $i$th macrocell, i.e. the number of fluid elements inside this macrocell at any fixed time $t$. The set of numbers $(n_1, n_2, \ldots, n_N)$, called a *macrostate*, can thus be viewed as a discretized realization of the coarse-grained distribution function of the system at the time $t$.

4. For any given macrostate, the mutual exchange of any two phase elements or the shift of an element in a different cell within the same macrocell leaves the macrostate unaltered. Thus, we can calculate the number $\Omega(n_1, n_2, \ldots, n_N)$ of all possible microscopic configurations that correspond to a given macrostate, and define a Boltzmann entropy $S = \ln \Omega$ for this particular macrostate. If we denote by $n = \sum_{i=1}^{N} n_i$ the total number of phase elements and by $\nu$ the (constant) number of microcells within each macrocell, the combinatorial calculation of $\Omega$ readily yields:

$$\Omega(n_1, n_2, \ldots, n_N) = \frac{n!}{n_1! n_2! \ldots n_N!} \prod_{i=1}^{N} \frac{\nu!}{(\nu - n_i)!} \qquad (11.44)$$

5. We finally seek to determine a statistical equilibrium state as the most probable macrostate, i.e. the one maximizing $S$ under the constraints

imposed by all preserved quantities of the phase flow. Besides mass conservation $n = \sum_{i=1,N} n_i$, we can assume conservation of the total energy of the system $E = \sum_{i=1}^{N} n_i \varepsilon_i$ (where $\varepsilon_i$ is the average energy of particles in the macrocell $i$), and perhaps of other quantities such as the total angular momentum (if spherical symmetry is preserved during the collapse) or any other 'third integral' of motion. In the simplest case of mass and energy conservation, we maximize $S$ by including the mass and energy constraints as Lagrange multipliers $\lambda_1$, $\lambda_2$ in the maximization process, namely:

$$\delta \ln \Omega - \lambda_1 \delta n - \lambda_2 \delta E = 0 . \tag{11.45}$$

We furthermore apply Stirling's formula for large numbers $\ln N! \approx N \ln N - N$. In view of (11.44), (11.45) then yields

$$F_i = \frac{\eta n_i}{\nu}\Big|_{S=\max} = \frac{\eta}{\exp(\lambda_1 + \lambda_2 \varepsilon_i) + 1} \tag{11.46}$$

where $\eta$ is the (constant) value of the phase space density inside each moving phase space element. Equation (11.46) is Lynden-Bell's formula for the value $F_i$ of the coarse-grained distribution function within the $i$th macrocell at statistical equilibrium. Following the conventions of thermodynamics, we interpret $\lambda_2$ as an inverse temperature constant $\lambda_2 \equiv \beta \propto 1/T$ and $\lambda_1$ in terms of an effective 'chemical potential' $\varepsilon_0 = -\lambda_1/\beta$ (or 'Fermi energy'). We thus rewrite (11.46) in a familiar form reminiscent of Fermi–Dirac statistics

$$F_i = \frac{\eta}{\exp[\beta(\varepsilon_i - \varepsilon_0)] + 1} = \frac{\eta \exp[-\beta(\varepsilon_i - \varepsilon_0)]}{1 + \exp[-\beta(\varepsilon_i - \varepsilon_0)]} \tag{11.47}$$

by recalling, however, that the energy and effective chemical potential in (11.47) have in fact dimensions of energy per unit mass, in accordance to our general treatment of orbits in $\mu$-space (Sect. 11.2.2). Therefore, contrary to the two-body relaxation, the process of violent relaxation cannot lead to mass segregation at the equilibrium state. At any rate, in the so-called non-degenerate limit $F_i << \eta$, (11.47) tends to the form of a Boltzmann distribution $F_i \simeq A \exp(-\beta \varepsilon_i)$, that is, the final state approaches the isothermal model.

The above exposition of Lynden-Bell's theory is simplified in many aspects. In particular: (a) the expression given for the constraint of the total energy is not precise. One should calculate the energy self-consistently by the gravitational interaction of the masses contained in each phase element. However, the final result turns out to be the same with this more precise calculation. (b) All phase elements in the above derivation are assumed to have the same value of the phase space density $\eta$, i.e. the same 'darkness' in Fig. 11.10.

A more general distribution function was derived by Lynden-Bell when the phase elements of Fig. 11.10 can be grouped into $K$ groups of distinct darkness $\eta_J$, $J = 1, \ldots K$. The final formula, derived also by the standard combinatorial calculation, reads:

$$F_i = \sum_{J=1}^{K} \frac{\eta_J \exp\left(-\beta_J(\varepsilon_i - \varepsilon_{0J})\right)}{1 + \sum_{J=1}^{K} \exp\left(-\beta_J(\varepsilon_i - \varepsilon_{0J})\right)} \quad (11.48)$$

that is, it depends on a set of $K$ pairs of Lagrange multipliers $\beta_J, \varepsilon_{0J}$, $J = 1, \ldots, K$. This more realistic formula links the initial conditions of formation of a system, parameterized by the values of $\eta_J$ which are conserved during the relaxation, to the final distribution function. In the non-degenerate limit, the latter is a superposition of nearly Boltzmann distributions, meaning that each group of phase elements is characterized by its own Maxwellian distribution of velocities which yields a different velocity dispersion in each group, depending on the value of $\eta_J$. This poses a problem as regards the possibility to express the overall distribution of velocities in the galaxy by a single Maxwellian function (see for example [274] and the debate [200, 275]). We return to this question in Sect. 11.3.5 where we discuss alternative formulations of the statistical mechanics of violently relaxing systems.

### 11.3.3 Incomplete Relaxation

The basic prediction of Lynden-Bell's theory, namely the possibility for a stellar system to settle down to a statistical equilibrium within a time comparable to a few system's mean dynamical times, has been completely verified in a series of numerical simulations over subsequent years [1, 3, 39, 41, 43, 76, 92, 102, 128, 129, 140, 141, 146, 169, 191, 206, 208, 210, 227, 305, 310, 311, 318, 320, 324, 334, 335]. However, the data of these experiments, as well as other considerations converge to the conclusion that Lynden-Bell's formula (11.47) is not applicable even in the simplest cases of realistic galactic systems [3, 75, 102, 126, 132, 145, 187, 200, 206, 273, 296, 320, 334].

There are many phenomena which obstruct the convergence of $F$ towards Lynden-Bell's prediction. We refer below to one of the most important factors considered in the literature: incomplete relaxation. This is a phenomenon that may happen even in the simplest case of systems relaxing via a monolithic collapse. The term 'incomplete' means that the process of mixing of phase elements in $\mu$-space, during the relaxation process, is not efficient enough so as to justify the assignment of equal a priori probability on a phase element to be in any of the macrocells of $\mu$. This also implies that some memory of initial conditions survives in the final equilibrium state. This phenomenon is commonly verified by N-body experiments [102, 206, 293, 305, 324, 325]. An example is shown in Fig. 11.11 [324] which shows a plot of the final versus initial energies (Fig. 11.11a) or angular momenta (Fig. 11.11b) for each particle in a N-body collapse experiment. The correlation between initial and final values of the angular momentum is obvious from the concentration of points near the diagonal.

We may quantify this correlation by calculating, in N-body collapse experiments, the time-dependence of the correlation coefficient defined by

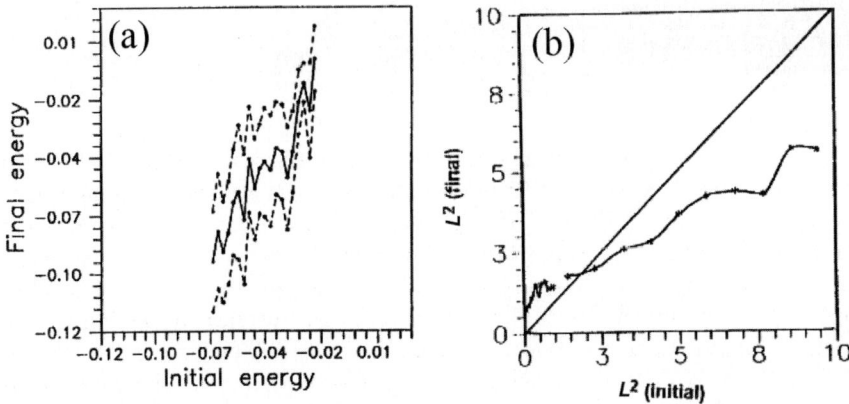

**Fig. 11.11.** (a) Initial versus final energies (*solid line* = mean, *dashed lines* = lower and upper limits) of the particles before and after the collapse. (b) Initial versus final angular momenta for the same particles (after [324])

$$\mathrm{CR}(t) = \frac{\sum_{i=1}^{N}(E_{0i} - \bar{E}_0)(E_{ti} - \bar{E}_t)}{\sqrt{\sum_{i=1}^{N}(E_{0i} - \bar{E}_0)^2 \sum_{i=1}^{N}(E_{ti} - \bar{E}_t)^2}} \quad (11.49)$$

where $E_{0i}$, $i = 1, \ldots, N$ are the energies of the $N$ particles at the initial snapshot of the experiment, $E_{ti}$ the energies of the same particles (each labelled by $i$) at a time $t$, and $\bar{E}_0$, $\bar{E}_t$ the mean energies, respectively (Fig. 11.12). In this and in subsequent plots we refer to a series of collapse experiments corresponding to the time evolution of the matter distributed in a spherical volume in the Universe containing one galactic mass, in which, at the moment of decoupling, we impose a field of density perturbations consistent with a standard $\Lambda$-CDM cosmological scenario. We furthermore distinguish between (a) experiments with a spherically symmetric field of initial density perturbations and (b) clumpy initial density perturbations (S and C experiments, see [102] for a detailed description of the initial conditions of the experiments). Finally, we examine various exponents $n$ of the power spectrum of density perturbations, that is, the r.m.s. dependence of a density perturbation on scale $r$ is given by

$$\frac{\delta\rho(r)}{\rho} \propto \frac{1}{r^{n+3/2}} \quad (11.50)$$

according to standard cosmological considerations [321]. In the case of S-experiments, (11.50) is viewed as the radial profile of a spherically symmetric density perturbation, while in the C-experiments the perturbation field inside the spherical volume is determined by a superposition of plane waves with power spectrum $P(k) \propto k^n$ and random phases. The resulting perturbation field is translated to perturbation of the particles' positions and velocities with respect to an ideal Hubble flow by means of Zel'dovich approximation [341].

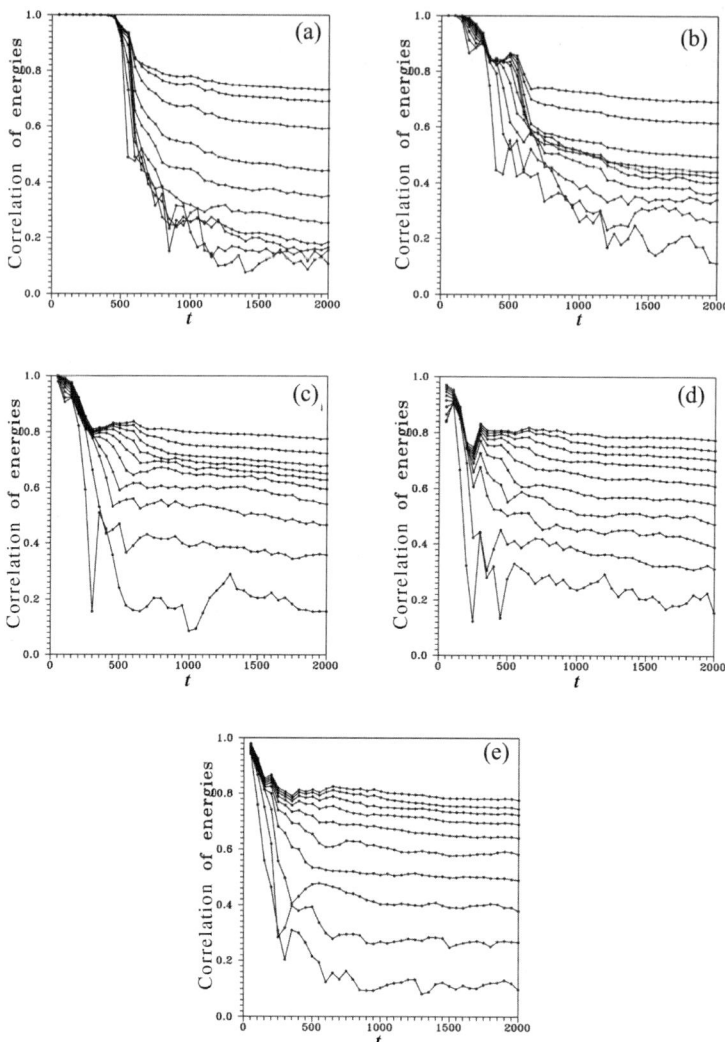

**Fig. 11.12.** The time evolution of the initial–final energy correlation coefficient (11.49) for five collapse experiments differentiated by the value of the power exponent $n$ of initial density perturbations, namely (**a**) $n = -2.9$, (**b**) $n = -2$, (**c**) $n = -1$, (**d**) $n = 0$, (**e**) $n = 1$. The curves in each panel from down to the top correspond to the values of the correlation coefficient for the innermost 10%, 20%,...90% of the bound matter

We choose different values of the exponent $n$ in the range $-3 < n \leq 1$, consistent with the hierarchical clustering scenario.

The value of $n$ is a parameter that regulates the violence of the collapse phase by affecting the distribution of power between perturbations of small and large scales. This can be seen by the following analysis, due to Palmer and Voglis [243]: if the r.m.s. profile of mass perturbation in a structure of scale $h$ at the moment of cosmological decoupling is, according to (11.50) taken to be $\mu^2(h) = \mu_0 h^{-(n+3)}$, then the total mass contained in the interior of a sphere of radius $h$, given by $M(h) = \frac{4\pi}{3}\rho_0 h^3[1+\mu(h)]$, where $\rho_0$ is the average density of the Universe at decoupling, will cause a gravitational attraction of the spherical shell at radius $h$ so that the expansion of the shell will gradually detach from the average Hubble expansion of the Universe. If $r(t)$ denotes the radius of the shell at the moment $t$, the solutions of the equations of motion in a $\Omega = 1$ Universe can be given parametrically in the form of cycloid motion

$$r = h\frac{1+\mu(h)}{2\mu(h)}(1-\cos u), \quad t = t_0 \frac{3[1+\mu(h)]}{4\mu(h)^{3/2}}(u - \sin u)$$

where $t_0$ is the time of decoupling, $r(t_0) = h$, and we use units in which $G = 1$ and the Hubble constant at decoupling is $H_0 = \sqrt{2}$. From these equations we find that a shell of radius $h$ will reach its maximum expansion at $t_{\max} \approx 3t_0\pi/4\mu(h)^{3/2}$, and from there on the shell will begin to collapse, the collapse time being almost equal to the expansion time. We may now use the form of the profile $\mu(h) \propto h^{-(n+3)/2}$ and find that the collapse time for a spherical shell including in its interior spherical volume a percentage $\frac{\Delta M}{M}$ of the total mass of the system is given by

$$t_{\text{collapse}} \propto \left(\frac{\Delta M}{M}\right)^{(n+3)/4}. \tag{11.51}$$

This power-law is well verified in $N$-body experiments. In Fig. 11.13 we show the evolution of the radii $r(t)$ of spherical shells containing a percentage $10\%, 20\%, \ldots, 90\%$ of the total mass of the collapsing object for different values of $n$, namely (a) $n = -2.9$, (b) $n = -2$, (c) $n = -1$, (d) $n = 0$ and (e) $n = 1$. It is immediately seen that in the limit $n \to -3$ (Fig. 11.13a), meaning a homogeneous profile of the initial density perturbation (11.50), all shells collapse at about the same time. This is the well-known spherical 'top-hat' model. On the other hand, as $n$ increases, the collapse becomes more synchronous, and in the other limit as $n \to 1$ (Fig. 11.13e) the outer shells collapse at a time which is an order of magnitude larger than the collapse time of the inner shells.

Figure 11.14 shows the value of the correlation coefficient (11.49) of the particles' energies at the initial and final snapshot of the experiment, as a function of the exponent $n$. There are nine curves in this diagram, corresponding to the value of the correlation coefficient for the innermost $10\%, 20\%, \ldots, 90\%$ of the matter. We see that, independent of the value of $n$, the innermost $20\%$ of the matter yields low correlation coefficients ($0.2$–$0.3$), meaning that we can speak about almost complete relaxation. In the case of the $n = -2.9$ experiment, this percentage rises to $40\%$. However, for the rest of matter the correlation coefficient has values that can be as high as $0.7$. This means that

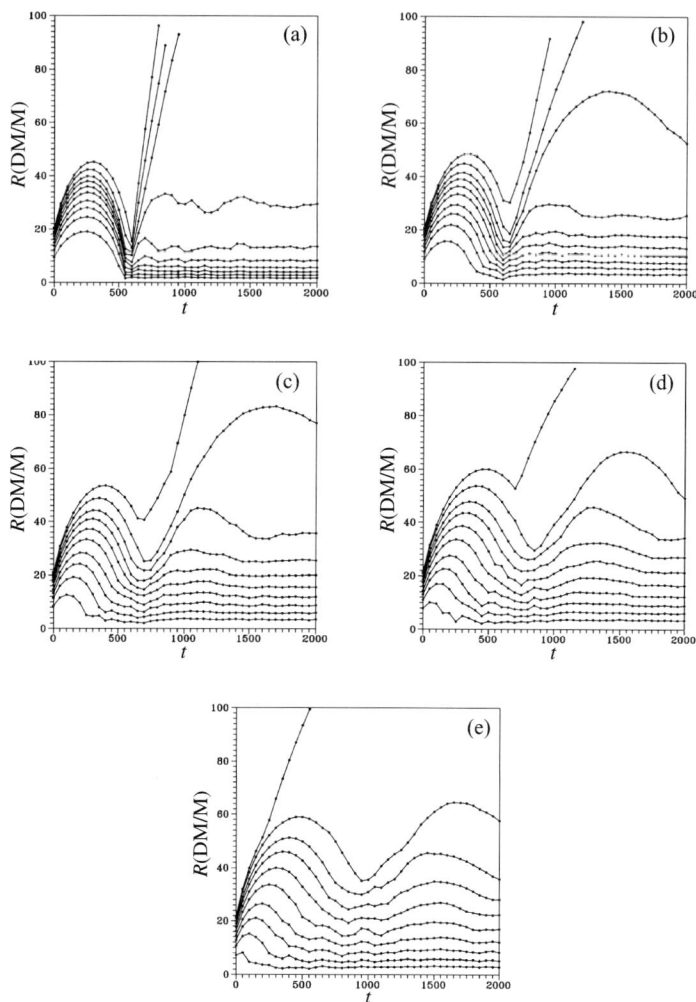

**Fig. 11.13.** The time evolution of the radii of spherical shells containing 10%, 20%,...90% of the matter in the same experiments as in Fig. 11.12

the mixing of energies is incomplete. Such high values of the correlation coefficient are observed in all the experiments, including the limit of the 'top-hat' model ($n = -2.9$).

This fact is remarkable and requires some further explanation. This is related to a problem regarding the very nature of violent relaxation that was posed by Miller (private correspondence with Lynden-Bell, see [214]). In the original approach of Lynden-Bell, the energies of stars are subject to stochastic changes caused by the time fluctuations of the self-gravitational potential of the system, since the rate of energy change of each star is given by

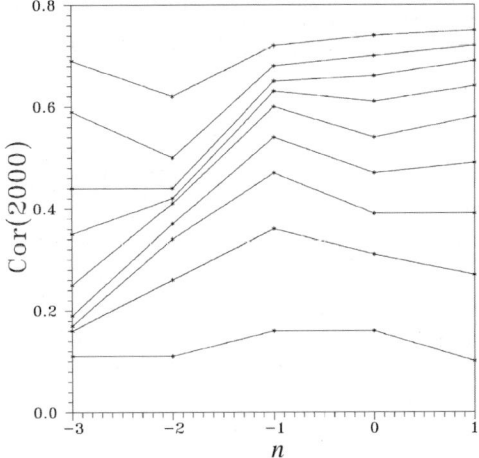

**Fig. 11.14.** The final value of the correlation coefficient (at $t = 2000$) for the 10%, 20%,...90% of matter (*from down to top*) as a function of the exponent $n$ of the initial density perturbations. The correlation coefficient has relatively high values for an important fraction of the matter of all the systems, indicating that the relaxation is incomplete

$$\frac{\mathrm{d}E}{\mathrm{d}t} = \frac{\partial \Phi}{\partial t} . \qquad (11.52)$$

The rate of relaxation is thus linked to the mean timescale of the time-dependent variations in the r.h.s. of (11.52), that is $T_{\mathrm{rel}} \sim < (\Phi/\dot\Phi)^2 >^{1/2}$. Lynden-Bell established that this timescale is of the order of the mean dynamical period of the system, hence the term 'violent' relaxation. Nevertheless, Miller notices that if we have an isolated galaxy and a mass $m$ which is uniformly distributed in a spherical shell surrounding the galaxy, then, if we let the mass $m$ vary in time $m \equiv m(t)$, the total gravitational potential $\Phi = \Phi_{\mathrm{galaxy}} + \Phi_{\mathrm{shell}}$ becomes time-dependent. As a result, the energy of each star in the galaxy changes, according to (11.52), but these changes are only due to the addition of a time-dependent uniform term to the energies of all stars and, in reality, they have *no* effect in the stars' orbits, since the shell does not exert any force to particles in its interior. Miller concludes that (11.52) cannot characterize the effectiveness, or timescale, of mixing of the energies in a violent relaxation process, but other criteria must be established in order to distinguish when and how fast such a mixing actually occurs.

The results for the 'top-hat' case $n = -2.9$ are in certain aspects similar to Miller's example. Since the shells all collapse at the same time, the variations of the energies of all the stars are *in-phase*, that is, all the stars gain or lose energy during collapse and rebound of the system, so that the mixing of energies is not very effective despite the fact that the rate of change of energies is very fast. On the other hand, in the limit $n \to 1$, the variations of energies of the stars are to a large extent out-of-phase, since the inner shells are at the rebound phase when the outer shells are still in the collapse phase (Fig. 11.13). This is caused by the decreasing profile of mass perturbation $\mu(h) \propto h^{-2}$. At the same time, this mechanism implies that the overall time fluctuations of the potential are less violent than in the 'top-hat' model. As a conclusion, in both limits $n \to -3$ and $n \to 1$ the relaxation cannot be complete, although the reasons for that are different in each case.

The question of more refined criteria characterizing the violence or effectiveness of the relaxation process is still unanswered to a large extent. A recent proposal in this direction was made by Kandrup [163, 166]. When the potential has strong time fluctuations, these fluctuations introduce chaos to the relaxing system through time-dependent terms of the Hamiltonian. This happens even in a spherically symmetric, but pulsating, or collapsing, system. For example, such chaos is found in models of spherical galaxies in which the galaxy undergoes stable periodic oscillations [192, 226, 282]. Now, in regions of phase space where chaos is prominent, the rate of mixing is determined by the Lyapunov times of the orbits of stars that move as ensembles within the phase space [164]. This so-called *chaotic mixing* process is much faster than the phase mixing process discussed already in Lynden-Bell [196]. In Kandrup's view the rate of chaotic mixing determines essentially the rate of approach of the system to equilibrium.

There is no direct experimental test so far, e.g. by $N$-body collapse simulations, of the validity of Kandrup's suggestion. One way to produce such tests is by a detailed exploration of plots from $N$-body collapse experiments showing in detail the spreading of particles in phase space during the relaxation phase. A schematic example is given in Fig. 11.15. If we consider a 'frozen' spherical potential corresponding to one snapshot of the collapse experiment, the invariant tori of the Hamiltonian of this momentary potential have the form shown schematically in Fig. 11.15. As long as the system is not in equilibrium, the phase flow is transverse to the direction determined by the foliation of these tori (arrows in Fig. 11.15). However, as the system approaches closer and closer to the equilibrium, the flow becomes more and more tangent to the directions defined by the foliation of the tori. This simple picture is not precise when chaotic mixing takes place. This causes irregularities of the flow both in the transverse and tangent directions that may be the dominant source of mixing. Such irregularities are distinguishable in some real plots of the phase flow in collapse experiments [39, 140, 141], but, to our knowledge, there has been no systematic qualitative or quantitative study of the time evolution of this flow so far.

### 11.3.4 Collective Instabilities

Stellar systems relaxing from different initial conditions cannot in general be expected to relax to the *same* equilibrium endstate, since the properties of the latter are determined, to a large extent, by dynamical instabilities affecting the system in the course of or after the relaxation process. The topic

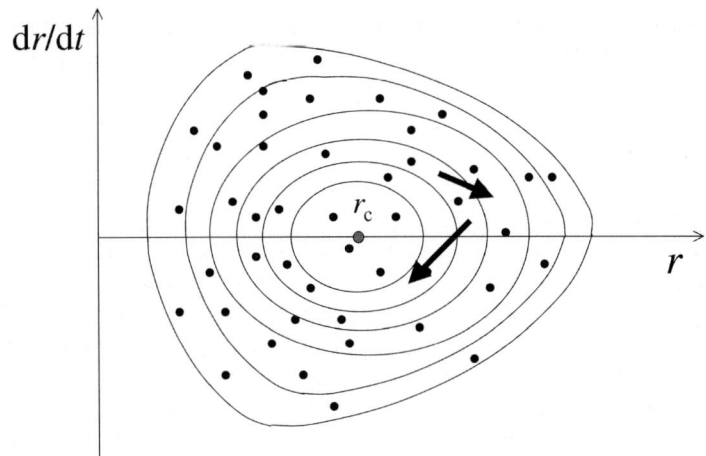

**Fig. 11.15.** Schematic representation of the theoretical invariant tori in the space $r, \dot{r}$ of a spherical system for a constant pair of energy–angular momentum values. If the system is collapsing, these tori correspond to a 'frozen' snapshot of the time-dependent spherical potential. The relaxation process continues for as long as the phase flow of the real particles (*bold arrows*) is transverse to the tori. The points represent the $N$-body sampling of the distribution function

of instabilities in collisionless stellar systems is a whole chapter of galactic dynamics (see [108, 241] for a review).

Collective instabilities in the simplest case of a spherical system were first discussed by Antonov [7]. Such instabilities may lead to interesting phenomena such as the 'gravothermal catastrophe' [199] that is believed to have played some role in dense systems such as the cores of spherical clusters. The main result of Antonov's studies is that a spherical isotropic system is stable against radial or non-radial instabilities if its distribution function is a monotonically decreasing function of the energy [33, p. 307],. Subsequent studies [89, 138] gave criteria for the stability of anisotropic systems under various types of radial perturbations. The analogue of such instabilities in the case of disks are axisymmetric instabilities [299].

A type of instability relevant to elongated galaxies is the 'radial orbit instability' [242, 251, 252]. If a galaxy contains initially many radial orbits, i.e. $\sigma_r \gg \sigma_t$ (Sect. 11.2.2), a small deviation of the angular distribution of these orbits from spherically symmetric creates a collective collaboration of the orbits, based on their mutual torques, that results in a large departure of the system from the spherical symmetry. The final states can be either axisymmetric (usually prolate) or triaxial. In the case of disks, Lynden-Bell [198] examined a similar collaboration of elongated orbits that can lead to the formation of a rotating bar inside the inner Lindblad resonance.

The general theory of the radial orbit instability is based on perturbative solutions to the collisionless Boltzmann equation. The final result can be cast in the form of Polyachenko's criterion: a system is stable against the radial orbit instability when

$$\frac{2T_\mathrm{r}}{T_\mathrm{t}} \leq 1.7 \pm 0.7 \qquad (11.53)$$

where $T_\mathrm{r} = <v_\mathrm{r}^2/2>$ and $T_\mathrm{t} = <v_\mathrm{t}^2/2>$ (in non-rotating galaxies), $v_\mathrm{r}$, $v_\mathrm{t}$ being the velocities of stars in the radial and transverse direction, respectively. The $\pm 0.7$ error in (11.53) is produced by a compilation of different values of the ratio $2T_\mathrm{r}/T_\mathrm{t}$ reported in the literature by use of different basic models used to study the instability [213, Sect. 6.2 and references there in]. For example, we may consider a spherical distribution function which is initially in steady state and find a somewhat different ratio $2T_\mathrm{r}/T_\mathrm{t}$ depending on what is the model chosen for the initial distribution function. Other, similar in spirit, criteria were proposed by different authors. For example, Merritt and Aguilar [216] proposed the criterion $2T/|U| \leq 0.1$, where $T$ is the initial kinetic energy and $U$ the initial potential energy of the system. Such criteria are verified in $N$-body studies of the radial orbit instability when we start with initial conditions which are perturbations to a spherical equilibrium [3, 18, 41, 216]. If, on the other hand, we consider initial conditions corresponding to a cosmological collapse scenario [43, 102], we find that when we start with a spherically symmetric collapsing object, which has an overpopulation of radial orbits, then the system relaxes to its final equilibrium only after the ratio $2T_\mathrm{r}/T_\mathrm{t}$, which is initially very large, settles down to a value near Polyachenko's value 1.7 (Fig. 11.16a). The resulting endstates are triaxial systems with axial ratios of long to short axis corresponding to E5–E6 galaxies. On the contrary, if we start with clumpy initial conditions (Fig. 11.16b), which are characterized by a more random initial distribution of the directions of the particles' velocities, the ratio $2T_\mathrm{r}/T_\mathrm{t}$ goes below the value 1.7 very quickly (at $t \simeq 150$ in Fig. 11.16b), at times smaller than the collapse time ($t_\mathrm{collapse} \approx 1000$). Then the systems do not exhibit a strong radial orbit instability, and the resulting endstates resemble to E2–E3 galaxies.

Other types of instabilities [241] are the 'bending' [108, 219, 300], 'tumbling bar' [6] instabilities and the bar instability in disks [15, 147, 240].

### 11.3.5 Alternative Formulations of the Statistical Mechanics of Violent Relaxation

Both the violent relaxation process and collective instabilities processes can be described by solutions of Boltzmann's equation (for the violent relaxation case see [140, 141, 146, 210] and see for the case of a collective instability references in subsection (3.4)). However, it is not clear how to distinguish between these two types of solutions, which essentially both describe excursions of a system in Liouville space, until the system settles down to a stable equilibrium state. We may say that when collective phenomena are present, these phenomena constitute the main factor determining the system's excursion in Liouville

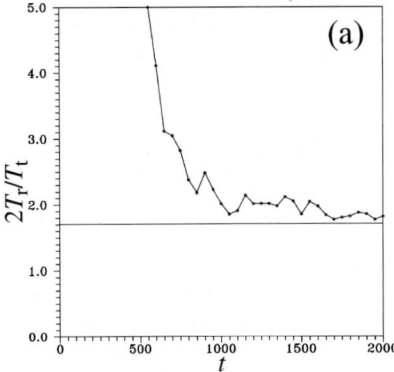 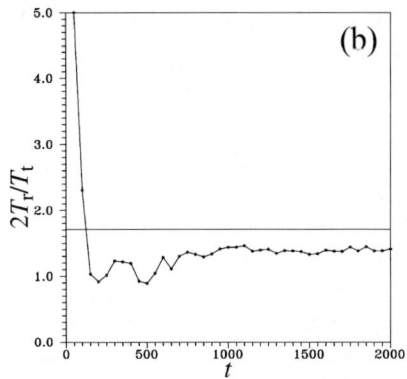

**Fig. 11.16.** The time evolution of the ratio $2T_r/T_t$ in two experiments of violent relaxation from (**a**) quiet (spherically symmetric) initial conditions and (**b**) clumpy initial conditions. For both systems, the final value stabilizes near Polyachenko's criterion $2T_r/T_t = 1.7$. Only for the system (**a**) the radial orbit instability is prominent

space. The extent that this happens determines also the limits of applicability of statistical mechanical considerations such as those forming the basis of the violent relaxation theory.

On the other hand, we can always say that a stable equilibrium state of a system of many particles should correspond to a local, or global, maximum of a kind of entropy functional $S[F]$ (where $F$ is the coarse-grained distribution function). The use of such functionals to describe the endstate of a system subject to dynamical instabilities was pioneered by Ipser [152] and Ipser and Horwitz [153]. In the case of violent relaxation, a debate was caused by a paper of Tremaine et al. [303], supporting the view that other functionals than the Boltzmann functional $S[F] = -\int F \log F \mathrm{d}^6\mu$ or its generalization by Lynden-Bell [196] can be used in the description of statistical equilibria. In particular, *any* functional $S[F]$ that is convex in $F$ will be an increasing function of time that reaches a maximum at the equilibrium state, that is, it can play the role of 'entropy' of a stellar system. This approach to equilibrium can be measured by quantities alternative to the entropy functionals of Tremaine et al. [204]. This point of view was immediately criticized by Kandrup [161], Sridhar [288] and Dejonghe [86] on the basis of the remark that Tolman's proof of H-theorem does not apply in the case of an arbitrary convex functional $S[F]$ and that the monotonic increase of such an 'entropy' cannot be established by elementary arguments. In order to resolve this issue, Soker [283] studied in detail the time evolution of a particular choice of functional $S[F]$ which is a variant of a functional proposed by Spergel and Hernquist [284]. He found that the relaxation process can be divided in two phases: during the first phase, which includes the first collapse and rebound, the Spergel–Hernquist entropy functional may increase or decrease with time. During the second phase (called the 'calm' phase), it is an increasing function of time. The calm phase can perhaps be identified with the so-called 'secondary infall' of matter [107] that characterizes the formation of dark halos, or with the process of progressive mixing in finer scales that takes place in the phase space during the late phase of relaxation.

Another class of modifications of Lynden-Bell's statistics aims at curing the problem of superpositions of Maxwellian velocity distributions with different dispersions when the phase space elements are divided in groups of different phase densities (Sect. 11.3.2). Kull et al. [178] suggested a statistical mechanics based on phase elements of unequal volumes but equal masses. They show that the resulting velocity distribution is again a superposition of Maxwellians, but this time they all have the same velocity dispersion. Nakamura [233] made a completely different proposal in order to address the same problem. He suggested to use a particle approach for collisionless systems and defined an entropy $S = -\sum P_{i,j} \log P_{i,j}$, where $P_{i,j}$ is the probability that if a particle is at the $i$th cell of the phase space at the initial time $t_0$, it will be at the $j$th cell in the end. Numerical simulations by Merrall and Henriksen [210] yielded the result that in collapse simulations the final velocity distribution appears to be a unique Maxwellian in the centre, but in merger simulations there were considerable deviations from such a unique distribution if the centres were initially well separated. On the other hand, Arad and Johansson [9] and Arad and Lynden-Bell [10] made a detailed comparison of Lynden-Bell's and Nakamura's theories both by numerical and analytical methods. The final conclusion is somewhat disappointing, since the authors support that both theories yield results not compatible with numerical experiments. Arad and Lynden Bell [10] conclude that a proper description of the violent relaxation process should rely on dynamical arguments for the evolution of the coarse-grained distribution function rather than on the classical statistical mechanical approach.

A yet different approach is based on the search for criteria that can characterize an equilibrium of Boltzmann's equation without reference to the concept of entropy, classical or modified. The basic proposal in this direction was made by Wiechen et al. [338] and Ziegler and Wiechen [242]. We notice first that Boltzmann's equation can be deduced from a Hamiltonian density function $H[f]$. Furthermore, an equilibrium state $f_0$ is a fixed point of $H[f]$. Ziegler and Wiechen [242] then define a 'dynamical energy function' $W[f]$ such that $f_0$ is a minimum of $W[f]$. The difference $W[f] - W[f_0]$ defines a kind of energy 'dissipated' during the relaxation process. The same authors proposed an algorithm for the calculation of the dynamical energy functional and of the state $f_0$. In a similar spirit, Kandrup [162] proposed to consider the stability character of 'orbits' in the so-called $\Gamma$-space, the space of all states $f$, by giving a suitable definition of Lyapunov characteristic number that is applicable to the case of the Hamiltonian density $H[f]$.

A final proposal has been to use the well-known Tsallis entropy [215]

$$S_q[f] = -\frac{1}{q-1} \int (f^q - f) \mathrm{d}^6 \mu \qquad (11.54)$$

as more relevant to the description of the relaxation process, since gravitational systems are, in general, non-extensive [250, 297, 298]. If the functional (11.54) is maximized under the usual constraints of mass and energy conservation, the resulting distribution function has the form of a *polytropic* distribution $f \propto |E|^p$, where the power index $p$ is related to the q-index of Tsallis' entropy. This approach was criticized by Chavanis [47], who points out the fact that the equilibria of galaxies are far from polytropic. Chavanis emphasizes that the use of the Tsallis entropy in stellar dynamics is somewhat ad hoc, because the Tsallis entropy applies when the phase space of a system is a fractal, or multifractal, while the phase space structure of stellar systems are not known. This is an open problem that requests more work to be clarified. On the other hand, Chavanis [45, 46] proposed a method to study the approach to equilibrium that is close in spirit to Arad and Lynden-Bell's call upon a dynamical description of violent relaxation. The proposal is to consider a Boltzmann-type equation that describes the time evolution of either the coarse-grained distribution function $F(\mathbf{x}, \mathbf{v}, t)$ [45] or a distribution function $\rho(\mathbf{x}, \mathbf{v}, \eta, t)$ that is different for each subset of phase elements with initial phase space density equal to $\eta$ (Sect. 11.3.2). In these studies, the analogue of the partial derivative $\partial f/\partial t$ in Boltzmann's equation (11.16), namely $\partial F/\partial t$ or $\partial \rho/\partial t$, is replaced by a diffusion-like term, the form of which is chosen on the basis of dynamical considerations.

### 11.3.6 The Number Density Function in the Space of Integrals of Motion: Stiavelli–Bertin Statistical Mechanics

The distribution function $f$ is a density function in the 6D phase space, i.e. it gives the mass of stars per unit values of the phase space coordinates. If, however, the orbits obey integrals of motion, the distribution function depends on these integrals $f \equiv f(I_1, I_2, \ldots, I_K)$, thus it can be expressed in terms of a different function, $N(I_1, I_2, \ldots, I_K)$ which yields the mass of stars $dm$ per unit value of each of the integrals $I_i$. The latter function, $N$, is called the *number density* function

$$N(\mathbf{I}) = \frac{dm}{d\mathbf{I}} \tag{11.55}$$

where $\mathbf{I}$ is the $K$-dimensional vector of integrals considered and $d\mathbf{I}$ is an infinitesimal volume in the space of integrals. The relation between $f$ and $N$ is specified by providing the *density of states function*

$$W(\mathbf{I}) = \frac{d\Omega(\mathbf{I})}{d\mathbf{I}} \tag{11.56}$$

where $d\Omega(\mathbf{I})$ is the elementary volume of phase space that comprises all phase space points $(\mathbf{x}, \mathbf{p})$ yielding values of the integrals in the range $\mathbf{I}, \mathbf{I} + d\mathbf{I}$.

There are indications that the number density function $N$ may be more fundamental than the distribution function $f$ in the characterization of particular properties of stellar dynamical systems. A first such suggestion was made by Binney [166] who found that in spherical isotropic galaxies obeying de Vaucouleurs law, the number density function $N(E)$ depends exponentially on the energy $N(E) \propto \exp(-\beta E)$, a fact that allows one to characterize these systems as 'isothermal after all' [166]. In order to find the isotropic spherical equilibrium associated with de Vaucouleurs' law (11.42), we recall that in isotropic systems only the energy $E$ appears as an argument of the distribution function $f$. We can then make use of a the well-known Eddington's inversion formula [96]:

$$f(E) = \frac{1}{\sqrt{8\pi^2}} \frac{\mathrm{d}}{\mathrm{d}E} \int_E^0 \frac{\mathrm{d}\rho}{\mathrm{d}\Phi} \frac{\mathrm{d}\Phi}{\sqrt{E-\Phi}} \qquad (11.57)$$

which allows one to find the unique isotropic distribution function $f(E)$ consistent with a given density–potential profile $\rho(r), \Phi(r)$ (by eliminating $r$ we use the function $\rho(\Phi)$ in the actual calculation). Since $\Phi$ is derived from (11.20), the only unknown of the problem is the density profile $\rho(r)$. However, we may also invert (11.10) and obtain $\rho(r)$ from a known surface brightness profile $\Sigma(R)$.

In the case of de Vaucouleurs' surface brightness profile (11.42) we find, numerically, a particular distribution function $f(E)$. The value of $f$ is the same at all the points of phase space which lie on the same hypersurface of constant energy $E$. We next consider an elementary phase space volume $\Delta\Omega(E) = \Delta^3\mathbf{x}\, \Delta^3\mathbf{p}$ by taking all the points of phase space corresponding to energies in a small interval $E, E + \Delta E$. The density of states function $W(E) = \Delta\Omega(E)/\Delta E$ is then calculated. Finally, we define the number density function

$$N(E) = \frac{\Delta N}{\Delta E} = \frac{\Delta N}{\Delta \Omega}\frac{\Delta \Omega}{\Delta E} = f(E)W(E) \qquad (11.58)$$

yielding the number of particles per unit energy of the system. We stress again that $N(E)$ represents a density in energy space, while $f(E)$ represents a density in phase space. The two functions can be linked only because the distribution of velocities is isotropic. Binney's numerical calculation showed that the number density function $N(E)$ for a system with de Vaucouleurs' profile is, to a good approximation, an exponential function $N(E) \simeq N_0 \exp(-\beta E)$. This suggests that a kind of statistical mechanics is applicable in these systems, which, however, should introduce a *non-uniform partition* of the phase space in terms of elementary volumes $\Delta\Omega$ corresponding to the energies in intervals $E, E + \Delta E$.

This approach can be generalized in the case of anisotropic systems. In that case we consider distribution functions of the form $f(E, L^2)$, and look for a number density distribution $N(E, L^2)$ in the space $(E, L^2)$, called the 'Lindblad space' [211]. The calculation of the elementary volume $\Delta\Omega(E, L^2) = \Delta^3\mathbf{x}\, \Delta^3\mathbf{p}$, corresponding to the volume of the union of invariant tori with energy and angular momentum values in the range $E, E + \Delta E, L^2, L^2 + \Delta L^2$ can be done as follows [239]: a phase space element corresponding to values of the phase space variables in the range $r, r + \mathrm{d}r,\ \theta, \theta + \mathrm{d}\theta,\ \phi, \phi + \mathrm{d}\phi,\ v_r, v_r + \mathrm{d}v_r,$

$v_t, v_t + \mathrm{d}v_t$ (where $v_r, v_t$ denote the modulus of the radial and transverse velocity, respectively) is given by

$$\mathrm{d}^3\mathbf{r}\,\mathrm{d}^3\mathbf{v} = r^2 \sin\theta\,\mathrm{d}r\,\mathrm{d}\theta\,\mathrm{d}\phi\,4\pi v_t\,\mathrm{d}v_t\,\mathrm{d}v_r\ . \tag{11.59}$$

Considering the transformation $r \to r$, $\theta \to \theta$, $\phi \to \phi$, $v_t^2 \to L^2/r^2$, $v_r^2 \to 2[E - \Phi(r)] - L^2/r^2$, we can take the determinant of the transformation's Jacobian matrix and write (11.59) in the form

$$\mathrm{d}^3\mathbf{r}\,\mathrm{d}^3\mathbf{v} = 2\pi \sin\theta\,\mathrm{d}\theta\,\mathrm{d}\phi\frac{\mathrm{d}r}{v_r}\mathrm{d}E\,\mathrm{d}L^2\ . \tag{11.60}$$

Then, the total phase space volume occupied by tori with energies in the interval $E, E + \mathrm{d}E$ and angular momenta $L^2, L^2 + \mathrm{d}L^2$ is given by

$$\mathrm{d}\Omega(E, L^2) = \mathrm{d}E\,\mathrm{d}L^2\pi\int_{r_\mathrm{p}(E,L^2)}^{r_\mathrm{a}(E,L^2)}\frac{2\mathrm{d}r}{v_r}\int_0^\pi \sin\theta\,\mathrm{d}\theta\int_0^{2\pi}\mathrm{d}\phi = \mathrm{d}E\,\mathrm{d}L^2 4\pi^2 T_\mathrm{r}(E,L^2) \tag{11.61}$$

where $r_\mathrm{p}(E, L^2), r_\mathrm{a}(E, L^2)$ are the radii of pericenter and apocenter, respectively, for given $(E, L^2)$, that is, the roots for the equation

$$E - \Phi(r) - \frac{L^2}{2r^2} = 0 \tag{11.62}$$

and $T_\mathrm{r}(E, L^2)$ is the radial period of orbits, i.e. the time needed to go from pericenter to apocenter and back to pericenter, given by

$$T_\mathrm{r}(E, L^2) = 2\int_{r_\mathrm{p}(E,L^2)}^{r_\mathrm{a}(E,L^2)} \frac{\mathrm{d}r}{\sqrt{2(E - \Phi(r)) - \frac{L^2}{r^2}}}\ . \tag{11.63}$$

We thus have that:

$$W(E, L^2) = \frac{\mathrm{d}\Omega}{\mathrm{d}E\,\mathrm{d}L^2} = 4\pi^2 T_\mathrm{r}(E, L^2)\ . \tag{11.64}$$

It is remarkable that for a wide class of galactic potentials the behaviour of the function $T_\mathrm{r}(E, L^2)$ is, to a very good approximation, independent of $L^2$, and very close to the Keplerian limit $T_\mathrm{r} \propto |E|^{-3/2}$. For example, Hénon [136] proved that the most general class of spherical potential functions for which the integral (11.63) is strictly independent of $L^2$ is the isochrone model:

$$\Phi(r) = -\frac{GM}{b + \sqrt{r^2 + b^2}} \tag{11.65}$$

and for this model (11.63) yields precisely the same result $T_\mathrm{r} \propto |E|^{-3/2}$ as in the Keplerian case. This is also verified in the monopole terms of the potential of $N$-body experiments [102, 320], and in the polytropic model [181].

In order, now, to generalize Binney's result $N(E) \propto \exp(-\beta E)$ in the anisotropic case, we request that the number density function $N(E, L^2)$ has exponential dependence on its arguments, that is

$$N(E, L^2) \propto \exp(-\beta(E + b'L^2)) \tag{11.66}$$

On the other hand, the generalization of (11.58) reads:

$$f(E, L^2) = \frac{N(E, L^2)}{W(E, L^2)}. \tag{11.67}$$

Thus, substituting the ansatz $W \propto |E|^{-3/2}$ in (11.67), (11.66) leads to

$$f(E, L^2) \propto |E|^{3/2} \exp(-\beta(E + b'L^2)). \tag{11.68}$$

The formula (11.68) was proposed by Stiavelli and Bertin [292, 293] as a candidate to fit the distribution function of spherically anisotropic systems. This can also be generalized to axisymmetric systems according to the formula

$$f(E, L_z, I_3) \propto |E|^{3/2} \exp\left(-\beta\left(E + b'\frac{L_z^2}{2} + cI_3\right)\right) \tag{11.69}$$

where we consider an axisymmetric potential

$$\Phi(r, \theta) = \frac{\zeta(r)}{r^2} + \frac{\eta \cos\theta}{r^2}$$

which yields an integrable system third integral

$$I_3 = \frac{(v_\theta^2 + v_\phi^2)r^2}{2} + \eta \cos\theta.$$

A more general formula involving axisymmetric Stäckel potentials is given in Stiavelli and Bertin [292].

The Stiavelli–Bertin distribution function can be derived on the basis of statistical mechanical considerations [293]. This is done by implementing the microcanonical approach of statistical mechanics, but assigning *unequal* a priori probabilities of a phase element to visit one of the macrocells of the $\mu$-space such as in Fig. 11.10, or partitioning this space to macrocells of unequal volume. The resulting entropy can be written in the form of a Boltzmann–Gibbs entropy functional defined in the Lindblad space:

$$S[N] = -\int N(E, L^2) \log N(E, L^2) dE\, dL^2 \tag{11.70}$$

under the usual constraints of mass and energy conservation, and one additional constraint regarding a combination of the energy and angular momentum that is quasi-preserved during the collapse [54]. The maximization of the entropy (11.70) leads then to an exponential law for the number density function $N(E, L^2)$ such as in (11.66). A partitioning of the phase space in terms of unequal volumes ($\propto |E|^{-3/2}$) seems quite justified by the fact that, in an integrable potential, the foliation of invariant tori creates a natural partition in phase space and that, when a system is in equilibrium, there are no motions of the phase flow transverse to these tori (see discussion of Fig. 11.15). In the next subsection we discuss the types of distribution functions found in $N$-body experiments by use of similar arguments.

Independently of whether the Stiavelli–Bertin formula for $N$ is the most convenient choice, the important point in the above analysis is the shift of emphasis from entropy functionals depending on $f$ to entropy functionals depending on $N$, which thus becomes the important quantity to study. This point is emphasized by Tremaine [302], see also Merritt et al. [220]. An entropy functional similar to (11.70) was proposed by Spergel and Hernquist [284] in the case of isotropic spherical systems. The resulting distributions were also found in good agreement with the results of numerical experiments.

### 11.3.7 The Distribution Function Found in $N$-Body Experiments of Violent Relaxation

The number of particles used in galactic $N$-body simulations has grown from $10^4$–$10^5$ in the 1990s to $N = 10^6$–$10^7$ today. Even so, it remains a hardly tractable task to obtain numerically the distribution function of a relaxed system by the counting method, i.e. by counting the number of particles in cells of the 6D $\mu$-space. Even a very coarse division of the phase space, say by 10 bins per dimension, would result in $10^6$ cells to consider, implying 1–10 particles per cell on average. Thus the signal would be hidden by the statistical noise.

On the other hand, with such a number of particles it is possible to do statistics in the space of integrals, or approximate integrals of motion, such as the Lindblad space $(E, L^2)$, which has dimension equal to two, thus allowing for a meaningful statistics. If the system has spherical symmetry, one can then pass from $N$ to $f$ according to the formulae of the previous subsection. Scatter plots of the positions of the particles in the space $(E, L^2)$ can be found in a number of papers [3, 206]. But the first systematic study of the resulting number density function $N(E, L^2)$ was made by Voglis [320], who proposed fitting formulae to represent the *contours* of $N$ in the space $(E, L^2)$ (Fig. 11.17). Similar figures were given by Natarajan et al. [234] and Trenti et al. [305].

Voglis' method gave three main results:

(a) A violently relaxed system in equilibrium is characterized by the existence of a *time-invariant* function $N(E, L^2)$ despite the fact that the arguments $(E, L^2)$ are not precise integrals of motion. In particular, the energy of particles has small fluctuations due to numerical fluctuations in the coefficients of the potential of the $N$-body code. On the other

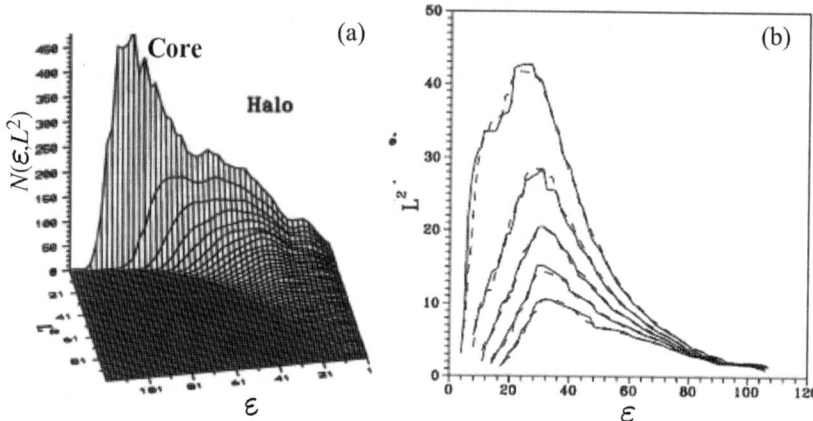

**Fig. 11.17.** (a) A typical form of the 'number density' distribution $N(E, L^2)$ of an N-body system after the relaxation. (b) The contours of the function $N(E, L^2)$ (*solid line*) together with the fitting by the model of [320] (*dashed line*). The quantity in the abscissa is $\mathcal{E} = -E$ (after [320])

hand, the modulus of the angular momentum $L^2$ is not even approximately preserved because the final system is not spherical. Nevertheless, the function $N(E, L^2)$ is found to remain invariant in time as a result of a 'detailed equilibrium' established in the space $(E, L^2)$, namely the numbers of exchanged particles between any two elementary cells of the space $(E, L^2)$ are equal in the course of the N-body run.

(b) The distribution $N(E, L^2)$ is characterized by the existence of two main loci of maximum of the distribution.

The first locus, called the 'core' is given by pairs of values $(E, L^2)$ which are very close to the locus of the energy of circular orbits $E_c(L^2)$ if we only consider the monopole term of the multipole potential expansion of the system. The function $N(E, L^2)$ near this maximum can be fitted by a modified Lynden-Bell's formula:

$$N(E, L^2) \propto \frac{|E|^p}{\exp\left(-\beta(E - E_c(L^2))\right) + 1} \quad (11.71)$$

where the function $E_c(L)$ plays the role of 'chemical potential'. The numerator $|E|^p$ represents a *polytropic* function. The polytropic index $p$ can be shown to depend monotonically on the power-exponent of the initial density perturbation $n$ that caused the system to collapse [102].

The second locus, called 'halo', is in mild energies but extends to high values of the angular momentum. The associated function $E_m(L_m^2)$ is given by two formulae relating the energy $E_m$ or angular momentum $L_m$ of the halo maximum with the value of the number density $N$ at the maximum, namely

$$\log |E|_m = \log \mathcal{E}_0 + \nu \log \log N, \quad L_m^2 = L_0^2(P - \log \log N) \quad (11.72)$$

with parameters $\mathcal{E}_0, \nu, L_0, P$ depending again monotonically on the exponent $n$ of density perturbations.

(c) The behaviour of the system near both loci indicates a local change of the sign of the temperature of the system from positive to negative. The concept of 'negative temperature' was introduced by Merritt et al. [220] who had proposed 'negative temperature' Stiavelli–Bertin like models

$$N(E, L^2) \propto \exp(\beta(E + b'L^2)), \quad (11.73)$$

i.e. with a positive factor $\beta$ appearing in the exponential dependence of $N$ on $E$. Merritt et al. [220] suggested that such models better fit the observed surface density profiles as well as the energy distributions of violently relaxed systems. The detailed fits to numerical experiments by Aguilar and Merritt [3] favoured the negative temperature models. However, even these models failed to reproduce the $N$-body distribution of particles $N(E, L^2)$, or $N(E)$, in the region of energies close to zero. The authors suggested that this might be attributed to incomplete relaxation in the outer parts of the systems. On the other hand, Voglis' study indicated that there is no fundamental reason to consider a unique sign of the constant $\beta$ throughout the whole available Lindblad space.

Efthymiopoulos and Voglis [102] presented a more fundamental understanding of these results in terms of modified Stiavelli–Bertin number density statistics. In the same time, they showed that the method is applicable to systems that deviate considerably from the spherical symmetry, i.e. triaxial systems corresponding to E5–E6 galaxies. The key remark is that if one considers a multipole expansion of the potential written in spherical coordinates:

$$\Phi(r, \theta, \phi) = \Phi_0(r) + \sum_{l=0}^{\infty} \sum_{m=-l}^{l} \Phi_{lm}(r) Y_l^m(\theta, \phi) \quad (11.74)$$

where $Y_l^m(\theta, \phi)$ are spherical harmonics, then the only part of the potential which is guaranteed to yield an integrable system is the monopole term $\Phi_0(r)$. One can then use this term to define tori of constant label values $E, L^2$ under the flow induced by the Hamiltonian $H_0$ corresponding to $\Phi_0$. These, of course, are not invariant tori of the full Hamiltonian of the system. They are, however, well-defined geometrical objects in the phase space, and, therefore, they can be used in order to produce a *partition* of the phase space in terms of volumes $d\Omega(E, L)$ given by (11.60), with $\Phi_0$ in the place of $\Phi$. This partition is a geometrical structure, not depending on the dynamics. One can then ask what is the value of the *coarse-grained* distribution function $F(d\Omega)$ within each elementary volume $d\Omega$. In the case of a spherical system, this is, by definition, equal also to the value of the fine-grained distribution function

$f(\mathbf{x}, \mathbf{v})$ at any point $\mathbf{x}, \mathbf{v}$ of $d\Omega$. In the case of an axisymmetric or triaxial system, however, $F$ is only an average value of $f$ throughout the volume $d\Omega$. We find nevertheless that $F$ can be used in the place of $f$ to reproduce the profile of the density and the profile of the anisotropy parameter $\beta$ of the system under study with very good accuracy [102].

As regards the functional form of $F$, this is given by (11.67) with $F$ in the place of $f$. The problem is thus again transferred to the determination of the number density function $N(E, L^2)$. It was found that if the system is divided into a number of spherical shells of radii $r$, width $dr$, then locally, within each shell, the number density function $\nu(E, L^2, r) = \Delta N/\Delta E \, \Delta L^2 \Delta r$ takes the form of a modified Stiavelli–Bertin's formula

$$\nu(E, L^2, r) = \frac{\exp(-\beta(r)E + B(r)L^2)}{\exp[-\beta_c(E - E_a(r, L^2))] + 1} . \tag{11.75}$$

This function fits well the numerical function $\nu(E, L^2, r)$ found in the $N$-body experiments. The latter has the same property as (a) above, i.e. it remains practically invariant in different time snapshots. The numerator of (11.75) has, precisely, the form of Stiavelli–Bertin statistics for the number density function (11.66). However, we find that the parameters $\beta$ and $B$, measuring the temperature and velocity anisotropy within the shell, are functions of the shell radius $r$ (Fig. 11.18). Since these parameters enter as Lagrange multipliers in the maximization of an entropy functional in the Lindblad space, such as the functional (11.70), the authors concluded that the results indicate a new type of statistics that incorporates the different degrees of mixing in phase space during relaxation between the inner and outer system's shells. We finally note that the denominator in (11.75) introduces again a cut-off of the shell number density function $\nu$ for energies lower than $E_a(r, L)$, the energy of an orbit reaching the shell at its apocenter:

$$E_a(r, L^2) = \frac{L^2}{2r^2} + \Phi_0(r) . \tag{11.76}$$

In the spherical case, (11.76) provides an absolute cut-off, i.e. no particle with energy $E < E_a$ can reach the shell. But in a triaxial system there is some tolerance around this cut-off introduced by the multipole terms of (11.74), which is measured by the value of the constant $\beta_c$.

The global number density function $N(E, L^2)$ found by integrating $\nu(E, L^2, r)$ over the radii of all shells

$$N(E, L^2) = \int_0^\infty \nu(E, L^2, r) dr \tag{11.77}$$

fitted quite nicely the numerical data in a series of experiments collapsing under either spherically symmetric or clumpy initial conditions (Fig. 11.19a–d). The goodness of the fit was also evident in the profiles of the density $\rho(r)$ and of the anisotropy parameter $\beta_{an}(r)$ of the same systems (Fig. 11.19e and Fig. 11.19f, respectively).

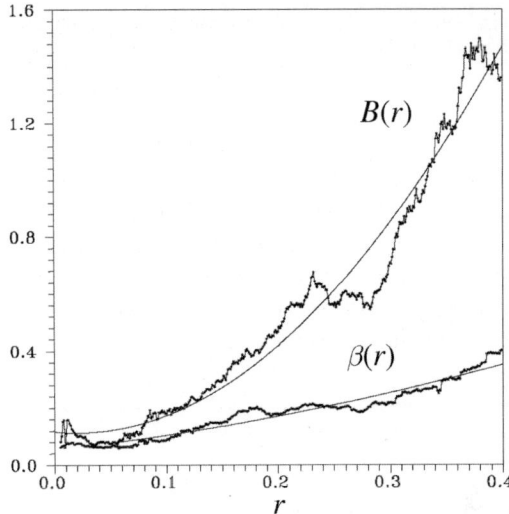

**Fig. 11.18.** The Lagrange multipliers $\beta(r)$ and $B(r)$ of the 'spherical shell' modified Stiavelli–Bertin statistics (11.75) for a relaxed system as functions of the distance $r$ from the centre (after [102])

## 11.4 The Orbital Approach: Global Dynamics and Self-Consistent Models of Galaxies

In the previous section, the focus was on studying the distribution function of galaxies on the basis of statistical mechanical considerations. However, a different approach to the same problem lies in studying the *orbital content* of stellar systems. An orbital study should give the main characteristics of the phase space structure and find which types have the dominant contribution in the self-consistency of the system. As a rule, a type of orbits is important if the form of the orbits supports the form of the galaxy.

In the sequel, we analyse the main types of orbits in spherical, axisymmetric and triaxial systems (we focus on non-rotating systems). We then refer to applications of 'global dynamics' in galaxies, based mostly on the frequency analysis of Laskar [93, 180, 181, 182, 183, 184, 185, 276]. Finally, we discuss the method of *self-consistent models* [266] which is widely used today in order to explore the relative contribution of various types of orbits in the composition of the distribution function of a galaxy.

### 11.4.1 Orbits in Spherical Systems

As already discussed in Sect. 11.2.3, the orbits in a spherical potential $\Phi(r)$ are confined to planes normal to their (constant) angular momentum vector $\mathbf{L} = \mathbf{r} \times \dot{\mathbf{r}}$. The modulus of $\mathbf{L}$ appears as a parameter in the effective one

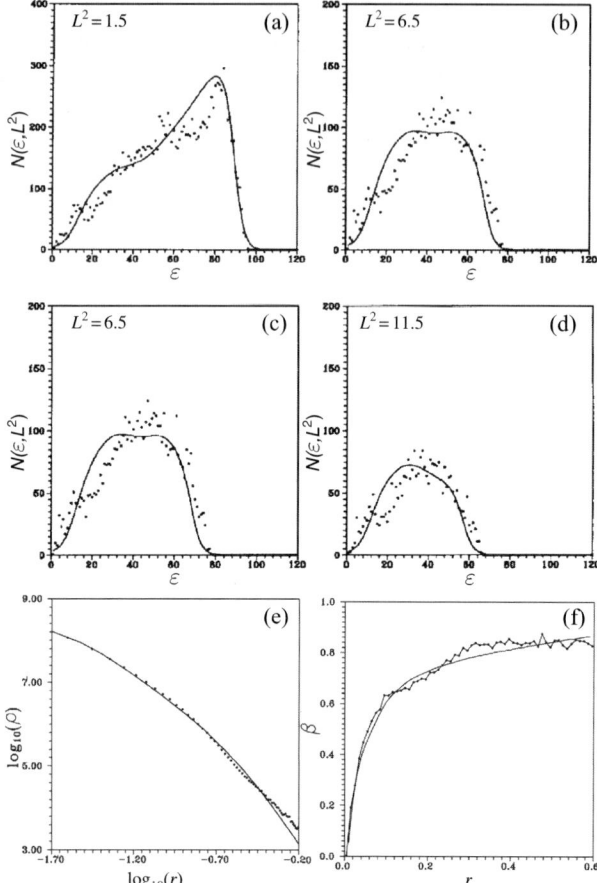

**Fig. 11.19.** (**a**–**d**) Four different slices of the distribution $N(\mathcal{E}, L^2)$ for four different constant angular momentum values $L^2$. The fitting by the model of Efthymiopoulos and Voglis [102] is shown as a *solid line*. (**e**) Reproduction of the $N$-body density profile $\rho(r)$ (*points*) by the model (*solid line*). (**f**) same as (**e**) for the anisotropy parameter profile $\beta_{\mathrm{an}}(r)$

degree of freedom Hamiltonian

$$H(r, p_r; L^2) = \frac{p_r^2}{2} + \Phi_{\mathrm{eff}}(r, L) = \frac{p_r^2}{2} + \frac{L^2}{2r^2} + \Phi(r) \tag{11.78}$$

with $p_r = \dot{r}$. The Hamiltonian (11.78) yields the radial motion on the orbital plane. The value $E_c = L^2/2r_c^2 + \Phi(r_c)$, where

$$\frac{\mathrm{d}\Phi(r_c)}{\mathrm{d}r} - \frac{L^2}{r_c^3} = 0 \tag{11.79}$$

yields the energy $E_c$ of the circular orbit with radius $r_c$. In galactic potentials, the radius $r_c$ corresponds to a minimum of the effective potential. As a result, the circular orbits are stable against *radial perturbations*. On the other hand, for any value of the energy $0 < E \leq E_c$ the orbits are confined between a minimum pericentric distance $r_p$ and a maximum apocentric distance $r_a$. These are the roots of (11.62). The forms of the orbits are rosettes (Fig. 11.20). The radial period is given by (11.63), while the azimuthal period is [33, p. 107]

$$T_\theta(E, L^2) = \frac{2\pi T_r(E, L^2)}{\Delta\phi} \quad (11.80)$$

with

$$\Delta\phi = 2L \int_{r_p(E,L^2)}^{r_a(E,L^2)} \frac{dr}{r^2 \sqrt{2(E - \Phi(r)) - L^2/r^2}}$$

If the orbit is close to circular ($r_a - r_p \ll r_c$), the radial period tends to the *epicyclic period* $T_\kappa = 2\pi/\kappa$, with

$$\kappa^2 = \frac{\partial^2 \Phi_{\text{eff}}(r_c)}{\partial r^2} = \frac{3L^2}{r_c^4} + \frac{\partial^2 \Phi}{\partial r_c^2} \ . \quad (11.81)$$

If the density $\rho(r)$ is a decreasing function of $r$, then $1 < T_\theta/T_r < 2$, that is, the angle $\Delta\phi$ covered within one radial period lies between $\pi$ and $2\pi$ [53]. Limiting cases are the Keplerian $\Phi(r) \propto -1/r, \rho(r) = \delta(r)$, where $\Delta\phi = 2\pi$, and the homogeneous $\Phi(r) \propto r^2, \rho(r) = $ const, where $\Delta\phi = \pi$. In these potentials, there are no rosettes but only closed (periodic) orbits. In any other case, we have closed orbits if

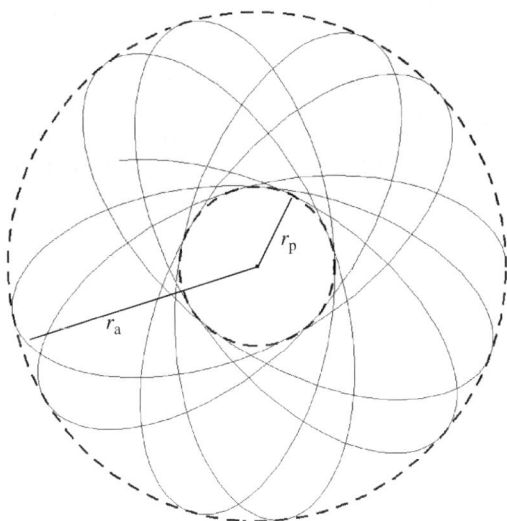

**Fig. 11.20.** A typical orbit (rosette) in a spherical potential

$$\Delta\phi = \frac{m}{n}2\pi \tag{11.82}$$

with $m, n$ as integers, $n \neq 0$.

### 11.4.2 Orbits in Axisymmetric Systems

The effective Hamiltonian of motion in the meridional plane of an axisymmetric galaxy is given by (11.23). We consider the potential symmetric on both sides of the plane $z = 0$, $(\Phi(R, z) = \Phi(R, -z))$. All orbits preserve, besides the energy, the angular momentum component $L_z = R^2\dot\varphi$, which is a parameter in the two degrees of freedom Hamiltonian (11.23). We call

$$\Phi_{\text{eff}}(R, z) = \Phi(R, z) + \frac{L_z^2}{2R^2}$$

the effective potential. The 3D orbit of a star is the result of the combination of the motion on the meridional plane $(R, z)$ and of the rotation about the $z$-axis with angular speed $\dot\varphi = L_z/R^2$ (which is not constant). If the orbit obeys a 'third integral', the orbit is called regular, otherwise it is called chaotic.

Figure 11.21a shows an example of regular orbit on the meridional plane with $L_z \neq 0$. The model used has a potential function that corresponds to a flat central density profile. The orbit appears as a deformed parallelogram in the meridional plane. However, as the orbit also rotates, there is a cylindrical hole around the $z$-axis that is created by the rotation of the left boundary of the parallelogram. Such regular orbits are called 'tubes' [85]. On the other hand, when $L_z = 0$ the orbit's left boundary touches the axis $z = 0$, and

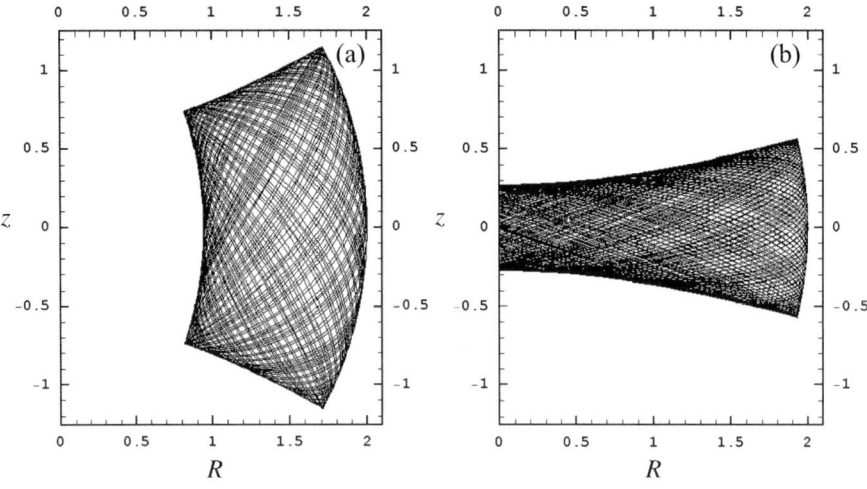

**Fig. 11.21.** (a) A tube-type orbit with $L_z \neq 0$ on the rotating meridional plane $(R, z)$ in the case of an axisymmetric potential. (b) When $L_z = 0$ we have a box orbit. This orbit is 2D and lies on the meridional plane $(R, z)$

the hole disappears (Fig. 11.21b). Furthermore, there is no rotation because $\dot{\varphi} = 0$. Thus, the orbits are 2D, and they are called 'box' orbits, because their shape on the meridional plane resembles a box with curvilinear sides.

The box orbits are quasi-periodic orbits associated with two independent oscillations with incommensurable frequencies, on the $R$ and $z$ axes, respectively. The limiting periodic orbits are stable orbits along the $z$-axis or the $R$-axis. In general, the $z$-axis orbit is stable for values of the energy close to the central potential value. At larger values the $z$-axis orbit becomes unstable and there can be no box orbits around it. At the transition to instability, a 1:1 stable periodic orbit bifurcates from the $z$-axis orbit. The 1:1 orbit forms a loop on the meridional plane. Such is the orbit of Fig. 11.28c below, that corresponds to the centre of the island of stability of the 1:1 resonance marked with (B) in Fig. 11.6a. Higher order periodic orbits can also exist that correspond to various ratios of the fundamental frequencies in the $z$ and $R$ axes.

In models with flat central profiles most orbits are regular [118]. An exploration of the phase space by means of Poincaré surfaces of section yields typically invariant curves corresponding to boxes or tubes, and only small secondary resonances with limited chaos. If, however, the galaxy has a central black hole or, more generally, a '*central mass concentration*' (CMC), the box orbits or tube orbits with low values of $L_z$ lose their regular character and they are converted to chaotic orbits (Sect. 11.4.4 below).

Finally, the orbits lying on the equatorial plane follow the same rules as the orbits in spherical potentials, since they feel a 2D axisymmetric potential $\Phi(R, 0)$.

### 11.4.3 Orbits in Triaxial Systems

In generic triaxial models of galaxies only the energy is a global integral of motion. An exception is the perfect ellipsoid (11.35) which yields an integrable Stäckel potential [80]. The regular orbits of this model have served as a basic guide for the form of regular orbits in generic triaxial potential models. Figure 11.22a–c shows the contour surfaces of the ellipsoidal coordinates $(\lambda, \mu, \nu)$, respectively (Sect. 11.2.3). The surfaces of constant $\lambda$ are ellipsoids, while the surfaces of constant $\mu$ and $\nu$ are hyperboloids, of one and two sheets, respectively [80]. The orbits can be of four types: (a) box, (b) inner long axis tube—ILAT, (c) outer long axis tube—OLAT and (d) short axis tube—SAT. Figure 11.23 shows the cross-sections of these orbits with the three principal planes as well as the limits of these orbits determined by the ellipsoidal coordinate lines.

Figure 11.24 shows the same orbits in 3D configuration space [291]. Box orbits fill a region that resembles a parallelepiped with curved surfaces. These orbits pass arbitrarily close to the system's centre. Three-dimensional boxes do not exist in an axisymmetric configuration. On the other hand, ILATs are tube orbits which fill an elongated region around the long axis, they have a hole along the same axis and they are compatible with triaxial or prolate

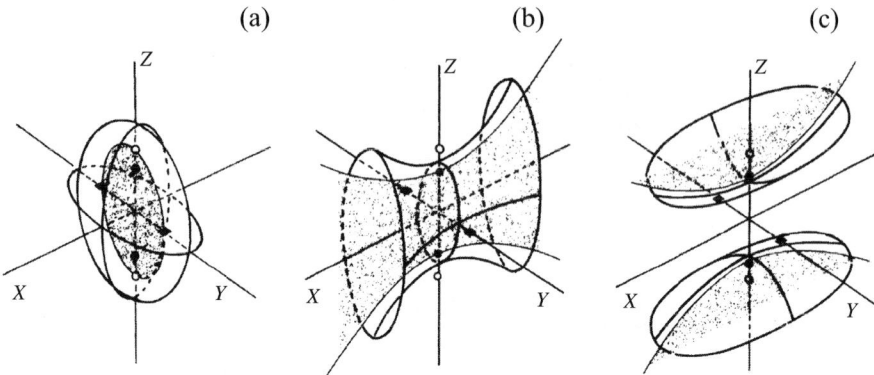

**Fig. 11.22.** (a–c) Contour surfaces of constant ellipsoidal coordinates $\lambda, \mu, \nu$, respectively. The surfaces of constant $\lambda$ are ellipsoids. The surfaces of constant $\mu$ are hyperboloids of one sheet and the surfaces of constant $\nu$ are hyperboloids of two sheets (after [80])

configurations. OLATs are tube orbits with a hole also around the long axis (like ILATs), which, however, do not approach close to the centre of the system. OLATs are also compatible with triaxial or prolate configurations. SATs resemble like OLATs except that their hole is around the small axis. SATs are compatible with triaxial or oblate configurations. Orbits such as in Fig. 11.21a are limiting cases of either a SAT in an oblate configuration, or an ILAT or OLAT in a prolate configuration.

Besides the above main families of orbits, in generic triaxial potentials there can be higher-order periodic orbits corresponding to different commensurabilities of the basic frequencies of oscillation in the three axes (see [213] for some examples of such orbits). When stable, these orbits are surrounded by quasi-periodic orbits which form thin tubes around the periodic orbits. We call the these orbits 'higher-order resonant tubes' (HORT).

### 11.4.4 Chaotic Orbits: The Role of Chaos in Galaxies

The role of chaos in galaxies is currently a very active field of research. In the case of elliptical galaxies, the successful construction of self-consistent models of triaxial galaxies composed practically only by regular orbits [266, 267, Subsection 4.6] suggested that galactic equilibria favour, for some reason, nearly integrable models with mostly regular orbits. An explanation was provided on the basis of Statler's [291] self-consistent models of the perfect ellipsoid. In these models (which are integrable) there was a clear predominance of box orbits, and it was naturally expected that such a predominance should be generic. Besides the usual box orbits, which are symmetric with respect to the three axes, Levison and Richtone's work [189] on self-consistent models of the logarithmic potential demonstrated that there were many 'tilted' box orbits

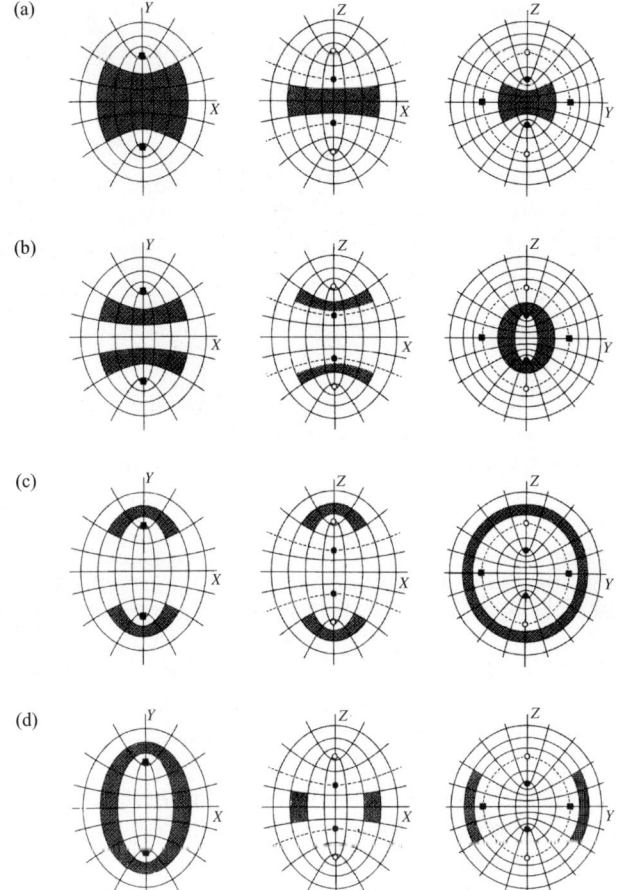

**Fig. 11.23.** The cross-sections of the four types of (regular) orbits with the three principal planes in the case of perfect ellipsoid (11.35). The coordinate surfaces confine the various types of orbits (after [80]). They are (**a**) box, (**b**) ILAT, (**c**) OLAT and (**d**) SAT orbits

that were probably not associated with the axial periodic orbits, but with other higher-order periodic orbits. On the other hand, Schwarzschild [268] studied triaxial models of galactic halos of the form $\varrho \propto r^{-2}$ (cuspy density profiles) and found a significant percentage of chaotic orbits indicating thereby the substantial role of chaos for systems with cuspy profiles. The role of chaos in such systems was emphasized in recent years mostly by Merritt and his collaborators (e.g. [217, 221, 222, 309], see [212, 215] for a review), supporting the view that the percentage of chaotic orbits in an elliptical galaxy with a central density cusp may raise up to 60%.

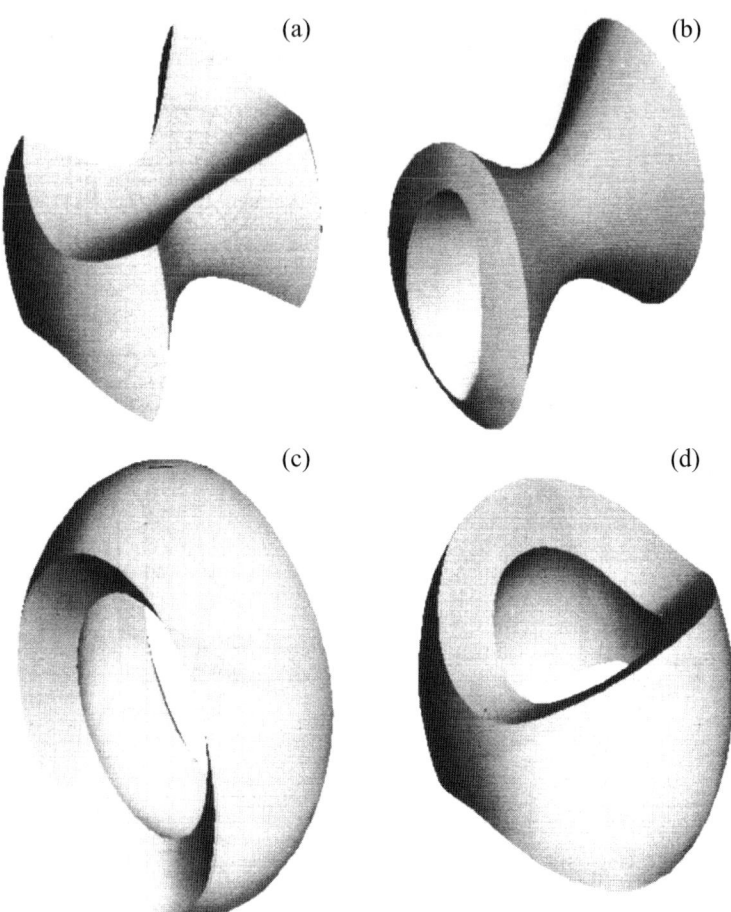

**Fig. 11.24.** (a) Box, (b) ILAT, (c) OLAT and (d) SAT. The four types of (regular) orbits in 3D space (for perfect ellipsoid). These orbits are very good guides for the form of regular orbits that exist in most galactic models (after [291])

Central black holes or CMCs are known also to contribute to the creation of a large percentage of chaotic orbits. From the early 1960s, it was known that black holes possibly exist at the centres of galaxies [197, 265, 340]. The presence of the black holes was proposed, initially, in order to explain the active galactic nuclei (AGN). During the last 10–15 years, in view of better quality observations (e.g. with the Hubble Space Telescope), many researchers [74, 115, 174, 175, 176, 201, 312, 313] found evidence of the existence of massive black holes at the centres of galaxies. The density of matter in many galaxies is not constant at the centre but it appears in a similar 'cuspy' form as in the models of Schwarzschild [1, 71, 106, 114, 186, 268]. Today, the dominant point of view is that practically all galaxies contain a massive black hole at their centre.

The presence of a CMC produces a significant number of chaotic orbits in galaxies that have triaxial form, by destroying the regular character of many regular orbits [109, 118, 148, 149, 157, 158, 165, 167, 217, 218, 221, 253, 254, 255, 278, 279, 309]. In particular, with the inclusion of a massive CMC, many (previously box) orbits acquire positive Lyapunov exponents that correspond to Lyapunov times much smaller than the age of galaxies in which they reside. The reason for this destabilization of the orbits is that, when approaching arbitrarily close to the centre, the box orbits are scattered by the CMC and become chaotic, tending to fill the whole available space inside the equipotential surface corresponding to the constant energy condition. As a consequence, the orbits cover a more spherical domain. The insertion of a CMC in a triaxial galaxy produces many chaotic orbits, which cannot, in general, support a triaxial equilibrium state. In reality, after such an insertion, the chaotic orbits cause a secular evolution of the system towards a different equilibrium state. We show below (Sect. 11.5.4) that, while under certain circumstances the final equilibrium can still be triaxial, more often it is very close to axisymmetric (oblate spheroid). In any case, the structure of the system in the final equilibrium state is mainly supported by regular orbits of the SAT type, which have a large amount of angular momentum, because the latter condition is required in order that an orbit avoids the (singular) centre.

Another example of the importance of chaos is the case of disk galaxies. Chaos is known to play an important role mostly near the *corotation* region [59, 68, 170]. In the case of barred galaxies, the chaos is prominent near corotation and it is considered as responsible for the termination of strong bars [61, Sect. 3.3.8]. On the other hand, recent findings from $N$-body experiments [328] suggest that the spiral structure beyond corotation is also composed almost entirely by chaotic orbits. A theoretical mechanism explaining this phenomenon was proposed by Voglis et al. [329].

### 11.4.5 Global Dynamics

In two degrees of freedom (DOF) Hamiltonian systems, an easy way to visualize the structure of the phase space is by means of Poincaré surfaces of section. In 3DOF cases, however, the surface of section is 4D and cannot be visualized. In such systems, an efficient method to study the phase space structure is by means of the analysis of the fundamental frequencies of the orbits. This is usually called the study of 'global dynamics' of galaxies. An early example of frequency analysis was given by Binney and Spergel [32], who used the Fourier transform to test the variability of the frequencies of orbits in a logarithmic potential model of a galaxy. But the most precise treatment of the same problem can be made by the frequency map method of Laskar [180, 181, 182]. The frequency map offers a clear representation of the picture of the phase space by providing a distinction of regular or chaotic domains in the space of *actions*, or of their associated frequencies. Thus, one may visualize the Arnold web of the various resonances and identify which resonances play the dominant role.

The distinction between the chaotic and regular orbits is based on the fact that the regular orbits have constant frequencies whereas the chaotic orbits show a variability of the frequencies calculated in different time windows. The calculation of the frequencies takes place with an advanced numerical technique that reduces, in general, the scaling of the error with respect to the width of the time window $T$ to $O(1/T^4)$, instead of $O(1/T)$ as in the fast Fourier transform. This method was implemented in galactic dynamics firstly by Papaphilippou and Laskar [244, 245]. The potential adopted was the logarithmic potential

$$V(x,y,z) = \ln\left(R_c^2 + x^2 + \frac{y^2}{q_1^2} + \frac{z^2}{q_2^2}\right) \tag{11.83}$$

that represents elliptical galaxies with flat density profiles at the centre. The parameter $R_c$ is a softening radius, and $q_1, q_2$ are two parameters that control the ellipticity and the triaxiality of the system.

Figure 11.25a–d shows a characteristic example of a frequency map of box orbits for four different sets of the parameters' values of the potential (11.83) and for one particular value of the energy [245]. Every *point* in these diagrams corresponds to one *orbit* in phase space. The axes give the *rotation numbers* $a_1, a_2$ of the orbits. The horizontal axis corresponds to the ratio $a_1 = \omega_x/\omega_z$ of the orbital frequency along the long axis $x$ over the frequency along the short axis $z$. Similarly, $a_2 = \omega_y/\omega_z$ is the ratio of the frequency of oscillation along the middle axis to the frequency along the short axis.

In these diagrams, areas filled with well-ordered points correspond to regular orbits, whereas areas with scattered points correspond to chaotic orbits. We also distinguish various resonance lines and resonance strips with borders covered by chaotic orbits. A resonance line is specified by a linear combination of the form $k_1 a_1 + k_2 a_2 + k_3 = 0$ with integer $k_1, k_2, k_3$. At the intersection of two resonance lines there are periodic orbits of various stability types. Inside each resonance strip, on the other hand, there are invariant tori of dimensionality lower than three. The orbits near the central resonance lines are usually on 2D elliptic tori, which cause a concentration of points along the resonant line. On the other hand, the orbits in resonance lines devoid of points are usually on tori which are at least partially hyperbolic. The study of Papaphilippou and Laskar demonstrated in a clear way the complexity of the phase space in 3D galactic systems by giving detailed information not only about the existence of periodic orbits but also about the interaction of resonances. They also confirmed that triaxial systems with a flat central density profile contain all the types of regular orbits found in the simple perfect ellipsoid model (see Fig. 11.24), but also many chaotic orbits that appear to play an important role in the system's global dynamics.

Wachlin and Ferraz-Mello [330] and Valluri and Merritt [309] used the same technique as Papaphillippou and Laskar [244] in order to study the dynamics

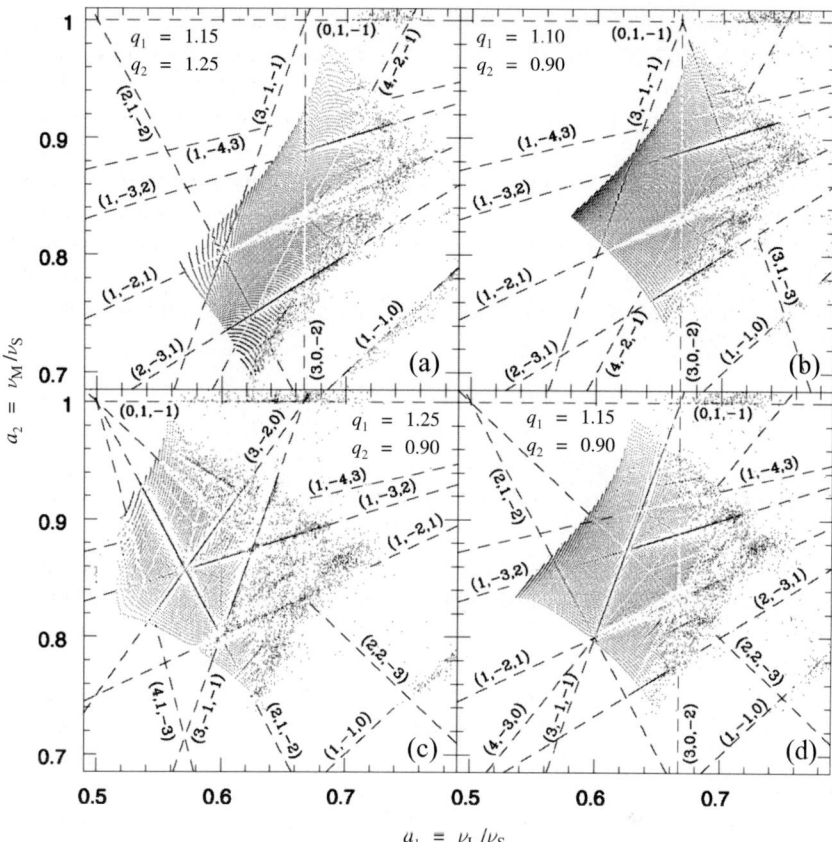

**Fig. 11.25.** Frequency maps (rotation numbers $(a_1, a_2)$) of the box orbits in the case of the logarithmic potential (11.83) for a fixed energy level and for various pairs of the parameters $q_1, q_2$. Each point in these diagrams corresponds to one orbit. We distinguish regions of regular orbits (*well-ordered points*), regions of chaotic orbits (*scattered points*) and various resonance lines (after [244])

of triaxial galaxies with cuspy central density profiles and massive black holes. These studies use the Dehnen (or $\gamma$) density model [83, 304]

$$\varrho(m) = \frac{(3-\gamma)M}{4\pi abc} m^{-\gamma}(1+m)^{-(4-\gamma)} \tag{11.84}$$

where

$$m^2 = \frac{x^2}{a^2} + \frac{y^2}{b^2} + \frac{z^2}{c^2}$$

with $a > b > c$. The total mass of the system is given by $M$ and the equidensity surfaces are stratified ellipsoids with axial ratio $a : b : c$. The parameter $\gamma$ specifies the form of the central density profile and can take values $0 \leq \gamma < 3$.

For $\gamma = 0$ we have a flat central density profile, while for $\gamma > 0$ we have a cuspy profile, with $\rho \to \infty$ as $m \to 0$, and $\gamma$ regulating the logarithmic slope of the density profile. The studied values of $\gamma$ were $0 \leq \gamma \leq 2$. Figure 11.26 shows the frequency maps in the domain of box orbits for different values of $\gamma$. These diagrams render obvious that, as the value of $\gamma$ increases, the total volume occupied by regular orbits decreases while that of chaotic orbits increases. Furthermore, the density of points near resonance lines increases.

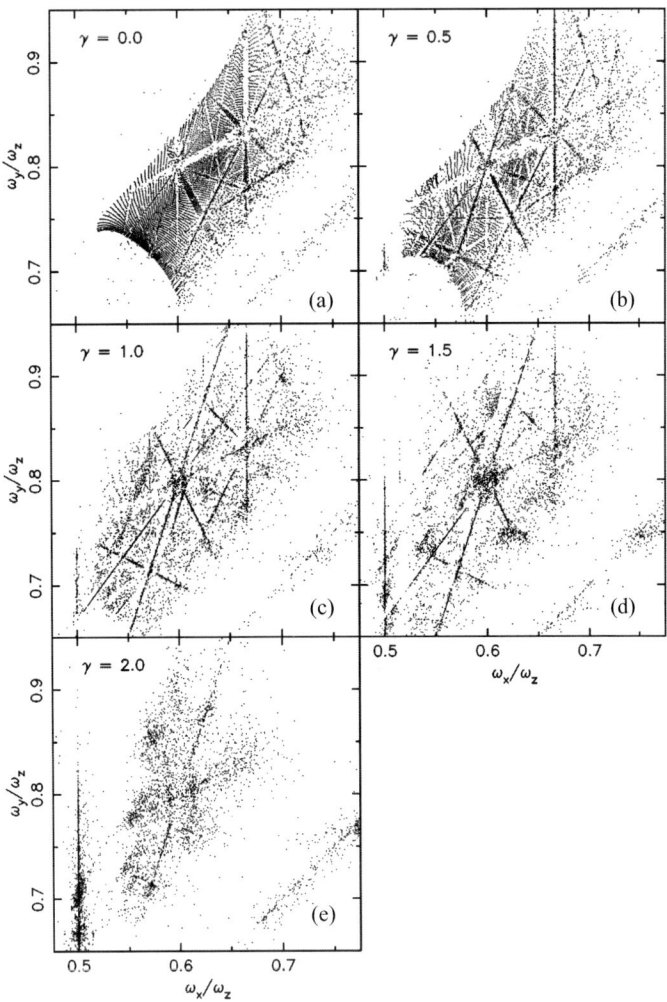

**Fig. 11.26.** As in Fig. 11.25 for the Dehnen (or $\gamma$) model (11.84), for various values of the parameter $\gamma$. Chaos becomes more prominent as the parameter $\gamma$ increases (after [309])

When $\gamma$ is large, the orbits near resonances are the only surviving regular orbits of the system.

Valluri and Merritt [309] studied the case with $\gamma = 0.5$ combined with massive central black holes. They found that chaos becomes even more prominent at the presence of the black hole, leading to full stochasticity when the mass of the black hole becomes of order $M_h \approx 0.03 M_{\text{galaxy}}$ (Fig. 11.27). The same authors noticed that the regular character of tube orbits is not greatly affected

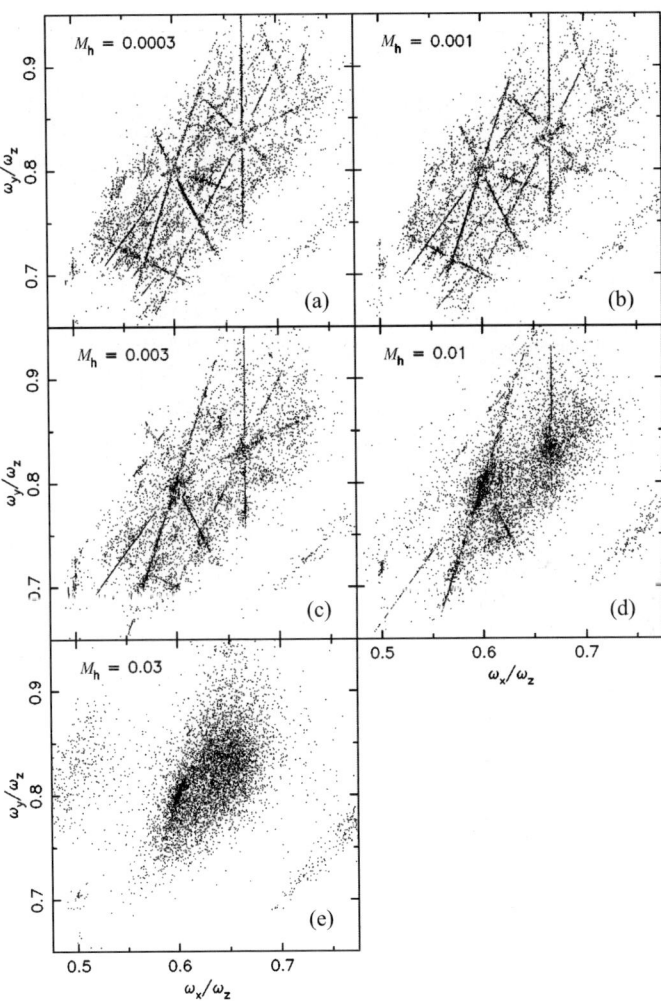

**Fig. 11.27.** As in Fig. 11.26, for $\gamma = 0.5$ and various mass values of a central black hole. The chaos in these cases is even more prominent than in the cases of Fig. 11.26 (after [309])

by the presence of a black hole or CMC, since, by definition, these orbits avoid anyway the centre of the system.

### 11.4.6 Self-Consistent Models: Schwarzschild's Method

The next step after a study of global dynamics is to examine the relative contribution of various types of orbits in supporting self-consistently the equilibrium state of the considered galactic model. The basic method towards such a study was introduced in stellar dynamics by Schwarzschild [266]. The main steps of Schwarzschild's method are the following:

1. A spatial density function $\rho(\mathbf{r})$ is initially selected and we pose the question whether this function can represent the density of a galaxy in steady-state equilibrium. Via Poisson's equation, the potential $\Phi(\mathbf{r})$ corresponding to $\rho(\mathbf{r})$ is obtained.
2. A grid of initial conditions is specified in a properly chosen subset of the phase space, e.g. on equipotential surfaces. The orbits with these initial conditions are integrated for sufficiently long time intervals. This creates a 'library of orbits' (typical number is of order $10^4$ orbits).
3. The configuration space is divided into small cells, and the time that each orbit spends inside each cell is recorded. Let $N_c$ be the number of cells, $N_o$ the number of orbits ($N_c \ll N_o$) and $t_{oc}$ the time that the orbit $o$ spends inside the cell $c$. We then assign statistical weights $w_o$ ($o = 1, \ldots, N_o$) to the orbits that represent the relative contribution of each orbit, i.e. percentage of stars that follow the same orbit, in the system. Based on these weights, it is possible to construct a *response* density, that is the density in ordinary space created by the superposition of the orbits with the above weights. The problem is then to find for which values of the weights the response density can be made to match the *imposed* density, namely $\rho(\mathbf{r})$. Mathematically, we look for solutions for $w_o$ of

$$\sum_{i=1}^{N_o} w_o t_{oc} = m_c \quad c = 1, \ldots, N_c \quad (11.85)$$

where $m_c$ is the total mass in the $c$th cell determined by the value of the imposed density $\rho(\mathbf{r_c})$ at the centre $\mathbf{r_c}$ of the cell times the volume of the cell. We furthermore impose the constraint $w_o \geq 0$, since the density contributed by each orbit can only be a positive quantity.

In most implementations of the above algorithm, a solution to (11.85), under the imposed constraints, is sought by reformulating the problem as an optimization problem. Namely, we look for weights that minimize the absolute difference, for all $c$, of the l.h.s and r.h.s of (11.85). This request takes the form of an objective function that is to be minimized. The problem is then solved by various algorithmic techniques such as linear programming solved, e.g. by the Simplex method [262, 266, 291], non-negative least squares (NNLS) [248, 263,

339], the Lucy algorithm [193, 237, 291], maximization of a suitably defined 'entropy' functional [261] and quadratic programming [217].

The main drawback in the above method is that the algorithm usually yields non-unique solutions. This is not physically unacceptable, since it is known that there can be more than one distribution functions compatible with a particular density function [248]. However, the non-uniqueness of Schwarzschild's solutions is also partly due to numerical reasons or to the fact that the problem's formulation is not sufficiently constrained. This means that the number of unknowns in the equations (i.e. the weights) is larger than the number of available equations. This is because in Schwarzschild's method we seek to match the distribution of matter only in configuration space but we ignore the velocity distribution which is an outcome of the method for any particular solution $w_o$. Thus, different solutions imply the same distribution in ordinary space but quite different distributions in velocity space. An alternative version of Schwarzschild's method has been applied [72, 73] that requests, besides the spatial density, to reproduce also the velocity profiles observed along the line of sight (Sect. 11.2.2). This new request reduces the number of solutions by increasing the number of equations, yet it still leaves some possibility for more than one solution.

Another selection criterion among different solutions is the *stability* properties of each solution. Actually this is also a drawback of Schwarzschild's method, since the latter cannot decide on whether a particular solution found by the method is stable or not. This can only be decided after an $N$-body realization of the system is prepared and run in the computer. Such a stability analysis for the original Schwarzschild [266] model by Smith and Miller [281] demonstrated that, despite the authors characterization of the model as 'robust', the model was actually *not* remaining in steady state (see also [278, p. 80]). This can be due to two reasons: (a) the self-consistent fitting is not perfect, i.e. we find weights that minimize the difference between the imposed and response density, but the difference is not actually equal to zero and (b) the model found is not stable. Smith and Miller [281] concluded that while the $N$-body system was evolving in time, the differences in time were seemingly not growing exponentially.

For most other self-consistent models of galaxies presented so far in the literature, there have been no accompanying $N$-body tests of the stability of the models. We believe, however, that such tests are indispensable, otherwise the conclusions drawn by the self-consistent modelling alone may not reflect real properties of the systems under study.

We finally mention a variant of Schwarzschild's method implemented in disk galaxies, i.e. spiral [246] or barred [170]. In these authors' approach, one has first to calculate the basic *families of periodic orbits* that (presumably) constitute the backbone of the galaxy. Then, one builds the library of orbits by considering initial conditions preferably in the neighbourhood of the periodic orbits (e.g. with a Gaussian distribution). This method can be advantageous over Schwarzschild's method in that one starts with some information as regards which orbits play the major role in the galaxy, rather than using a uniform grid of initial conditions. On the other hand, this can also turn to be a disadvantage in case the leading hypothesis about which orbits are important is wrong.

A further complication is due to the fact that there are cases in which the distribution of matter has a *minimum* rather than maximum near some particular stable periodic orbits. Such an example is shown in Fig. 11.28, referring to the same $N$-body experiment as in Figs. 11.4–11.7. In Fig. 11.28a we see that, for a particular value of the energy, there is a large island of stability near a 1:1 stable periodic orbit of the system. If, however, we find the number of $N$-body particles as a function of the distance from the central periodic orbit (Fig. 11.28b), this plot has a *minimum* rather than maximum at the position of the periodic orbit. This is because the form of this orbit is an elongated ellipse (Fig. 11.28c), but the elongation is at right angle with the elongation of the galaxy (Fig. 11.4a,b). This means that the distribution of matter has to have a minimum near the particular periodic orbit, because the shape of this orbit, and of other nearby quasi-periodic orbits, cannot support the shape of the galaxy.

### 11.4.7 The Importance of Chaos Through Self-Consistent Models of Galaxies

Schwarzschild [266] constructed self-consistent models of triaxial elliptical galaxies using only regular orbits (box and SAT). In the decade following his paper, Schwarzschild's method was implemented for the construction of self-consistent models using almost exclusively regular orbits [189, 238, 259, 260, 262, 267, 291]. These systems were either integrable (e.g. spherical or the perfect ellipsoid), or nearly integrable (e.g. axisymmetric with a flat central

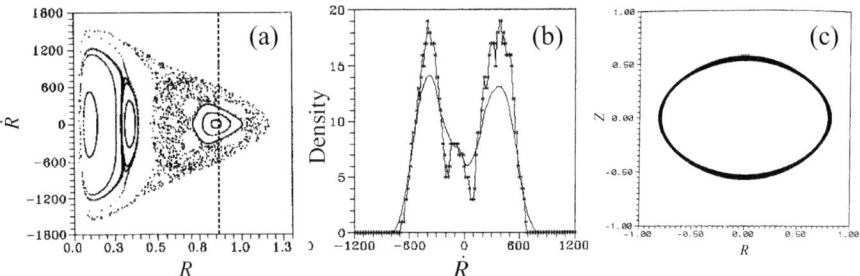

**Fig. 11.28.** (a) Same as in Fig. 11.7a. (b) The density of real particles of the $N$-body system along a vertical line passing from the centre of the 1:1 resonance (*dashed line* in (a)) has a *minimum* at the centre of the resonance. (c) The form of a very thin tube orbit in the meridional plane around the stable 1:1 periodic orbit. The orbit is elongated perpendicularly to the $Z$-axis, while the system as a whole is elongated along the $Z$-axis (Fig. 11.4)

density profile). Goodman and Schwarzschild [127] pointed out that a large number of the orbits that were considered as regular in their models were in reality weakly chaotic (called 'semistochastic'). However, these orbits exhibited a nearly regular behaviour for times comparable to the galaxy's lifetime. This means that the orbits obey an approximate integral of motion (Sect. 11.2.3), or that they exhibit the 'stickiness phenomenon' (Sect. 11.2.3) associated with slow Arnold diffusion in phase space.

On the other hand, as we have seen above, Schwarzschild [268] searched for self-consistent solutions of cuspy halo models of the form $\varrho \propto r^{-2}$ near the centre. In these models, he found a large number of chaotic orbits. For systems with a moderate ellipticity (smaller than E5), it was in fact possible to find multiple solutions that contained either a mixed population of regular and chaotic orbits or only regular orbits. But for more elongated models (ellipticities larger than E5) it was not possible to find any solution composed entirely by regular orbits. It thus became evident that cuspy models of elliptical galaxies render necessary the presence of chaotic orbits.

Merritt and Fridman [217] studied the construction of self-consistent models in systems with the potential function corresponding to the $\gamma$ model for the values $\gamma = 1$ (weak cusp) and $\gamma = 2$ (strong cusp). In both models the axial ratios were $c/a = 0.5$ and $b/a = 0.79$. These correspond to a case of a maximally triaxial E5 galaxy, i.e. a galaxy with triaxiality

$$T = \frac{a^2 - b^2}{a^2 - c^2} \tag{11.86}$$

equal to $T \simeq 0.5$ ($T = 0$ in oblate systems ($a = b$), and $T = 1$ in prolate ($b = c$) systems). Furthermore, the same authors quantified the distinction of orbits into regular or chaotic by means of the Lyapunov characteristic number. They concluded that efforts to construct self-consistent models without the presence of chaotic orbits were unsuccessful in both cases of weak or strong cusps. If instead chaotic orbits are included in the same manner as regular ones (one weight assigned to each chaotic orbit), then it becomes possible to find successful models. In fact, these models are *not* precisely stationary, because the chaotic orbits exhibit observable diffusion in phase space, especially at low energy levels (where diffusion times are much smaller than the Hubble time). As a result, the superposition of orbits creates a model that changes shape in time. Furthermore, many chaotic orbits do not manage to fill ergodically the whole available connected chaotic domain during an integration time $\approx 100 T_{\text{dyn}}$, which corresponds to the age of the system. Nevertheless, by ignoring chaotic orbits at low energy levels (which are responsible for the fastest diffusion), it was possible to produce quasi-stationary solutions that provided good self-consistent models, especially in the case of weak cusp.

The next step in the same study was the search of *fully mixed* solutions. These are solutions in which all the chaotic orbits of the library that correspond to the same value of the energy can essentially be considered as different pieces of only *one* orbit. If this can be established, the library representatives of this one orbit do not cause a macroscopic change in the shape of the response density field due to chaotic diffusion. This means that the resulting models are guaranteed to be stationary. Merritt and Fridman [217] concentrated on such solutions applicable to the chaotic orbits of low energy levels. This was completely successful in the case of weak cusps, but only partly successful in the case of strong cusps. The so-found 'fully mixed' solutions yielded large percentages of chaotic orbits in the final orbital composition, up to 45% in the case of weak cusps, and 60% in the case of strong cusps.

## 11.5 The $N$-Body Approach

### 11.5.1 Numerical Integration of the $N$-Body Problem

The simplest method to integrate the $N$-body problem, with $N$ large, is the so-called *direct* method, namely the direct numerical solution of the softened equations of motion

$$\ddot{\mathbf{r}}_\mathbf{i} = -\sum_{j=1, j\neq i}^{N} \frac{Gm_j(\mathbf{r_i} - \mathbf{r_j})}{(|\mathbf{r_i} - \mathbf{r_j}|^2 + \varepsilon^2)^{3/2}} \qquad (11.87)$$

for each particle $i = 1, \ldots, N$. The softening parameter $\varepsilon$ is introduced in order to avoid the singular behaviour of the Keplerian force whenever two particles have a close approach. The value of $\varepsilon$ should be such that the accelerations implied by (11.87) 'mimic' the collisionless character of the system under study (Sect. 11.2.1), without, however, introducing a large error to Newton's law. Typical values are a fraction of $D/N^{1/3}$, where $D$ is the typical scale length of the integrated system. Special softening techniques are often combined with numerical implementations of regularization techniques (in the case of two-body encounters) or approximate regularization (for triple encounters). A review of these techniques is made by Aarseth [2], a leading scientist in this field over decades.

The direct method is very accurate, but its algorithmic complexity is $O(N^2)$, hence prohibitive for $N$ large. For this reason, Aarseth has developed codes that use a direct summation only for neighbouring particles, while they use a multipole expansion of the potential, or the force, for groups of distant particles. The algorithm used to split the particles into neighbouring or distant was introduced by Ahmad and Cohen [4]. The introduction of this scheme reduces the algorithmic complexity from $O(N^2)$ to about $O(N^{1.5})$.

Another direction followed in order to reduce the computational cost of $N$-body calculations was the construction of special hardware (GRAPE, e.g. [202, 203, 294]) based on specially designed chips to perform the sum (11.87) with speed exceeding by orders of magnitude any program written in conventional programming languages.

In simulations of galactic systems we are usually not interested in having an accuracy comparable to that requested in Celestial Mechanics. In the latter case the integration must often extend over billions of periods of the solar system bodies, while in the former case we are interested in integration times of order $10^2$–$10^3$ periods of the stars. Furthermore, in galactic dynamics we usually pose questions regarding the collective behaviour of the system, which do not require an accuracy of integration of individual orbits as high as in celestial mechanical calculations.

Such differences have led to the consideration of special techniques to integrate the $N$-body problem with $N$ large. Besides traditional 'particles-in-cell' or 'grid' methods (see the review by Sellwood [269]) that are applicable also to plasma physics, there are two methods that fit particularly the nature of the gravitational $N$-body problem: the TREE method [19, 144, 209] and the smooth potential field method [5, 51, 52, 143, 333]. The so-called 'spherical harmonics' method [208, 216, 317] is a hybrid method similar to the smooth field code but with a 'stepwise' numerical calculation of the radial part of the spherical harmonics expansion of the potential of the system [269], that requires sorting of the particles with respect to their distances from the centre. There have also been simulations in which the collisionless Boltzmann equation (11.16) is solved directly. This is numerically tractable in systems forced to retain a particular symmetry (usually spherical). The Boltzmann equation can be integrated either through calculation of its moments [146] or by its characteristic system of ordinary differential equations [39, 137, 140].

The main idea of the TREE algorithm [18] is to define a criterion by which, considering, say, the $i$th particle of the system, the other particles can be divided in groups of close or distant particles with respect to the position of the $i$th particle. The forces on the $i$th particle by close particles are added by direct summation. However, the forces by distant particles are added by considering one term for each group rather than for each particle. This effectively reduces the algorithmic complexity of the code from $O(N^2)$ to $O(N \log N)$.

In the algorithm of Barnes and Hut [18] all the particles are set initially in one cubic cell of volume $s_0$. This cell is divided by consecutive bisections into subcells of volume $s_k = s_0/2^{3k}$, where the index $k = 1, 2, \ldots$ denotes the order of division. If, for some order $k$, one subcell contains no more than one particle, this subcell is not further divided, otherwise the subdivision continues at order $k+1$. The data structure storing the hierarchy of all subcells is called the 'tree'.

A 'tolerance parameter' $\theta_{\text{tol}}$ is also defined to distinguish close cells from distant cells with respect to the position of one particle (suggested values are around $\theta_{\text{tol}} \approx 1$, [144]). Considering the $i$th particle, a cell is called distant if the following condition holds:

$$r \geq \frac{s_k^{1/3}}{\theta_{\text{tol}}} \tag{11.88}$$

where $r$ is the distance of the cell from the $i$th particle. If condition (11.88) is true, the particles in the cell are viewed as one group of mass $M_c$ (the sum of the particles' masses). The contribution of this group to the force on the $i$th particle is calculated by a multipole expansion (usually up to quadrupole terms).

The TREE method is very efficient. While the integration time is drastically reduced, there are many different cases of $N$-body experiments that can be effectively treated with TREE. Examples are (a) collapsing galaxies [41, 76, 320, 324] (b) merging galaxies [16, 17, 40, 142, 319], (c) multiple merger events [102] and (d) cosmological simulations (where TREE is often combined with a particle-mesh (PM) algorithm, [38, 177]). This flexibility is due to the fact that the TREE code can follow simultaneously the evolution of different parts of a system that may have large density contrasts or a rapidly varying spatial distribution. For these reasons, TREE codes or hybrid TREE–PM codes have been developed continually over the years, resulting in drastic improvements of the $O(N \log N)$ scaling [84] and in parallel implementations of the algorithm for either galactic or cosmological simulations (e.g. [21, 37, 91, 177, 228, 319, 331]). The TREE code can also be combined with the special hardware GRAPE [14, 112].

On the other hand, the main disadvantage of the TREE code is that it does not allow one to have the potential function of the system $\Phi(\mathbf{x}, t)$ in a closed analytical form. This means that one cannot easily make global dynamical studies with TREE, such as, e.g. the calculation of orbits, variational equations, Poincaré sections, frequency maps, etc. (Sect. 11.4).

The class of *self-consistent field codes* is particularly suited to global dynamical studies of galaxies. The main idea in such codes is that a spatial distribution of particles represents a Monte Carlo realization of an ideally smooth density field. The smooth density $\rho$ is given by (11.6), i.e. in terms of a smooth distribution function $f$. If the system's geometry is not very peculiar, the smooth function $\rho(\mathbf{x})$ can be expanded in a truncated series of basis functions. Different basis functions can be chosen tailored to the particular properties of the system under study.

We shall follow the formalism of Weinberg [333] in order to show the method to obtain suitable sets of basis functions for triaxial systems. If we anticipate that the average density profile of the system that is to be simulated will not be very different from a model function $\rho_{00}(r)$, we express the monopole term of the density as a truncated series of the form:

$$\rho_{\text{monopole}}(r) = \rho_{00}(r) \sum_{n=0}^{n_{\max}} b_{n00} u_{n00}(r) \qquad (11.89)$$

The sum in the r.h.s. represents the residuals of the fit of the monopole term of the real density of the system by the model density $\rho_{00}(r)$. The coefficients $b_{n00}$ are unknown and the main task of the $N$-body code is to find their values. The functions $u_{n00}(r)$, on the other hand, are known functions which are eigenfunctions of a Sturm–Liouville problem specified below. We can similarly express all multipole contributions to the density, i.e. we fix some model functions $\rho_{ml}(r)$ and express the density as:

$$\rho(r,\theta,\phi) = \sum_{l=0}^{l_{max}} \sum_{m=-l}^{l} \sum_{n=0}^{n_{max}} b_{nml}\rho_{ml}(r)u_{nml}(r)Y_l^m(\theta,\phi) \tag{11.90}$$

with functions $u_{nml}(r)$ specified by a Sturm–Liouville problem and coefficients $b_{nml}$ calculated by the $N$-body code.

The Sturm–Liouville problem for $u_{nml}(r)$ is formulated as follows: writing the potential in a form similar to (11.90)

$$\Phi(r,\theta,\phi) = \sum_{l=0}^{l_{max}} \sum_{m=-l}^{l} \sum_{n=0}^{n_{max}} c_{nml}\Phi_{ml}(r)u_{nml}(r)Y_l^m(\theta,\phi) \tag{11.91}$$

we couple equations (11.90) and (11.91) via Poisson equation (11.7) in spherical coordinates. After the separation of variables, this leads to:

$$-\frac{d}{dr}\left(r^2\Phi_{ml}^2 \frac{du_{nml}}{dr}\right) + \left[l(l+1)\Phi_{ml}^2 - \Phi_{ml}\frac{d}{dr}\left(r^2\frac{d\Phi_{ml}}{dr}\right)\right]u_{nml} =$$
$$-(4\pi G\lambda_{nml}r^2\Phi_{ml}\rho_{ml})u_{nml} \tag{11.92}$$

with $\lambda_{nml} = b_{nml}/c_{nml}$. This equation, supplemented with appropriate boundary conditions at two particular radii $r_a$ and $r_b$ is a case of the Sturm–Liouville eigenvalue problem

$$-\frac{d}{dx}\left[p(x)\frac{dy}{dx}\right] + q(x)y = \lambda w(x)y \tag{11.93}$$

with

$$p(x) = x^2\Phi_{ml}^2(x)$$
$$q(x) = l(l+1)\Phi_{ml}^2(x) - \Phi_{ml}(x)\frac{d}{dx}\left(x^2\frac{d\Phi_{ml}(x)}{dx}\right). \tag{11.94}$$
$$w(x) = -4\pi Gx^2\Phi_{ml}(x)\rho_{ml}(x)$$

The functions $u_{nml}(r)$ are eigenfunctions of a differential operator acting on $u_{nml}$ in the l.h.s. of (11.92). Since this operator does not depend on $n$, the index $n$ can be identified to the serial index of successive eigenvalues and eigenvectors, starting from the ground state value $n = 0$. The problem is well defined if boundary conditions are given in the form

$$a_1 u - a_2\left(p(r)\frac{du}{dr}\right) = \lambda\left(a'_1 u - a'_2\frac{du}{dr}\right) \quad \text{at} \quad r = r_a$$

$$b_1 u + b_2\left(p(r)\frac{du}{dr}\right) = 0 \quad \text{at} \quad r = r_b \tag{11.95}$$

In galactic problems, the radii $r_a, r_b$ are set equal to $r_a = 0$ (centre of the system), and $r_b = R_p$ or $r_b \to \infty$. If $r_b = R_p$, the radius $R_p$ is set to represent the size of the system, and the boundary conditions at $R_p$ are obtained by the request of continuity, and continuous derivative, of the potential function at the point $R_p$ where we pass from Poisson to Laplace equation.

If the model functions $\Phi_{ml}(r)$, $\rho_{ml}(r)$ satisfy Poisson's equation, they are called *potential–density pair* functions. The eigenfunctions of the Sturm–Liouville problem (11.92) are mutually orthogonal with respect to the inner product definition:

$$<f|g> = \int_{r_a}^{r_b} f(r)g(r)w(r)dr \qquad (11.96)$$

The above equations provide the general framework of the self-consistent field method. In order to have a concrete $N$-body implementation we proceed by the following steps:

(a) Specify a set of potential–density model functions $\Phi_{ml}(r)$, $\rho_{ml}(r)$. These are arbitrary functions which may or may not really depend on the indices $l$ or $m$.
(b) Substitute the functions $\Phi_{ml}(r)$ and $\rho_{ml}(r)$ in (11.92) and solve the Sturm–Liouville problem. This will specify the eigenfunctions $u_{nml}$ and eigenvalues $\lambda_{nml}$. Although a numerical solution of (11.92) is in principle possible for any choice of $\Phi_{ml}(r), \rho_{ml}(r)$, we prefer to use sets of functions for which the solution of (11.92) is reduced to known functions from the literature. Examples are:
   – The Clutton-Brock [52] set:

$$\rho_{ml}(r) = \left(\frac{1}{\sqrt{4\pi}}\right) \frac{r^l}{(1+r^2)^{l+5/2}} (2l+1)(2l+3)$$

$$\Phi_{ml}(r) = -\sqrt{4\pi} \frac{r^l}{(1+r^2)^{l+1/2}} \qquad (11.97)$$

in which the eigenfunctions $u_{nlm}(r)$ are Gegenbauer polynomials of the form $C_n^{l+1}(\xi)$ where $\xi = (r^2-1)/(r^2+1)$
   – The set of Allen et al. [5]

$$\rho_{ml}(r) = -1$$
$$\Phi_{ml}(r) = 1 \qquad (11.98)$$

where the eigenfunctions $u_{nml}(r)$ are spherical Bessel functions
   – The Hernquist–Ostriker [143] set

$$\rho_{ml}(r) = \sqrt{4\pi} \frac{1}{2\pi} \frac{(2l+1)(l+1)}{r} \frac{r^l}{(1+r)^{2l+3}}$$

$$\Phi_{ml}(r) = -\sqrt{4\pi} \frac{r^l}{(1+r)^{2l+1}} \qquad (11.99)$$

where the eigenfunctions $u_{nml}$ are Gegenbauer polynomials of the form $C_n^{2l+\frac{3}{2}}(\xi)$ where $\xi = (r-1)/(r+1)$.

In the case of galactic disks, potential–density pairs were proposed by Clutton-Brock [51], Kalnajs [159], Aoki and Iye [8] and Earn [94]. We note that such pairs are also extremely useful in the study of the stability properties of galaxies [241].

(c) Given the positions of the $N$ particles, we calculate the coefficients $b_{nml}$ and $c_{nml}$ of the full density and potential expansions (11.90) and (11.91). We determine first the coefficients $b_{nml}$ by exploiting the orthogonality of the functions $u_{nml}$, i.e. $<u_{nml}|u_{n'ml}> = \delta_{n,n'}$, as well as the orthogonality of the spherical harmonic functions. If we multiply both sides of (11.90) by $w(r)u_{nml}(r)Y_l^m(\theta,\phi)$ and take the integral over all positions we find:

$$b_{nml} = \int_{r_a}^{r_b} \int_0^\pi \int_0^{2\pi} -4\pi\phi_{ml}(r)r^2\rho(r,\theta,\phi)u_{nml}(r)Y_l^{m*}(\theta,\phi)\mathrm{d}r\,\mathrm{d}\theta\,\mathrm{d}\phi \qquad (11.100)$$

Assuming that the positions of the $N$ particles provide a Monte Carlo sampling of the function $\rho(r,\theta,\phi)$, the triple integral in (11.100) can be approximated by a sum over particles:

$$b_{nml} \simeq \sum_{i=1}^{N} -4\pi\phi_{ml}(r_i)u_{nml}(r_i)Y_l^{m*}(\theta_i,\phi_i) \qquad (11.101)$$

The potential coefficients $c_{nml}$ are finally determined via the relation $c_{nml} = b_{nml}/\lambda_{nml}$.

The use of the Monte Carlo integration method implies that, contrary to what the term 'smooth field' might signify, there is some sort of noise in the system introduced by its discreteness. This noise appears in a quite different way than in the direct or TREE algorithm. Namely, the uncertainties, or inevitable small fluctuations of the coefficients $b_{nml}$ during the simulation of even an 'equilibrium' system result in a relaxation of this system which is essentially due to discreteness effects (see [332] for a detailed discussion). However, this noise is reduced, in general, as the number of particles $N$ increases. As $N$ increases we may also use a larger number of basis functions to fit the density, or the potential.

Since the calculation of the sum (11.101) has an $O(N)$ algorithmic complexity, the overall complexity of a smooth field code scales linearly with $N$. This, in combination with the fact the parallelization of the sum (11.101) is straightforward, renders such codes very powerful even in small computers or computer clusters, with applications reaching $N = 10^7$–$10^8$.

Besides its linear algorithmic complexity, the main power of a smooth field code lies in that the outcome of the potential evaluation can be expressed analytically in terms of a series of basis functions. This allows one to have the Hamiltonian of the system in closed form, a fact which greatly facilitates the orbital or global dynamical study of the system. The use of smooth field codes has been very fruitful in galactic dynamics so far, and we can anticipate only better prospects for the future.

### 11.5.2 The Global Dynamics of $N$-Body Systems

We have already discussed Schwarzschild's method for the construction of self-consistent models of galaxies as well as the limitations of this method (Sect. 11.4.5). Such limitations are not present if one uses $N$-body simulations. The equilibria of such simulations are by definition self-consistent and stable. Thus, questions like what is the relative importance of ordered or chaotic orbits in a galaxy are better answered within the framework of global dynamical studies of $N$-body systems. The $N$-body method allows one to deal also with systems exhibiting secular evolution such as, e.g. systems with a CMC or central black hole.

An early example of orbital analysis in triaxial systems resulting from $N$-body simulations was given by Udry and Martinet [307]. These authors presented histograms yielding the distribution of particles with respect to the ratios of the fundamental frequencies of their orbits. They subsequently discuss the link between the orbital structure and the shape of the $N$-body systems.

A extended study of global dynamics in $N$-body systems was made by Voglis et al. [327], Kalapotharakos et al. [158] and Kalapotharakos and Voglis [157]. This is a study of $N$-body systems in equilibrium resulting from the cosmological simulations described in Sect. 11.2.5. We have seen that spherically symmetric (or 'quiet') initial conditions lead to very elongated galaxies as a result of the radial orbit instability. We call such an experiment the 'Q-system'. On the other hand, clumpy initial conditions lead to a less elongated final state (the 'C-system'). These systems remain in a steady state for long time periods ($\gg 1 t_{\text{Hubble}}$). The smooth field code of Allen et al. [5] is used to calculate a smooth analytic potential and the corresponding density. These are given by:

$$\Phi(r,\vartheta,\varphi) = -\frac{G}{R_0} \sum_{l=0}^{\infty} \sum_{n=0}^{\infty} \sum_{m=-n}^{n} b_{lmn} j_n \left( a_{ln} \frac{r}{R_0} \right) P_n^{|m|}(\cos\vartheta) e^{im\varphi} \text{ if } r \leq R_0$$

(11.102)

$$\Phi(r,\vartheta,\varphi) = -\frac{G}{R_0} \sum_{l=0}^{\infty} \sum_{n=0}^{\infty} \sum_{m=-n}^{n} b_{lmn} j_n (a_{ln}) \left( \frac{R_0}{r} \right)^{n+1} P_n^{|m|}(\cos\vartheta) e^{im\varphi} \text{ if } r > R_0$$

(11.103)

$$\rho(r,\vartheta,\varphi) = s \frac{1}{4\pi R_0^3} \sum_{l=0}^{\infty} \sum_{n=0}^{\infty} \sum_{m=-n}^{n} a_{ln}^2 b_{lmn} j_n \left( a_{ln} \frac{r}{R_0} \right) P_n^{|m|}(\cos\vartheta) e^{im\varphi} \text{ if } r \leq R_0$$

(11.104)

where $b_{lmn}$ are the coefficients of the expansion and $R_0$ is a parameter of the code fixing the outermost radius inside which Poisson's equation is solved.

The functions $j_n(r)$ are spherical Bessel functions and $P_n^m$ are Legendre polynomials.

When calculating orbits, care is needed as regards the characterization of an orbit as regular or chaotic. The usual criterion of the Lyapunov characteristic number

$$\text{LCN} = \lim_{t \to \infty} \frac{1}{t} \ln \left| \frac{\xi(t)}{\xi(0)} \right| \qquad (11.105)$$

cannot be applied in a straightforward manner in the case of galaxies. The reason has to do with the (inevitably) finite time of numerical integration of the orbits (in order to obtain an estimate of the LCN), that has to be compared with the lifetime of the system. In fact, the periods of the orbits of the stars in the inner and outer parts of a galaxy differ by about three orders of magnitude. This suggests that the Lyapunov times of the orbits (inverse of the LCNs) should be normalized with respect to the periods of orbits. On the other hand, the lifetime of the galaxy defines a second relevant timescale, i.e. an orbit is effectively chaotic only if its Lyapunov time is of the order of the galaxy's lifetime or smaller. This second timescale is uniform for all orbits. The question then is what is a proper normalization of Lyapunov times (or Lyapunov exponents) that provides a fair measure of the 'chaoticity' of an orbit.

In order to address this question, Voglis et al. [327] introduced a new type of calculation based on the combination of two methods: (a) the 'specific time Lyapunov number' $L_j$, normalized with respect to the orbit's inverse of the period and (b) the alignment index (AI). The alignment index is a numerical method based on certain properties of the time evolution of deviation vectors [280, 322, 323]. Along regular orbits, the index AI has a value close to unity while along chaotic orbits it tends exponentially to zero.

In order to make the distinction of the orbits, one considers a particular snapshot of the system and uses the potential expansion (11.102–11.103) as a time-independent potential in which orbits can be calculated. The orbits with initial conditions given by the positions and velocities of the N-body particles are integrated. The distinction of the orbits in the case of the Q-system, after an integration for $N_{\text{rp}} = 1200$ radial periods of each orbit, is shown in Fig. 11.29. The triangular group of points in the down-right part of the diagram corresponds to regular orbits with $\log(\text{AI}) \gtrsim -3$ and Lyapunov numbers decreasing in time as $t^{-1}$. The group of points in the up-left part ($\log(\text{AI}) < -12$) are chaotic orbits. The index AI of such orbits reaches, after a fast decrease, the accuracy limit of the computer program. Furthermore, their calculation of the LCN stabilizes to a positive limit. We also distinguish a lane of points connecting the two groups. These are particles in weakly chaotic orbits. In this case the time evolution of the index AI shows a relatively slow decrease, compared to that of strongly chaotic orbits.

Based on the above method, the bodies found in chaotic orbits are 23% and 32% of the total mass in the C- and Q-systems, respectively. When the whole procedure was repeated after 100 half-mass crossing times ($T_{\text{hmct}}$), the above

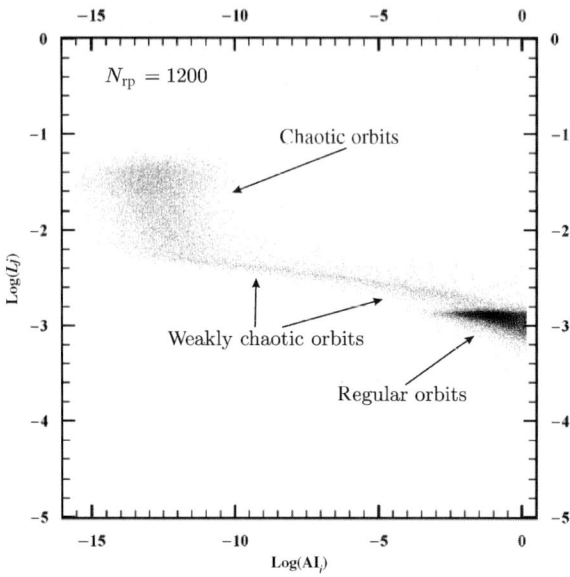

**Fig. 11.29.** The distinction of regular and chaotic orbits on the $(\log L_j - \log \mathrm{AI}_j)$ plane. The orbits with small values of $\mathrm{AI}_j$ and stabilized values of $L_j$ are identified as chaotic orbits. The orbits with large values of $\mathrm{AI}_j$ ($> 10^{-3}$) and decreasing values of $L_j$ (as $t^{-1}$) are identified as regular orbits. The orbits on the lane joining the two regions are weakly chaotic orbits (after [327])

percentages remained essentially unaltered. Keeping track of the identities of particles that were characterized in regular or chaotic orbits, there was no change of character except for a 3% of the particles.

It should be stressed that only a fraction of the particles characterized chaotic can, in fact, develop appreciable as chaotic diffusion within a Hubble time. The estimated percentage of such particles is less than 8%. Although most orbits are only weakly chaotic, the spatial distribution of the chaotic mass component is very different from that of the regular component (Fig. 11.30). Namely, the chaotic component is distributed rather spherically, i.e. isotropically, while the regular component has a spatial distribution elongated in the direction of the long axis of the system. Moreover, contrary to the regular component, the chaotic component has a flat central surface density profile (Fig. 11.31). The superposition of the two profiles creates a hump in the total surface density profile roughly at the point where the two profiles cross each other. In the case of the Q-system, this transition is manifested also in the ellipticity profile (Fig. 11.32) which has an abrupt decrease by two units from the inner region, where regular orbits dominate, to the outer region, where chaotic orbits dominate.

11 Special Features of Galactic Dynamics    373

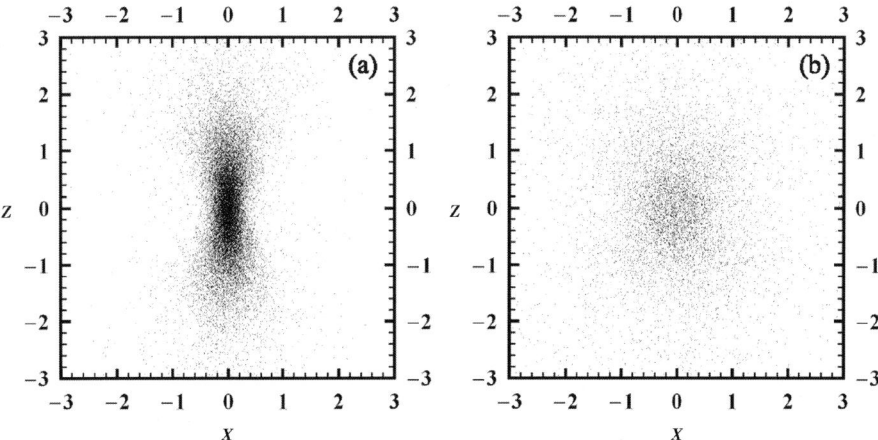

**Fig. 11.30.** Projection of the particles in (**a**) regular and (**b**) chaotic orbits on the plane of short–long axis. The distribution of chaotic orbits is almost spherical while the distribution of regular orbits is strongly elongated along the long axis of the system (after [327])

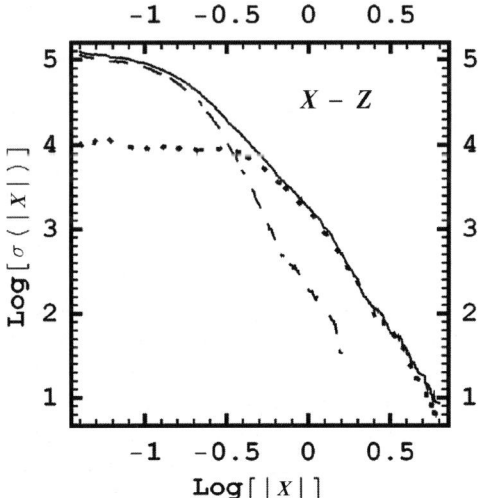

**Fig. 11.31.** Profiles of the surface density on the projected plane of short–long axis for the regular component (*dashed line*), chaotic component (*dotted line*) and the overall system (*solid line*). The superposition of the two profiles (regular and chaotic) creates a hump in the overall surface density profile roughly at the point where the two partial profiles cross each other (after [327])

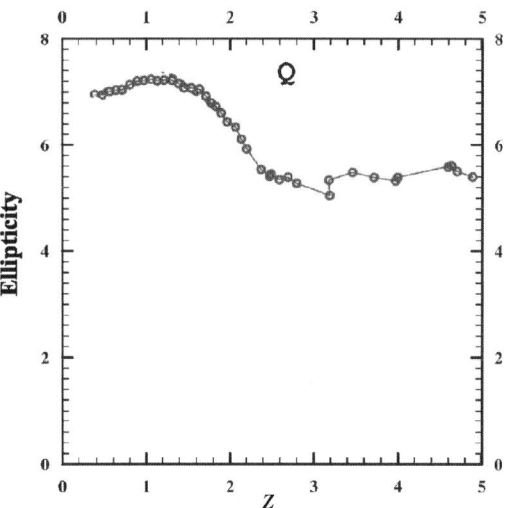

**Fig. 11.32.** The ellipticity of the Q-system on the short–long axis plane as a function of the distance from the centre along the long axis $z$. The abrupt decrease of the ellipticity at about $r_{rc} = 2$ marks the transition from a region where regular orbits dominate (inside $r_{rc}$) to a region where chaotic orbit dominate (outside $r_{rc}$) (after [327])

The conclusion is that there is an upper limit to the ellipticity of galaxies containing chaotic orbits, while systems with only regular orbits can reach much larger values of the ellipticity.

In a similar study, Muzzio et al. [231] found that the fraction of mass in chaotic motion in their system (similar to the Q-system above) is about 53%. This larger fraction is due to two reasons: (a) the use of a smaller threshold in the Lyapunov number for the characterization of an orbit as chaotic and (b) a less flat density profile near the centre. However, the different spatial distribution of the particles in regular or chaotic orbits (as in Fig. 11.30), was also found in the experiments of Muzzio et al.

There is a variety of different orbital structures that are able to support systems with smooth centres [70, 155, 156, 157]. This seems to apply both to regular and chaotic orbits. The self-consistency condition imposes some restrictions on the permissible orbital distributions. For example, the Q-system, being very elongated, is formed by particles moving almost exclusively in box orbits, and it has only a small fraction of particles in tube types (SAT or LAT). This, despite the fact that the SAT and LAT types of orbits are very stable and occupy an extended domain in phase space. On the other hand, in the C-system, which is more spherical, the particles move preferentially in tube orbits, especially SAT. This fact clearly shows that even if a global dynamical analysis establishes the existence of large domains of stability in phase space, this does not imply that the real particles of the $N$-body system will fill these domains. The preferential domains in phase space are only partly determined by the regular or chaotic character of the orbits. The other determining factor is the request for self-consistency.

Figure 11.33 shows the orbital content of the Q-system in the frequency space (the fundamental frequencies are calculated by the algorithm of [276]). The points correspond to particles moving on regular orbits (Fig. 11.33a), and chaotic orbits (Fig. 11.33b). In Fig. 11.33a we can distinguish the distribution of particles in different domains of the frequency space, according to whether an orbit is of the box type or one of the tube subtypes. The points in Fig. 11.33b show a scatter due to their chaotic character, which implies variability of the frequencies. The orbits with large variability of frequencies have also large Lyapunov characteristic numbers. Nevertheless, there are also chaotic orbits that remain localized along the resonance lines of other, regular, orbits. These are weakly chaotic orbits which are temporarily trapped in particular resonances, diffusing mostly along the resonance lines and only marginally across these lines.

### 11.5.3 Secular Evolution Under the Presence of a CMC: Self-Organization

Secularly evolving models can be created by inserting a black hole, or CMC, to a Q- or C-system [158]. The evolution and the properties of these systems depend on the value of the relative mass parameter $m = M_{\rm cmc}/M_{\rm galaxy}$. We consider values of $m$ in a range [0.0005, 0.01]. Just after the insertion of the CMC, the fraction of mass in chaotic motion increases suddenly to the level

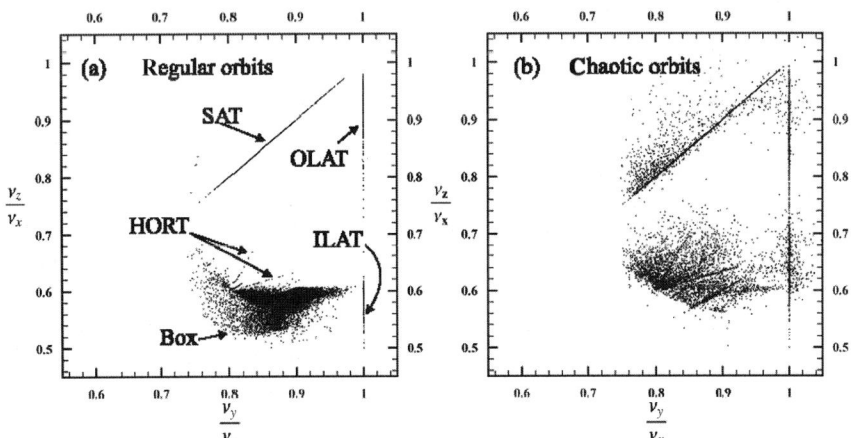

**Fig. 11.33.** Frequency maps (rotation numbers) for (**a**) the regular and (**b**) the chaotic orbits of the Q-system. The regions of various types of regular orbits are marked in (**a**). In (**b**) many chaotic orbits are concentrated in particular resonance lines and they diffuse mainly along these lines (after [157])

of 80% in systems generated by the Q-system, and 50% in systems generated by the C-system. The sudden rise of the chaotic component causes a secular evolution in these systems (Sect. 4.6). This is mainly due to the anisotropic, i.e. *non-mixed* distribution of the chaotic orbits caused by the fact that, before the insertion of the CMC, these were mostly regular orbits (boxes) of the original system. Voglis and Kalapotharakos [326] found that the mean rate of exponential divergence (or mean level of LCN) of this chaotic component has a narrow correlation with $m$, scaling as $m^{1/2}$. Furthermore, in order to measure the effectiveness of chaotic diffusion, these authors defined a parameter called 'effective diffusion momentum' $\mathcal{L}$ as the product of the anisotropically distributed chaotic mass times the mean logarithmic divergence of the orbits of this mass. Numerically, it is found that the parameter $\mathcal{L}$ measures the ability of secular evolution of the system. Namely, if $\mathcal{L} \lesssim 0.0045$ there is negligible secular evolution due to chaotic diffusion even for times longer than a Hubble time. On the other hand, if $\mathcal{L} \gtrsim 0.0045$ the models evolve following a process of self-organization that converts chaotic orbits to regular. The resulting reduction of entropy is partly balanced by the increase of the mean level of exponential divergence of the remaining chaotic orbits. During the whole process, the fraction of chaotic mass distributed anisotropically decreases in time, resulting in smaller values of $\mathcal{L}$. The evolution ceases when $\mathcal{L}$ goes below the value 0.0045.

Figure 11.34 shows the distributions of the Lyapunov numbers for all the systems after half a Hubble time from the CMC insertion. Smaller CMCs produce in general smaller Lyapunov numbers (the Lyapunov number at the peak of the distribution scales with $m$ as $m^{1/2}$). This also implies a slower rate of secular evolution. In fact, the morphology of systems with small CMCs ($m \leq 0.001$) remains close to the morphology of the original Q or C systems for at least a Hubble time, as indicated by a plot of the time evolution of the triaxiality index $T$ (Fig. 11.35). The fact that regular orbits of the original systems are now characterized as weakly chaotic does not have serious consequences in the resulting morphology of the systems.

In systems with $m \geq 0.005$ the secular evolution is faster, and it leads from a prolate, or maximally triaxial shape to a final equilibrium which is characterized either by almost zero triaxiality (oblate), or moderate triaxiality, depending on the size of $m$ and on the initial orbital distribution of the system at the time when the CMC is inserted. Larger CMCs and small initial percentages of tube orbits (like in the Q-system) favour oblate final equilibria.

During the secular evolution of the systems, the fraction of mass in chaotic motion decreases in time, and in the final equilibrium it reaches a range 12–25%. As already mentioned, the systems present strong indications of *self-organization*. This means that in the course of secular evolution, many chaotic orbits are gradually converted into regular orbits of the SAT type. This process can be understood with the help of Figs. 11.36 and 11.37. Figure 11.36a–d shows projections of the 4D Poincaré sections at successive snapshots of the secular evolution of a Q-system with a CMC $m = 0.01$. The phase portraits

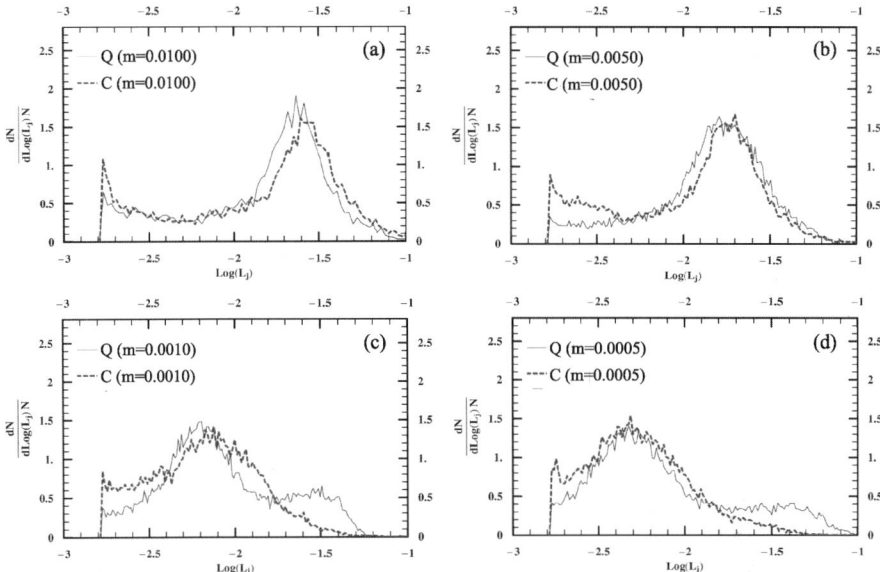

**Fig. 11.34.** The distribution of the real particles with respect to values of $\log L_j$ in four models with CMCs. For larger values of the mass $m$, the maximum of the distributions is shifted at larger values of $\log L_j$ (after [158])

in the background are obtained by integrating many reference orbits in a potential frozen at the time corresponding to each snapshot. On the other hand, the orbits of the real particles of the $N$-body system are integrated in an evolving $N$-body potential and superposed on the phase portraits at different snapshots. Figure 11.36 shows the successive Poincaré consequents (stars or dots) of one orbit of a real particle. Initially, before the insertion of the CMC, this is box orbit. Thus, immediately after the insertion this orbit becomes a chaotic orbit yielding Poincaré consequents on the chaotic domain of the surface of section (stars). However, as the volume of the regular domain progressively increases, the character of the orbit is converted at a particular moment from chaotic to regular (open circles). The new regular orbit is of the SAT type. As a result of many such orbits, the triaxiality parameter $T$ of the system decreases.

Figure 11.37a–d shows the particles of the Q-system (with $m = 0.01$) on the plane of rotation numbers at four different snapshots ($t = 30, 90, 150, 210$, a Hubble time corresponds to 300 time units). Initially the box orbits of the original Q-system are converted to chaotic orbits or they are trapped along the various resonance lines of HORT orbits. As the chaotic orbits diffuse in the phase space the system's geometry changes. Namely, the system becomes less elongated and its triaxiality parameter $T$ decreases. Due to this evolution, some particles are trapped in orbits confined on SAT tori. During this self-consistent evolution, the areas of HORT and ILAT orbits move upwards,

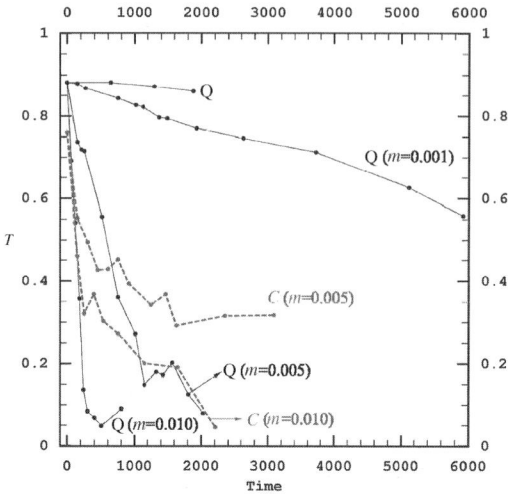

**Fig. 11.35.** The time evolution of the triaxiality parameter $T$ in various systems. One Hubble time corresponds to $t_{\text{Hub}} \approx 300$. The systems (Q, $m = 0.01$), (Q, $m = 0.005$) and (C, $m = 0.01$) reach an oblate ($T = 0$) final equilibrium state. Only the system (Q, $m = 0.01$) achieves this equilibrium within a Hubble time. The (C, $m = 0.005$) system reaches an equilibrium with modest triaxiality. Systems with smaller CMCs ($m \leq 0.001$) do not appear to evolve significantly within a Hubble time (e.g. system (Q, $m = 0.001$) (after [272])

approaching the line of SAT orbits. The number of SAT orbits increases while the number of all other orbital types decreases. At the equilibrium position there are only regular orbits of SAT type and chaotic orbits.

Kalapotharakos et al. [158] concluded that a system must have a CMC of at least $m \approx 0.01$, in order to complete its evolution and reach a new equilibrium state within a Hubble time. For smaller mass parameters the evolution to equilibrium is prolonged over many Hubble times. The exact evolution rate of the systems with CMCs depends mainly on two factors: (a) the fraction of non-mixed chaotic orbits and (b) the mean Lyapunov number of these orbits [326]. For example, the C-system with $m = 0.01$ needs longer time than the Q-system with the same $m$ in order to evolve towards the final oblate configuration (Fig. 11.35). This is because the C-System, as we have seen above, had initially a smaller fraction of box orbits than the Q-system. Therefore, after the insertion of the CMC, the C-system has a smaller fraction of chaotic orbits than the Q-system.

These results are in agreement with previous studies of Merritt and Quinlan [218] and Holley-Bockelmann et al. [149]. Merritt and Quinlan [218] studied the evolution of systems for various values of the mass of the central black hole. The black hole is inserted at the centre of a maximally triaxial E5 elliptical galaxy in equilibrium. They found that black holes with $m = 0.01$ are

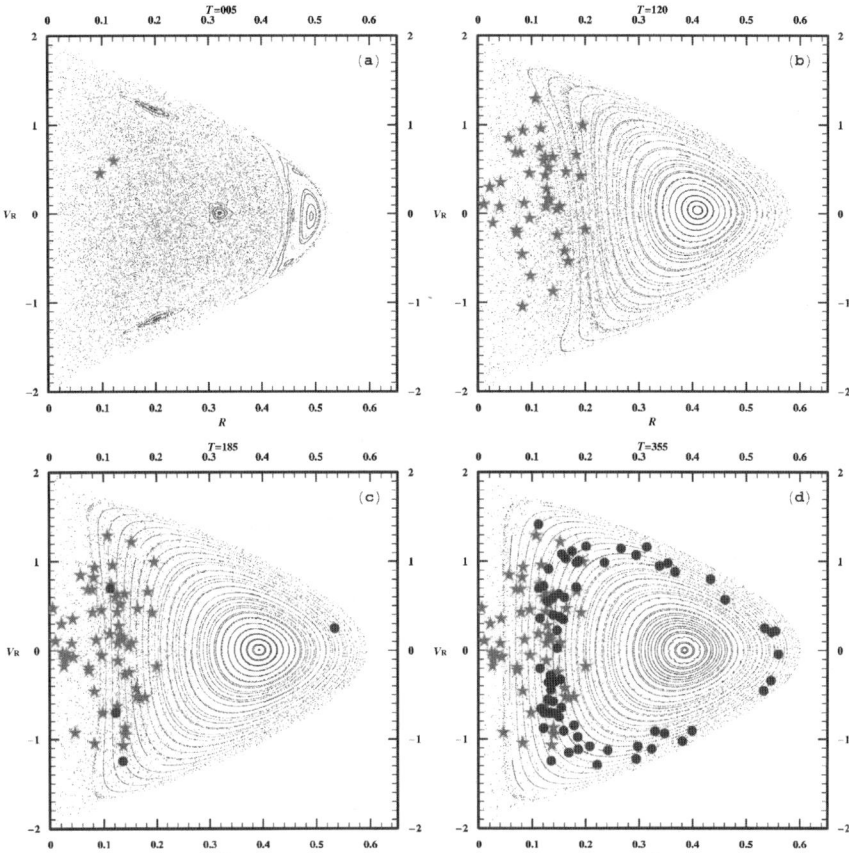

**Fig. 11.36.** Projections on the $(R, V_R)$ plane of the 4D Poincaré section $z = 0, \dot{z} > 0$ at four different snapshots of the evolution of the $(Q, m = 0.01)$ system, at times (**a**) $t = 5$, (**b**) $t = 120$, (**c**) $t = 185$ and (**d**) $t = 355$. The island of stability to the right corresponds to SAT orbits and its size increases as a result of changes in the self-consistent potential. The *stars* or *dots* give the successive Poincaré consequents of an orbit which was a box in the original Q-system (before the insertion of the CMC), up to the time corresponding to each panel. A star is plotted as long as the orbit falls in the chaotic domain of the surface of section. Part of the portrait. A dot is plotted after the moment when the orbit is captured in the regular domain (after [272])

capable to make the system evolve towards an oblate axisymmetric configuration within a Hubble time. On the other hand, Holley-Bockelmann et al. [149] found that the insertion of a black hole with mass $m = 0.01$ left their system unchanged (especially in the external parts). However, the original system that they used was a triaxial E2 elliptical galaxy, while the original system of Merritt and Quinlan [218] is similar to the Q-system considered above (many

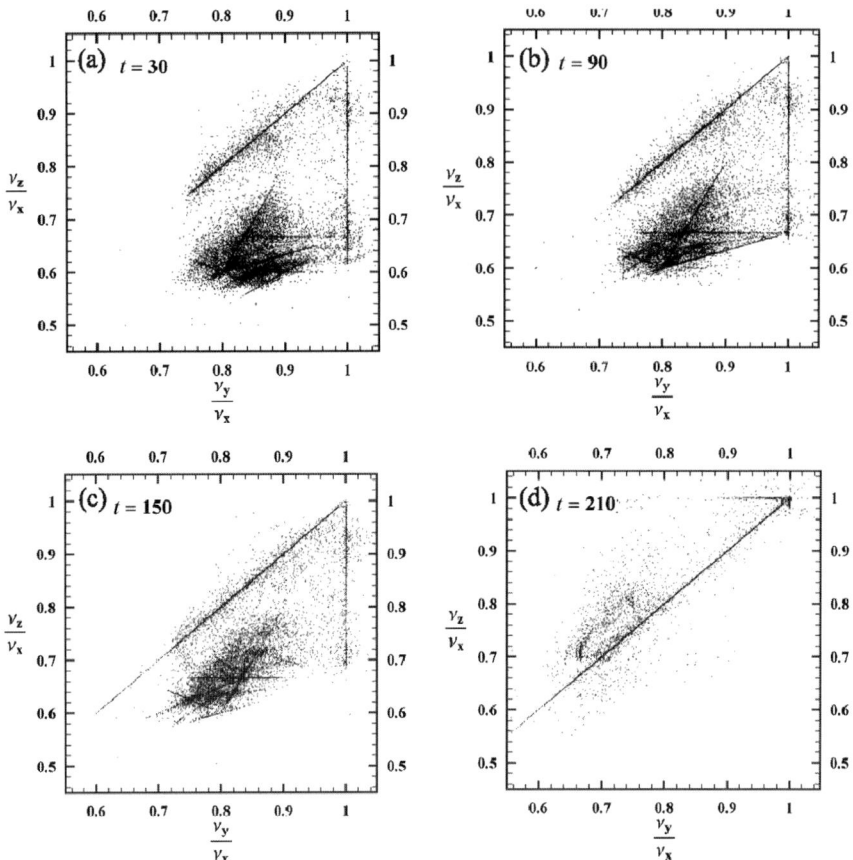

**Fig. 11.37.** Frequency maps for the orbits of all the particles of the (Q, $m=0.01$) system at the snapshots: (**a**) $t = 30$, (**b**) $t = 90$, (**c**) $t = 150$ and (**d**) $t = 210$. Initially there are many (previously box) chaotic orbits, trapped in various resonance lines (**a**). During the self-consistent evolution of the system, the population of SAT orbits increases while the population of all other types of orbits decreases (**b,c**). At equilibrium all orbits are either regular (of SAT type) or chaotic (**d**) (after [157])

box orbits). As explained above, given that box orbits become chaotic after the insertion of the black hole, such a system evolves rapidly towards a new equilibrium. On the contrary, the system of Holley-Bockelmann et al. [149] (similar to the C-system above) has a smaller initial fraction of box orbits and it evolves at a slower rate.

# References

1. Aarseth, S. and Binney, J., 1978. *Mon. Not. R. Astron. Soc.* **185**, 227.
2. Aarseth, S. J.: *Direct methods for N-Body Simulations*, Lect. Notes Phys. **433**, 275–312. Springer-Verlag, New York (1994).
3. Aguilar, L. and Merritt, D., 1990. *Astrophys. J.* **354**, 33.

4. Ahmad, A. and Cohen, L., 1973. *J. Comput. Phys.* **12**, 389.
5. Allen, A., Palmer, P., and Papaloizou, J., 1990. *Mon. Not. R. Astron. Soc.* **242**, 576.
6. Allen, A., Palmer, P., and Papaloizou, J., 1992. *Mon. Not. R. Astron. Soc.* **256**, 695.
7. Antonov, V., 1960. *Sov. Astron.* **4**, 859.
8. Aoki, S. and Iye, M., 1978. *Publ. Astron. Soc. Jpn* **30**, 519.
9. Arad, I. and Johansson, P., 2005. *Mon. Not. R. Astron. Soc.* **362**, 252.
10. Arad, I. and Lynden-Bell, D., 2005. *Mon. Not. R. Astron. Soc.* **361**, 385.
11. Arnold, V., 1963. *Russ. Math. Surv.* **18**, 5.
12. Arnold, V., 1964. *Sov. Math. Dokl.* **5**, 581.
13. Arnold, V., 1978. *Mathematical Methods of Classical Mechanics*, Springer, New York.
14. Athanassoula, E., Bosma, A., Lambert, J., and Makino, J., 1998. *Mon. Not. R. Astron. Soc.* **293**, 369.
15. Athanassoula, E. and Sellwood, J., 1986. *Mon. Not. R. Astron. Soc.* **221**, 213.
16. Barnes, J., 1988. *Astrophys. J.* **331**, 699.
17. Barnes, J., 1992. *Astrophys. J.* **393**, 484.
18. Barnes, J., Goodman, J., and Hut, P., 1986. *Astrophys. J.* **300**, 112.
19. Barnes, J. and Hut, P., 1986. *Nature* **324**, 446.
20. Bazzani, A. and Marmi, S., 1991. *Nuovo Cimento B* **106**, 673.
21. Becciani, U. and Antonuccio-Delogu, V., 2001. *Comput. Phys. Commun.* **136**, 54.
22. Belmonte, C., Boccaletti, D., and Pucacco, G., 2006. *Celest. Mech. Dyn. Astron.*, in press.
23. Benettin, G., Galgani, L., and Giorgilli, A., 1985. *Celest. Mech.* **37**, 2.
24. Berry, M., 1978. *AIP Conf. Proc.* **46**, 16.
25. Bertin, G., 2000. *Dynamics of Galaxies*, Cambridge Univ. Press, Cambridge.
26. Bertin, G. and Trenti, M., 2004. *AIP Conf. Proc.* **703**, 318.
27. Bertola, F. and Capaccioli, M., 1975. *Astrophys. J.* **338**, 723.
28. Binney, J., 1976. *Mon. Not. R. Astron. Soc.* **177**, 19.
29. Binney, J., 1978. *Mon. Not. R. Astron. Soc.* **183**, 501.
30. Binney, J., 1982. *Mon. Not. R. Astron. Soc.* **201**, 15.
31. Binney, J., 1982. *Mon. Not. R. Astron. Soc.* **200**, 951.
32. Binney, J. and Spergel, D., 1982. *Astrophys. J.* **252**, 308.
33. Binney, J. and Tremaine, S., 1987. *Galactic Dynamics*, Princeton University Press, New Jersey.
34. Birkhoff, G., 1927. *Dynamical Systems*, Am. Math. Soc., Providence, RI.
35. Bishop, J. L., 1987. *Astrophys. J.* **322**, 618.
36. Boccaletti, D. and Pucacco, G., 1996. *Theory of Orbits*, Springer, Berlin.
37. Bode, P. and Ostriker, P., 2003. *Astrophys. J. Suppl. Ser.* **145**, 1.
38. Bouchet, F. and Hernquist, L., 1988. *Astrophys. J. Suppl. Ser.* **68**, 521.
39. Burkert, A., 1990. *Mon. Not. R. Astron. Soc.* **247**, 152.
40. Burkert, A. and Naab, T.: *Major Mergers and the Origin of Elliptica Galaxies*, Lect. Notes Phys. **626**, 327. Springer, Heidelberg (2003).
41. Canizzo, J. and Hollister, T., 1992. *Astrophys. J.* **400**, 58.
42. Cappellari, M., Verolme, E., van der Marel, R., Verdoes Kleijn, G., Illingworth, G., Franx, M., Carollo, C., and de Zeeuw, T., 2002. *Mon. Not. R. Astron. Soc.* **578**, 787.

43. Carpintero, D. and Muzzio, J., 1995. *Astrophys. J.* **440**, 5.
44. Chandrasekhar, S., 1942. *Principles of Stellar Dynamics*, Chicago University Press, Chicago.
45. Chavanis, P., 1998. *Mon. Not. R. Astron. Soc.* **300**, 981.
46. Chavanis, P., 2002. astro-ph/0212205.
47. Chavanis, P., 2006. *Physica A* **365**, 102.
48. Cherry, T., 1924. *Proc. Camb. Phil. Soc.*. **22**, 325.
49. Cherry, T., 1924. *Proc. Camb. Phil. Soc.* **22**, 510.
50. Chirikov, B., 1979. *Phys. Rep.* **52**, 263.
51. Clutton-Brock, M., 1972. *Astrophys. Space Sci.* **16**, 101.
52. Clutton-Brock, M., 1973. *Astrophys. Space Sci.* **23**, 55.
53. Contopoulos, G., 1954. *Z. Astrophys.* **35**, 67.
54. Contopoulos, G., 1960. *Z. Astrophys.* **49**, 273.
55. Contopoulos, G., 1963. *Astron. J.* **68**, 763.
56. Contopoulos, G., 1966. *The Theory of Orbits in the Solar System and in Stellar Systems*, IAU Symp. **25**, Academic Press, London.
57. Contopoulos, G., 1966(7). In Hénon, M. and Nahon, F. (eds), *Les Nouvelles Méthodes de la Dynamique Stellaire*, Bull. Astron., 2e Ser., **2**, CNRS, Paris, p. 223.
58. Contopoulos, G., 1971. *Astron. J.* **76**, 147.
59. Contopoulos, G., 1983. *Astron. Astrophys.* **117**, 89.
60. Contopoulos, G.: *Order and Chaos*, Lect. Notes Phys. **433**, 33–68. Springer, Heidelberg (1994).
61. Contopoulos, G., 2004. *Order and Chaos in Dynamical Astronomy*, Springer, New York.
62. Contopoulos, G., 2004b. *Adventures in Order and Chaos*, Kluwer Academic Publishers, Dordrecht.
63. Contopoulos, G., Efthymiopoulos, C., and Giorgilli, A., 2003. *J. Phys. A* **36**, 8639.
64. Contopoulos, G., Efthymiopoulos, C., and Voglis, N., 2000. *Celest. Mech. Dyn. Astron.* **78**, 243.
65. Contopoulos, G., Galgani, L., and Giorgilli, A., 1978. *Phys. Rev. A* **18**, 1183.
66. Contopoulos, G. and Grosbol, P., 1988. *Astron. Astrophys.* **197**, 83.
67. Contopoulos, G. and Moutsoulas, M., 1965. *Astron. J.* **70**, 817.
68. Contopoulos, G. et al.: *Orbits in Barred Galaxies*, Lect. Notes Phys. **474**, 19. Springer, Heidelberg (1996).
69. Contopoulos, G., Voglis, N., Efthymiopoulos, C., and Grousouzakou, E., 1995. In Hunter, J. and Wilson, R. (eds), *Waves in Astrophysics*, N. Y. Acad. Sci. Proc., **773**, pp. 145–168.
70. Contopoulos, G., Voglis, N., and Kalapotharakos, C., 2002. *Celest. Mech. Dyn. Astron.* **83**, 191.
71. Crane, P., Stiavelli, M., King, I., Deharveng, J., Albrecht, R., Barbicri, C., Blades, J., Boksenberg, A., Disney, M., Jakobsen, P., Kamperman, T., Machetto, F., Mackay, C., Paresce, F., Weigelt, G., Baxter, D., Greenfield, P., Jedrzejewski, R., Nota, A., and Sparks, W., 1993. *Astron. J.* **106**, 1371.
72. Cretton, N., de Zeeuw, P. T., van der Marel, R. P., and Rix, H. W., 1999. *Astrophys. J. Suppl. Ser.* **124**, 383.
73. Cretton, N., Rix, H. W., and de Zeeuw, T. P., 2000. *Astrophys. J.* **536**, 319.
74. Cretton, N. and van den Bosch, F., 1999. *Astrophys. J.* **514**, 704.

75. Cupperman, S., Goldstein, S., and Lecar, M., 1969. *Mon. Not. R. Astron. Soc.* **146**, 161.
76. Curir, A., Diaferio, A., and de Felice, F., 1993. *Astrophys. J.* **413**, 70.
77. Davies, R., Efstathiou, G., Fall, S., Illingworth, G., and Schechter, P., 1983. *Astrophys. J.* **266**, 41.
78. de Vaucouleurs, G., 1948. *Ann. d'Astrophys.* **11**, 247.
79. de Vaucouleurs, G. and Capaciolli, M., 1979. *Astrophys. J.* Suppl. **40**, 699.
80. de Zeeuw, T., 1985. *Mon. Not. R. Astron. Soc.* **216**, 279.
81. de Zeeuw, T. and Lynden-Bell, D., 1985. *Mon. Not. R. Astron. Soc.* **215**, 713.
82. de Zeeuw, T. and Merritt, D., 1983. *Astrophys. J.* **267**, 571.
83. Dehnen, W., 1993. *Mon. Not. R. Astron. Soc.* **265**, 250.
84. Dehnen, W., 2000. *Astrophys. J.* **536**, L39.
85. Dehnen, W. and Gerhard, O., 1993. *Mon. Not. R. Astron. Soc.* **261**, 311.
86. Dejonghe, H., 1987. *Astrophys. J.* **320**, 477.
87. Dejonghe, H. and de Bruyne, V.: *Dynamics of Galaxies: From Observations to Distribution Functions*, Lect. Notes Phys. **626**, 243. Springer, Heidelberg (2003).
88. Dejonghe, H., de Bruyne, V., Vauterin, P., and Zeilinger, W., 1996. *Astron. Astrophys.* **306**, 363.
89. Dejonghe, H. and Merritt, D., 1988. *Astrophys. J.* **328**, 93.
90. Deprit, A., 1969. *Celest. Mech.* **1**, 12.
91. Dubinski, J., 1996. *New Astron.* **1**, 133.
92. Dubinski, J. and Calberg, R., 1991. *Astrophys. J.* **378**, 496.
93. Dumas, H. S. and Laskar, J., 1993. *Phys. Rev. Lett.* **70**, 2975.
94. Earn, D., 1996. *Astrophys. J.* **465**, 91.
95. Eddington, A., 1915. *Mon. Not. R. Astron. Soc.* **76**, 37.
96. Eddington, A., 1916. *Mon. Not. R. Astron. Soc.* **76**, 572.
97. Efthymiopoulos, C., 1999. PhD Thesis, University of Athens.
98. Efthymiopoulos, C., 2005. *Celest. Mech. Dyn. Astron.* **92**, 29.
99. Efthymiopoulos, C., Contopoulos, G., and Giorgilli, A., 2004. *J. Phys. A* **37**, 10831.
100. Efthymiopoulos, C., Contopoulos, G., Voglis, N., and Dvorak, R., 1997. *J. Phys. A* **30**, 8167.
101. Efthymiopoulos, C. and Sandor, Z., 2005. *Mon. Not. R. Astron. Soc.* **364**, 253.
102. Efthymiopoulos, C. and Voglis, N., 2001. *Astron. Astrophys.* **378**, 679.
103. Eggen, O., Lynden-Bell, D., and Sandage, A., 1962. *Astrophys. J.* **136**, 748.
104. Faber, S. M., Tremaine, S., Ajhar, E. A., Byun, Y., Dressler, A., Gebhardt, K., Grillmair, C., Kormendy, J., Lauer, T. R., and Richstone, D., 1997. *Astron. J.* **114**, 1771.
105. Fassò, F., Guzzo, M., and Benettin, G., 1998. *Commun. Math. Phys.* **197**, 347.
106. Ferrarese, L., van den Bosch, F., Ford, H., Jaffe, W., and O'Connell, R., 1994. *Astron. J.* **108**, 1598.
107. Filmore, J. and Goldreich, P., 1984. *Astrophys. J.* **281**, 1.
108. Fridman, A. and Polyachenko, V., 1984. *Physics of Gravitating Systems*, Springer, New York.
109. Fridman, T. and Merritt, D., 1997. *Astron. J.* **114**, 1479.
110. Froeschlé, C., Guzzo, M., and Lega, E., 2000. *Science* **5487**, 2108.
111. Froeschlé, C. and Scheidecker, J., 1973. *Astron. Space Sci.* **25**, 373.
112. Fukushige, T., Ito, T., Ebisuzaki, J. M. T., Sugimoto, D., and Umemura, M., 1991. *Publ. Astron. Soc. Jpn* **43**, 841.

113. Gebhardt, K., Lauer, T., Kormendy, J., Pinkney, J., Bower, G., Green, R., Gull, T., Hutchings, J., Kaiser, M., Nelson, C., Richstone, D., and Weistrop, D., 2001. *Astrophys. J.* **122**, 2469.
114. Gebhardt, K., Richstone, D., Ajhar, E. A., Lauer, T. R., Byun, Y., Kormendy, J., Dressler, A., Faber, S. M., Grillmair, C., and Tremaine, S., 1996. *Astron. J.* **112**, 105.
115. Gebhardt, K., Richstone, D., Kormendy, J., Lauer, T. R., Ajhar, E. A., Bender, R., Dressler, A., Faber, S. M., Grillmair, C., Magorrian, J., and Tremaine, S., 2000. *Astron. J.* **119**, 1157.
116. Gerhard, O., 1981. *Mon. Not. R. Astron. Soc.* **197**, 179.
117. Gerhard, O. and Saha, P., 1991. *Mon. Not. R. Astron. Soc.* **251**, 449.
118. Gerhard, O. E. and Binney, J., 1985. *Mon. Not. R. Astron. Soc.* **216**, 467.
119. Giordano, C. and Cincotta, P., 2004. *Astron. Astrophys.* **423**, 745.
120. Giorgilli, A., 1979. *Comput. Phys. Commun.* **16**, 331.
121. Giorgilli, A., 1988. *Ann. Inst. H. Poincaré* **48**, 423.
122. Giorgilli, A., 2002. *Notes on Exponential Stability of Hamiltonian Systems*, Centro di Ricerca Matematica Ennio de Giorgi, Pisa.
123. Giorgilli, A. and Galgani, L., 1978. *Celest. Mech.* **17**, 267.
124. Giorgilli, A. and Skokos, C., 1997. *Astron. Astrophys.* **317**, 254.
125. Goldstein, H., 1980. *Classical Mechanics*, Addison-Wesley Series in Physics, Addison-Wesley, Reading, MA.
126. Goldstein, S., Cupperman, S., and Lecar, M., 1969. *Mon. Not. R. Astron. Soc.* **143**, 209.
127. Goodman, J. and Schwarzschild, M., 1981. *Astrophys. J.* **245**, 1087.
128. Gott, J., 1973. *Astrophys. J.* **186**, 481.
129. Gott, J., 1975. *Astrophys. J.* **201**, 296.
130. Gurzandyan, V. and Savvidy, G., 1986. *Astron. Astrophys.* **160**, 203.
131. Gustavson, F., 1966. *Astron. Astrophys.* **71**, 670.
132. Guzzo, M., Fassò, F., and Benettin, G., 1998. *Math. Phys.* Electronic 4, paper 1.
133. Guzzo, M., Lega, E., and Froeschlé, C., 2002. *Physica D* **163**, 1.
134. Guzzo, M., Lega, E., and Froeschlé, C., 2005. *Discrete Contin. Dyn. Syst.* **5**, 687.
135. Hagihara, Y., 1970. *Celsetial Mechanics*, MIT Press, Cambridge.
136. Hénon, M., 1959. *Ann. d'Astrophys.* **22**, 126.
137. Hénon, M., 1964. *Ann. d'Astrophys.* **27**, 83.
138. Hénon, M., 1973. *Astron. Astrophys.* **24**, 229.
139. Hénon, M. and Heiles, C., 1964. *Astron. Astrophys.* **69**, 73.
140. Henriksen, R. and Widrow, L., 1997. *Phys. Rev. Lett.* **78**, 3426.
141. Henriksen, R. and Widrow, L., 1999. *Mon. Not. R. Astron. Soc.* **302**, 321.
142. Hernquist, L., 1992. *Astrophys. J.* **400**, 460.
143. Hernquist, L. and Ostriker, J., 1992. *Astrophys. J.* **386**, 375.
144. Herquist, L., 1987. *Astrophys. J.* Suppl. **64**, 715.
145. Hjorth, J. and Madsen, J., 1991. *Mon. Not. R. Astron. Soc.* **253**, 703.
146. Hoffman, Y., Shlosman, Y., and Shavin, G., 1979. *Mon. Not. R. Astron. Soc.* **189**, 737.
147. Hohl, F., 1971. *Astrophys. J.* **168**, 343.
148. Holley-Bockelmann, K., Mihos, J. C., Sigurdsson, S., and Hernquist, L., 2001. *Astrophys. J.* **549**, 862.
149. Holley-Bockelmann, K., Mihos, J. C., Sigurdsson, S., Hernquist, L., and Norman, C., 2002. *Astrophys. J.* **567**, 817.

150. Hori, G., 1966. *Publ. Astron. Soc. Jpn* **18**, 287.
151. Illingworth, G., 1977. *Astrophys. J.* **218**, 43.
152. Ipser, J., 1974. *Astrophys. J.* **193**, 463.
153. Ipser, J. and Horwitz, G., 1979. *Astrophys. J.* **232**, 863.
154. Jeans, J., 1915. *Mon. Not. R. Astron. Soc.* **76**, 70.
155. Jesseit, R., Naab, T., and Burkert, A., 2005. *Mon. Not. R. Astron. Soc.* **360**, 1185.
156. Kalapotharakos, C., 2005. PhD Thesis, University of Athens.
157. Kalapotharakos, C. and Voglis, N., 2005. *Celest. Mech. Dyn. Astron.* **92**, 157.
158. Kalapotharakos, C., Voglis, N., and Contopoulos, G., 2004. *Astron. Astrophys.* **428**, 905.
159. Kalnajs, A., 1976. *Astrophys. J.* **205**, 745.
160. Kaluza, M. and Robnik, M., 1992. *J. Phys. A* **25**, 5311.
161. Kandrup, H., 1987. *Mon. Not. R. Astron. Soc.* **225**, 995.
162. Kandrup, H., 1998. *Astrophys. J.* **500**, 120.
163. Kandrup, H.: *Chaos and Chaotic Phase Mixing in Galaxy Evolution and Charged Particle Beams*, Lect. Notes Phys. **626**, 154. Springer, Heidelberg (2003).
164. Kandrup, H. and Mahon, M., 1994. *Astron. Astrophys.* **290**, 762.
165. Kandrup, H. and Sideris, I., 2002. *Celest. Mech. Dyn. Astron.* **82**, 61.
166. Kandrup, H., Vass, I., and Sideris, I., 2003. *Mon. Not. R. Astron. Soc.* **341**, 927.
167. Kandrup, H. E. and Siopis, C., 2003. *Mon. Not. R. Astron. Soc.* **345**, 727.
168. Karney, C., 1983. *Physica D* **8**, 360.
169. Katz, N., 1991. *Astrophys. J.* **368**, 325.
170. Kaufmann, D. E. and Contopoulos, G., 1996. *Astron. Astrophys.* **309**, 381.
171. Kent, S. and de Zeeuw, T., 1991. *Astron. J.* **102**, 1994.
172. King, I., 1962. *Astron. J.* **67**, 471.
173. Kolmogorov, A., 1954. *Dokl. Akad. Nauk. SSSR* **98**, 527.
174. Kormendy, J., Bender, R., Evanst, A. S., and Richstone, D., 1998. *Astron. J.* **115**, 1823.
175. Kormendy, J., Bender, R., Magorrian, J., Tremaine, S., Gebhardt, K., Richstone, D., Dressler, A., Faber, S. M., Grillmair, C., and Lauer, T. R., 1997. *Astrophys. J. Lett.* **482**, 139.
176. Kormendy, J. and Richstone, D., 1995. *Annu. Rev. Astron. Astrophys.* **33**, 581.
177. Kravtsov, A., Klypin, A., and Khokhlov, A., 1997. *Astrophys. J. Suppl. Ser.* **73**, 94.
178. Kull, A., Treumann, R., and Böhringer, H., 1997. *Astrophys. J.* **484**, 58.
179. Kuzmin, G., 1956. *Astron. Zh.* **33**, 27.
180. Laskar, J., 1990. *Icarus* **88**, 266.
181. Laskar, J., 1993. *Physica D* **67**, 257.
182. Laskar, J., 1993. *Celest. Mech. Dyn. Astron.* **56**, 191.
183. Laskar, J., 1999. In Simó, C. (ed.), *HamiltonianSystems with Three or More Degrees of Freedom*, NATO ASI Series,**C:533**, Kluwer Academic Publishers, Dordrecht, p. 134.
184. Laskar, J., 2003. math/0305364
185. Laskar, J., Froeschle, C., and Celletti, A., 1992.*Physica D* **56**, 253.
186. Lauer, T., Ajhar, E., Byun, Y., Dressler, A., Faber, S.,Grillmair, C., Kormendy, J., Richstone, D., and Tremaine, S., 1995.*Astron. J.* **110**, 2622.

187. Lecar, M. and Cohen, L., 1972. In Lecar, M. (ed.),*Gravitational N-Body Problem*, Reidel, Dordrecht.
188. Lega, E., Guzzo, M., and Froeschlé, C., 2003.*Physica D* **182**, 179.
189. Levison, H. and Richstone, D., 1987. *Astrophys. J.***314**, 476.
190. Lochak, P., 1992. *Russ. Math. Surv.* **47**, 57.
191. Londrillo, P., Messina, A., and Stiavelli, M., 1991.*Mon. Not. R. Astron. Soc.* **250**, 54.
192. Louis, P. and Gerhard, O., 1988. *Mon. Not. R. Astron.Soc.* **233**, 337.
193. Lucy, B., 1974. *Astron. J.* **79**, 745.
194. Lynden-Bell, D., 1962. *Mon. Not. R. Astron. Soc.***124**, 1.
195. Lynden-Bell, D., 1962. *Mon. Not. R. Astron. Soc.***124**, 95.
196. Lynden-Bell, D., 1967. *Mon. Not. R. Astron. Soc.***136**, 101.
197. Lynden-Bell, D., 1969. *Nature* **223**, 669.
198. Lynden-Bell, D., 1979. *Mon. Not. R. Astron. Soc.***187**, 101.
199. Lynden-Bell, D. and Wood, R., 1968. *Mon. Not. R.Astron. Soc.* **138**, 495.
200. Madsen, J., 1987. *Astrophys. J.* **316**, 497.
201. Magorrian, J., Tremaine, S., Richstone, D., Bender, R., Bower,G., Dressler, A., Faber, S. M., Gebhardt, K., Green, R., Grillmair,C., Kormendy, J., and Lauer, T., 1998. *Astron. J.***115**, 2285.
202. Makino, J. and Funato, Y., 1993. *Publ. Astron. Soc.Jpn* **45**, 279.
203. Makino, J., Taiji, M., Ebisuzaki, T., and Sugimoto, D., 1997.*Astrophys. J.* **480**, 432.
204. Mathur, S., 1988. *Mon. Not. R. Astron. Soc.***231**, 367.
205. Matthias, M. and Gerhard, O., 1999. *Mon. Not. R.Astron. Soc.* **310**, 879.
206. May, A. and van Albada, T., 1984. *Mon. Not. R. Astron.Soc.* **209**, 15.
207. McGill, C. and Binney, J., 1990. *Mon. Not. R. Astron.Soc.* **244**, 634.
208. McGlynn, T., 1984. *Astrophys. J.* **281**, 13.
209. McMillan, S. and Aarseth, S., 1993. *Astrophys. J.***414**, 200.
210. Merrall, T. and Henriksen, R., 2003. *Astrophys. J.***595**, 43.
211. Merritt, D., 1985. *Mon. Not. R. Astron. Soc.***214**, 25P.
212. Merritt, D., 1996. *Science* **271**, 337.
213. Merritt, D., 1999. *Proc. Astr. Soc. Pac.***111**(756), 129.
214. Merritt, D., 2005. In Buchler, J., Gottesman, S., and Mahon, M.(eds), *Nonlinear Dynamics in Astronomy and Physics*, N. Y.Acad. Sci. Proc.
215. Merritt, D., 2006. *Rep. Prog. Phys.* **69**, 2513.
216. Merritt, D. and Aguilar, L., 1985. *Mon. Not. R. Astron.Soc.* **217**, 787.
217. Merritt, D. and Fridman, T., 1996. *Astrophys. J.***460**, 136.
218. Merritt, D. and Quinlan, D., 1998. *Astrophys. J.***498**, 625.
219. Merritt, D. and Sellwood, J., 1994. *Astrophys. J.***425**, 551.
220. Merritt, D., Tremaine, S., and Johnstone, D., 1989.*Mon. Not. R. Astron. Soc.* **236**, 829.
221. Merritt, D. and Valluri, M., 1996. *Astrophys. J.***471**, 82.
222. Merritt, D. and Valluri, M., 1999. *Astron. J.***118**, 1177.
223. Michie, R., 1963. *Mon. Not. R. Astron. Soc.***125**, 127.
224. Milani, A. and Nobili, A., 1985. *Astron. Astrophys.***144**, 261.
225. Milani, A. and Nobili, A., 1992. *Nature* **357**,569.
226. Miller, R. and Smith, B., 1994. *Celest. Mech. Dyn.Astron.* **59**, 161.
227. Mineau, P., Feix, M., and Rouet, J., 1990. *Astron.Astrophys.* **228**, 344.
228. Miocchi, P. and Capuzzo-Dolcetta, R., 2002. *Astron.Astrophys.* **382**, 758.

229. Morbidelli, A. and Guzzo, M., 1997. *Celest. Mech. Dyn.Astron.* **65**, 107.
230. Moser, J., 1962. *Nachr. Akad, Wiss. Gottingen II Math.Phys.* **1**, 1.
231. Muzzio, J. C., Carpintero, D. D., and Wachlin, F. C., 2005.*Celest. Mech. Dyn. Astron.*, in press.
232. Naab, T., Burkert, A., and Hernquist, L., 1999.*Astrophys. J. Lett.* **523**, 133.
233. Nakamura, T., 2000. *Astrophys. J.* **531**, 739.
234. Natarajan, P., Hjorth, J., and van Kampen, E., 1997.*Mon. Not. R. Astron. Soc.* **286**, 329.
235. Negroponte, J. and White, S., 1983. *Mon. Not. R.Astron. Soc.* **205**, 1009.
236. Nekhoroshev, N., 1977. *Russ. Math. Surv.***32**(6), 1.
237. Newton, A. and Binney, J., 1984. *Mon. Not. R. Astron.Soc.* **210**, 711.
238. Niederman, L., 1998. *Nonlinearity* **11**, 1465.
239. Ogorodnikov, K., 1965. *Dynamics of Stellar Systems*,Pergamon Press, New York.
240. Ostriker, J. and Peebles, P., 1973. *Astrophys. J.***186**, 467.
241. Palmer, P., 1995. *Instabilities in CollisionlessStellar Systems*, Cambridge University Press, Cambridge.
242. Palmer, P. and Papaloizou, J., 1987. *Mon. Not. R.Astron. Soc.* **224**, 1043.
243. Palmer, P. and Voglis, N., 1983. *Mon. Not. R. Astron.Soc.* **205**, 543.
244. Papaphilippou, Y. and Laskar, J., 1996. *Astron.Astrophys.* **307**, 427.
245. Papaphilippou, Y. and Laskar, J., 1998. *Astron.Astrophys.* **329**, 451.
246. Patsis, P. A., Contopoulos, G., and Grosbol, P., 1991.*Astron. Astrophys.* **243**, 373.
247. Petrou, M., 1983. *Mon. Not. R. Astron. Soc.***202**, 1195.
248. Pfenniger, D., 1984. *Astron. Astrophys.* **141**,171.
249. Pfenniger, D., 1986. *Astron. Astrophys.* **165**,74.
250. Plastino, A. and Plastino, A., 1993. *Phys. Lett. A***174**, 384.
251. Polyachenko, V., 1981. *Sov. Astron. Lett.* **7**,79.
252. Polyachenko, V. and Shukhman, I., 1984. *Sov. Astron.***25**, 533.
253. Poon, M. Y. and Merritt, D., 2001. *Astrophys J.***549**, 192.
254. Poon, M. Y. and Merritt, D., 2002. *Astrophys. J.***568**, 89.
255. Poon, M. Y. and Merritt, D., 2004. *Astrophys. J.***606**, 774.
256. Pöshel, J., 1993. *Math. Z.* **213**, 187.
257. Ratcliff, S., Chang, K., and Schwarzschild, M., 1984.*Astrophys. J.* **279**, 610.
258. Richstone, D., 1980. *Astrophys. J.* **238**, 103.
259. Richstone, D., 1982. *Astrophys. J.* **252**, 496.
260. Richstone, D., 1984. *Astrophys. J.* **281**, 100.
261. Richstone, D., 1987. In de Zeeuw, T. (ed.), *Structureand Dynamics of Elliptical Galaxies*, Proceedings of the IAUSymposium, **127**, Reidel, Dordrecht, p. 261.
262. Richstone, D. and Tremaine, S., 1984. *Astrophys. J.***286**, 27.
263. Rix, H., de Zeeuw, T., van der Marel, N. C. R., and Carollo,M., 1997. *Astrophys. J.* **488**, 702.
264. Rosenbluth, M., Sagdeev, R., Taylor, J., and Zaslavskii, M.,1966. *Nucl. Fusion* **6**, 217.
265. Salpeter, E. E., 1964. *Astrophys. J.* **140**,796.
266. Schwarzschild, M., 1979. *Astrophys. J.* **232**,236.
267. Schwarzschild, M., 1982. *Astrophys. J.* **263**,599.
268. Schwarzschild, M., 1993. *Astrophys. J.* **409**,563.
269. Sellwood, J., 1987. *Ann. rev. Astron. Astrophys.***25**, 151.

270. Sersic, J., 1963. *Bolet. de la Asoc. Argentina deAstron.* **6**, 41.
271. Sersic, J., 1968. *Atlas de galaxias australes*,Observatorio Astronomico, Cordoba, Argentina.
272. Servizi, G., Turchetti, G., Benettin, G., and Giorgilli, A.,1983. *Phys. Lett. A* **95**, 11.
273. Severne, G. and Luwel, M., 1986. *Astron. Space Sci.* **122**, 299.
274. Shu, F., 1978. *Astrophys. J.* **225**, 83.
275. Shu, F., 1987. *Astrophys. J.* **316**, 502.
276. Sidlichovsky, M. and Nesvorny, D., 1997. *Celest. Mech.Dyn. Astron.* **65**, 137.
277. Siegel, C., 1941. *Ann. Math.* **42**, 806.
278. Siopis, C., 1999. PhD Thesis, University of Florida.
279. Siopis, C. and Kandrup, H. E., 2000. *Mon. Not. R.Astron. Soc.* **319**, 43.
280. Skokos, C., 2001. *J. Phys. A* **34**, 10029.
281. Smith, B. and Miller, R., 1982. *Astrophys. J.***257**, 103.
282. Smith, H. and Contopoulos, G., 1995. *Astron.Astrophys.* **314**, 795.
283. Soker, N., 1996. *Astrophys. J.* **457**, 287.
284. Spergel, D. and Hernquist, L., 1992. *Astrophys. J.***397**, 75.
285. Spitzer, L., 1987. *Dynamical Evolution of GlobularClusters*, Princeton University Press, New Jersey.
286. Spitzer, L. and Hart, M., 1971. *Astrophys. J.***164**, 399.
287. Springer, V., Yoshida, N., and White, S., 2000. *NewAstron.* **6**, 79.
288. Sridhar, S., 1987. *J. Astrophys. Astron.* **8**,257.
289. Stäckel, P., 1890. *Mathemat. Ann.* **35**, 91.
290. Stäckel, P., 1893. *Mathemat. Ann.* **42**,537.
291. Statler, T. S., 1987. *Astrophys. J.* **321**, 113.
292. Stiavelli, M. and Bertin, G., 1985. *Mon. Not. R.Astron. Soc.* **217**, 735.
293. Stiavelli, M. and Bertin, G., 1987. *Mon. Not. R.Astron. Soc.* **229**, 61.
294. Sugimoto, D., Chikada, Y., Makino, J., Ito, T., Ebisuzaki, T.,and Umemura, M., 1990. *Nature* **345**, 33.
295. Szebehely, V., 1967. *Theory of Orbits*, Academic Press,New York.
296. Takizawa, M. and Inagaki, S., 1997. astroph p. 2002T.
297. Taruya, A. and Sakagami, M., 2002. *Physica A***307**, 185.
298. Taruya, A. and Sakagami, M., 2003. *Physica A***318**, 387.
299. Toomre, A., 1964. *Astrophys. J.* **139**, 1217.
300. Toomre, A., 1966. *Geophys. Fluid Dyn.* **66-46**,111.
301. Toomre, A. and Toomre, J., 1972. *Astrophys. J.***178**, 623.
302. Tremaine, S., 1987. In de Zeeuw, T. (ed.), *Structureand Dynamics of Elliptical Galaxies*, IAU Symposium, **127**,Reidel, Dordrecht, p. 367.
303. Tremaine, S., Hénon, M., and Lynden-Bell, D., 1986.*Mon. Not. R. Astron. Soc.* **219**, 285.
304. Tremaine, S., Richstone, D., Byun, Y., Dressler, A., Faber, S.,Grillmair, C., Kormendy, J., and Lauer, T., 1994. *Astron.J.* **107**, 634.
305. Trenti, M., Bertin, G., and van Albada, T., 2005.*Astron. Astrophys.* **433**, 57.
306. Tsallis, C., 1988. *J. Stat. Phys.* **52**, 479.
307. Udry, S. and Martinet, L., 1994. *Astron. Astrophys.***281**, 314.
308. Udry, S. and Pfenniger, D., 1988. *Astron. Astrophys.***198**, 135.
309. Valluri, M. and Merritt, D., 1998. *Astrophys. J.***506**, 686.
310. van Albada, T., 1982. *Mon. Not. R. Astron. Soc.***201**, 939.
311. van Albada, T., 1987. In *Structure and Dynamics ofElliptical Galaxies*, IAU Symposium, **127**, Reidel,Dordrecht, p. 291.

312. van der Marel, R. P., de Zeeuw, P. T., and Rix, H. W., 1997.*Astrophys. J.* **488**, 119.
313. van der Marel, R. P. and van den Bosch, F. C., 1998.*Astron. J.* **116**, 2220.
314. Varvoglis, H. et al.: *The "Third" Integral in the Restricted Three-Body Problem Revisited*, Lect. Notes Phys. **626**, 433–441. Springer,Heidelberg (2003).
315. Verhulst, F., 1979. *Philos. Trans. R. Soc. Lond.***290**, 435.
316. Verolme, E., Cappellari, M., Copin, Y., van der Marel, R.,Bacon, R., Bureau, M., Davies, R., Miller, B., and de Zeeuw, T.,2002. *Mon. Not. R. Astron. Soc.* **335**, 517.
317. Villumsen, J., 1982. *Mon. Not. R. Astron. Soc.***199**, 493.
318. Villumsen, J., 1984. *Astrophys. J.* **284**, 75.
319. Vitturo, H. and Carpintero, D., 2000. *Astron.Astrophys.* Suppl. **142**, 157.
320. Voglis, N., 1994. *Mon. Not. R. Astron. Soc.***267**, 379.
321. Voglis, N.: *Galaxy Formation*,Lect. Notes Phys. **433**, 365. Springer-Verlag, New York,(1994).
322. Voglis, N., Contopoulos, G., and Efthymiopoulos, C., 1998.*Phys. Rev. E* **57**, 372.
323. Voglis, N., Contopoulos, G., and Efthymiopoulos, C., 1999.*Celest. Mech. Dyn. Astron.* **73**, 211.
324. Voglis, N., Hiotelis, N., and Harsoula, M., 1995.*Astron. Space Sci.* **226**, 213.
325. Voglis, N., Hiotelis, N., and Hoeflich, P., 1991.*Astron. Astrophys.* **249**, 5.
326. Voglis, N. and Kalapotharakos, C., 2006. In Fridman, A., Marov,M., and Kovalenko, I. (eds), *Astrophysical Disks: Collectiveand Stochastic Phenomena*, Astrophysics and Space Science Library,**337**, Springer, Netherlands, p. 157.
327. Voglis, N., Kalapotharakos, C., and Stavropoulos, I., 2002.*Mon. Not. R. Astron. Soc.* **337**, 619.
328. Voglis, N., Stavropoulos, I., and Kalapotharakos, C., 2006.*Mon. Not. R. Astron. Soc.*, in press, astro-ph/0606561.
329. Voglis, N., Tsoutsis, P., and Efthymiopoulos, C., 2006.*Mon. Not. R. Astron. Soc.*, in press, astro-ph/0607174.
330. Wachlin, F. C. and Ferraz-Mello, S., 1998. *Mon. Not. R.Astron. Soc.* **298**, 22.
331. Warren, M. and Salmon, J., 1993. *Proc. SuperComput,* p.12.
332. Weinberg, M., 1998. *Mon. Not. R. Astron. Soc.***297**, 101.
333. Weinberg, M., 1999. *Astron. J.* **117**, 629.
334. White, S., 1976. *Mon. Not. R. Astron. Soc.***177**, 717.
335. White, S., 1978. *Mon. Not. R. Astron. Soc.***189**, 831.
336. White, S. and Narayan, R., 1987. *Mon. Not. R. Astron.Soc.* **229**, 103.
337. Whittaker, E., 1916. *Proc. R. Soc. Edinb* **37**,95.
338. Wiechen, H., Ziegler, H., and Schindler, K., 1988. *Mon.Not. R. Astron. Soc.* **232**, 623.
339. Wozniak, H. and Pfenniger, D., 1997. *Astron.Astrophys.* **317**, 14.
340. Zel'dovich, Y., 1964. *Sov. Astron.***8**, 13.
341. Zel'dovich, Y., 1970. *Astron. Astrophys.* **5**,89.
342. Ziegler, H. and Wiechen, H., 1989. *Mon. Not. R. Astron.Soc.* **238**, 1261.

# 12

# On the Difficulty to Foresee Solar Cycles: A Non-Deterministic Approach

Jean-Pierre Rozelot

Observatoire de la Côte d'Azur, Dpt Fizeau, & Université de Nice-Sophia
Antipolis Avenue Copernic, 06130 Grasse (F)
rozelot@obs-azur.fr

**Abstract.** The so-called "sunspot number" is one of the longest running time series available in astronomy. It forms an observed data set providing an index of solar activity allowing to trace its variation on a time scale of about 400 years since around AD 1600. This long period of time covers a wide range of activity, from vanishing sunspots during great minima to very high peaks during these last 50 years. However, forecast of solar activity is still a matter of debate. This paper gives a review of the recent achievements and findings in long-term activity. We emphasize determinism and chaos in sunspot cyclicity, giving the new "standard mapping" of the Sun. We indicate that prolonged periods of suppressed activity can be interpreted in terms of phase changes. We describe the "phase catastrophe" during the Dalton minimum, likely leading to a lost numbered cycle. Based on this non-linear physics, the behavior of solar activity is tackled in a new way, and a review of the different methods (rescaled range analysis, Grassberger and Procaccia algorithms, Shugira and May methods) is presented. A consistent scenario is given for the next grand minimum, scheduled between 2010 and 2050. Such predictions are of high importance, not only to put constraints on solar dynamo theories, but also in a new field of research, the Space Weather Physics.

## 12.1 Introduction

Since about 15 years, new views have emerged in physics concerning the conceptual ideas of determinism. It is perhaps not useless to point out how this philosophical notion is translated into science and to give again its primitive definition stated by Claude Bernard in 1865: "once known the conditions of existence of a phenomenon, it must be always and necessarily reproduced with the will of the experimenter; the negation of this property is the negation of science". In fact, such a concept fertilized the ideas of physicists for a very long time because it was convenient to think that for an event having had a cause in the past, the same causes producing the same effects, the future would be predictable. All should be only a question of formulation of the equations, necessarily approximated in a first step, and a question of an excellent definition of the initial conditions. At the time of Laplace, a partisan of this philosophy, and up to the end of the nineteenth century, it was believed that with a progressive progress of our knowledge, the set of equations would be better defined, the initial conditions would be increasingly exact, so that the numerical errors would be reduced; the system would be then integrable (one recalls that a system is integrable if it presents an integral for all its degrees of freedom). Moreover, at the end of this century, mathematicians thought that if one cannot find an integral directly, the solution of the system would border the exact solution on the condition of approximating this solution by series and by increasing the number of their terms indefinitely, because it was believed that the precision grew with the number of terms. Since with two terms the result is better than with only one and three better than with two, one can reasonably envisage that each new term improves the estimate; thus a perturbative series converges toward the exact solution. When the system evolves with time, it becomes possible to determine its future evolution, the determinism postulating that this future proceeds in a single way from the present and from the past. All these hopes appear now, unfortunately, to be badly founded.

These theories of causality and determinism collapsed when one realized that completely different events could be obtained when changing the initial conditions by an infinitesimal quantity (very small causes may produce great effects). This sensitivity to the initial conditions has been largely described elsewhere. As an example, one may consult the Lorentz model which is used now as an academic case (see [2, p. 327], [47, p. 187] or [3]).[1]

Poincaré (1854–1912) was the first who understood that it was not sufficient to study the local variations of a phenomenon, but that it was necessary to apprehend it in its totality to understand the long-term effects. He is also the first to realize that the precision of the initial conditions played a significant role in the evolution of the medium-term systems. However, the real implications of his work was developed only 70 years later, by the Russian mathematical school. Resuming the work of Poincaré, Kolmogorov, Arnold and Moser were able to achieve considerable progress in what it is agreed now to call the "chaos theory". Let us note that this last term remains a bit unsuitable, because it does not mean in contemporary physics (and particularly for the resolution of the planets and asteroids trajectories) a complete disorder, but lets imply a certain idea of organization in a complex structure, where the key will be precisely to find, in an apparently diffuse expansion, some zones of relative stabilities (orbits for example). One calls "dynamic chaos" the non-periodic, irregular behavior, caused by the non-linear

---

[1] The title of a conference given by Edward Lorentz in 1972 was "Predictability: does the flap of a butterfly's wings in Brazil set off a tornado in Texas?" Such a title was so ear-catching that it was taken again by other authors as if it was true, whereas Lorentz wanted to illustrate the difficulties of weather forecasting and did not claim to say that this was possible.

nature inherent to the system. Locally, chaos is almost always present in the vicinity of resonances.

It is particularly attractive to apply to the solar case, the results developed in celestial mechanics, a nearby discipline. Some authors have already tested the method; albeit the approach seems promising, there is no considerable development yet. The goal of this article is to give a broad outline of what can be developed and to show some results.

## 12.2 Solar Cycle Activity

If we were inhabitants of a planet in orbit around a remote star, say for example "51 Pegasi-b", and that from over there we were observing our Sun, it would seem a variable star, a fact not revealed by the common sense. With regularity, the solar luminosity varies, passing by alternative maxima and minima. This periodicity is appreciated on time scales ranging from very short term (day–hour–minute) to middle term (year–month–week) and long term (1000–100–10 years). Observations of solar luminosity made by ground-level radiometry, radiometry from balloons, aircraft and rockets, continuous radiometry from space probes and satellites, measurements of reflected light from solar system bodies, measurements of solar line depths and limb darkening demonstrate that solar luminosity variations are less than $1.2°/_{oo}$. Temporal variations of total solar irradiance (TSI: integrated flux over all wavelengths), measured from several spacecraft since 1978, show a modulation on approximatively 11 years (Fig. 12.1). Thus, the "true" solar cycle, on physical grounds, is the TSI cycle, but as TSI variations were associated as early as 1962 to the strong influence of magnetic activity coming from both sunspots and facula, and as it was discovered that the number of sunspots varied in time, the so-called "solar cycle" was reduced to the sunspot cycle, of approximately 11 years.

Today, "solar cycles" refer to a wide variety of frequencies and features: sunspots, flares, coronal mass ejections, irradiance, etc. The way they are linked together is still unknown (or poorly known). However, the study of time evolution of sunspot activity is of great interest for solar physics and solar-terrestrial physics since it reflects many processes including heat flux perturbations due to local variations in thermal impedance, variations in convective heat transport efficiency, energy storage in magnetic fields and variations in wave heating at the photosphere. The main feature of sunspot activity is its undecennal cycle, modulated by long-term effects. Sometimes sunspot activity is nearly suppressed, such as during the years ranging from about 1790 to 1820 [27] or dramatically suppressed, leading to a so-called great minimum. The most recent great minimum was the Maunder minimum (MM) lasting from 1645 to 1715, when sunspot activity almost vanished [9]. Other periods of time of very low activity exist (see Table 12.1), so that it is tempting to see in such a large-amplitude modulation a regular cycle (one can deduce

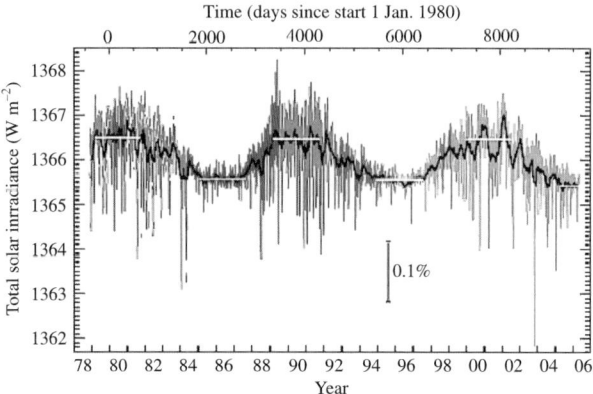

**Fig. 12.1.** The Sun as a variable star. This picture shows daily measurements of the total solar irradiance (TSI) with a 81-day running average. The TSI increases around the maxima of sunspot number that occurred near 1980, 1990 and 2001. The TSI variation amplitudes of the three sunspot cycles shown (horizontal lines) are 0.92, 0.89 and 0.90 Wm$^{-2}$, respectively, with an average minimum of 1365.52 ± 0.009 Wm$^{-2}$ and differences of the minima from this average of +0.051, +0.037 and −0.089 Wm$^{-2}$. From Foukal et al. [12] Indeed, ≈ 95% of TSI can be modelized through a weighted function of spots and faculae, but one is today still unable to give an accurate prediction for a next cycle

a general long-term periodic oscillation of about 120 years [40]). It is thus common to describe sunspot activity as a multiharmonic process with several fundamental harmonics superposed to each other. But, on the other hand, looking at the sunspot activity series, one can see a random component which is larger than the observational uncertainties. Hence, since the early 1990s, several authors have studied solar activity as an example of low-dimensional deterministic chaos described by a strange attractor (see, e.g., [24, 30, 37]). This approach has been criticized because the analyzed data set is too short [4] and disturbed by filtering [31]. Indeed, the real difficulty is to determine

**Table 12.1.** Periods of low sunspot activity that lasted from about AD 1000, based on evidence such as carbon-14 in tree rings and records of auroral activity

| Period of time (Years) | Name | |
|---|---|---|
| 1010–1050 | Oort | |
| 1280–1340 | Wolf | |
| 1420–1550 | Spörer | (may span also from 1450 to 1570) |
| 1645–1715 | Maunder | |
| 1790–1820 | Dalton | |

from a "chaotic" approach, the underlying physics of solar activity. In this scope, the work remains to be done, as resonant effects have not yet been fully studied.

### 12.2.1 Observations

The observations of the Sun started in a regular way around AD 1610, thanks to Galileo. With the telescope of his own invention, he was delighted to show the sunspots to philosophers in Rome's Monte Pincio gardens. However, people in more ancient days had already announced from time to time black spots on the Sun, which they had seen with the naked eye. Chinese were in this matter very true precursors. For example, the work of Ma-Twan-Lin indexes 45 observations made between the years 301 and 1205, that is to say during 904 years.[2] But nobody had highlighted a specified cycle. One had to wait until 1848 for the Swiss astronomer Rudolph Wolf, resuming the observations of an astronomer amateur, Heinrich Schwabe, to characterize the solar activity by a quantity $\Re$, which is calculated by first counting the number of sunspot groups and then the number of individual sunspots indexed on the Sun. The "sunspot number" is then given by the sum of the number of individual sunspots and 10 times the number of groups. Since most sunspot groups have, on average, about 10 spots, this formula for counting sunspots gives reliable numbers even when the observing conditions are less than ideal and small spots are hard to see.

This quantity $\Re$ is nowadays determined daily, on an international basis: astronomers have thus one of the longest time series of astronomical nature.[3] It must be noted that such a number has no true physical meaning. It is because $\Re$ correlates with the radio flux at 10.7 cm or with an index combining a ratio of the core/wing of the MgII line, two parameters based on the physics, that we can use this estimate $\Re$ as a true indicator of solar activity (Fig. 12.2).

A critical study of this database reveals a basic periodicity at 11.1 years (the Schwabe cycle) on the average, as well as other periodicities, in particular a 88.1-year period (the Gleissberg cycle), a longer one of 211.5 years (the Suess cycle), and anomalies. Thus, as previously seen, for the period going from 1645 to 1715, practically no group of spots are listed on the Sun: this is a prolonged minimum, called the Maunder minimum (after the work of Edward Walter Maunder and Annie Maunder). It is known today, by various methods, that other prolonged minima (or maxima) have existed in the past (see also [33, 41, pp. 5–24]).

The existence of such grand minima of solar activity is a key stone for solar/stellar theories. The reconstitution of the old cycles can be used in different studies, not only in solar dynamo theories, but also in other fields such

---

[2] Theophrastus (370 to about 285 BC) first identified sunspots in the year 325 BC.
[3] The International Sunspot Number is compiled by the Sunspot Index Data Center at the Uccle Observatory in Belgium. The NOAA Sunspot Number is compiled by the US National Oceanic and Atmospheric Administration in Boulder (CO), USA.

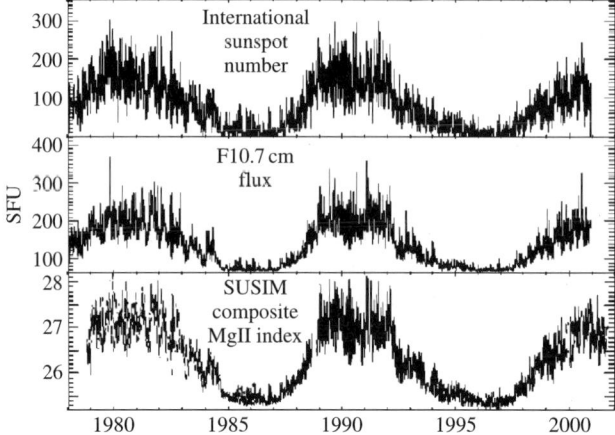

**Fig. 12.2.** Indexes of solar activity: the sunspot number (**top**) is well correlated with two other physical parameters, the radio flux at 10.7 cm (**middle**) and the ratio (core-to-wing) of the MgII UV line (**bottom**). So, the past database of counting sunspots can be used as a "true" indicator of solar activity

as in paleoclimatology. We will see further what one can await of the physical nature of these cycles; are they of different nature or not?

### 12.2.2 Solar Activity Forecasting

The foreseeability of the solar cycle constitutes an important chapter of current research in solar physics. The revival is also mainly due to the impulse given by the development of Space Weather, for which forecasting solar activity is crucial: impact on electronics on board satellites, impact on future inhabited spacecrafts, etc. The nature of the cycle, purely deterministic or chaotic, is an important feature to know, as it conditions physics. It seems that solar theoreticians today continue to ignore the difference and are still trying to solve the problem of solar activity by sets of equation and well-determined initial conditions. If it is sure that a precise knowledge of the internal solar dynamo would lead to considerable progresses to establish the morphology of cycles (Dikpati et al. [5], Dikpati and Gilman [6], who show that a cycle is determined by the four previous cycles), we are yet to know which information is really carried by a well-determined cycle.[4]

---

[4] Damon and Jirikowic, 1994 (IAU Colloquium 143, Pap ed., p. 301), have shown that $\Re$ can be approached by a function of the type $\Re = \sum_{i=1}^{4} \alpha_i \cos^2(\omega_i t + \phi_i) \alpha_c \cos^2(\omega_c t + \phi_c)$, where the $\omega_i$ are four overtones of 211.5, 105.9, 88 and 52.9 years and $\omega_c$ the modulated 11.1-year carrier signal. However, no particularly physical meaning can be attributed to the 105.9- and 52.9-year periods, which looks like harmonics.

In "traditional" physics, the simplest way is to determine by Fourier analysis all the periodicities contained in the signal. With about 30 sine functions, one could think that the signal could be reconstructed with a good approximation. The extrapolation of the model to later temporal values should then provide a good forecast. Unfortunately, the comparison with the observed values always gives a relatively significant discrepancy. A predicted point diverges from the observed one after 3 or 4 years ahead. This is the limit of predictability. In theory of chaos, this is called the Lyapunov exponent: it measures the time put by a non-linear system (for example a pendulum or a planet orbit) to find its equilibrium, time which can be finished or not. In this last case, the solutions diverge exponentially with time [47]. Space thus studied is characterized by its geometry, whose dimension can be a non-integer. This space is said of fractal dimension and the signature of the deterministic chaos is an attractor; an estimate of its dimension in the solar case can be put at 3.3.

The result can be summarized in a very concise way: as many authors, as many methods, as many periodicities. Even the basic 11-year period may fluctuate from 10.9 to 11.3 years according to the authors. In fact the analysis is very sensitive to the length of the sample. As the analysis is made over different periods of time, a dispersion is obtained on the results. Moreover, if the physical identification of the undecennal cycle can be done by the magnetic inversion of the spots polarity (this is called the Hale cycle, that is to say a 22 years cycle, which has thus a physical meaning) one understands badly why the Sun would be an oscillator having a fundamental well explained (on physical grounds) and on which a multiplicity of other periods would be superposed, without particular physical meaning (or not yet known?).

Briefly speaking, the failure of all traditional methods (technique known as "fast Fourier transform", "wavelet" techniques, methods of the maximum of entropy, generalized least squares or discrete methods, cyclogrammes as well as the approach by neuronal networks which could be undoubtedly very promising – [23]), indicates that

- The forecast is good but the prediction is poor (inaccuracy on the date, inaccuracy on the value of the level of the signal).
- The "observed" is not enough to deduce the "observable".

Let us explain briefly these two points. For the first one, as an example, in May 1996 (at the time of the minimum), one is able to forecast that the next solar maximum and the next solar minimum will take place, respectively, around 2001 and 2007. But, afterward, it can be seen that cycle 23 was peaked by two high values, of 169.1 (July 2000) and 150.7 (September 2001) surrounded by two smaller ones of 80.1 (February 2001) and 82.2 (July 2001). This double structure is completely unexplained and would have been never forecast. In the same way, it was claimed as early as June 2006 that the onset of solar cycle 24 has been observed (precursor signs of reversed magnetic polarity in sunspots), but the activity remains rather large (14.5 in September 2006 and 10.4 in October 2006); this cycle will be longer than it was first expected.

As far as the second point above mentioned is concerned, one can say in other words that it is not possible to deduce a future observable value from the observed past values, even if the data set lasts from now till nearly 257 years.

This is why it is necessary to jump to other methods. We will only focus here on deterministic chaos techniques.

## 12.3 Tools

Thus the solar cycle of activity remains still largely misunderstood and no valid theory is to date advanced to explain its oscillatory character and its amplitude modulation. One will find for instance in Soon [43] a rather complete summary of this fundamental question of the physics of the Sun: how to model the fluctuations observed on the surface of spots, radiance and perhaps of the diameter, when it is known that such features vary on scales of time going from a few seconds to several hundred years?

From another point of view, the analysis of the time series showed that the reconstruction of the cycles could not be carried out with an excellent precision, neither from a Fourier analysis or wavelets analysis, i.e., starting from cycles of regular periodicities, nor starting from purely random sequences, i.e., starting from cycles of perfectly irregular periodicities. On this subject, one may refer usefully to the model of Barnes et al. [1], who were able to simulate many cycles of variable amplitude, but obtained in a purely fortuitous manner, i.e., by adjustment of a free parameter. However, if it is now proved that the solar activity cycle can be rather well modeled with a small number of periods (Rozelot, 1994), [26], on a time ranging less than 100 years (see Fig. 12.3), inevitably, for longer periods of time, a progressive drift appears. The basic model remains valid, but must be shifted by a few years. This aspect of the problem raises the question of the solar internal clock. If the solar cycles have a memory under the form of a phase–amplitude correlation, do they need, or not, to be controlled by a clock [8, 17]? These models deduced from the observations are useful "to get an idea", but are insufficient for an accurate forecast, for giving a physical explanation to the studied phenomenon, and cannot account for the fine observed fluctuations.

A third way remains to be explored, which appears to be intermediate. It starts from the observed phenomenon, then passes, in the space phases; the dynamic evolution of the system is obtained *in fine*. By doing this, significant parameters related to the physical state of the system are determined, which define what one could call its "metric": dimension $d$ of the attractor, weak or strong dimension of the chaos, Lyapunov exponents $\lambda$, entropy $K2$. The point which remains to be developed is to pass from the numerical experimentation, upstream or downstream from the theory, to the account of these quantities in the theoretical approaches.

This work remains to be made. And as Bergé wrote so well [1, p. 234], "one must remember that the observer goes on the same way as the theorist, but in a diametrically opposite direction". Thus, applying the first return principle (Fig. 12.4, see also Sect. 12.5.) should render possible the experimenter to

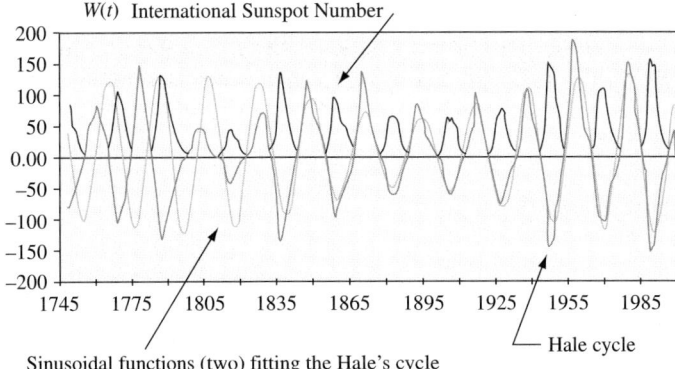

**Fig. 12.3.** The Hale cycle of solar activity: odd cycles reversed to be negative. For the years 1834–1998, this cycle can be adjusted by a function composed of two sinusoidal functions of periods $T_1 = 21.5$ years and $T_2 = 11.3$ years. The agreement is rather satisfactory (average quadratic deviation of 23.6 only). From 1749 to 1833, the beat of these two waves can again be used to model the Hale cycle, with the proviso of shifting the series by $-6.2$ years. This phase anomaly, well known under the name of "great solar anomaly" (cf. [17] for example), is at the basis of discussions on internal mechanisms of the solar clock

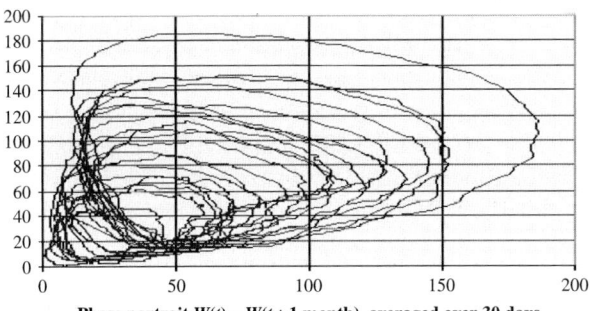

**Fig. 12.4.** The "solar standard mapping". Projection in the plane of the trajectories described by the solar activity cycle in the phase space of dimension 3. The $W(t)$ values are averaged on the number of days deduced from the first zero of the autocorrelation function $W(t + \tau)$. The topology of these trajectories reveals a limit cycle (also well visible in Fig. 12.5) and orbits of increasing areas with increasing activity. When the activity reaches its maximum value, a phase catastrophe occurs, which brings back the trajectory toward low levels of amplitude. By analogy with theoretical dynamic systems, one can speak of a very strong dissipation of energy

rebuild the iteration function, which is the unknown of our problem (through which analytical function(s) one can model the solar cycle?). On the contrary, the theorist would start from known functions to determine the successive iterates.

## 12.4 Some Results

The results presented here come mainly from work of several authors for which we may briefly describe hereafter their original contribution. We give only one restricted list, non-exhaustive, and the reader will find in these articles all the useful references for a more detailed examination of the question. Once more, let us recall that the terms used (chaos, bifurcation, attractor...) are those of the vocabulary introduced by Bergé et al. [2] and are not those which the common sense gives to them. The analysis are made using the three main methods: the algorithm of Grassberger and Procaccia [14, 15], the Shugira and May method [44] and the rescaled range analysis (Hurst, 1965 method developed in Feder [10]). The Hurst exponent is an empirical mean to determine if the studied phenomena is random, chaotic or regular.[5]

Weiss [48] pointed out that one can evaluate the dimension of the solar attractor to 2.1, but he estimated that this value is hardly convincing because of a too short sample of data, and he concluded that one cannot establish certainly that the aperiodic modulation of the 11-year cycle is the consequence of a deterministic chaos or a stochastic disturbance. Let us note that he estimated to 4096 the minimum number of points of the considered sample to draw a valid conclusion. Let us specify that the daily series of the values of the sunspot number, available since 1749, provides more than 90,000 points of measurements and the monthly values, approximately 3000.

Ostryakov and Usoskin [30] determined the value of solar attractor, using the algorithm of Grassberger and Procaccia on 2868 values of the sunspot number (1749–1987). Analyzing the average over a year (a) and over a month (b) of solar activity and radiocarbon data C14 (1524–1900)—(c) they obtained the following dimensions $d$ of the solar attractor: 3.3 (a), 4.3 (b) and 4.7 (c). During the Maunder minimum such a dimension turns out to be significantly higher: 8.0 (c), whereas during the period of a phase catastrophe (1792–1828) the sunspot number averaged over a month yields $d = 3.0$ (b). They have also investigated the sensitivity of their inferences to the number of available experimental points. Positive values of the Kolmogorov entropy and first Lyapunov exponent explicitly show the stochastic behavior of the Sun (on a scale time lower than 5 years).

---

[5] The Hurst exponent is symmetrically distributed, for many natural phenomena, about a mean of 0.73, with a standard deviation of about 0.09. Statistical results show that $H = 0.5$ for a noisy random phenomena. If $0 < H < 0.5$, the temporal series is less correlated than the random noise; if $0.5 < H < 1.0$, the series is more correlated, thus is called persistent.

Gizzatullina et al. [13] analyzing the C14 temporal data from 4300 BC to 1950 AD using the algorithm of Grassberger and Procaccia, found a dimension of 3.3 of the attractor responsible for the solar activity.

Feynman and Gabriel [11], considering that the periods of time when the solar activity was suppressed show a phase shift (see Fig. 12.3; including other prolonged minima lead to the similar shift), interpreted this phenomenon like the proof of a chaotic solar dynamo and showed that this dynamo functions near to the bifurcation mode (frequencies doubling observed: 11, 22 and 88 years for instance).

Carbonnell et al. [4] approached the analysis of solar activity starting from the algorithm of Grassberger and Procaccia on daily values, from 1818 to 1990 (that is to say 62,942 points), and compared the results obtained with filtered series (however, no indication is given on the theoretical reconstruction of this new data set leading to perfect regular curves, from where the noise is completely excluded). The authors do not conclude positively in favor of a deterministic chaos of low dimension, due to the fact that the Ruelle criterion would not be satisfied in the solar case.

Kremliovsky [19] developed a topological analysis of solar activity (SA), based on three assumptions: (i) SA is a chaos of weak dimension, (ii) SA has a strong dissipation, (iii) SA presents temporal intermittencies. The principal conclusions are then the following ones: (i) SA is a deterministic process, with a basic cycle at 11 years, unstable in amplitude, phase and period; (ii) the chaotic mode is forced by a competition between the linear instability of growth and the effects of resonance having a tendency to remove this last one. If the conditions of resonance are almost satisfied, then SA tends to be removed (as in the Maunder minimum); if the instability of growth gains, then the cycle rises monotonously in amplitude to turn over to a chaotic mode; (iii) cycles of 80 and 1000 years can be highlighted like a "chaotic walk" of the whole SA.

Kurths and Ruzmaikin [20] applied the space phase reconstruction as in Fig. 12.4 to reconstruct the annual sunspot relative numbers. They proved that the embedding dimension is $m = 3$–$4$, with a delay time $\tau = 2$–$5$ years. This rather simple approach yields relatively good results for short-term forecasts ($< 11$ years). However, this approach must be modified in the vicinity of a grand minimum.

Ruzmaikin et al. [42] analyzed the persistence of the time series of the solar activity by the method known as "the rescaled range analysis" (R/S, see Feder [10]) and found a Hurst exponent close to 0.8, thus indicating a phenomenon of "long memory" in solar activity.

Note that in his book, Feder found, for the ranging time 1749–1948, $H = 0.96 \pm 0.04$, an estimate first given by Mandelbrodt and Wallis [22]. We are unable to compute such a high value through the daily or monthly sunspot data; we found $H = 0.84 \pm 0.06$ on the ranging time 1749–1996. Oliver and Ballester [29] found $H = 0.717 \pm 0.002$ (January 1874–July 1993, computed through the north–south asymmetry of daily area of spots (43,558 values); however, the 12.1 periodic component was removed.

Rozelot [37], using the method of Shugira and May [44] to separate chaos from the noise, showed that solar activity was controlled by a deterministic chaos of low dimension and a Lyapunov exponent of $0.25\,\mathrm{yr}^{-1}$ (impossibility to predict with a good precision the value of the solar activity beyond 4 years, which explains why the models of the Fourier type do not act). Let us note that the method theoretically does not need a minimum number of points; however, it can be implemented with a sufficient length of the data estimated at approximately 500 (250 points in each series, one for the starting library and one for the predicted library). In this case, the number of monthly values of SA is amply sufficient.

Usoskin et al. [45] compared the mappings obtained from 1953 to 1965 both for solar activity and the records of the cosmic radiation (the Climax neutron monitor) and concluded for a phase anomaly of the solar activity during the last years of cycle 20.

Oliver and Ballester [29] suggested that the asymmetry of the sunspots (NS) of solar activity could be represented by a model of three components constructed by means of a term of long tendency (modeled by a polynomial of degree 3), of a periodic component equal to 12.1 years, of low amplitude, on which is superimposed a stochastic signal, persistent in the Hurst meaning. The analysis of this last term was led by the method known as "the rescaled range analysis" (R/S)—see above Ruzmaikin, 1994.

Zhang Qin [49], using a sample of 1709 values (monthly values of solar activity from 1850 to 1992) and using the algorithm of Grassberger and Procaccia, showed that solar activity was controlled by a weak chaos beyond 8 years and by a strong chaos (i.e., stochastic) below this characteristic time.

## 12.5 Portrait of the Solar Activity Cycle in the Phases Space

The method used is that described originally by Packard et al. (1980) and Ruelle [35], and is explained for example in Rozelot [38]. Figure 12.4 gives the result for the solar cycles ranging from 1749 to 1998, where one has projected in a two-dimension space (the plan [ $W(t+1)$, $W(t)$ ]) the evolution of the trajectories described by the cycle of solar activity in the space phases of dimension 3. The $W(t)$ values were averaged over the number of days deduced from the first zero of the autocorrelation function $W(t) \otimes W(t+\tau)$, which is 35 days. The topology of these trajectories reveals a limit cycle (better visible in Fig. 12.5) and orbits of increasing areas as the activity grows. When this one reaches its maximum value, a phase catastrophe occurs, which brings back the trajectory towards low levels of amplitude. By analogy with theoretical dynamic systems, one can speak of a very strong dissipation of energy.

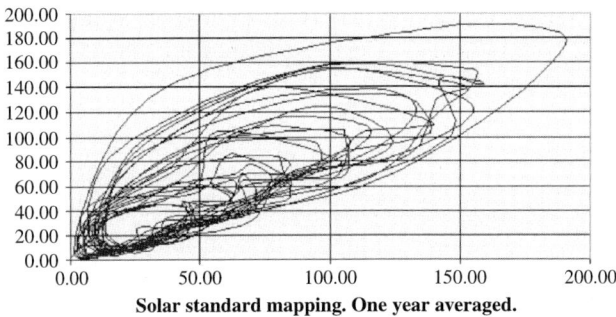

**Fig. 12.5.** Same as Fig. 12.4, but values of the solar cycle were smoothed over 1 year. The limit cycle appears and the abrupt return from the large to the low amplitudes shows very well the phase catastrophe

On Fig. 12.5, the $W(t)$ values were simply averaged over 1 year, while Fig. 12.6 makes a "zoom" over the period going from the years 1750 to 1833. The abrupt return of the trajectory of the points of co-ordinates of approximately $W(150-140)$ to those of co-ordinates $W(10-20)$ clearly indicates an instability, known under the name of great anomaly, and discovered by other processes. It should be noted that this same method applied to the periodic function fitting the Hale cycle (Fig. 12.3) leads to traditional Lissajous. The difference between the representation of the trajectories $W(t)$ and $F(t)$ indicates (that is already known) that the frequencies $\omega_k$, components of the solar activity signal, are not directly commensurable. By analogy with the studies on the dynamical chaos, we propose to call the curve described in Fig. 12.4, the *standard solar mapping*.

Looking at Fig. 12.6, Usoskin et al. [46] have suggested that one solar cycle was lost in the beginning of the Dalton minimum (during the 1790s), due to

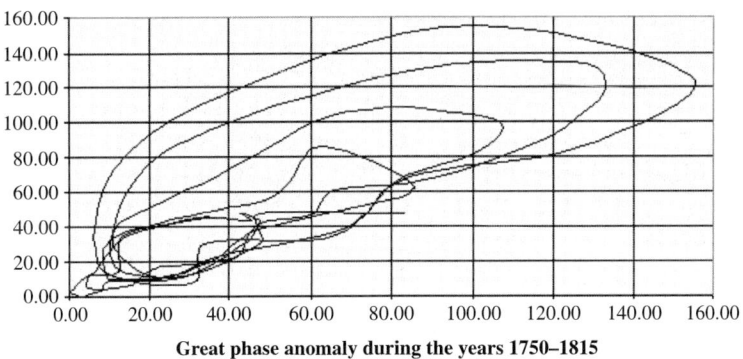

**Fig. 12.6.** Mapping of the great anomaly of phase of the years 1750–1815

this exceptionally long solar cycle 4 in 1784–1789, with such an irregular phase anomaly.

## 12.6 Conclusion

It seems that a significant work was carried out since the year 1990. The disturbing character of some results, for example related to the limit of predictability, is not really worrying if one correctly applies the basic definitions to the statements. To this purpose, it is perhaps not useless to point out the ideas of Poincaré [32] concerning the unpredictability of certain physical systems: "a very small cause, which escapes to us, determines a considerable effect that we cannot see, and then we say that this effect is due to chance... It may happen that small differences under defined initial conditions generate very large ones in the final results". Perhaps that applies in the solar case, whose cycle is very sensitive to the initial beginning of a new cycle, i.e., at the precise moment when a new minimum starts.

The magnetohydrodynamic approach was privileged so far and continues to be, in increasingly complex space–time models, including the implementation of 3D models, where ingredients of small scale are increasingly difficult to be taken into account (perhaps because of the fractal dimension of the structures[6]). Confrontation of the models thus suggested, with the observations, is certainly increasingly satisfactory, but at a price of increasingly large complications, and the agreement is still not always attractive. For example, if one succeeds rather well in modeling "afterward" an observed eruptive phenomenon, of which one can succeed in obtaining the chart of intensity, polarization, etc., it is still unfortunately impossible to know why this phenomenon occurred. The solar forecast, as a whole, remains extremely fragile.

From the "solar standard mapping" (as defined from Fig. 12.4), one can deduce the following:

(1) Consolidated back to Galileo's observations in 1610, the solar cycle of activity follows a deterministic process, with a basic cycle of $10.9 \pm 1.2$ years, but unstable in amplitude, phase and period. The cycle is asymmetric, with a fast rise and a slow decay; the rise time decreases with cycle amplitude, and large-amplitude cycles are preceded by short-period cycles; the secular amplitudes increase since the Maunder minimum and about hemispheric symmetries [16]. Subcycles with periods of 152–158 days were also noted from flare rates [7] or other solar indices [18]. Such subcyles are also called Rieger periodicities and were even discovered on stars (e.g., with a 294-day cycle on UX Arietis), believed to be caused by equatorially trapped Rossby-type waves modulating the emergence of magnetic flux at the surface [25]. The centennial increase in global geomagnetic activity

---

[6] A special topical issue of *Sol. Phys.* (Vol. 228) was devoted to this subject: fractal and multifractal analysis of photospheric features.

was, however, considerably smaller than the secular increase in solar activity (Mursula et al. 2004). Predictions include a strong next cycle (24) with a sunspot number of 145 ± 30 in 2010 and a weak following cycle (25) with 70 ± 30 spots and a long cycle peaking in 2023 [16]. Thus, the next large minima is awaited for the years 2010–2050 (see also [21]).

On such basic cycles would be superimposed a chaotic mode forced by a competition between a linear instability of growth and effects of resonance tending to remove it. When the conditions of resonance are almost satisfied, then the cycle of activity tends to be removed (as in the case of the Maunder minimum); when the instability of growth gains, then the cycle rises monotonously in amplitude until reaching a maximum value, which makes it to turn over to a chaotic mode. This chaos of weak dimension is governed by a Lyapunov exponent of about 0.02–0.03 for the whole of the cycles and about 0.2–0.4 from one cycle to another (in $yr^{-1}$).

(2) The dimension of the attractor lies between 2.1 [48] to 2.8 [49] and 3.0 [13] to 3.3 [30] according to the modalities of calculation. This concept must be specified again in the future.

The last point which may continue to pose problem is that of the sensitivity of the results regarding the Ruelle criterion [34, 36]. This one stipulates that the minimum number of data necessary to validate the results using the algorithm of Grassberger and Procaccia is $N_{min} \leq 42^M$, where $M$ is the greatest integer lower than the fractal dimension of the attractor. For $M = 2$, $N_{min}$ is 1764 and for $M = 3$, $N_{min}$ is 74,088. In the first case, the results are validated as soon as monthly data are used; in the second case, it is necessary to treat the daily data, particularly noised. One can also consider that for $\tau = 35$, the number of data $N'$ using $\tau$ as a sample is about 50 in the case of the monthly values (that is to say $1750/\tau$); for $\tau \geq 10$, Ruelle indicates that the dimension of attractor must be $D \leq 2\ lg\ N'$, which gives $D \leq 3.4$.

In spite of perhaps still contrasted results, the methods of non-deterministic physics lead to a lot of concordant results, which necessarily would have to be intensified in the future.

## References

1. Barnes, J. A., Sargent, H. H., and Tryon, P. V., 1980. In Pepin, R. O., Eddy, J. A. and Merrill, R. B. (eds), *The Ancient Sun*, Pergamon Press, New York.
2. Bergé, P., Pommeau, Y., and Vidal, C., 1988. *l'Ordre dans le chaos*, Herman, Paris.
3. Bois, E., 1996. Comptes-Rendus de l'Ecole Thématique du CNRS *Chaos et Fractales dans le cycle d'activité solaire*, p. 8.
4. Carbonnel, M., Oliver, R., and Ballester, J. L., 1994. *Astron. Astrophys.* **290**, 983.
5. Dikpati, M., de Toma, G., Gilman, P., Arge, C., and White, O., 2004. *Astrophys. J.* **601**(2), 1136.

6. Dikpati, M. and Gilman, P., 2005. *Astrophys. J.* **635**, L193–L196.
7. Díaz, A. J., Oliver, R., Ballester, J. L., and Roberts, B., 2004. *Astron. Astrophys.* **424**, 1055.
8. Dicke, R. H., 1978. *Nature* **276**, 676.
9. Eddy, J. A., 1976. *Science*, **192**, 1189.
10. Feder, J., 1988. *Fractals*, Plenum Press, New York, p. 149.
11. Feynman, J. and Gabriel, S. B., 1990. *Sol. Phys.* **127**, 393.
12. Foukal, P., Fröhlich, C., Spruit, H., and Wigley, T. M. L., 2006. *Nature* **443**, 161.
13. Gizzatullina, S. M., Rukavishnikov, V. D., Ruzmaikin, A. A. and Tavastsherna, K. S., 1990. *Sol. Phys.* **127**, 281.
14. Grassberger, P. and Proccacia, I., 1982. *Phys. Rev. Letters* **50**, 346.
15. Grassberger, P. and Proccacia, I., 1983. *Physica D, 13*, 34.
16. Hathaway, D. H. and Wilson, R. M., 2004. *Sol. Phys.* **224**, 5.
17. Hoyng, P., 1996. *Sol. Phys.* **169**, 253.
18. Kane, R. P., 2005. In Sripathi Acharya, B., Gupta S., Jagadeesan, P., Jain, A., Karthikeyan, S., Morris, S., and Tonwar, S. (eds), *Proceedings of the 29th International Cosmic Ray Conference*, August 3–10, 2005, Pune, India, **2**, Tata Institute of Fundamental Research, Mumbai, p. 293.
19. Kremliovsky, N. M., 1994. *Sol. Phys.* **151**, 351.
20. Kurths, J. and Ruzmaikin, A. A., 1990. *Sol. Phys.* **126**, 407.
21. Livingston, W., Gray, D., Wallace, L., and White, O. R., 2004. In Sankarasubramanian, K., Penn, M., and Pevtsov, A. (eds), *Large-Scale Structures and Their Role in Solar Activity*, ASP Conference Series, **346**, Proceedings of the Conference held October 18–22, 2004 in Sunspot, New Mexico, USA, p. 353.
22. Mandelbrodt, B. and Wallis, J., 1969. *Water Resour. Res.* **5**, 321.
23. Maris, G., 2006. Second International Symposium and Space Climate, Sinaia, Romania, 13-16/09/2006
24. Mundt, M. D., Maguire, W. B., and Chase, R. P., 1991. *J. Geophys. Res.* **96**, 1705.
25. Massi, M., Neidhöfer, J., Carpentier, Y., and Ros, E., 2005. *Astron. Astrophys.* **435**, L1–L4.
26. Merzlyakov, V. L., 1997. *Sol. Phys.* **170**, 425.
27. Mursula, K., Usoskin, I. G., and Nevanlinna, H., 2003. *Geophys. Res. Abstr.* **5**, 10361.
28. Oliver, R. and Ballester, J. L., 1996. *Solar Phys.* **169**, 215.
29. Oliver, R. and Ballester, J. L., 1997. *Solar Phys.* **169**, 215
30. Ostryakov, V. M. and Usoskin, I. G., 1990. *Solar Phys.* **127**, 495.
31. Price, C. P., Prichard, D., and Hogenson, E. A., 1992. *J. Geophys. Res.* **97**(A12), 19113.
32. Poincaré, H., 1927. In *Science et Méthode*, Flammarion (ed.), p. 68.
33. Rezelot, J. P. and Lefebvre, S., 2006. in Advance in Understanding Elements of the Sun–Earth Links, *Lect. Notes Phys.* **699**, 5
34. Ruelle, D. and Takens, F., 1971. *Commun. Math. Phys.* **20**, 167.
35. Ruelle, D., 1980, *La Recherche*, **11**, 132.
36. Ruelle, D., 1990. *Proc. Roy. Soc., Lond.* **A427**, 241.
37. Rozelot, J. P., 1995. *Astron. Astrophys.* **297**, 45.
38. Rozelot, J. P., 1997. Comptes-Rendus de l'Ecole Thématique du CNRS, *Reconstruction d'attracteurs et mesure du chaos à partir de l'analyse d'un signal: applications au cas du Soleil*, p. 17.

39. Rozelot, J. P., 1996. Comptes-Rendus de l'Ecole Thématique du CNRS, *Chaos et Fractales dans le cycle d'activité solaire*, p. 100.
40. Rozelot, J. P., 2001. *J. Atmos. Sol. Terrestr. Phys.* **63**, 375.
41. Rozelot, J. P.: Solar and Heliospheric Origins of Space Weather Phenomena, Lect. Notes Phys. **699**. Springer, Heidelberg (2006).
42. Ruzmakin, A. A., Feyman, J., and Robinson, P., 1994. **149**, 395.
43. Soon, W., 1997. Comptes-Rendus de l'Ecole Thématique du CNRS, *Chaos et Fractales dans le cycle d'activité solaire*, p. 110.
44. Shugira, G. and May, R. M., 1990. *Nature* **344**, 734.
45. Usoskin, I. G., Kovaltsov, G. A., Kananen, H., Mursula, K., and Tanskanen, P. J., 1997. *Sol. Phys.* **170**, 447.
46. Usoskin, I. G., Mursula, K., and Kovaltsov, G. A., 2002. *Geophys. Res. Lett.* **29**, 2183.
47. Tel, T. and Gruiz, M., 2006. *Chaotic Dynamics*, Cambridge University Press, 393p.
48. Weiss, O., 1990. *Phil. Trans. R. Soc. Lond.* **A 330**, 617.
49. Zhang Qin, 1997. *Chin. Phil. Trans.*

# Index

1:1 resonance, 115–117, 223, 319–321, 351, 362
47Uma, 213
51Peg, 209

Action-angle coordinates, 4
Adiabatic invariants, 23, 68, 76, 77
Apsidal alignment, 245, 251–252
Arnol'd web, 8
Arnold, 62, 315, 392
Arnold's mechanism, 30–33, 62
Arnold's model, 31, 33–36, 62
Arnold diffusion, 29, 33, 36, 38, 55–56, 59–61, 315, 320, 363
Arnold web, 29–30, 37–38, 42, 48–49, 52, 54–56, 315, 317–318, 356
Asteroid families, 111, 114–115, 118, 123–126, 145–146
Asteroids, 10, 20, 22, 111–148, 151, 164–165, 225, 234, 237, 313, 392
Attractor, 68–69, 71–74, 76, 394, 397–398, 400–401, 405
Averaged Hamiltonian, 130, 146, 274, 283–284, 287
Axisymmetric system, 306–310, 342, 350–351

Binary star, 234, 236, 238, 240
Binary systems, 211, 214–222, 238, 240
Boltzmann equation, 300, 335, 365
Brown dwarf, 210, 240
Bulirsch-Stoer, 224

Capture, 68–70, 141, 214, 234, 238–239, 379
Capture into resonance, 68, 70, 80–83, 238
Catalogue, 114, 169, 230
Chaos, 20, 22–25, 92, 113–114, 124–125, 128, 132, 145–146, 298, 307, 313, 315, 334, 351–353, 355, 358–359, 362, 392–394, 397–398, 400–403, 405
Chaotic diffusion, 24, 111–148, 316, 364, 372, 376
Chaotic mixing, 334
Chirikov's diffusion, 29
Classical impulse approximation, 259, 262
Collisionless, 299–300, 303, 305, 335, 338, 364–365
Comets, 165, 257–271, 273–295
Comet showers, 258–259
"Connectance", 91–108
Constant acceleration, 240, 243–245, 247–248
Corotation, 114, 237, 355

Darwin's Torque, 68, 79–80
Degenerate systems, 6, 19
Dehnen model, 357–358
Delaunay elements, 4, 274, 286
Determinism, 391–392
De Vaucouleurs law, 323, 340
Differential fast Lyapunov indicator, 68, 72–74

Diffusion coefficient, 30, 32, 35–36, 47–48, 50, 56–60, 61–63, 125–128, 136–138, 146–147
Dissipative effects, 67
Dissipative parameter, 68–69, 75
Dissipative standard map, 68–70, 80
Distribution function, 127, 129, 139, 298–299, 301–303, 304–307, 315–320, 322, 324–328, 335–340, 342–343, 345, 347, 361, 366
Doppler, 186, 198, 210, 235, 240
Doppler shift, 210
Dybczyński's impulse approximation, 262–264
Dynamical friction, 237

Eccentricity
  Eccentricities Escape orbit, 247
  Eccentricity observations, 235–236
  Excitation process, 240
Ellipsoidal coordinates, 313, 351–352
Encounter parametric plane, 266
Escape lines, 52
Euler-Poinsot system, 5
Excitation
  Excitation frequency, 244, 246
  Extrasolar planets, 233, 234, 235, 236, 238, 239, 240–253, 254
  galaxy gaseous disk, 237, 238
  gravitational instability, 240, 254
  gravitational torque, 170, 199, 237
  impulse, 239, 259, 262, 263, 264, 265, 396
  integrable, 243–247
  sudden excitation, 242, 251, 252
Exocatalogue, 230
Extra-solar planet, 209–210, 216
Extra-solar planetary system, 209–230

"Fast web", 55–59, 62
Fast drift, 9, 15–18, 43–44, 54–55, 60–63
Fast Lyapunov Indicator, 34–35, 38–40, 62, 68, 70, 72–74
FLI, 38–45, 48–49, 51, 55–57, 72, 216, 220–221, 230
Fokker-Planck equation, 127, 136
"Food web", 93–95
Forecasting activity, 396–398

Four-Dimensional Standard Mapping, 75
Frequency analysis, 8, 68, 70–72, 75, 201, 247, 355
Froeschlé's mapping, 33
Fully mixed, 363–364

Galactic tide, 258, 273–295
Galileo, 395, 404
$\gamma$ Cephei, 217, 219–220
Geometry of resonances, 16–17, 33, 38, 41–43
"Geršgorin theorem", 93, 97
Giant molecular clouds, 257
Gl 777A, 224
Gleissberg cycle, 395
Gliese, 86, 219
Global diffusion, 30, 50, 55, 59, 62
Global diffusion process, 50
Global dynamics, 138, 299, 320, 347–364, 370–375
GRAPE, 364, 366
Grassberger algorithm, 400–402, 405
Gravitational lensing, 210

Habitability, 209, 211, 225
Habitable zone, 209, 211, 222, 224–225, 230
Hamiltonian perturbation theory, 3, 7–13, 39
HD108874, 225–227
HD160691, 212
HD41004, 216, 219–220
HD82943, 211–213
Hubble space telescope, 210, 322, 354
Hurst exponent, 400–401

Impact frame, 263–264
Impulse approximation, 259, 262, 264–265
Initial conditions, 8, 23–24, 30, 33, 35–36, 40–43, 45–50, 52–53, 55–61, 73–76, 86, 114, 134, 138, 140, 141, 144, 146–147, 196, 215, 221, 224, 228–230, 237, 281, 292, 318, 324–325, 328–329, 334, 336–337, 346, 360–362, 370–371, 392, 396, 404
Instabilities, 234, 254, 298, 334–336, 337

Index    411

Integrable system, 1–4, 6, 7–13, 17, 19–20, 29–63, 67–68, 83, 310, 314–315, 342, 345
Integrable systems, 3, 6
Integration method, 195, 224, 369
Integrator, 258, 261, 273–276, 279, 280–282, 286, 292–295
Invariant curves
  strange attractors, 69, 71, 72, 73, 394
Invariant curves, 69, 71, 132, 319–320, 351
Iso-energetically non-degenerate, 43, 52
Isolating integrals, 32, 304–305

Jeans theorem, 299, 303–305, 307, 313, 315–322
Jet
  bipolar jets, 241
  jet angle, 246, 247, 253
  jet-induced acceleration, 242, 243, 245, 247, 250, 251, 252, 253
  jet's inclination, 246
  migration planet-disk interaction, 236, 237–239
  planetesimal accumulation, 240, 254
  planet-planet scattering, 236–237, 254
  precession reflex velocity, 235, 240
  relaxation, 239–240, 299–301, 306, 322, 323, 325, 327, 328, 331, 332, 333, 334, 336–339
  stellar jets, 241–242, 246
Resonance
  Andromedae, 250–251
  mean motion resonances, 13, 20, 22, 112, 114, 225, 236, 237, 238–239, 253
  secular resonance, 13, 112, 113–115, 119, 121, 146, 184, 193, 202, 212, 237, 238, 245
  Sedna Solar system, 250, 252
  stardust, 298
  stellar encounters, 236, 239, 259, 260–261, 265–266, 267
  temporal crisis, 254
  transit, 210, 235, 240
  truncation, 171–172, 199, 242, 280, 291, 307, 308, 309, 310, 312, 313
  T Tauri stars, 238, 241, 246

Jupiter, 22, 112, 115–119, 121, 129–130, 132, 136, 138–139, 141, 143, 145, 147, 152, 184, 192, 197, 199, 201–202, 210, 217, 222–223, 225–226, 233–234, 236–241, 251–252, 254

KAM theorem, 7
Kepler system, 4
Kolmogorov, 392, 400
Kuiper belt, 116, 202, 222, 250, 252
Kuustanheimo-Stiefel transformation, 276, 294

$L_4$, 10, 223, 227–230
$L_5$, 10, 223
Lagrangian points, 224
Laplace, 13, 20, 132, 201, 287, 295, 368, 392
LARKS, 281–282, 292–295
LCN, 371, 376
Lie integration, 224
Lie–Poisson, 274, 286–291, 295
Lindblad space, 340, 342–343, 345–346
Line-of-sight velocity distribution, 302
Liouville-Arnold theorem, 3, 6
Local diffusion, 30, 50, 60
LPV2, 291–295
L-type motion, 215
Lyapunov, 114, 118, 120, 125, 127, 133, 138, 140, 145–146, 300, 316, 334, 338, 355, 363, 371, 374–376, 378
Lyapunov Characteristics Indicator, 38
Lyapunov exponent, 11, 32, 38, 50–51, 69, 72, 133, 224, 301, 314, 355, 371, 397–398, 400, 402, 405

MacDonald's Torque, 78–79
Main belt, 10, 21, 30, 113, 115, 123, 132, 138–140, 146, 225, 234
MAPP, 286, 292–295
Mappings, 33, 68–69, 71, 75, 258, 273–274, 285–286, 309, 312, 402
Mass-ratio, 21, 215–219, 222, 224, 230
Matese elements, 274, 284–286
Mean motion resonances, 13, 20, 22, 112, 114, 225, 236–239, 253
Migration, 115–116, 123, 145, 202, 234, 236, 238–239, 242, 247–250, 252–254

Moser, 392
Multi-planet system, 211–214

N-body problem, 299, 316, 364–370
Nekhoroshev theorem, 10, 11
Nekhoroshev theorems, 2, 7, 29, 37, 40
Nekhoroshev theory, 62, 312
Neptune, 13, 22, 152, 164, 201–202, 222, 252
Non-convex, 51, 54, 56, 58
Non convex functions, 30
Non-uniform partition, 340
Number density, 122–123, 339–340, 342–346
Numerical integrations, 8, 11, 116, 119, 136–137, 147, 200, 224, 258–259, 274, 315

Observational errors, 226
Oort cloud, 250, 257–271, 273–295
Oort cloud comets, 257–271, 273–295
Oort constants, 274–275

Pendulum, 23, 31, 68, 76–77, 119, 132, 137, 146–147, 397
Perfect ellipsoid, 314, 351, 352–355, 356, 362
Periodic orbits, 68, 69, 71, 72, 73, 74, 83, 84–85, 87–88, 114, 132, 136, 141, 146, 349, 351, 352, 353, 356, 361, 362
Phase anomaly, 399, 402, 403–404
Phase shift, 401
Pisot–Vijayaraghavan number, 75
Pluto, 164, 222
Poincaré, 7, 13, 192, 312, 319, 320, 351, 355, 366, 376, 377, 379, 392, 404
Polyachenko criterion, 336–337
Prediction, 328, 394, 397, 405
Priori unstable system, 30, 32
Probability of capture, 82, 83
P-type motion, 215, 221–222
"Quasi-diagonal dominance", 93, 95–96, 98–100, 108

Quasi–integrable maps, 37
Quasi–integrable systems, 2, 7–13, 17, 29–63

"Rationally convex", 54–56, 58
Regularized, 274, 275–283
Relaxation time, 299, 300, 301
Resonance overlapping, 38, 59, 141
Resonances, 13, 14, 112, 113, 116, 117, 119, 132, 139, 141, 144, 145, 146, 147, 200, 225, 236, 237, 238–239, 253, 309, 314, 315, 320, 351, 356, 359, 375, 393
Restricted three body problem, 20, 21, 22, 129, 137, 138, 146, 220, 221, 230
RLI, 230
Rotation number, 68, 74, 356, 357, 375, 377
Ruelle, 401, 402, 405
Runge-Kutta, 224

Saturn, 139, 146, 147, 152, 184, 201, 202, 222, 237, 238, 240, 254
Schwabe cycle, 395
Schwarzschild method, 360–362, 370
Secular evolution, 355, 370, 375–376
Secular resonances, 13, 112, 113–115, 119, 145, 146, 147, 237
Self-consistent, 31, 299, 304, 316, 318, 327, 347–364, 366, 368, 370, 377, 379, 380
Self-consistent field, 318, 366, 368
Self-organization, 375–376
Sersic law, 322
Solar clock, 399
Solar cycle, 393, 396, 397, 398, 400, 402, 403, 404
Solar luminosity, 393
Solar standard mapping, 399, 403, 404
Space phase, 398, 401, 402
Space Weather, 391, 396
Spherical system, 305–306, 314, 316, 335, 343, 345, 347–350
Spin-Orbit Model, 77–78
Spin-orbit problem, 67, 80, 84–85
Spin-orbit resonance, 77, 186, 187–192, 193, 200
"Stability recovery phenomenon", 98, 100–105, 107
Stability limit, 224
Stable zone, 216, 217, 218, 219, 221, 222

Statistical mechanics, 301, 323, 328, 336–338, 340, 342, 399
Stellar perturbations, 257, 258, 259, 260, 261, 265, 268
Stiavelli–Bertin model, 345, 346, 347
Stickiness, 128, 316, 363
Strange attractors, 69, 71, 72, 73, 394
S-type motion, 215–216
Sun-like stars, 210, 211, 233
Sunspots
  sunspots activity, 393, 394
  sunspots number, 395, 396, 399, 400, 405
Maunder minimum
  Dalton minima, 391, 403
  Ort, 394
  Spörer, 394
  Wolf, 394, 395
Symplectic, 1, 3, 5, 6, 29, 30, 37, 40, 62, 92, 133, 274, 275–283, 309, 312

Taylor development, 274, 285, 286, 295
Terrestrial planets, 222–223, 224, 225–226, 227, 230, 237
The Dissipative Standard Map, 68–70, 80
The Sequential Impulse Approximation, 264–265
Third integral, 210, 299, 303, 306, 309, 313, 314, 315, 318, 319, 320, 327, 342, 350
Three-body problem, 21, 67, 83–84, 85, 113, 115, 129, 137, 138, 140, 146, 147, 215, 224, 230

Three-body resonances, 67, 83–84, 85
Through frequency analysis, 68
Transit, 38, 42, 45, 61, 62, 91, 94, 140, 210, 235, 240, 351, 372, 374
TREE, 365, 366, 369
Triaxiality index, 376
Triaxial system, 299, 313–315, 336, 345, 346, 347, 351–352, 356, 366, 370
Trojan, 115–123, 145, 223, 226, 227, 313
Tsallis entropy, 338–339

Uranus, 152, 201–202, 222

Velocity
  residual velocity, 242, 249, 250
  velocity dispersion, 242, 302, 328, 338
  velocity pulse, 249
  viscosity, 239, 241
Velocity ellipsoid, 302, 303, 306, 307, 315
Violent relaxation, 299, 306, 322, 323–327, 332, 333, 336–339, 343–347

"Walras law", 95
Weak dimension chaos, 401, 405
Whiskered torus, 32

Yarkovsky, 115, 116, 118, 122, 123, 145

Zel'dovich approximation, 329
Zwicky paradox, 323

Printing: Krips bv, Meppel, The Netherlands
Binding: Stürtz, Würzburg, Germany